METHODS IN MOLECULAR BIOLOGY

For further volumes:
http://www.springer.com/series/7651

For over 35 years, biological scientists have come to rely on the research protocols and methodologies in the critically acclaimed *Methods in Molecular Biology* series. The series was the first to introduce the step-by-step protocols approach that has become the standard in all biomedical protocol publishing. Each protocol is provided in readily-reproducible step-by-step fashion, opening with an introductory overview, a list of the materials and reagents needed to complete the experiment, and followed by a detailed procedure that is supported with a helpful notes section offering tips and tricks of the trade as well as troubleshooting advice. These hallmark features were introduced by series editor Dr. John Walker and constitute the key ingredient in each and every volume of the *Methods in Molecular Biology* series. Tested and trusted, comprehensive and reliable, all protocols from the series are indexed in PubMed.

High Performance Computing for Drug Discovery and Biomedicine

Edited by

Alexander Heifetz

In Silico Research and Development, Evotec UK Ltd, Abingdon, UK

Humana Press

Editor
Alexander Heifetz
In Silico Research and Development
Evotec UK Ltd
Abingdon, UK

ISSN 1064-3745 ISSN 1940-6029 (electronic)
Methods in Molecular Biology
ISBN 978-1-0716-3448-6 ISBN 978-1-0716-3449-3 (eBook)
https://doi.org/10.1007/978-1-0716-3449-3

Cover Illustration Caption: Photograph of the supercomputer MareNostrum, which is managed by Barcelona Supercomputing Center.

This Humana imprint is published by the registered company Springer Science+Business Media, LLC, part of Springer Nature.
The registered company address is: 1 New York Plaza, New York, NY 10004, U.S.A.

Paper in this product is recyclable.

Dedication

This book is dedicated to my brilliant wife, Diana, who made me believe in myself, always inspires, helps, and supports, and without whom this book would not be written. Thank you, my dear!

Preface

Over the past two decades, significant advances in cloud and high-performance computing (HPC) technologies have led to remarkable achievements in both computational drug discovery (CDD) and biomedicine. HPC is a cross-disciplinary field that employs computational models to extract precise, accurate, and reproducible predictions that enable novel insights into the physical world. This approach has yielded a series of new approaches in drug design, as well as new algorithms and workflows for biomedicine. HPC-based algorithms are the subject of substantial research themselves and their efficient implementation requires an understanding of computer architecture, parallel computing, and software engineering. To continue to benefit from these approaches and to evolve them further, there is a clear need for multidisciplinary collaborations between key experts from the scientific community, including biologists, medicinal chemists, drug designers and clinicians, and specialists in the fields of HPC, programming, artificial intelligence, and machine learning (AI/ML). Successful integration of these biomedical models into the drug design process will result in safer, more efficient medicines whose effects on the human body are able to be simulated (for example, through virtual clinical trials) before they are given to patients. The capacity to build a tool chain that simulates treatment outcomes for many diseases and disorders will be of great benefit in managing our health and wellbeing.

This book provides an overview of the state of the art in the development and application of HPC to CDD and computational biomedicine. It comprises two major sections. The first section is dedicated to CDD approaches that, together with HPC, can revolutionize and automate drug discovery process. These approaches include knowledge graphs, natural language processing (NLP), Bayesian optimization, automated virtual screening platforms, alchemical free energy workflows, fragment-molecular orbitals (FMO), HPC-adapted molecular dynamic simulation (MD-HPC), and how it can be integrated with machine learning (ML) to accelerate COVID-19 drug discovery and the potential of cloud computing for drug discovery. The second section is dedicated to computational algorithms and workflows for biomedicine, including an HPC framework to assess drug-induced arrhythmic risk, digital patient applications relevant to the clinic, virtual humans simulations, cellular and whole body blood flow modelling for stroke treatments, and prediction of the femoral bone strength from CT data. Also covered are an introduction to HPC, quantum computing, efficient and safe data management for biomedical applications, the potential future applications of HPC, and the current path toward exascale computing as applied to healthcare. The review of these topics will allow a diverse audience, including computer scientists, computational and medicinal chemists, biologists, clinicians, pharmacologists, and drug designers, to navigate the complex landscape of what is currently possible and to understand the challenges and future directions that HPC-based technologies can bring to drug development and biomedicine.

Abingdon, UK | *Alexander Heifetz*

Contents

Contributors

JAZMIN AGUADO-SIERRA • *Barcelona Supercomputing Center, Barcelona, Spain; Elem Biotech S.L., Barcelona, Spain*

ZAINAB ALTAI • *School of Sport Rehabilitation and Exercises Sciences, University of Essex, Colchester, UK*

SIMON ARSÈNE • *Novadiscovery SA, Lyon, France*

APOLLO K. BARON • *Elem Biotech S.L., Barcelona, Spain*

MIKE J. BODKIN • *Evotec (UK) Ltd., Abingdon, Oxfordshire, UK*

JEAN-PIERRE BOISSEL • *Novadiscovery SA, Lyon, France*

BHUSHAN BONDE • *Evotec (UK) Ltd., Dorothy Crowfoot Hodgkin Campus, Abingdon, Oxfordshire, UK; Digital Futures Institute, University of Suffolk, Ipswich, UK*

KARIM ZOUAOUI BOUDJELTIA • *Laboratory of Experimental Medicine (ULB222), Faculty of Medicine, Université libre de Bruxelles, CHU de Charleroi, Charleroi, Belgium*

RENEE BRIGHAM • *Visible Heart® Laboratories, Department of Surgery and the Institute for Engineering in Medicine, University of Minnesota, Minneapolis, MN, USA*

HELMUT BRÜCHLE • *Leibniz Supercomputing Centre of the Bavarian Academy of Sciences and Humanities (LRZ-BAdW), Munich, Germany*

LARA BRUEZIÈRE • *Novadiscovery SA, Lyon, France*

CONSTANTINE BUTAKOFF • *Elem Biotech S.L., Barcelona, Spain*

FRANCESC CARRERAS • *Hospital de la Santa Creu i Sant Pau, Universitat Autònoma de Barcelona, CIBERCV, Barcelona, Spain*

BASTIEN CHOPARD • *Scientific and Parallel Computing Group, Computer Science Department, University of Geneva, Carouge, Switzerland*

BHASKAR CHOUBEY • *Digital Futures Institute, University of Suffolk, Ipswich, UK; Chair of Analogue Circuits and Image Sensors, Siegen University, Siegen, Germany*

LIONEL COLLIANDRE • *Evotec SAS (France), Toulouse, France*

MATTHEW CORNEY • *Evotec (UK) Ltd., in silico Research and Development, Abingdon, Oxfordshire, UK*

EULALIE COURCELLES • *Novadiscovery SA, Lyon, France*

CLAIRE COUTY • *Novadiscovery SA, Lyon, France*

PETER V. COVENEY • *Department of Chemistry, Centre for Computational Science, University College London, London, UK; Advanced Research Computing Centre, University College London, London, UK; Computational Science Laboratory, Institute for Informatics, Faculty of Science, University of Amsterdam, Amsterdam, the Netherlands*

BENJAMIN CZAJA • *SURF Cooperation, Amsterdam, The Netherlands*

CONSTANTINO DIAZ GONZALEZ • *Evotec SE, Integrated Drug Discovery, Molecular Architects, Campus Curie, Toulouse, France*

DAVID FILGUEIRAS-RAMA • *Fundación Centro Nacional de Investigaciones Cardiovasculares (CNIC), Instituto de Investigación Sanitaria del Hospital Clínico San Carlos (IdISSC), CIBERCV, Madrid, Spain*

NATACHA GO • *Novadiscovery SA, Lyon, France*

PAULA DOMINGUEZ GOMEZ • *Elem Biotech S.L., Barcelona, Spain*

SOLÈNE GRANJEON-NORIOT • *Novadiscovery SA, Lyon, France*

JOSE M. GUERRA • *Hospital de la Santa Creu i Sant Pau, Universitat Autònoma de Barcelona, CIBERCV, Barcelona, Spain*
STEPHAN HACHINGER • *Leibniz Supercomputing Centre of the Bavarian Academy of Sciences and Humanities (LRZ-BAdW), Munich, Germany*
MOHAMAD HAYEK • *Leibniz Supercomputing Centre of the Bavarian Academy of Sciences and Humanities (LRZ-BAdW), Munich, Germany*
ALEXANDER HEIFETZ • *In Silico Research and Development, Evotec UK Ltd., Abingdon, UK*
HOLGER HENNIG • *Evotec SE, Hamburg, Germany*
ANNA M. HERZ • *EaStChem School of Chemistry, Joseph Black Building, University of Edinburgh, Edinburgh, UK*
GUILLAUME HOUZEAUX • *Barcelona Supercomputing Center, Barcelona, Spain*
PAUL A. IAIZZO • *Visible Heart® Laboratories, Department of Surgery and the Institute for Engineering in Medicine, University of Minnesota, Minneapolis, MN, USA*
TINEN L. ILES • *Department of Surgery, Medical School, University of Minnesota, Minneapolis, MN, USA*
TIM JAMES • *Evotec (UK) Ltd., Abingdon, Oxfordshire, UK; Evotec (UK) Ltd., in silico Research and Development, Abingdon, Oxfordshire, UK*
FERDINAND JAMITZKY • *Leibniz Supercomputing Centre of the Bavarian Academy of Sciences and Humanities (LRZ-BAdW), Munich, Germany*
SARAH JEMAI • *Novadiscovery SA, Lyon, France*
J. CHARLES G. JEYNES • *Evotec (UK) Ltd., in silico Research and Development, Abingdon, Oxfordshire, UK*
RIAD KAHOUL • *Novadiscovery SA, Lyon, France*
TAHSIN KELLICI • *Evotec (UK) Ltd., In Silico Research and Development, Abingdon, Oxfordshire, UK; Merck & Co., Inc., Modelling and Informatics, West Point, PA, USA*
JULIE KLEINE-SCHULTJANN • *Novadiscovery SA, Lyon, France*
MARTIN KOTEV • *Evotec SE, Integrated Drug Discovery, Molecular Architects, Campus Curie, Toulouse, France*
ALEXANDER KULESZA • *Novadiscovery SA, Lyon, France*
XINSHAN LI • *Department of Mechanical Engineering, Insigneo institute for in silico medicine, University of Sheffield, Sheffield, UK*
BASTIEN MARTIN • *Novadiscovery SA, Lyon, France*
REUBEN L. MARTIN • *Research Department of Structural & Molecular Biology, Division of Biosciences, University College London, London, UK; Evotec (UK) Ltd., Abingdon, Oxfordshire, UK*
GERALD MATHIAS • *Leibniz Supercomputing Centre of the Bavarian Academy of Sciences and Humanities (LRZ-BAdW), Munich, Germany*
JULIEN MICHEL • *EaStChem School of Chemistry, Joseph Black Building, University of Edinburgh, Edinburgh, UK*
ERICA MONTEFIORI • *Department of Mechanical Engineering, Insigneo institute for in silico medicine, University of Sheffield, Sheffield, UK*
CLAUDIO MONTEIRO • *Novadiscovery SA, Lyon, France*
INAKI MORAO • *Evotec (UK) Ltd., In Silico Research and Development, Abingdon, Oxfordshire, UK*
CHRISTOPHE MULLER • *Evotec SAS (France), Toulouse, France*
YVES PARÈS • *Novadiscovery SA, Lyon, France*
PRATIK PATIL • *Evotec (UK) Ltd., Oxfordshire, UK; Digital Futures Institute, University of Suffolk, Ipswich, UK*

REMY PETKANTCHIN • *Scientific and Parallel Computing Group, Computer Science Department, University of Geneva, Carouge, Switzerland*
EMMANUEL PHAM • *Novadiscovery SA, Lyon, France*
JULIE PITRAT • *Novadiscovery SA, Lyon, France*
IVAN PRIBEC • *Leibniz Supercomputing Centre of the Bavarian Academy of Sciences and Humanities (LRZ-BAdW), Munich, Germany*
GAVIN J. PRINGLE • *EPCC, University of Edinburgh, Bayes Centre, Edinburgh, UK*
FRANCK RAYNAUD • *Scientific and Parallel Computing Group, Computer Science Department, University of Geneva, Carouge, Switzerland*
BRITT VAN ROOIJ • *Philips Medical Systems, Best, The Netherlands*
CHRISTIAN SPIEKER • *University of Amsterdam, Amsterdam, Netherlands*
VLADIMIR JOSEPH SYKORA • *Evotec (UK) Ltd., Oxfordshire, UK*
CARLOS TEIJEIRO BARJAS • *SURF BV, Amsterdam, the Netherlands*
ELIOTT TIXIER • *Novadiscovery SA, Lyon, France*
ANDREA TOWNSEND-NICHOLSON • *Research Department of Structural & Molecular Biology, Division of Biosciences, University College London, London, UK*
MARIANO VAZQUEZ • *Barcelona Supercomputing Center, Barcelona, Spain; Elem Biotech S. L., Barcelona, Spain*
MARCO VERDICCHIO • *SURF BV, Amsterdam, the Netherlands*
SHUNZHOU WAN • *Department of Chemistry, Centre for Computational Science, University College London, London, UK*
GABOR ZAVODSZKY • *University of Amsterdam, Amsterdam, Netherlands*

Chapter 1

Introduction to Computational Biomedicine

Shunzhou Wan and Peter V. Coveney

Abstract

The domain of computational biomedicine is a new and burgeoning one. Its areas of concern cover all scales of human biology, physiology, and pathology, commonly referred to as medicine, from the genomic to the whole human and beyond, including epidemiology and population health. Computational biomedicine aims to provide high-fidelity descriptions and predictions of the behavior of biomedical systems of both fundamental scientific and clinical importance. Digital twins and virtual humans aim to reproduce the extremely accurate duplicate of real-world human beings in cyberspace, which can be used to make highly accurate predictions that take complicated conditions into account. When that can be done reliably enough for the predictions to be actionable, such an approach will make an impact in the pharmaceutical industry by reducing or even replacing the extremely laboratory-intensive preclinical process of making and testing compounds in laboratories, and in clinical applications by assisting clinicians to make diagnostic and treatment decisions.

Key words Molecular modeling, Machine learning, Binding affinity, Computer-aided drug design, Clinical decision support systems, Digital twin

1 Introduction

Technologies originating with the digital revolution over the past 20 years have dramatically evolved multiple paradigms in the biomedicine and healthcare sectors. The use of computer models and simulation is now widespread to accelerate drug development in the pharmaceutical industry and to aid in diagnosis and treatment of diseases in clinical application. One of the major advantages of computational modeling is that it provides insight into the physical, chemical, and biological bases of clinical interventions. These could be the underlying molecular interactions and mechanisms of drugs to their target biomolecules or a virtual three-dimensional view of the region of a patient's anatomy on which surgeons may operate. Molecular-level insights are often inaccessible experimentally, while details of organs may not be visible without invasive procedures.

Alexander Heifetz (ed.), *High Performance Computing for Drug Discovery and Biomedicine*, Methods in Molecular Biology, vol. 2716, https://doi.org/10.1007/978-1-0716-3449-3_1,

As these computational models become more reliable, one would hope to use these methods to quantitatively predict the outcome of experiments or clinical operations prior to performing them. The computational methods may come to replace many experimental procedures. In this way, computational techniques should reduce time and cost in the industrial processes of drug discovery, which takes on average of more than 10 years and \$2.6 billion to bring a new drug to market [1]. As far as clinicians are concerned, they can invoke clinical decision support systems (CDSS) to enhance decision-making in the clinical workflow. CDSS encompass a variety of tools, some based on knowledge, while others on models and computation. While the use of knowledge management is still the main feature of CDSS, the components based on modeling and computation are starting to make an impact on decision-making [2]. A recent study showed that a personalized CDSS may be created by constructing a computational pipeline at the molecular level with individual patients' genomic data to select the optimal drugs for a given patient, based on the predicted ranking of the binding efficacy of the particular drugs to variant proteins [3].

Computational modeling can also convert images obtained from medical scans into highly realistic virtual 3D models which can be used for training and for rehearsing complex cases prior to surgery. Due to these potential benefits, computer-based techniques have been adopted as routine techniques by the scientific community, along with the pharmaceutical industry and the healthcare sector.

The relentless enhancement in the performance of high-end computers, especially with advances in emerging embedded and many-core GPU accelerators of increasing diversity, is another key factor accounting for the increasing adoption of computer-based methods in science and industry in general, and in biomedicine in particular, over recent decades. In 1977, a "supercomputer" of that time—an IBM System/370 Model 168—was used for the first molecular dynamics simulation of a small protein [4] which had a top speed of a million (10^6) floating point operations per second (FLOPS), or a MegaFLOPS. In June 2022, the first machine to officially reach the exascale was announced, which achieved a peak performance of 1.102 exaFLOPS, or 1.102 quintillion (i.e., a billion billion or 10^{18}) FLOPS. The machine is Frontier, built at Oak Ridge National Laboratory (ORNL) in Tennessee, USA. Just 1 s of job execution on the entirety of Frontier would have needed more than 31,000 years to complete for a "supercomputer" from 45 years ago.

This increase in computing power enables us to perform complex calculations at very high speeds, with the simulation settings closer to real-world situations. The first molecular dynamics (MD) simulation of a protein mentioned above [4] simulated a small protein in vacuo for 9.2 picoseconds. Today, it is routine to run simulations for systems consisting of tens to hundreds of thousands of atoms for tens of nanoseconds or longer. During the COVID-19 pandemic, computational modeling and simulation play an important role in the understanding of the SARS-CoV-2 virus and in the early drug development to find potential therapeutic compounds. One of the studies, which won the ACM Gordon Bell award in 2020, used Summit supercomputer to simulate the spike protein and viral envelope with a model consisting of 305 million atoms [5]. As one of the authors stated: "we are giving people never-before-seen, intimate views of this virus, with resolution that is impossible to achieve experimentally." The molecular systems we study today are much more realistic as to their physiological conditions, with proteins being surrounded by water and ions, frequently including portions of cell membranes if required. The force fields used to describe the interactions between all the atoms in these simulations have also significantly improved, helping in some respects to make the simulations more accurate.

It should be noted that, while we can run such spatially enormous structures, it does not mean we can run them for long enough to extract anything useful from them. In the particular case of spike protein and viral envelope simulation [5], a total of ~702 ns trajectories were collected. The biological processes such as receptor binding and membrane fusion, however, occur over much longer timescales, from microseconds to hours. The larger the simulation domain, the longer the timescale one must run over; owing to the serial nature of time evolution of almost all codes employed on supercomputers, this unfortunately often means that the processes of interest get further and further out of reach for the computers available. This is why we need to use multiscale methods, to bridge between timescales. During the COVID-19 pandemic, for example, multiple mathematical models were used to evaluate the virus spread, to control the infectious disease, and to define an optimal strategy for vaccine administration [6].

The speed with which such simulations are executed is itself of critical concern as, if it can be done quickly enough, it enables decision-makers to take appropriate actions in a restricted "time-critical" window of time. Pharmaceutical companies need to quickly weed out compounds that may have toxicity issues or poor pharmacokinetics in preclinical studies, the so-called "fail fast, fail early" strategy, before making expensive late-stage investments during clinical trials. It is hoped that such work can be accomplished in silico rather than in actual experimental assays. In healthcare, a computerized CDSS need to deliver actionable recommendations to clinicians at the point of care, on timescales that are sufficiently rapid to be used in a decision-making context. The ability of clinicians to determine correct patient-specific interventions quickly should improve the efficiency of healthcare, providing better patient outcomes while eliminating unnecessary costs.

2 Methods and Protocols

Computational approaches have been widely used in biomedicine. Here we will summarize a few areas where the impacts are representative.

2.1 Drug Development

A major step in drug discovery is to identify lead compounds which have high binding affinities with a protein target. This happens at the early preclinical stage. The binding affinity, also known as the free energy of binding, is the single most important initial indicator of drug potency, and the most challenging to predict. It can be determined experimentally by a number of methods, with various accuracies but is generally expensive, error-prone, and time-consuming. Alternatively, one may seek to calculate the binding energy computationally. Here, methods drawn from computational chemistry offer a route forward; these are primarily based on in silico molecular dynamics (MD), for which several approaches to determining the free energy are possible.

The most common approaches for free energy predictions are the endpoint and the alchemical approaches [7]. As the name indicates, the endpoint methods are based on simulations of the final physical states of a system, namely, the bound and unbound states in the drug binding case. The most well-known endpoint methods are molecular mechanics Poisson–Boltzmann surface area (MM/PBSA) and molecular mechanics generalized Born surface area (MM/GBSA) [8]. The alchemical approach computes the free energy change by a "alchemical" path which is nonphysical. The process transforms, for example, one ligand into another, or a given amino acid residue into a mutated form, by morphing one chemical species of group directly into another. As the free energy is a state function, it does not matter which path, physical or nonphysical, is used to compute changes in it. MD simulations are performed using alchemical transformations to accelerate otherwise much more expensive and less reliable physical pathways to compute free energies. To further improve the precision and to generate reproducible predictions, ensemble-based approaches have been development, both for the endpoint and alchemical methods. The formal is called *enhanced sampling of molecular dynamics with approximation of continuum solvent* (ESMACS) [9], and the latter is called *thermodynamic integration with enhanced sampling* (TIES) [10].

The use of endpoint and alchemical approaches is not mutually exclusive but indeed can be even more powerful when performed in tandem. The former is less accurate but also less computationally expensive than the latter, which is well suited for use in the initial hit-to-lead activities within drug discovery. The alchemical approaches, on the other hand, are most relevant to lead optimization following the identification of promising lead compounds. A combination of the two methods, along with machine learning approaches, in a workflow has been proposed to accelerate COVID-19 drug discovery on high-performance computers [11]. While the currently ongoing COVID-19 pandemic has a negative impact in many areas, the computer-aided drug discovery market does experience a boost, in which the aforementioned approaches have been extensively applied, separately or jointly, to find novel drug candidates and to reposition existing drugs. An exciting global collaboration, called the COVID Moonshot, has come together for the discovery of new, urgent drug treatment for COVID-19 [12]. We ourselves are currently participating in a large-scale collaboration in which ML, docking, endpoint, and alchemical approaches are applied interactively to find promising drug candidates from libraries consisting of billions of compounds. The most attractive drug candidates are subsequently being studied experimentally with some under consideration for inclusion in possible clinical trials.

2.2 Personalized Medicine

In the post-genomic era, we are all very familiar with the notion that one-size-fits-all has limited validity in medicine. A drug that works well in one person might not be effective for another at all. While percentage of patients responding fall in the range of 50–75% for many drugs, it can be as low as 25% for cancer chemotherapy [13]. This is because the way that a particular medication works depends on how it interacts with other molecules in our bodies. A small variation, one amino acid mutation in one protein, for example, can determine a drug as an effective medication or a medication that doesn't work. In the past decade, the field of oncology has witnessed substantial changes in the way patients with cancer are managed, with increasing focus on personalized medicine based on genomic variants of individual patients. Pharmacogenomics, a combination of pharmacology and genomics, provides invaluable information on how an individual's genetic profile influences the response to medication. It enables personalized medicine which uses therapeutics to a subset of patients based on their specific genetic and molecular feature, who are expected to benefit. Two mutations in the kinase domain of epidermal growth factor receptor (EGFR), exon 19 deletions and L858R substitution in exon 21, have been commonly reported in patients with non-small cell lung carcinoma (NSCLC) tumors. The kinase inhibitor drugs gefitinib and erlotinib are effective in patients with these mutations, while inefficient for patients without. Even having these mutations, drug resistances can arise through other mutations such as the so-called gatekeeper mutation T790M.

One main reason for the occurrence of drug resistance arises from the binding affinity changes caused by mutations in the primary sequence of amino acid residues within the protein target. We have shown in our own previous publications how the ensemble-based binding free energy methods can be used to assess functional and mechanistic impacts of mutations in the case of EGFR [14], FGFR1 (fibroblast growth factor receptor 1) [15], and ER (estrogen receptor) [3] variants. The provision of extensive genomic sequencing technologies and the rising number of large-scale tumor molecular profiling programs across institutions worldwide have revolutionized the field of precision oncology. The treatment of patients with breast cancer, for example, has shifted from the standard of care, in which all patients receive similar interventions (such as mastectomy, axillary dissection, radiotherapy, and chemotherapy), to personalized medicine, where molecular characteristics of individual patients are used for prognostics, risk assessment, and selection of medical interventions [16]. Realistically, however, we still have a long way to go to deliver patient-specific treatments comprehensively across the healthcare system, whereas the stratification of the population into clusters/groups which have closer similarities might well help get finer grained differentiation so as to provide better treatments to these distinct groups. As a potential component in CDSS [2], the computation of binding free energies can be used to provide recommendations for more accurate and personalized treatment based on patients' specific genomic variances. In the longer term, a related approach could be used to design new drugs which are resistant to such mutations (*see* Subheading 2.1, Drug Development).

2.3 Medical Diagnosis with Machine Learning

While artificial intelligence (AI) is the broad science of computer algorithms designed to mimic human cognitive functions such as learning and problem-solving, machine learning is a specific subset of AI that uses datasets to train a computer to learn observable phenomena [17]. Machine learning is particularly powerful for image-based pattern recognition and has become integral to feature detection in biomedical imaging. This is frequently of considerable benefit in the initial stages of segmentation and reconstruction of complex three-dimensional geometries. Machine learning also has considerable promise in classifying categories of observed behavior and as a less computationally demanding surrogate for inclusion within clinically based decision support systems. While the availability of diagnostic images is rapidly increasing, they are underutilized in many countries and regions due to the lack of trained diagnostic specialists. Machine learning therefore brings considerable promise to clinical practice with medical images. A survey of the diagnostic performance based on medical imaging shows that deep learning models perform on a par with healthcare professionals [18]. By adding causal reasoning into machine leaning, the prediction can achieve expert clinical accuracy, i.e., an accuracy placing in the top 25% of doctors in the cohort [19]. An AI technology, known as InnerEye, has been used in some NHS hospitals to automate lengthy radiotherapy preparations for image-guided radiotherapy. The UK government has recently set a pioneering £100 m NHS consortium to use a groundbreaking artificial intelligence (AI) tool to speed up the diagnosis of breast cancer.

It should be noted, however, that the biggest drawback in A I/ML is its total lack of explanatory power. Medical decision-making can and will never be made on the basis that an AI system advised someone to take an action. Mechanistic understanding is essential. That is why, in a new book co-authored by one of us (PVC) and Roger Highfield [20], we talk about "Big AI"—meaning AI which pays attention to our scientific understanding and conforms with the laws of nature (e.g., of physics and chemistry) and understands the structural characteristics of the problems being studied. There is vast amount of biology and medicine for which mechanistic understanding is lacking (exacerbated by the perpetual pressure to "translate" any basic medical findings to the clinic as soon as possible), so there is plenty of room for improvement. AI methods are riddled with biases of many kinds which are often implicit and may easily be missed by those producing such solutions; and there are lots of open problems to face once similar data is integrated from different sources, even if nominally being from the same measurements. In short, there are often big but hidden uncertainties in these systems.

To derive the greatest benefit from working with AI, especially in healthcare, it is best to think of them like very junior colleagues. They are good at handling data and can learn quickly but are prone to making mistakes. The indispensable expert-in-the-loop in AI-based models allows clinicians and scientists to respond to output of the software, explaining what and why, while themselves making the key clinical decisions. These responses from clinicians can in turn become training data for the AI to learn from.

2.4 Human Digital Twins

As defined by the National Cancer Institute at Frederick, "A digital twin is a virtual model used to make predictions and run simulations without disrupting or harming the real-world object." It originates in engineering around 20 years ago and is only now being seriously applied to medicine and biology. Here we focus on biomedicine, where the definition of digital twins covers different scales, from the cellular, organ, and whole human to population levels. For the purpose of personalized medicine, the digital twin, also known by different names including digital patient, "avatar," virtual human, and so on, is built by integrating and processing vast amounts of data from individual patients to create the virtual human body. It is an evolving model which keeps integrating new data and predicting future statuses.

The ultimate purpose of human digital twins is to use patient-specific modeling to provide support for actionable medical and clinical decisions. High-resolution models can be evaluated virtually for many possible kinds of medical interventions in order to find the best option for the patient. This includes the optimal drugs which benefit most for the patient with minimal side effects or the optimal clinical operation strategies. Although digital twinning is still at an early stage of development and adoption in the healthcare sector, the concept is already finding some actual as well as potential application areas including patient monitoring, disease prediction, personalized treatment, and population health studies as well as in silico clinical and other kinds of health trials. Companies like ELEMBio (https://www.elem.bio/) that spun out from Barcelona Supercomputing Center (BSC) are offering such services already; the Virtual Humans predictive modeling platform from ELEMBio allows biotech companies to run supercomputer-based in silico trials to assess the cardiac safety ranges of drugs.

A prime example of the use of digital twins in medicine is the virtual human heart (Fig. 1), which is the culmination of research that stretches back more than half a century to experiments in the 1950s on the conveniently large nerves from a squid. Today, remarkable progress has been made in creating realistic human digital heart. Here we provide but a few examples. Blanca Rodriguez's team in Oxford makes human virtual heart predictions

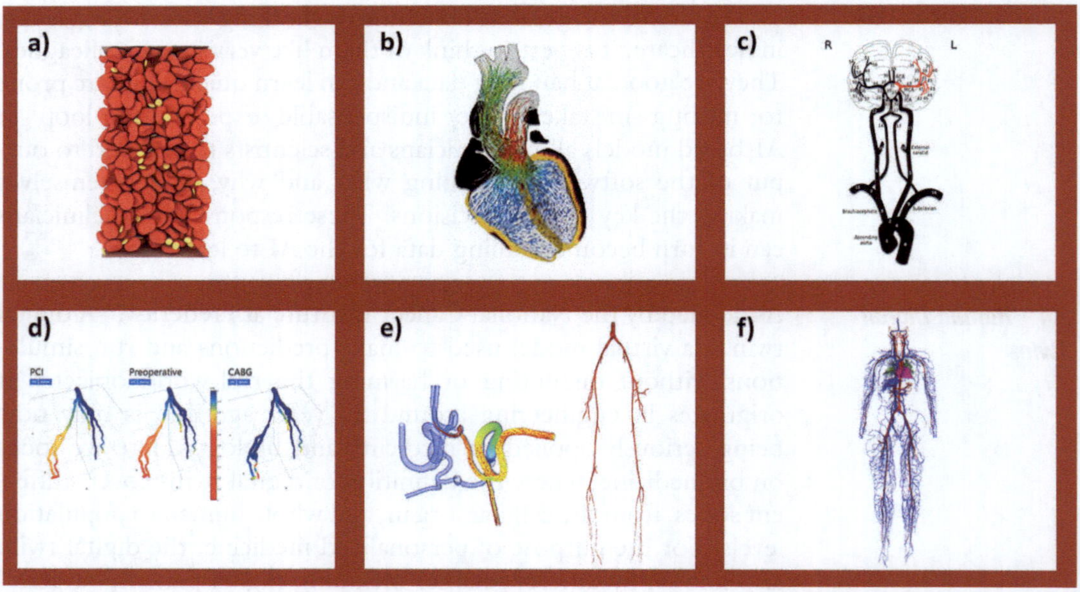

Fig. 1 Digital twin models for the cardiovasculature, which have paved the way for the dawn of precision cardiology. (**a**) "Digital Blood" and flow-diverting stents within Palabos (University of Amsterdam, Université de Genève, ATOS); (**b**) Alya Red virtual heart (Barcelona Supercomputing Center, University of Oxford); (**c**) OpenBF for vascular networks (University of Sheffield, SURFsara); (**d**) AngioSupport for coronary artery disease (LifeTec Group, SURFsara); (**e**) HemeLB model of blood flow in arteries (University College London); and (**f**) HemeLB-Alya coupling for cardiovascular flow in virtual human scale geometries (University College London, Barcelona Supercomputing Center)

which are more accurate than comparable animal studies [21], offering a way to reduce vivisection. Our colleagues in the Barcelona Supercomputing Center have developed the Alya Red heart model [22], which typically takes 10 h to simulate ten heart beats. Working with the company Medtronic, their simulations can help position a pacemaker, fine-tune its electrical stimulus, and model the effects of an innovative design called the micra [23]. It assesses the drug dosage and potential interactions between antimalarial drugs to provide guidance for their use in the clinic. Personalized virtual heart models, based on patient data, have been created by a team led by Reza Razavi at King's College London to predict tachycardia [24], while in Johns Hopkins University, a team led by Natalia Trayanova is creating digital replicas of the heart's upper chambers to guide the treatment of irregular heartbeats by the carefully targeted destruction of tissue [25]. In France, Dassault Systèmes has created a cohort of "virtual patients" to help test a synthetic artificial heart valve for regulators, working with the US Food and Drug Administration [26]. This represents another important milestone for *Virtual You* because, until recently, regulatory agencies have relied on experimental evidence alone.

There are other real and potential clinical applications of digital twins. CT2S (Computer Tomography to Strength) is a digital twin workflow that allows prediction of the risk of hip fracture for an individual based on CT scans [27]. The clinical application of this code is to provide a more accurate intervention strategy for elderly people with weaker bones, such as those who are clinically defined as osteopenic but not receiving any treatments. The workflow provides a complete assessment of bone strength in 3D and analyze the risk factor of that particular individual in sustaining a fall in the future. One of us (PVC) has worked with an international team to create a digital twin of a 60,000-mile-long network of vessels, arteries, veins, and capillaries using billions of data points from digitized high-resolution cross sections of a frozen cadaver of a 26-year-old Korean woman, Yoon-sun. By taking over a German supercomputer, SuperMUC-NG for several days, they could show in a realistic way how virtual blood flowed for around 100 s through a virtual copy of Yoon-sun's blood vessels down to a fraction of a millimeter across. The team is now charting variations in blood pressure throughout her body and simulating the movement of blood clots. Meanwhile, remarkable progress has been made in creating simple virtual cells [28] by Markus Covert at Stanford University, while at the Auckland Bioengineering Institute, Peter Hunter and colleagues are working on an array of organs. Even the most complex known object, the human brain, is being simulated, for instance, to plan epilepsy surgery in a French clinical trial. At last, the virtual human is swimming into view.

And there is the remarkable story of how modeling and simulation of the spread of the COVID-19 pandemic was common currency in UK's national newspapers and media for the best part of 2 years, being the only means we had available on which to take rational decisions about the kinds of NPI (non-pharmaceutical interventions) government could apply to reduce the spread of COVID and the resulting deaths of tens of thousands of citizens. Our own work [29] has revealed the substantial level of uncertainty in one of the most influential models used in Britain (CovidSim).

Quantitative methods and modeling, such as physiologically based pharmacokinetic (PBPK) modeling, have been involved in the regulatory authorities, which is already well advanced—particularly in the FDA. A recent example is the FDA approval of a generic diclofenac sodium topical gel, based on a totality of evidence including a virtual bioequivalence assessment instead of a comparative clinical endpoint bioequivalence study. What matters most here is how one can define in silico protocols that pass muster. It is already well established that one must conform with specified verification, validation, and uncertainty quantification (VVUQ) procedures. V&V has long been required, e.g., by the American Institute of Mechanical Engineers, for the design of safety critical devices which engender the safety of human beings. To be used for regulatory decision-making, the models and simulations need to be sufficiently verified and validated (V&V) for their intended purpose.

3 Summary and Perspective

Computational medicine will lead to better understanding of mechanisms for not only disease development and drug treatment but greatly improved healthcare. It should also contribute substantially to reduce the cost of healthcare. It can also accelerate drug development and reduce costs in pharmaceutical industry. Since around 2020, R&D across the pharmaceutical industry has no longer been a profitable activity. The success rate over all drug discovery projects in the industry is no more than about 4% of all those initiated. Computational medicine has potential to change the way drugs are developed in both speeding up the discovery process and reducing the costs and thus to stimulate drug discovery efforts. Clinically implementing computational models, simulations, and digital twins will make medicine truly personalized and predictive.

We still have a long way to go toward developing an AI that comes close to the one between our ears, but a good way to start is to place less blind faith in pure data and to give machines more scientific understanding of the world they inhabit. AI, in this "smarter" incarnation that we call "Big AI," features in a new book we have written, *Virtual You*, the first general account of how to make medicine truly predictive and personal by the use of digital twins—computer models that behave just like an individual person's body. AI will play a role in this venture, since it has come a long way in the past decade. But we should never forget that we have seen several false dawns during its rise over the past half century (aficionados talk of "AI winters"). Today, we are enjoying an AI summer, where machines excel over humans in very narrow domains, from playing Go to figuring out the structure of proteins in the body.

But, amid all the hype, we should remember that computers have always been our superior in a narrow way: even the very first such devices could multiply and divide numbers with an ease beyond the ability of most people. Indeed, the current generation of AI is only as good as the data it has been trained on. Give it biased data, and you get biased answers because it makes statistical inferences, glorified curve fitting of these data.

There is also a risk of what is called overfitting: you can think of it as over-learning, so while AI can be so in thrall to what it was trained on, it can't reliably make sense—or generalize—its predictions to new data and circumstances, such as a new patient it has never encountered before. Even when AI works, after expending a lot of energy and compute power, we have little idea *how* it works: a machine learning algorithm trained to distinguish a chihuahua from a muffin would have no idea what either is.

We need Big AI because, when our virtual twin gives insights into the right drug, lifestyle, or diet, we need to understand how it came to its conclusions. In *Virtual You* [20], we discuss various examples of a new generation of "Big AI," where the brute force of machine learning is augmented with physics-based models, mathematical theories which describe how the world actually works. Big AI will play its role in the virtual human and also marks another step toward general artificial intelligence, when an AI agent is finally able to match us in any intellectual task.

Acknowledgments

We are grateful for funding from the UK MRC Medical Bioinformatics project (grant no. MR/L016311/1), the EPSRC-funded UK Consortium on Mesoscale Engineering Sciences (UKCOMES grant no. EP/L00030X/1), the European Commission for EU H2020 CompBioMed2 Centre of Excellence (grant no. 823712), and EU H2020 EXDCI-2 project (grant no. 800957). We thank Dr. Roger Highfield for his helpful discussions with PVC which led to some of the notions we refer to in this article.

References

1. DiMasi JA, Grabowski HG, Hansen RW (2016) Innovation in the pharmaceutical industry: new estimates of R&D costs. J Health Econ 47:20–33. https://doi.org/10.1016/j.jhealeco.2016.01.012
2. Wright DW, Wan S, Shublaq N, Zasada SJ, Coveney PV (2012) From base pair to bedside: molecular simulation and the translation of genomics to personalized medicine. Wiley Interdiscip Rev Syst Biol Med 4(6):585–598. https://doi.org/10.1002/wsbm.1186
3. Wan S, Kumar D, Ilyin V, Al Homsi U, Sher G, Knuth A, Coveney PV (2021) The effect of protein mutations on drug binding suggests ensuing personalised drug selection. Sci Rep 11(1):13452. https://doi.org/10.1038/s41598-021-92785-w
4. McCammon JA, Gelin BR, Karplus M (1977) Dynamics of folded proteins. Nature 267(5612):585–590. https://doi.org/10.1038/267585a0
5. Casalino L, Dommer AC, Gaieb Z, Barros EP, Sztain T, Ahn S-H, Trifan A, Brace A, Bogetti AT, Clyde A, Ma H, Lee H, Turilli M, Khalid S, Chong LT, Simmerling C, Hardy DJ, Maia JDC, Phillips JC, Kurth T, Stern AC, Huang L, McCalpin JD, Tatineni M, Gibbs T, Stone JE, Jha S, Ramanathan A, Amaro RE (2021) AI-driven multiscale simulations illuminate mechanisms of SARS-CoV-2 spike dynamics. Int J High Perform Comput Appl 35(5):432–451. https://doi.org/10.1177/10943420211006452
6. Libotte GB, Lobato FS, Platt GM, Silva Neto AJ (2020) Determination of an optimal control strategy for vaccine administration in COVID-19 pandemic treatment. Comput Methods Prog Biomed 196:105664. https://doi.org/10.1016/j.cmpb.2020.105664
7. de Ruiter A, Oostenbrink C (2020) Advances in the calculation of binding free energies. Curr Opin Struct Biol 61:207–212. https://doi.org/10.1016/j.sbi.2020.01.016
8. Kollman PA, Massova I, Reyes C, Kuhn B, Huo S, Chong L, Lee M, Lee T, Duan Y, Wang W, Donini O, Cieplak P, Srinivasan J, Case DA, Cheatham TE 3rd (2000) Calculating structures and free energies of complex molecules: combining molecular mechanics and continuum models. Acc Chem Res 33(12):889–897. https://doi.org/10.1021/ar000033j
9. Wan S, Knapp B, Wright DW, Deane CM, Coveney PV (2015) Rapid, precise, and reproducible prediction of peptide-MHC binding affinities from molecular dynamics that correlate well with experiment. J Chem Theory Comput 11(7):3346–3356. https://doi.org/10.1021/acs.jctc.5b00179
10. Bhati AP, Wan S, Wright DW, Coveney PV (2017) Rapid, accurate, precise, and reliable relative free energy prediction using ensemble based thermodynamic integration. J Chem Theory Comput 13(1):210–222. https://doi.org/10.1021/acs.jctc.6b00979
11. Bhati AP, Wan S, Alfe D, Clyde AR, Bode M, Tan L, Titov M, Merzky A, Turilli M, Jha S, Highfield RR, Rocchia W, Scafuri N, Succi S, Kranzlmuller D, Mathias G, Wifling D, Donon Y, Di Meglio A, Vallecorsa S, Ma H, Trifan A, Ramanathan A, Brettin T, Partin A, Xia F, Duan X, Stevens R, Coveney PV (2021) Pandemic drugs at pandemic speed: infrastructure for accelerating COVID-19 drug discovery with hybrid machine learning- and physics-based simulations on high-performance computers. Interface Focus 11(6):20210018. https://doi.org/10.1098/rsfs.2021.0018
12. von Delft F, Calmiano M, Chodera J, Griffen E, Lee A, London N, Matviuk T, Perry B, Robinson M, von Delft A (2021) A white-knuckle ride of open COVID drug discovery. Nature 594(7863):330–332. https://doi.org/10.1038/d41586-021-01571-1
13. Spear BB, Heath-Chiozzi M, Huff J (2001) Clinical application of pharmacogenetics. Trends Mol Med 7(5):201–204. https://doi.org/10.1016/s1471-4914(01)01986-4
14. Wan S, Coveney PV (2011) Rapid and accurate ranking of binding affinities of epidermal growth factor receptor sequences with selected lung cancer drugs. J R Soc Interface 8(61):1114–1127. https://doi.org/10.1098/rsif.2010.0609
15. Bunney TD, Wan S, Thiyagarajan N, Sutto L, Williams SV, Ashford P, Koss H, Knowles MA, Gervasio FL, Coveney PV, Katan M (2015) The effect of mutations on drug sensitivity and kinase activity of fibroblast growth factor receptors: a combined experimental and theoretical study. EBioMedicine 2(3):194–204. https://doi.org/10.1016/j.ebiom.2015.02.009
16. Ellsworth RE, Decewicz DJ, Shriver CD, Ellsworth DL (2010) Breast cancer in the personal genomics era. Curr Genomics 11(3):146–161. https://doi.org/10.2174/138920210791110951
17. Choudhary A, Fox G, Hey T (eds) (2023) AI for science. World Scientific Press, Singapore

18. Liu X, Faes L, Kale AU, Wagner SK, Fu DJ, Bruynseels A, Mahendiran T, Moraes G, Shamdas M, Kern C, Ledsam JR, Schmid MK, Balaskas K, Topol EJ, Bachmann LM, Keane PA, Denniston AK (2019) A comparison of deep learning performance against health-care professionals in detecting diseases from medical imaging: a systematic review and meta-analysis. Lancet Digit Health 1(6):e271–e297. https://doi.org/10.1016/S2589-7500(19)30123-2
19. Richens JG, Lee CM, Johri S (2020) Improving the accuracy of medical diagnosis with causal machine learning. Nat Commun 11(1): 3923. https://doi.org/10.1038/s41467-020-17419-7
20. Coveney PV, Highfield R (2023) Virtual you: how building your digital twin will revolutionize medicine and change your life. Princeton University Press, Princeton
21. Passini E, Britton OJ, Lu HR, Rohrbacher J, Hermans AN, Gallacher DJ, Greig RJH, Bueno-Orovio A, Rodriguez B (2017) Human in silico drug trials demonstrate higher accuracy than animal models in predicting clinical pro-arrhythmic cardiotoxicity. Front Physiol 8:668
22. Vázquez M, Arís R, Aguado-Sierra J, Houzeaux G, Santiago A, López M, Córdoba P, Rivero M, Cajas JC, Alya Red CCM (2015) HPC-based cardiac computational modelling. In: Klapp J, Ruíz Chavarría G, Medina Ovando A, López Villa A, Sigalotti LDG (eds) Selected topics of computational and experimental fluid mechanics. Springer, Cham, pp 189–207
23. Roberts PR, Clementy N, Al Samadi F, Garweg C, Martinez-Sande JL, Iacopino S, Johansen JB, Vinolas Prat X, Kowal RC, Klug D, Mont L, Steffel J, Li S, Van Osch D, El-Chami MF (2017) A leadless pacemaker in the real-world setting: the micra transcatheter pacing system post-approval registry. Heart Rhythm 14(9):1375–1379. https://doi.org/10.1016/j.hrthm.2017.05.017
24. Chen Z, Cabrera-Lozoya R, Relan J, Sohal M, Shetty A, Karim R, Delingette H, Gill J, Rhode K, Ayache N, Taggart P, Rinaldi CA, Sermesant M, Razavi R (2016) Biophysical modeling predicts ventricular tachycardia inducibility and circuit morphology: a combined clinical validation and computer modeling approach. J Cardiovasc Electrophysiol 27(7):851–860. https://doi.org/10.1111/jce.12991
25. Boyle PM, Zghaib T, Zahid S, Ali RL, Deng D, Franceschi WH, Hakim JB, Murphy MJ, Prakosa A, Zimmerman SL, Ashikaga H, Marine JE, Kolandaivelu A, Nazarian S, Spragg DD, Calkins H, Trayanova NA (2019) Computationally guided personalized targeted ablation of persistent atrial fibrillation. Nat Biomed Eng 3(11):870–879. https://doi.org/10.1038/s41551-019-0437-9
26. Baillargeon B, Rebelo N, Fox DD, Taylor RL, Kuhl E (2014) The living heart project: a robust and integrative simulator for human heart function. Eur J Mech A/Solids 48:38–47. https://doi.org/10.1016/j.euromechsol.2014.04.001
27. Benemerito I, Griffiths W, Allsopp J, Furnass W, Bhattacharya P, Li X, Marzo A, Wood S, Viceconti M, Narracott A (2021) Delivering computationally-intensive digital patient applications to the clinic: an exemplar solution to predict femoral bone strength from CT data. Comput Methods Prog Biomed 208: 106200. https://doi.org/10.1016/j.cmpb.2021.106200
28. Maritan M, Autin L, Karr J, Covert MW, Olson AJ, Goodsell DS (2022) Building structural models of a whole mycoplasma cell. J Mol Biol 434(2):167351. https://doi.org/10.1016/j.jmb.2021.167351
29. Edeling W, Arabnejad H, Sinclair R, Suleimenova D, Gopalakrishnan K, Bosak B, Groen D, Mahmood I, Crommelin D, Coveney PV (2021) The impact of uncertainty on predictions of the CovidSim epidemiological code. Nat Comput Sci 1(2):128–135. https://doi.org/10.1038/s43588-021-00028-9

Chapter 2

Introduction to High-Performance Computing

Marco Verdicchio and Carlos Teijeiro Barjas

Abstract

Since the first general-purpose computing machines came up in the middle of the twentieth century, computer science's popularity has been growing steadily until our time. The first computers represented a significant leap forward in automating calculations, so that several theoretical methods could be taken from paper into practice. The continuous need for increased computing capacity made computers evolve and become more and more powerful. Nowadays, high-performance computing (HPC) is a crucial component of scientific and technological advancement. This book chapter introduces the field of HPC, covering key concepts and essential terminology to understand this complex and rapidly evolving area. The chapter begins with an overview of what HPC is and how it differs from conventional computing. It then explores the various components and configurations of supercomputers, including shared memory systems, distributed memory systems, and hybrid systems and the different programming models used in HPC, including message passing, shared memory, and data parallelism. Finally, the chapter discusses significant challenges and future directions in supercomputing. Overall, this chapter provides a comprehensive introduction to the world of HPC and is an essential resource for anyone interested in this fascinating field.

Key words High-performance computing, Supercomputer, HPC, Compute node, Parallel programming, Batch system

1 Introduction

High-performance computing (HPC) involves using highly powerful and specialized computers, called supercomputers, to solve complex problems and perform intensive calculations and very diverse tasks [1]. Supercomputers are designed to handle large amounts of data, process them quickly with high accuracy, and carry out multiple computations simultaneously.

Supercomputers differ from conventional computers in several ways: first, in terms of hardware, and in particular because of the aggregation of compute power. Even if a regular individual processor (also called "core" or, popularly by extension, central processing unit or "CPU," as it is the central calculating part of any computer) or memory cards in a supercomputer are typically

Alexander Heifetz (ed.), *High Performance Computing for Drug Discovery and Biomedicine*, Methods in Molecular Biology, vol. 2716, https://doi.org/10.1007/978-1-0716-3449-3_2,

not extremely different from the ones in a regular laptop, supercomputers have many more processors and a very large amount of storage capacity (in terms of nonpersistent storage, which is generally referred to as "memory," and persistent storage, which is usually referred to as "disk storage" or even just "storage") than regular computers. Moreover, it is very common that supercomputers include additional accelerator devices such as powerful graphics processing units (GPUs), allowing them to have better support for complex and heterogeneous tasks. Another difference is the requirement of specific software: supercomputers can only work efficiently with software that has been designed with parallel processing in mind, so that they can perform multiple computations by many different users simultaneously. However, the key difference is the engineering perspective: supercomputers are designed in a scalable way, so that from small pieces, it is possible to construct a large and complex system. In fact, supercomputers are generally designed based on a large amount of building blocks (i.e., compute nodes) that include a high amount of computing hardware, and then every node in the system is connected to each other by means of special high-performance interconnection network. The aggregation of different nodes is effectively possible via an efficient design of the interconnection network (at hardware level) and the use of an operating system with specific applications (at software level) which are aware of all the available compute power and facilitate an efficient use of it. As a result of this, the application developers need to use different programming techniques designed to take advantage of these systems' unique architecture and capabilities, such as message passing, shared memory, and data parallelism.

Overall, HPC represents a significant advance in computing power and has enabled breakthroughs in scientific research, engineering, and other fields that would not have been possible with conventional computing. However, at the same time, science and engineering have been continuously pushing the boundaries of human knowledge and development, so that they have challenged state-of-the-art computing systems and motivated further developments. Therefore, there is a very important synergy and interdependence between compute power and all the different knowledge fields, and HPC is at the forefront of the support for the largest and more complex problems proposed by human knowledge.

2 Supercomputer Architectures

Supercomputers are designed to perform multiple calculations simultaneously, and this is possible, thanks to the way that the different elements (hardware) are working together (thanks to the suitable software). At the hardware level, there are thousands of processing nodes connected by a high-speed network, but it is important to understand how these hardware elements can interact with each other, and this is the essential commitment of software development for HPC systems.

There are two main concepts in computer science that must be introduced here. Just like in natural language that there are verbs that express an action and nouns that represent entities (which perform or receive the action of the verb), in computers there are also "instructions" and "data" with analogous behavior [2]. Instructions are all the actions that a computer performs, and data are the elements that receive the effects of the instructions: therefore, every software application running in a computer consists of a large set of instructions and an associated set of data that is processed and modified by the instructions. As a result, the software needs to be implemented with the most suitable set of instructions that can exploit the available hardware in the most efficient way, so that the instructions perform the required actions over the available set of data and eventually solve the target computational problem.

In 1966, Michael J. Flynn proposed a classification for computer architectures based on how instructions are applied on the data [3, 4]. The four categories in Flynn's classification are the following:

1. SISD (single instruction, single data): This architecture is characterized by a single processing unit that operates on a single stream of data at a time. It is the simplest form of computation which is the basis of usual personal computers, thus effectively not considered as HPC.
2. SIMD (single instruction, multiple data): This architecture is characterized by a single processing unit operating simultaneously on multiple data streams using the same instruction. SIMD is commonly used in applications that require processing large amounts of data in parallel, such as graphics and video processing.
3. MISD (multiple instruction, single data): This architecture is characterized by multiple processing units operating simultaneously on a single data stream using different instructions. MISD architectures are rarely used in practice due to their complexity and low efficiency, but they are convenient in special systems that require high reliability (i.e., fault tolerance).
4. MIMD (multiple instruction, multiple data): This architecture is characterized by multiple processing units operating simultaneously on multiple data streams using different instructions. MIMD architectures, such as supercomputers, are commonly used in high-performance computing applications that require a high degree of parallelism.

Based on these concepts of instructions and data, we can identify two general strategies to scale applications' performance on a supercomputer: "scale out" or "scale up."

Scaling up involves adding more processing power and memory to a computing system, such as a more powerful processor, adding more memory, or using faster disk storage devices. Scaling up is typically used when a single computer can handle the workload, but the system needs more resources to handle the workload efficiently. Supercomputers are usually equipped with high-performance computing processors, accelerators, memories, and storage elements, leading to better performance without minor workflows or application adaptations.

Scaling out, on the other hand, refers to adding more computing resources, such as additional computers, to a system to increase its capacity and performance. Scaling out is typically used when a single computer or processing unit cannot handle the workload, and better performances are achieved by distributing the problem's instructions to many computing elements.

In supercomputing, scaling out and scaling up are used to increase a system's processing power and memory available to the application or workflow to handle more extensive data streams and complex computational problems. Users of supercomputers often use a combination of both scaling out and scaling up to achieve high performance and fully exploit the capabilities of these large machines.

On top of the previous classification, the hardware components (e.g., memory hierarchy, interconnection networks) are also defining different types of processing architectures for computers. Here are some examples:

1. Vector architectures: These architectures use vector processors, which can simultaneously perform the same operation on multiple data elements and are optimized for scientific and engineering applications. This type of architecture is characterized by its ability to perform a single operation on multiple pieces of data simultaneously, which makes it ideal for applications that require the processing of large amounts of data, such as weather forecasting or molecular modeling. However, vector processors are not well-suited for tasks that require complex decision-making processes (e.g., when the instructions that are executed depend on data values). Popular examples of this type of architecture are the general-purpose GPUs used as accelerators in HPC systems [5].
2. Shared memory architectures: In these architectures, all processing units share a common pool of memory, and communication between processors is typically facilitated through a shared bus or interconnect. Shared memory systems allow multiple processors to access the same data simultaneously,

which can result in significant performance gains for specific applications. This architecture is ideal for applications that require frequent communication between processors, such as image processing or video rendering. However, these architectures can be expensive, and the performance gains from adding additional processors can diminish as the number of processors increases, because of the complexity in the coordinated access to data by many instructions executed by different processors.

3. Distributed memory architectures: Each processing unit has its private memory, and communication between processors is facilitated through a network connecting multiple computers to form a single system. Each computer has its memory, and communications between computers are achieved through a network. Distributed memory architectures are ideal for applications that require a large amount of memory or involve large amounts of data. However, these architectures can be challenging to program, and communication between different computing elements can be a bottleneck.
4. Hybrid architectures: These architectures combine aspects of both shared and distributed memory systems. Hybrid architectures are ideal for applications that require a specific combination of processing power, memory, and communication capabilities, and it is the most common configuration for modern supercomputers. For example, most of the current HPC systems use a distributed memory architecture for most of its processing nodes but also include a shared memory within each node for efficient communication between processors.

In conclusion, there are several different architectures of supercomputers, each with its strengths and weaknesses. Choosing the correct architecture for a particular application, instruction workflow, and data stream requires careful consideration of factors such as the type of computations that need to be performed, the amount of memory required, and the communication requirements and the topology of the system.

3 Supercomputer Components

As indicated in the previous sections, supercomputers consist of hundreds (or thousands) of compute nodes connected by fast and high-performance interconnects, where each node has one processing unit (or a group of processors), a memory block (RAM), storage, and a networking system between the different components (I/O bus system). Each of these components is an essential functional part that facilitates calculations, data storage, and interaction inside the computer, but the implementation and complexity of each of these functional parts will be very different in each supercomputer.

In current modern CPU architectures, there are usually several compute cores, which have access to relatively small but high-speed memory on the chip itself, which is named "cache memory." There are typically several numbered levels of cache memory that will be used to keep data that is continuously used by the CPUs, and the lower the number of the cache level is, the closer this memory is to the processor, and thus the faster and more expensive it is. For example, a usual cache setup has three levels (L1, L2, and L3), where the first and second levels are specific to the individual core and the third is shared among the cores of a CPU. In addition to regular CPUs, compute nodes can include accelerators or specialized hardware for specific tasks, with their memories and interconnect systems.

Figure 1 shows a general schematic of a multi-socket compute node equipped with multicore CPUs. A multi-socket compute node is a system that contains more than one CPU socket on the motherboard, allowing multiple CPUs to work together as a single entity. Each CPU socket can host a separate CPU, and these CPUs communicate with each other via a high-speed interconnect. Multi-socket compute nodes are widely used in HPC systems, where large-scale data processing or scientific simulations require high computational power. In addition to the multiple CPU sockets, multi-socket compute nodes typically have a high amount of memory and other hardware components that can handle large datasets and complex computations.

One essential factor in compute nodes is the topology and access to the memory for each CPU. In modern multiprocessor systems, the most common case is to have a NUMA (non-uniform memory access) architecture [6]. In a NUMA system, the memory is divided into multiple local memory banks attached to each CPU socket, so that each CPU can access its local memory bank with low latency, but accessing the memory banks of other CPUs can incur a higher latency. When a CPU requests data from its local memory bank, the response time is faster than if the same CPU were to request data from a remote memory bank attached to another CPU. This nonuniform access time can cause performance bottlenecks when CPUs frequently access data from remote memory banks, leading to a slowdown in the overall system performance. Modern compute nodes often employ a NUMA-aware operating system and software to optimize data access patterns based on the system's architecture to mitigate this issue. For example, some applications can be designed to allocate instructions and data to specific CPUs and memory banks respectively, minimizing cross-CPU communication and improving overall system performance. However, it is still a very important task for software developers to understand how NUMA systems behave and design software that

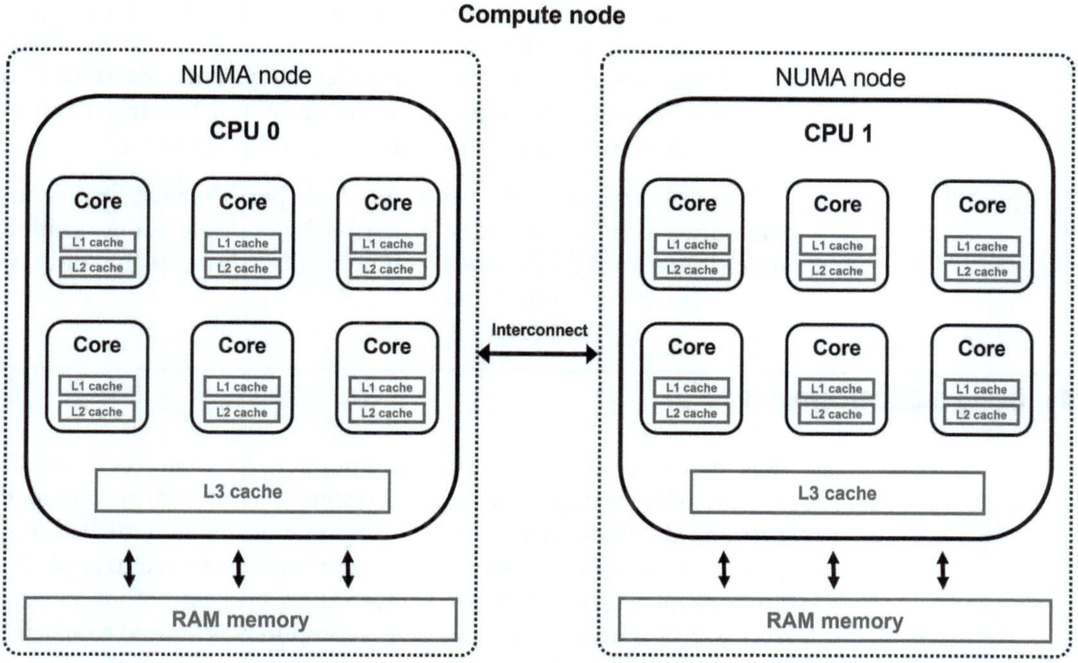

Fig. 1 Sketch of a dual socket multicore compute node

can deal with the different types of access, as well as it is important for HPC users to determine the correct amount of compute resources to efficiently run a parallel application on a given system.

Regarding disk storage, supercomputers rely on specialized file systems to support the complex computations demanded by HPC applications. HPC systems typically have two types of filesystems:

1. Parallel filesystems: These filesystems are designed to support high-performance I/O operations on large datasets distributed across multiple storage devices. These filesystems are usually implemented as distributed file systems, meaning that they can span multiple servers or storage devices and provide a unified view of data to the users: effectively, every data written to a file by a given processor in a given node will be almost immediately visible and readable by any other processor, ensuring data consistency and availability. The most common parallel filesystems used in HPC systems are Lustre [7], GPFS (IBM Spectrum Scale) [8], and BeeGFS [9]. These filesystems provide a scalable and high-bandwidth storage solution that can be used by multiple compute nodes in parallel, thus providing excellent I/O performance.
2. Local filesystems: These other filesystems are used to store data and files on individual compute nodes or storage devices. They are usually implemented as standard file systems (similar to those available on consumer-grade computers) and are

mounted on individual compute nodes. They store local copies of data required for processing by the compute nodes. Local filesystems are faster than parallel filesystems for small- to medium-sized datasets but do not scale well for large datasets that need to be accessed by multiple compute nodes.

Each compute node can access high-performance filesystems shared across different nodes (accessible by multiple nodes simultaneously) or local to the node itself (and directly accessible only by the cores on the compute node).

4 Using a Supercomputer

The first step in using a supercomputer is to gain access to the system. Depending on the HPC system, access can be obtained through different means, such as academic or research institutions, government agencies, or private companies. Users need to apply for an account and go through a registration process before gaining access. In some cases, access to supercomputers is limited to specific projects or research areas, and users must be affiliated with an organization that has been granted access. An HPC system usually has many users accessing and using it simultaneously.

Once the login credentials and account are granted on the HPC system, users need to log in to the remote cluster to access the resources. For this, it is usually required to use SSH (Secure Shell) [10]. SSH is a protocol for secure remote access to a computer or server over a network. It provides a secure encrypted connection between two computers, enabling secure communication and data transfer and preventing unauthorized access and eavesdropping. It is widely used by system administrators, developers, and other users who need to remotely access servers, run commands, and transfer files securely. SSH uses public key cryptography to authenticate the server to the client and the client to the server: this ensures that the connection is secure and that both parties can trust each other. SSH is available on most operating systems and can be used for various purposes, including remote login, file transfers, and remote command execution.

Figure 2 shows a schematic representation of the user environment when accessing a supercomputer. Once logged in, the user does not interact directly with the compute nodes but is connected to a type of specialized node, usually called login node (or head node), which is presented with an environment very similar to the one available on the compute nodes (similar operating system, access to the shared filesystem, etc.), but with restricted functionality allowed. The login nodes are the main access points to the cluster, and they are not meant to be used to perform heavy computations. Instead, they work as a gateway and are designed

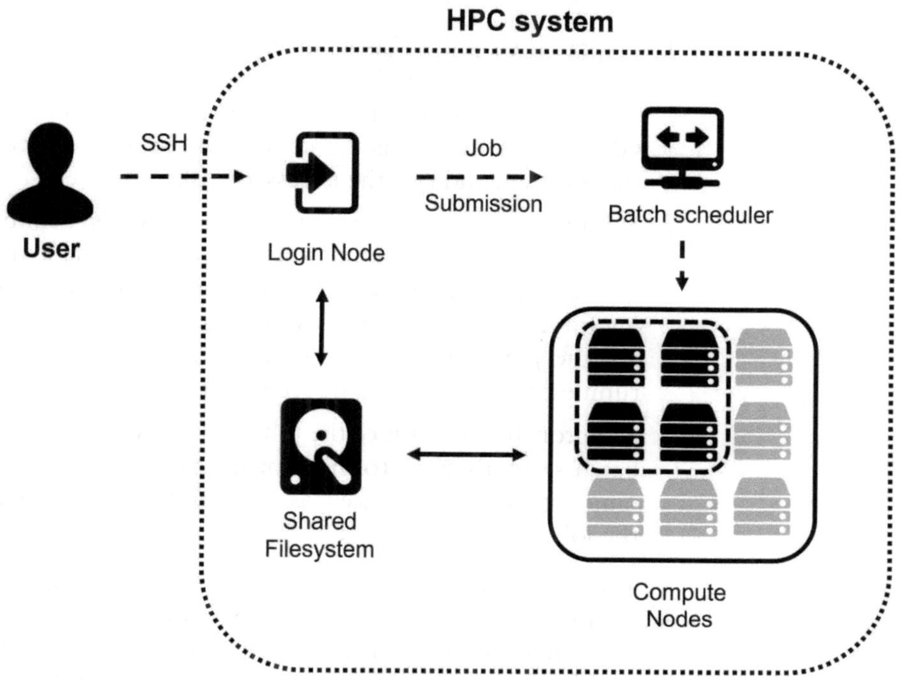

Fig. 2 Schematic representation of the user environment in a supercomputer

to set up the HPC simulation (upload the data, prepare inputs, monitor and control the status of the run) and compile and install applications and for lightweight pre−/post-processing and testing tasks.

Users do not have interactive access to the compute nodes on the login node; instead, they prepare and submit their jobs to the batch system scheduler. A batch scheduler is a complex software application that manages the scheduling and execution of batch jobs on an HPC system. It allows users to submit jobs to the system, which are then queued and scheduled for execution on the available compute resources.

Batch schedulers are critical in managing and scheduling compute-intensive jobs on clusters: initially, all users will submit their jobs to a queue, and then the scheduler decides the optimal order of execution of these jobs based on the available resources, job priority, and dependencies. As a result, this scheduling system keeps continuously updated information on the current status of the system and can take an immediate decision on whether any job will start to run or wait in the job queue. In general, HPC batch schedulers are highly optimized to ensure the best resource utilization and guarantee the fair access to cluster resources for all users in the system.

In order to use a batch system on an HPC system, users typically need to do the following:

1. Prepare their jobs: This involves creating a script that specifies the resources needed for the job (e.g., number of cores, amount of memory, runtime, use of specific hardware like GPUs, etc.), as well as the commands to be executed to load the program and run the necessary tasks.
2. Submit their job: Using the batch system's command-line interface, users can submit the job script to the system's job queue. The exact syntax for submitting jobs can vary depending on the batch system. However, it typically involves specifying the job script and any required options (e.g., job name, running time, type of nodes required).
3. Monitor their job: Once the job is submitted, users can use the batch system's tools to monitor its status (e.g., queued, running, completed), as well as view any output generated by the job.

There are several implementations of batch systems, each using its directives to request resources and commands to submit, monitor jobs, and check available hardware. Examples of currently available batch scheduling systems are SLURM [11], PBS [12], SGE [13], and LSF [14].

In addition to the batch scheduler, different software and tools are available for the users on a supercomputer. All these tools are running over the key controller application, which is operating system (OS). The OS is the ultimate responsible for the allocation of resources, scheduling, and use of the hardware provided by the system. Modern HPC systems use almost exclusively UNIX-based operating systems [15], which are well-designed OS for multiuser environments, allow multitasking, and are designed to be machine-independent. Moreover, UNIX-based operating systems support a wide range of tools that are suitable for scientific and engineering applications, as well as software development or many other topics.

Apart from the OS and the batch scheduler, HPC systems may usually have a set of commonly used software and computing libraries to perform numerical simulations, data analysis, molecular dynamics simulations, weather forecasting, and more: this set of applications is commonly referred to as "software stack." These applications are often requiring other specialized applications or libraries to be installed in advance to support their full functionality, and moreover every installed software or library must be optimized for its use in the HPC environment. In order to manage the various software components and dependencies between different applications, HPC systems typically use a module environment. A module environment is a system that allows users to dynamically load (i.e., activate) the use of given applications in the system and define a consistent environment with a customized set of software and libraries with their respective dependent modules. Using modules, users can load and make available within their job different types and versions of applications. Modules allow users to easily switch between different software and dependencies versions without manually installing and configuring each one, thus favoring flexibility, portability, and reproducibility (which are essential for scientific research and collaboration). Moreover, at the administration level, the use of modules allows the installation of any application in an HPC system regardless of whether it is compatible or not with other application: when the module for a given software is not loaded, the software is inactive and effectively nonexisting for the user, but after loading the module, the software becomes fully operational. In any case, the potential incompatibilities between applications must be handled by HPC users correctly, so that the selected modules are consistent and can work together when loaded at the same time.

5 Basics of Parallel Programming

When a job is submitted to a supercomputer with the right resource information and the necessary software modules, the batch scheduler allocates the resources for the job and it launches the program on the reserved nodes. However, the efficient use of a given parallel application (and, or course, its efficient development) is always determined by the parallel programming model used for the implementation of the HPC application.

Traditional computer programs are written for serial computations, which means that the instructions within a single code are executed sequentially, one after the other, on a single processor, and only one instruction at a time can be executed over a given piece of data (assuming a simple processing pipeline). Nevertheless, parallel computing allows for concurrent execution of multiple instructions, so that the computational problem is divided into multiple parts that can be executed simultaneously, and each of these different instances of the program uses a different processor to run. Parallel programming also requires some coordination mechanism which manages the different processes and controls the execution and data access within the parallel session of the application.

Parallel computing offers several advantages over traditional serial computing. By using multiple processors to execute a program or computation in parallel, parallel computing can significantly reduce the time taken to complete the computation, leading to faster execution times and increased throughput. Parallel applications can also scale with the number of computing elements to handle more complex and more extensive datasets. Since, within parallel applications, the workload is distributed across multiple processors or nodes, adding more resources can increase efficiency, reduce memory usage, and improve the system's overall performance. Thus, parallel computing requires a different approach to programming than traditional serial computing, and depending on the architecture of the parallel machine used (*see* Subheading 2), different programming models may bring the optimal implementation on different systems.

The basis of all parallel applications, just like for regular single-core applications, is in the concept of instructions and data. Therefore, parallel programming models are defining combinations for the execution of instructions on data, and these combinations rely on two important definitions: "process" and "thread." Using rough definitions, a process can be identified as a bundle of a set of instructions together with a set of data on which the given instructions operate, whereas a thread will be identified as a set of instructions alone (without associated data by default). From these descriptions, it should be assumed that a process contains one thread by default.

On the basis of processes and threads, the most traditional types of parallel programming models can be defined as follows:

1. Shared memory parallelism, where a single process is containing multiple threads that share the access to the memory pool of the hosting process, where its associated data is stored. In this model, each processor can access any data in the shared memory space, allowing multiple processors to work on the same data simultaneously. This can lead to significant performance gains, especially for algorithms that require frequent data access and sharing among multiple threads or processes. OpenMP [16] is a popular shared memory programming model that provides a set of compiler directives and library routines for developing multithreaded parallel applications. OpenMP, commonly used in multiprocessor systems, such as multicore processors, can provide high performance and low overhead parallelization. Other programming models for shared memory parallelism typically include using locks, semaphores, and other synchronization mechanisms to ensure that data access is adequately coordinated and managed. The major risk for a shared memory application is a race condition, where a thread operates over inconsistent data because of ill-coordinated data access with respect to other threads.
2. Distributed memory parallelism, where multiple processes work together to solve a problem, each having its local memory. In this model, each processor operates independently on its subset of the problem data, communicating with other processors to exchange information or coordinate their work. Distributed memory parallelism is commonly used in large-

scale computing systems, where the amount of data to be processed exceeds the capacity of a single processor or node. By distributing the problem by running multiple processes across multiple nodes, it is possible to achieve high levels of performance and scalability. However, every process is defined as independent from each other, and therefore the distributed memory parallelism involves the use of explicit communication mechanisms between processes, so that they can be synchronized among them. The most extended approach is the message-passing model, where processors communicate with each other by sending and receiving messages, and the de facto standard interface for this type of communication and synchronization between distributed processes is MPI (message passing interface) [17]. Distributed memory parallelism is a powerful tool for tackling large-scale computational problems and is widely used in scientific computing, machine learning, and big data analytics, and the communication mechanisms between processes may be very involving. Among the major risks to avoid with distributed memory programming are deadlocks, where a badly coordinated communication protocol causes the execution to stall because of messages that are not sent/received at the correct time.

3. Hybrid shared/distributed parallelism, where multiple processes (that communicate with each other via some message-passing model) are running multiple threads (which share the access to their respective hosting process' data). This approach is the most flexible and powerful to adapt to heterogeneous hardware architectures, but at the same time it's the most complex to program, as it inherits all the benefits and risks of the two previous models. The usual approach is a combination of MPI + OpenMP, but there may be several alternatives depending on the type of system architecture.

Despite its benefits, parallel computing also poses several challenges and limitations. Developing parallel applications is generally more complex than developing serial applications due to the need to manage concurrency, communication, and synchronization between multiple processors or nodes. This part will generally make parallel programming much more challenging and time-consuming than traditional serial programming. Moreover, although parallel computing can improve the scalability of computations, there are limits to how much scalability can be achieved: that is, for a given application, there may be parts of it that can be parallelized, but some other processing may be inherently sequential, and therefore no parallelism will be possible. As a result, when the number of processors or nodes increases, the overhead of communication and synchronization needs to be contained to ensure the maximum benefit from parallelism, and this is generally done by defining a reasonable amount of compute tasks that every single compute core must perform in parallel with other cores [1]. In this aspect, load balancing should also be named as an important aspect to consider when developing parallel applications: load balancing refers to the problem of distributing the workload evenly among multiple processes/threads in a parallel application, so that an inefficient load balancing can lead to idle processors or nodes, reducing the system's overall performance.

In general, parallel computing is a rapidly evolving field, with ongoing research and development aimed at improving parallel systems' performance, scalability, and ease of use. Parallel libraries and applications are continuously being developed, extended, and optimized, but the introduced concepts represent the essential basis to begin working on parallel application usage and/or development.

6 Conclusions

The introduction of high-performance computing (HPC) has revolutionized scientific research by enabling scientists to perform complex computations that were once impossible. In this chapter, we have explored the various aspects of HPC, starting with the definition of supercomputing and its role in scientific research. We then delved into the different architectures of supercomputers and when they are most suitable for scientific research. Finally, we discussed the importance of efficient utilization of HPC systems and the workflow for using them.

By understanding the fundamental concepts of HPC, researchers can take full advantage of the power and speed of supercomputers to accelerate scientific discoveries. However, it is essential to note that utilizing HPC systems effectively requires a deep understanding of the underlying architecture and programming techniques. As such, researchers must continually educate themselves on the latest developments to stay at the forefront of HPC research.

HPC has opened up new possibilities for scientific research and helped push the boundaries of what is possible in the field. As technology advances, we can expect even more significant breakthroughs and discoveries to emerge, further highlighting the importance of HPC in modern scientific research. Just like in science and engineering, the evolution of architectures may bring new paradigms and challenges in the future that may even be hardly imaginable nowadays.

References

1. Hager G, Wellein G (2010) Introduction to high performance computing for scientists and engineers. CRC Press, Boca Raton. https://doi.org/10.1201/9781420078779
2. Hennessy JL, Patterson DA (2018) Computer architecture: a quantitative approach, 6th edn. Morgan Kaufmann Publishers, Cambridge
3. Flynn MJ (1966) Very high-speed computing systems. Proc IEEE 54(12):1901–1909. https://doi.org/10.1109/PROC.1966.5508
4. Flynn MJ (1972) Some computer organizations and their effectiveness. IEEE Trans Comput C-21(9):948–960. https://doi.org/10.1109/TC.1972.5009071
5. Stone JE, Gohara D, Shi G (2007) Accelerating large-scale scientific computations with GPUs. Comput Sci Eng 9(3):14–21. https://doi.org/10.1109/MCSE.2007.55
6. Hennessy JL, Patterson DA (2017) "NUMA: multicore and beyond," computer architecture: a quantitative approach, 6th edn. Morgan Kaufmann Publishers, Cambridge, pp 603–612
7. Braam, P. J., Zahir, R. (2002). Lustre: A scalable, high performance file system. Cluster File Systems, Inc, 8(11): 3429–3441
8. Schmuck FB, Haskin RL (2002) IBM's general parallel file system. IBM Syst J 41(4):685–698
9. Herold M, Behling M, Brehm M (2017) BeeGFS – a high-performance distributed file system for large-scale cluster computing. In: International conference on computational science. Springer, Cham, pp 882–891
10. Ylonen T, Lonvick C (2006) The Secure Shell (SSH) protocol architecture. RFC 4251:1–30
11. Yoo AB, Jette MA, Grondona M (2003) SLURM: Simple Linux Utility for Resource Management. In: Proceedings of the 2003 ACM/IEEE conference on supercomputing. IEEE, pp 1–11
12. Bryson M (1997) PBS: the portable batch system. In: Proceedings of the sixth workshops on enabling technologies: infrastructure for collaborative enterprises (WET ICE '97). IEEE, pp. 273–278
13. Grewal R, Collins J (2012) Sun grid engine: a review. Int J Adv Res Comput Sci 3(2):38–41
14. IBM (2018) IBM spectrum LSF. IBM. https://www.ibm.com/products/spectrum-lsf
15. Ritchie DM, Thompson K (1978) The UNIX time-sharing system. Commun ACM 21(7): 365–376
16. Chapman B, Jost G, Van der Pas R (2008) Using OpenMP: portable shared memory parallel programming, vol 1. MIT Press, Cambridge
17. Gropp W, Lusk E, Skjellum A (1996) MPI: a message-passing interface standard. Int J Supercomput Appl High Perform Comput 10(4):295–308

Chapter 3

Computational Biomedicine (CompBioMed) Centre of Excellence: Selected Key Achievements

Gavin J. Pringle

Abstract

CompBioMed is a Centre of Excellence for High Performance Computing Applications, funded by the European Commission's Horizon 2020 program, running from October 1, 2017, to April 1, 2024. CompBioMed develops computer-based tools to simulate each human body in health and disease. The author provides a general overview and then presents his personal opinion on key achievements: collaborations between industry and academia, the two IMAX short films, training to foster a culture of HPC among biomedical practitioners, our free service to port and tune biomedical applications to HPC, providing future supercomputers access to surgery, and bringing FDA-endorsed credibility to biomedical simulations.

Key words Centre of Excellence, Horizon 2020, Computational biomedicine, HPC, Supercomputing, Cardiovascular medicine, Molecular medicine, Neuro-musculoskeletal medicine, Digital twin, Virtual human, Healthcare value chain, Vascular disease, Drug discovery, Medical devices, Training, Urgent computing, FDA approval

1 Introduction

I have been a consultant in supercomputing for over 30 years, and I've worked at EPCC [1] for the last 25 years. EPCC is a world-renowned, self-funded department of the University of Edinburgh, providing solutions for big computing and big data. I was lucky enough to join the CompBioMed project at the start, on October 1, 2017, and have been actively involved with its day-to-day running for the last 6.5 years.

The CompBioMed [2] project is a Centre of Excellence for High Performance Computing Applications, funded by the European Commission's Horizon 2020 program [3]. The project consists of two phases: our first phase for 3 years, grant agreement no. 675451, H2020-EINFRA-2015-1, and then our second phase for 4 years, grant agreement no. 823712, H2020-INFRAEDI-

Alexander Heifetz (ed.), *High Performance Computing for Drug Discovery and Biomedicine*, Methods in Molecular Biology, vol. 2716, https://doi.org/10.1007/978-1-0716-3449-3_3,

2018-1 which, thanks to a 6-month no-cost extension, will end on April 1, 2024.

For the first phase of our Centre of Excellence (CoE), my leading role was as the work package leader for sustainability and innovation, while for the second phase, I mainly coordinate both our e-seminar series and our scalability service, wherein we enable and/or optimize biomedical applications for supercomputers.

I was asked to write this chapter to give an overview of the CoE and present some of our key achievements. This is, of course, subjective and I shall present the achievements which are most evident to me.

Much of the text employed from hereon in has been adapted from our public website [2] and, as such, the list of contributing authors is easily over 50 from over 15 countries, and I wish to thank everyone who, unbeknownst to them, have tacitly contributed to this chapter.

The chapter continues as follows. First, I present a three-part overview of the CompBioMed project. The first part is high level which we use to target the public, wherein I also give details of the CompBioMed partners and their roles in both phases. The second part is a lower-level discussion on the healthcare value chain which we employed to appeal to the press, while the third part is more detailed to target both clinicians and academics. The closing section lists a number of key achievements, namely, the verdant collaborations between industry and academia fostered by the CoE, our two IMAX short films, how we began a culture of HPC among biomedical practitioners, our free service to optimize biomedical applications for HPC, a workflow providing future supercomputers into the operating theatre, and finally bringing FDA-endorsed credibility to biomedical simulations.

1.1 CompBioMed Overview

Our bodies are exquisitely complex machines; thus, getting the best out them and fixing their problems is a challenge that we seek to master, all of us. So, how can we get the best out of our bodies? Imagine a virtual human, not made of flesh and bone, one made of bits and bytes and not just any human, but a virtual version of you, accurate at every scale from the way your heart beats down to the letters of your DNA code.

Many drugs only work well on some people and can cause serious side effects in others. The reason is variations in DNA, our genetic differences. We understand how these DNA differences change the building blocks of your body, the proteins, and so through a virtual human, we could simulate in a computer how drugs interact with your unique protein makeup. By testing drugs on your virtual body, your doctor may eventually be able to assess a wide range of drugs and select precisely the right one to suit you.

Virtual humans could help doctors to plan risky surgery too. They could be employed to help determine how to reach an aneurysm deep in the brain that is at risk of rupture, which could cause a stroke. Surgeons can then try out the best treatment or implant to suit the location and shape of that particular aneurysm, and they could even double-check that the implant would not cause problems such as clotting before they try it out on you.

With the power of virtual humans, the medical possibilities are limitless.

1.1.1 Digital Twins

It is worth pointing out that the term digital twin is used across many fields of computer simulation to refer to any computing model that an application employs to emulate a real-world object. As such, the virtual human, the ultimate goal of CompBioMed, is an example of a digital twin, but not all digital twins are biomedical in nature.

1.2 Computational Biomedicine

Computational biomedicine is the name given to the use of computer-based tools and approaches to simulate and model the human body in health and disease. In the European Union, this new science has become synonymous with the concept of the virtual physiological human [4], an initiative that focuses on a methodological and technological framework that, once established, will enable collaborative investigation of the human body as a single complex system.

1.3 CompBioMed Centre of Excellence

CompBioMed is a CoE that is focused on the use and development of computational methods for biomedical applications. We have users within academia, industry, and clinical environments who are working to train more people in the use of our products and methods.

The cutting edge of computational biomedicine harnesses computer simulations that are conducted on massively powerful supercomputers. The tremendous power of these machines allows larger and more complex biological systems to be simulated, yielding better, more accurate, and more meaningful output.

Computational methods based on human biology are now reaching maturity in the biomedical domain, rendering predictive models of health and disease increasingly relevant to clinical practice by providing a personalized aspect to treatment. Computer-based modeling and simulation is well established in the physical sciences and engineering, where the use of high-performance computing (HPC) is now routine.

1.3.1 HPC and Supercomputers

It is worth noting here that any cluster of multicore computers can be described as HPC; however, only the current top 500 fastest HPC platforms get to be called supercomputers.

1.4 The Healthcare Value Chain

1.4.1 Entrepreneurial Opportunities

Our CoE has innovation at the forefront of its aims, promoting interdisciplinary entrepreneurial opportunities driven by our users' needs. Our industrial partners participate fully in the centre's activities, with the number of associate partners growing continuously over the lifetime of the centre.

1.4.2 Activities

We support and facilitate modeling and simulation activities and provide education and training for a diverse set of communities. We target research scientists from physical, computer, and biomedical sciences; software and infrastructure developers, from industry; and medical end users, including clinicians. Indeed, we actively invest in community building to spread knowledge, tools, and best practice to students and researchers across this domain.

1.4.3 Software

Our CoE provides a focal point for the development and sustainability of software tools and services capable of delivering high-fidelity three- and four-dimensional (including time) modeling and simulation of all aspects of the human body, from the genomic level to the whole human and beyond, in health and disease. Our core applications are listed below.

1.4.4 Transforming Industry

HPC has the potential to enhance industries in the healthcare sector including pharmaceuticals and medical device manufacturers and underpins a range of emerging sectors, such as those concerned with e-health and personalized medicine. The innovative modeling and simulation techniques we develop and promote within our centre have become of great interest and relevance to industrial researchers, HPC hardware vendors, and independent software vendors around the world.

1.5 Research

This section briefly outlines CompBioMed's three main pillars of research.

1.5.1 Cardiovascular Medicine

Cardiovascular disease accounts for half of sudden deaths in Europe; improvements in patient risk stratification and prediction of clinical intervention are both urgent and challenging. In this area, we study two critically important disease areas: firstly, cardiovascular diseases having a direct effect on the function of the heart itself, be it on the electrophysiology, mechanics, or blood flow (and, ultimately, the combination of all three), and secondly, arterial disorders, be they aneurysms in the abdominal aorta or in intracranial arteries, or stenosis in carotid or coronary arteries.

1.5.2 Molecular Medicine

Computational models are becoming immensely powerful at various stages in drug discovery and development, from molecular design to assessment of toxicity. Computational models can also be used for repositioning and targeting therapies for precision medicine, through rapid and accurate assessment of drug efficacy in specific disease cases. They can also provide added value to medical device measurement data, for example, as acquired by various imaging modalities.

1.5.3 Neuro-musculoskeletal Medicine

Despite a common perception that most neuro-musculoskeletal diseases are not life-threatening, around 30% of the elders who face an osteoporotic fracture of the hip joint will die of related complications within 12 months. We actively investigate applications such as skeletal muscle, cartilage, and connective tissue modeling of the deformation of tissues under biomechanical loads.

1.6 The Clinic

1.6.1 To and from the Clinic

We manage and provision access to personal (patient-specific)-derived medical data in a research environment. We also perform high-fidelity 3D and 4D HPC-based simulations and, ultimately, aim to provide clinical decision support within short periods (often minutes, hours, or a few days).

1.6.2 Medical Data

The rapid rise of detailed medical imaging, genomic data, and abundant proteomic, metabolomic, biological, and physiological data on all levels of biological organization (macromolecules, cells, tissues, organs, organ systems, the whole body, and, indeed, epidemiology) permits the development of mathematical, mechanistic, and predictive multiscale models of human health and disease. The software tools and techniques that we develop within CompBioMed help healthcare providers make sense of the vast array of data now available and will have a major impact on the clinical decision-making process.

1.7 CompBioMed Partners

1.7.1 Core Partners

Peter Coveney of UCL is the principal investigator of the CompBioMed CoE. We had/has around 15 core partners from both academic and industrial institutions, where the academic partners include universities, clinical partners, international partners, and HPC centres, while the industrial partners include SMEs.

1.7.2 Phase 1

The details for each core partner can be found at [5]:

1. The University College London (UCL) led the project and took a substantial role in "management," "biomedical research activities," and "workflow and performance."
2. The University of Amsterdam (UvA) brought in expertise in computational biomedicine, modeling and simulation, and HPC and led "training and outreach."
3. EPCC, the University of Edinburgh's supercomputing centre, brought its experience in collaborating on large-scale projects to CompBioMed and led "innovation and sustainability." They also provided access to two of their supercomputers, ARCHER2, and a GPU cluster, Cirrus.

4. SARA is the National Supercomputing and e-Science Support Centre in the Netherlands and used their experience in porting and scaling of applications and led "usage and operations." They also provided access to their supercomputer, Cartesius, and HPC clusters.
5. The Barcelona Supercomputing Center (BSC) led "biomedical research activities" with a substantial contribution to "empowering biomedical applications." They also provided access to their supercomputer, MareNostrum.
6. The University of Oxford (UOXF) played a leading role in cardiovascular exemplar research and played a substantial role in the adaptation to commodity high-end infrastructure.
7. The University of Geneva (UNIGE) worked within "biomedical research activities" on simulation and modeling and "empowering biomedical applications."
8. The University of Sheffield (USFD) led "empowering biomedical applications" and headed the neuro-musculoskeletal exemplar as part of "biomedical research activities."
9. CBK Sci Con Limited is a consultancy that offers technical and management advice to business in e-science domains and provided commercial management for the project and supported UCL in management and, with a comprehensive network, consulted with industrial partners.
10. Universitat Pompeu Fabra (UPF) played a substantial role in "biomedical research activities" within the molecular medicine exemplar and in "empowering biomedical applications."
11. LifeTec Group is a medical technology-driven SME offering high-tech R&D services and smart customized solutions to accelerate innovations in healthcare and played a substantial role in "innovation and sustainability"; however, their main tasks were the development and application of cardiovascular and orthopedic in silico models and tools in collaboration with other partners.
12. Acellera is a Barcelona-based company that focused on providing modern technologies for the study of biophysical phenomena and provided technology and know-how for large-scale molecular simulations in "biomedical research activities," "innovation and sustainability," and "empowering biomedical applications."
13. Evotec is an international drug discovery solutions company and, as a leading industrial application partner, was responsible for four key objectives: adaptation of modeling protocol to HPC platform, developing of new HGMP-HPC-based tools, testing and application of HGMP-HPC integrated technology, and dissemination of the results.

14. Bull (AtoS) is the trusted partner for enterprise data and was heavily involved in "upscaling of CompBioMed production applications for future HPC platforms" and in "innovation and sustainability." They also contributed to (big) data management aspects of the project within "integrating compute and data infrastructure."
15. Janssen Pharmaceutica NV is the Belgian affiliate of Janssen, pharmaceutical companies of Johnson & Johnson, and were heavily involved in "biomedical research activities" and worked closely in tasks relating to free energy binding calculations and GPCR modeling.

1.7.3 Phase 2

The details for each core partner can be found in [6]:

1. UCL leads the overall project and takes a substantial role in "project management, dissemination, and innovation" and "research and applications."
2. UvA leads "research and applications," bringing in two biomedical applications, and contributes to multiscale UQ algorithms and UQPs, to exascaling of codes, to training and dissemination activities.
3. EPCC's major participation is in "operations and services," along with "dissemination and innovation." They also provide access to two of their supercomputers, ARCHER2, and a GPU cluster, Cirrus.
4. SURF is using their experience in porting and scaling of applications while leading "operations and services." They also provide access to their supercomputer, Snellius, and HPC clusters.
5. BSC is leading "incubator applications" with their representative acting as application manager in the executive board.
6. UOXF plays a leading role under cardiovascular and molecular medicine exemplar research and will play a substantial role in high-performance data and for in silico trials.
7. UNIGE is working within "research and applications" on simulation and modeling, their implementation on HPC platforms, and their coupling with other applications.
8. USFD is leading "engagement, training, and sustainability" and is working within the neuro-musculoskeletal exemplar as part of "research and applications."
9. CBK Sci Con Ltd. takes a significant role in "project management, dissemination, and innovation," leading the dissemination and innovation activities. It also contributes to engagement and sustainability tasks, as well as those related to commercialization.

10. UPF plays a substantial role in "research and applications," particularly in the molecular medicine exemplar. In "incubator applications," UPF provides expertise in high-throughput MD simulations and machine learning (ML) methods for drug discovery.
11. The Leibniz Supercomputing Centre (Leibniz-Rechenzentrum, LRZ), new to CompBioMed, is leading "data management and analytics" and provides access to their supercomputer, SuperMUC-NG.
12. Acellera plays a substantial role in research and development of innovative applications and their commercialization as well as in data and analytics.
13. Evotec is responsible for three key objectives: adaptation of hierarchical GPCR modeling protocol (HGMP) to HPC platform, development of new HGMP-HPC-based tools/plugins, and testing and application of HGMP-HPC integrated technology.
14. Bull (AtoS) is strongly involved in co-design activities on the road to the future of supercomputing, so-called exascale. Indeed, Bull is especially active in optimizing CompBioMed2 applications.
15. Janssen has a substantial role in "research and applications," particularly tasks relating to development of science for molecular medicine.
16. The University of Bologna, new to CompBioMed, contributes to "engagement, training, and sustainability" and the development of computational medicine solutions for neuro-musculoskeletal conditions.

International Partners

1. The Argonne National Laboratory is a multidisciplinary science and engineering research centre that supports CompBioMed2 by directly supporting the application of machine learning techniques to drug design, drug resistance, and other challenging biomedical research problems.
2. Rutgers University strengthens CompBioMed by directly supporting the use and adaptation of RADICAL-Cybertools within the project.

1.7.4 Associate Partners

The CompBioMed CoE has built strong and fruitful collaborative relationships between our core and associate partners by sharing resources and ideas. Our CoE focuses on the end user and brings in institutions that benefit from and are of benefit to our CoE. The associate partner scheme allows these institutions to interact more proactively with us, and us with them. We have almost 50 associate partners [7] that are keen to continue growing. Partners include academic and other research institutes and industrial partners. The former includes HPC centres, while the latter includes cloud providers and SMEs.

The list of opportunities and benefits for associate partners includes access to and provision of project resources including HPC facilities, software, and training materials; invitations to project meetings, workshops, and training events; participation in our visitor program; incubator coordination; potential inclusion in our External Expert Advisory Board; and a listing of software services on the CompBioMed website.

1.8 Biomedical Software: Core Applications

We present all our core applications within our software hub [8], wherein the biomedical community can access the resources developed, aggregated, and coordinated by CompBioMed. This includes access to the application itself, user documentation, training materials, links to scientific publications, and user support. Information on each of the following core applications can be found in [8] and are summarized below.

1.8.1 Alya

Alya, developed at BSC, performs cardiac electromechanics simulations, from tissue to organ level. The simulation involves the solution of a multiscale model using a FEM-based electromechanical coupling solver, specifically optimized for the efficient use of supercomputers.

1.8.2 HemeLB

HemeLB, developed at UCL, is a 3D macroscopic blood flow simulation tool that has been specifically optimized to efficiently solve the large and sparse geometries characteristic of vascular geometries. It has been used to study flow in aneurysms, retinal networks, and drug delivery, among many other cases.

1.8.3 HemoCell

HemoCell, developed at UvA, is a parallel computing framework for simulation of dense deformable capsule suspensions, with special emphasis on blood flows and blood-related vesicles (cells). The library implements validated mechanical models for red blood cells and can reproduce emergent transport characteristics of such complex cellular systems.

1.8.4 Palabos

Palabos, developed at UNIGE, is a general software library for computational fluid dynamics using the lattice Boltzmann method. In biomedical research, the main domains of applications are cardiovascular, including the simulation of blood flow in arteries, the investigation of the effect of medical devices such as stents, and cell-level blood simulations to investigate fundamental blood properties.

1.8.5 Binding Affinity Calculator

The binding affinity calculator (BAC), developed at UCL, is a workflow tool that runs and analyzes simulations designed to assess how well drugs bind to their target proteins and the impact of changes to those proteins. Use of ensemble simulations enables robust, accurate, and precise free energy computations from both alchemical and end-point analysis methodologies. BAC uses high-level python object abstractions for defining simulations, physical systems, and ensemble-based free energy protocols which wrap around common molecular dynamics codes.

1.8.6 CT2S, ARF10, and BoneStrength

BoneStrength, CT2S, and ARF10 are a set of tools to model the risk of hip fracture developed by the USFD and University of Bologna. The Computed Tomography to Strength (CT2S) workflow is a digital twin solution, used for the prediction of the risk of hip fracture for an individual based on CT scans. ARF10 is a multiscale model for the prediction of 10-year hip fracture which uses CT2S. Based on the ARF10 workflow, the University of Bologna is developing the in silico trial solution BoneStrength.

1.8.7 openBF

openBF, developed at the Insigneo Institute at USFD, is an open-source 1D blood flow solver based on the MUSCL finite-volume numerical scheme. The software allows the wave propagation problem in 1D models of the human cardiovascular system to be solved and the prediction of flow rate, pressure, and other hemodynamic variables across the network.

1.8.8 PlayMolecule

PlayMolecule is a drug discovery web service developed and maintained by Acellera. It contains a variety of applications that allow users to accelerate and improve their drug discovery workflows using novel machine learning methods (such as binding affinity predictors) or through molecular dynamics simulations to elucidate biological structures and binding modes.

1.8.9 TorchMD

TorchMD, developed at UPF, is used for the development and usage of machine learning potentials for molecular simulations. TorchMD is divided into three different parts/applications: TorchMD itself, which is an end-to-end differentiable molecular dynamics code; TorchMD-NET, which is a software for training and creation of neural network potentials; and TorchMDCG, which is a specific application to learn coarse-grained potentials, currently specialized in protein folding.

1.8.10 Virtual Assay

Virtual Assay, licensed through Oxford University Innovation, is used to perform simulations of drug effects in populations of human ventricular cells and to predict potential side effects of the drugs on the human heart. Virtual Assay has been designed to be accessible by everyone, including users with little or no expertise in computer modeling and simulations.

2 Selected Key Achievements

2.1 Collaborations

One of the most impressive achievements of CompBioMed is how well the core partners have worked with each other to produce such high-quality results. The willingness to contribute, the quality of the work, the ability to hit deadlines, and the overall camaraderie have been most rewarding. Through these collaborations, we have produced a number of exemplars of applying supercomputers and HPC for biomedicine.

Thanks to collaborations with CompBioMed's HPC centre partners, the applications HemeLB, Palabos, Alya, and BAC have all now demonstrated excellent preparedness for exascale supercomputers.

Collaborations have also produced a number of exemplars in computational biomedicine. Exemplars in cardiovascular medicine, from the end of phase 1, were as follows:

- "Digital blood" and flow-diverting stents within the Palabos application (UvA, UNIGE, Bull): digital blood in massively parallel CPU/GPU systems (under experimental validation) and flow-diverting devices for intracranial simulations (under clinical validation for patient-specific scenarios)
- Alya cardiac computational model using the Alya application (BSC, UOXF): fluid-electromechanical cardiac model of anti-arrhythmic drug cardiotoxicity and cardiac resynchronization (under clinical assessment with 12 patients).
- Coupling of the two applications HemeLB and Alya, for cardiovascular flow in virtual human-scale geometries (UCL, BSC): 3D flow in arterial networks (under clinical validation).
- The OpenBF application for vascular networks (USFD, SARA): cerebral vasospasm (under verification and validation assessment).
- The AngioSupport application for coronary artery disease (LTG, SARA): developed an interactive tool to support coronary interventions (under clinical assessment with 72 patients).

Exemplars in molecular medicine, from the end of phase 1, were as follows:

- Machine learning and molecular dynamics (UPF, Acellera, Pfizer, Janssen, Argonne, Rutgers, BNL).
- Molecular dynamics for drug discovery programs (Janssen, UCL, UPF, Acellera): affinity prediction, docking, and drug discovery, via a computational triage of similar molecules, used to improve drug discovery on Janssen portfolio.
- Supercomputers and binding affinities (UCL, Janssen, GSK): computational methods for initial stages in the drug discovery,

with a diverse range of drug discovery targets, especially in cancer.

- The development of computational methods for modeling the structures of G-protein-coupled receptors (GPCRs) is being carried out collaboratively by Evotec and UCL. These methods are intended for use in the initial stages of drug discovery, aiming to enhance the understanding of GPCRs' structures, their activation processes and their interactions with small drug like molecules.

Around 30% of both our core and associate partners are from industry. New collaborations created between both industrial and academic partners have enabled us to hone our incubation activities, and our visitor program funds visits to industrial companies to and from academic centres.

I helped form and run our External Expert Advisory Board along with key influencers from both hospitals and industry, particularly SMEs. Membership is currently listed [9] as follows: Peter Coveney (Chair), University College London; Paul Best, CBK; Philippe Bijlenga, Geneva University Hospitals; Alexander Heifetz, Evotec; Brendan Bouffler, Amazon AWS; Enrico Gianluca Caiani, Politecnico di Milano; Francesc Carreras, Hospital de la Santa Creu i Sant Pau; David Filgueiras, the Centro Nacional de Investigaciones Cardiovasculares Carlos III (CNIC); Malcolm Finlay, NHS and UCL; Neil Chue Hong, Software Sustainability Institute; Mahmood Mirza, Neuravi, Ireland; Ana Paula Narata, Tours Hospital; Mark Palmer, Medtronic; Gavin Pringle, EPCC; Frits Prinzen, Maastricht University; Brad Sherborn, MERCK, USA; Sarah Skerratt, Vertex Pharmaceuticals, Stanford University, and Oxford University; Derek Sweeney, CADFEM Ireland; and Kenji Takeda, Microsoft Azure.

2.1.1 ELEM Biotech

ELEM Biotech [10] is a biomedical company which develops a simulation environment for tissue and organ levels. ELEM Biotech is a start-up from BSC, spun out with CompBioMed assistance, and was incorporated in July 2018. The Barcelona-based company works with hospitals to model heart conditions for real patients, providing realized personalized medicine. They provide supercomputer-based simulations of the cardiovascular system for medical device manufacturers, pharmaceutical industry, CROs, and academia and have projects with Medtronic [11] (cardiac mechanics and morphine pumps for spinal fluid) and iVascular [12] (stent manufacturing and deployment).

2.2 IMAX Films

2.2.1 The Virtual Humans IMAX Film

CompBioMed dedicated considerable effort toward the "Virtual Humans" IMAX film, available via our YouTube page [13], released in September 2017, which has reached over 2.5 million viewers to date.

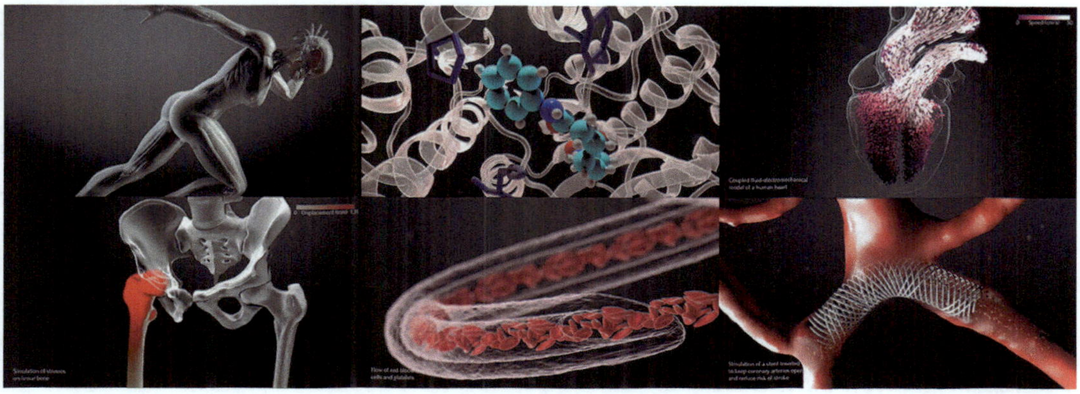

Fig. 1 Screenshots from the Virtual Humans IMAX film

The Virtual Humans film describes the core concept of computational biomedicine, looking toward a future where digital avatars are used to inform medical decisions.

The film itself was produced using structural data taken directly from simulations and enhanced using advanced graphics techniques and then rendered in 4 k resolution on MareNostrum at BSC. Six screenshots from the film are shown in Fig. 1.

2.2.2 The Next Pandemic IMAX Film

A sequel IMAX film to the Virtual Humans IMAX film was recently released, named The Next Pandemic, which explores supercomputers as a new weapon against the next pandemic, harnessing supercomputers to explore the spread of the virus through the air, the effectiveness of masks, simulations revealing key insights for antiviral drug development, predictions related to vaccines, treatments, or public health campaigns. Six screenshots from the film are shown in Fig. 2.

2.3 Creating a Culture of HPC Among Biomedical Practitioners

There are only a few medics using HPC in Europe, which necessitates strategies for engaging healthcare professionals in a technology that has brought such significant benefits to other sectors. CompBioMed introduces the concept of a virtual human to practitioners in the medical and life sciences not only via our dissemination activities, such as our IMAX short films; our regular e-seminar series (that I coordinate) [14], workshops, and conferences; and—the topic of this section—through university lectures and courses. Via teaching, we create a culture of HPC among biomedical practitioners, which is key to expediting their use of HPC, and embedding HPC biomedicine in university curricula is effective in achieving this.

The student selected component (SSC) of any medical school curriculum worldwide is a taught course that allows students to choose a specific topic in a medically relevant area of interest to

Fig. 2 Screenshots from The Next Pandemic IMAX film

them, and this SSC provides an ideal opportunity to formally integrate HPC into the medical curriculum.

In 2017, I worked with UCL and SURF to create an SSC course entitled HPC for Medics. I co-wrote and presented the course, and we helped the UCL medical students, who had no command-line computing experience whatsoever, to efficiently run parallel applications on remote HPC clusters in EPCC and SURF. This proved so popular that it has since expanded to run on HPC systems at BSC, and USHEFF, and on cloud HPC from our associate partner Alces Flight and via the three modern-day delivery modalities: in-person, remotely, and hybrid. The course is currently successfully embedded in curriculum at three universities: UCL for 6 years, USFD for 3 years, and UPF for 2 years, and work is underway with UvA and UOXF to deliver training at these institutions. To date, we have delivered HPC training to ca 2000 students in (bio)medical fields who would not have typically received this.

The course itself provides students with a theoretical understanding of the importance of the relationship between human microbiomes—the microorganisms present in and on the human body—and human health and provides them with the practical opportunity to use state-of-the-art computational resources to run a metagenomics pipeline. These courses have run with students collecting their own microbiome data and performing experimental work to determine the microbial sequences that are analyzed computationally. During the COVID-19 lockdown, microbiome data from databases replaced the student's own sequences, which added data management to the HPC work.

An expanded version of the metagenomics SSC has been used since 2017–2018 for a third year undergraduate research project for students taking MSc/BSc Biochemistry and BSc Molecular Biology at UCL, to complete research that integrates both experimental and computational work.

Currently, UCL, USFD, and UPF are collaborating with the associate partner Dassault Systèmes [15], to develop an HPC-based SSC in cardiovascular research, and there are plans to extend further to neuro-musculoskeletal research.

The use of our training sessions helps to promote engagement with medical students and biomedical science practitioners to incorporate HPC in their future clinical work. As such, CompBioMed is helping to establish a culture of integrating HPC with experimental/clinical methods to inform university biomedical education and, subsequently, professional practice.

2.4 Free Support to Enable and/or Optimize Applications for HPC

CompBioMed offers free support, via our so-called scalability service [16], to organizations in their initial steps toward either parallelizing their existing computational biomedical applications or improving the performance of those applications already parallelized and thereafter deploying them on HPC platforms.

The service boosts the performance of biomedicine applications via a range of support routes, such as informal discussions about efficient use of parallel platforms in general and code reviews, to porting and profiling applications and suggesting improvements, or working closely with the client and adapting the source code on their behalf. The service has published our agreed SLA and an internal helpdesk to ensure no request for support is missed.

Our website provides links to a group email of experts in both HPC and biomedical applications, and access to the public Slack channel #scalability, hosted by "In Silico World" Community of Practice [17], which provides a safe space to share scaling questions. The website also proffers both a detailed application form and a more informal web-form version.

Notably, we present a renowned overview for programmers with ideas on improving the performance of parallel applications for supercomputers [18], developed in collaboration with HPC experts from CompBioMed and another CoE, EXCELLERAT [19].

Many biomedical applications deal with sensitive data; thus, clients are assured that great care is taken when adapting their applications and managing the associated data: the website provides our data policies, which cover data privacy, data security, and research data management.

To date, the service has been exploited by both experts and students from across the EU. Further, we have improved the performance and portability of our own key applications through a combination of both this service and our own co-design efforts.

2.5 Providing HPC to Surgery

In the field of personalized medicine, one example of urgent computing is the placement of a stent (or flow diverter) into a vein in the brain: once the stent is inserted, it cannot be moved or replaced. Surgeons of the future will use large-scale simulations of stent placement, configured for the individual being treated. These simulations will use live scans to help identify the best stent, along with its location and attitude.

CompBioMed prepares biomedical applications for future exascale supercomputers, where these machines will have a remarkably high node count. Individual nodes have a reasonable mean time to failure; however, when you collect hundreds of thousands of nodes together, the overall mean time to failure is much higher. Moreover, given exascale applications will employ MPI, and a typical MPI simulation will abort if a single MPI task fails, and then a single node failure will cause an entire simulation to fail.

Computational biomedical simulations may well employ time- and safety-critical simulations, where results are required at the operating table faster than real time. Given we will run these urgent computations on machines with an increased probability of node failure, one mitigation is to employ resilient HPC workflows.

Two classes of such workflows employ replication. The first resilient workflow replicates computation, where the same simulation is launched concurrently on multiple HPC platforms. Here the chances of all the platforms failing are far less than any individual; however, such replication can prove expensive, especially when employing millions of cores. The second resilient workflow replicates data, where restart files are shared across a distributed network of data and/or HPC platforms. Then, a simulation that finds its host HPC fail will simply continue from where it left off on another platform.

In both classes, the simulation's results will be available as if no failure had occurred; however, the first class (replicated computation) will produce results quicker, as the simulation will not have to wait in a second batch system. On the other hand, from a personal experience, the time to coordinate multiple HPC platforms to simulate concurrently feels like it increases as the cube of the number of platforms. As such, the total turnaround time of the second class (replicated data) is more likely to be faster. Turnaround might be reduced further via the use of batch reservations.

The LEXIS project [20] has built an advanced engineering platform which leverages large-scale geographically distributed resources from the existing HPC infrastructure, employs big data analytics solutions, and augments them with cloud services.

LEXIS has the first class of reliant workflow in its arsenal. EPCC is working with LEXIS using the UvA's HemoFlow application [21], to create an exemplar of the second class: resilience via data replication.

For our workflow, we have created both a data network and an HPC network. The data network is based on nodes of CompBioMed and of the LEXIS "Distributed Data Infrastructure," relying on the EUDAT-B2SAFE system [22]. The HPC network includes five HPC systems at EPCC, LRZ, and IT4I. Both these networks are distributed across countries to mitigate against a centre-wide failure, e.g., power outage. The application is ported to all the HPC systems in advance. The input data resides on the LEXIS Platform and replicated across the data nodes. The essential LEXIS Platform components are set up across their core data centres (IT4I, LRZ, ICHEC, and ECMWF), with redundancy built in for failover and/or load balance. These components include an advanced workflow manager, namely, the LEXIS Orchestrator System.

The test workflow will progress as follows. The workflow manager launches the HPC simulation. Restart files are created at regular intervals and the workflow manager replicates these across all the data nodes. A node failure is then emulated, which causes the entire MPI simulation to fail. This failure triggers the workflow manager to restart the simulation on one of other remote HPC platforms using the latest restart files, pre-staged from the nearest data node. An automated choice of platform ensures the fastest turnaround as performed by the LEXIS Platform's broker.

The staging of data for exascale simulations must naturally consider both the amount of data to be moved and the bandwidth required; however, given biomedical simulations can contain patient-sensitive data, and data staging must also ensure the staging follows all relevant legal requirements, including the common FAIR principles of research data management.

Exascale supercomputers bring new challenges and our resilient HPC workflow mitigates the low probability but high-impact risk of node failure for urgent computing. These supercomputers are emerging and present an exciting opportunity to realize personalized medicine via ab initio computational biomedical simulations which, in this case, will provide live, targeted guidance to surgeons during lifesaving operations.

2.6 FDA-Endorsed Credibility to Biomedical Simulations

Finally, a key achievement is our work in verification, validation, and uncertainty quantification (VVUQ), which is required to augment the software credibility, which may then establish FDA-endorsed biomedical models and simulations.

Validation means that the simulation software correctly reproduces the multiple physics of the question of interest for a determined context of use. This not only requires correctly solving the programmed model, but that the model effectively models the physics. To do so, experimental data is required to compare ex vivo, in vitro, and in vivo data against in silico results. This stage requires a detailed description of the variables of the physical problem which are, in most of the cases, difficult to obtain with a high accuracy.

The goal of the risk-informed credibility assessment framework is to empower the medical device industry and the regulatory agencies to determine and justify the appropriate level of credibility for using a computational model to inform a decision which is internal to an organization or part of a regulatory activity.

Standardization of the VVUQ process for medical devices has been addressed by the ASME V&V 40 Subcommittee, with the release of "Assessing Credibility of Computational Modeling through Verification and Validation: Application to Medical Devices" [23], augmenting V&V 10 and V&V 20 methodologies.

The risk-informed framework initiates with the definition of a context of use which details the specific role and scope of the computational model. The model's risks depend on its influence and the decision consequences which, in turn, leads to the establishment of the credibility goals for each of the model's V&V stages.

CompBioMed is developing a unified strategy and analysis to perform VVUQ techniques for biomedical device models. To date, code-specific strategies and risk analyses have been established for OpenBF, CT2S, BBCT/BoneStrength, and Alya Red.

Executing these V&V methodologies for complex simulation codes that reproduce physical behaviors is a challenging task by itself that then drives innovation in each research field. Credible numerical software is mandatory for industrial adoption and clinical translation of simulation code, and this makes VVUQ mandatory to engage with a range of industries operating across the entire healthcare value chain, from healthcare providers to pharmaceutical and medical device manufacturers.

References

1. EPCC (2023). https://www.epcc.ed.ac.uk. Accessed 5 Apr 2023
2. CompBioMed (2023). https://www.compbiomed.eu. Accessed 5 Apr 2023
3. Horizon2020 (2020). https://research-and-innovation.ec.europa.eu/funding/funding-opportunities/funding-programmes-and-open-calls/horizon-2020_en. Accessed 5 Apr 2023
4. VPH Institute (2023). https://www.vph-institute.org. Accessed 5 Apr 2023
5. CompBioMed (2021) CompBioMed phase 1 core partners. https://www.compbiomed.eu/about/core-partners-new. Accessed 5 Apr 2023
6. CompBioMed (2023) CompBioMed phase 2 core partners. https://www.compbiomed.eu/about/compbiomed2-core-partners. Accessed 5 Apr 2023
7. CompBioMed (2023) CompBioMed associate partners. https://www.compbiomed.eu/associate-partners. Accessed 5 Apr 2023
8. CompBioMed (2023) CompBioMed software hub. https://www.compbiomed.eu/compbiomed-software-hub. Accessed 5 Apr 2023
9. CompBioMed (2021) CompBioMed external expert advisory board. https://www.compbiomed.eu/innovation/external-expert-advisory-board. Accessed 5 Apr 2023
10. Elem biotech (2023). https://www.elem.bio. Accessed 5 Apr 2023
11. Medtronic (2023). https://www.medtronic.com/uk-en/index.html. Accessed 5 Apr 2023
12. iVascular (2023). https://ivascular.global. Accessed 5 Apr 2023
13. CompBioMed (2018) CompBioMed virtual humans film. https://www.youtube.com/watch?v=1FvRSJ9W734. Accessed 5 Apr 2023
14. CompBioMed (2023) CompBioMed e-seminar series. https://www.compbiomed.eu/compbiomed-e-seminars. Accessed 5 Apr 2023
15. Dassault Systèmes (2023). https://www.3ds.com. Accessed 5 Apr 2023

16. CompBioMed (2023) CompBioMed scalability service. https://www.compbiomed.eu/scalability. Accessed 5 April 2023
17. In Silico World (2023). https://insilico.world/community. Accessed 5 Apr 2023
18. Pringle GJ et al (2023) Rough guide to preparing software for Exascale. https://www.compbiomed.eu/rough-guide-to-preparing-software-for-exascale. Accessed 5 Apr 2023
19. EXCELLERAT (2023) EXCELLERAT P2 the European Centre of Excellence for Engineering Applications. https://www.excellerat.eu. Accessed 5 Apr 2023
20. LEXIS Project (2023). https://www.lexis-project.eu. Accessed 5 Apr 2023
21. Závodszky G (2023) HemoFlow. https://www.zavodszky.com. Accessed 5 Apr 2023
22. EUDAT (2020) EUDAT B2SAFE. https://www.eudat.eu/b2safe. Accessed 5 Apr 2023
23. The American Society of Mechanical Engineers (2018) Assessing credibility of computational modeling through verification and validation: application to medical devices, V&V 40. https://www.asme.org/codes-standards/find-codes-standards/v-v-40-assessing-credibility-computational-modeling-verification-validation-application-medical-devices. Accessed 5 Apr 2023

Chapter 4

In Silico Clinical Trials: Is It Possible?

Simon Arsène, Yves Parès, Eliott Tixier, Solène Granjeon-Noriot, Bastien Martin, Lara Bruezière, Claire Couty, Eulalie Courcelles, Riad Kahoul, Julie Pitrat, Natacha Go, Claudio Monteiro, Julie Kleine-Schultjann, Sarah Jemai, Emmanuel Pham, Jean-Pierre Boissel, and Alexander Kulesza

Abstract

Modeling and simulation (M&S), including in silico (clinical) trials, helps accelerate drug research and development and reduce costs and have coined the term "model-informed drug development (MIDD)." Data-driven, inferential approaches are now becoming increasingly complemented by emerging complex physiologically and knowledge-based disease (and drug) models, but differ in setup, bottlenecks, data requirements, and applications (also reminiscent of the different scientific communities they arose from). At the same time, and within the MIDD landscape, regulators and drug developers start to embrace in silico trials as a potential tool to refine, reduce, and ultimately replace clinical trials. Effectively, silos between the historically distinct modeling approaches start to break down. Widespread adoption of in silico trials still needs more collaboration between different stakeholders and established precedence use cases in key applications, which is currently impeded by a shattered collection of tools and practices. In order to address these key challenges, efforts to establish best practice workflows need to be undertaken and new collaborative M&S tools devised, and an attempt to provide a coherent set of solutions is provided in this chapter. First, a dedicated workflow for in silico clinical trial (development) life cycle is provided, which takes up general ideas from the systems biology and quantitative systems pharmacology space and which implements specific steps toward regulatory qualification. Then, key characteristics of an in silico trial software platform implementation are given on the example of *jinkō.ai* (nova's end-to-end in silico clinical trial platform). Considering these enabling scientific and technological advances, future applications of in silico trials to refine, reduce, and replace clinical research are indicated, ranging from synthetic control strategies and digital twins, which overall shows promise to begin a new era of more efficient drug development.

Key words Modeling and simulation (M&S), Knowledge, In silico trials, Clinical trials, Systems biology, Systems pharmacology, Software platform, Drug development, Drug regulation, Health technology assessment, Synthetic control arms, Digital twins

Alexander Heifetz (ed.), *High Performance Computing for Drug Discovery and Biomedicine*, Methods in Molecular Biology, vol. 2716, https://doi.org/10.1007/978-1-0716-3449-3_4,

Abbreviations

AB	Absolute benefit
ADME	Absorption, distribution, metabolism, and excretion
ANOVA	Analysis of variance
BMI	Body mass index
CHMP	Committee for Medicinal Products for Human Use
CM	Computational model
CMAES	Covariance matrix adaptation evolution strategy
CoU	Context of use
COVID-19	Coronavirus disease
DPM	Disease progression model
EMA	European Medicines Agency
FDA	US Food and Drug Administration
GSA	Global sensitivity analysis
HTA	Health technology assessment
ICH	International Council for Harmonisation
K-M	Kaplan–Meier
KM	Knowledge model
LDLc	Low-density lipoprotein cholesterol
M&S	Modeling and simulation
MIDD	Model-informed drug research and development
MIDDD	Model-informed drug research, development, and discovery
NASH	Nonalcoholic steatohepatitis
NLEM	Nonlinear mixed effect modeling
NLP	Natural language processing
nova	Novadiscovery
NPE	Number of prevented events
NPI	Nonpharmaceutical intervention
ODE	Ordinary differential equation
PBPK	Physiologically based pharmacokinetic
PD	Pharmacodynamics
PK	Pharmacokinetic
popPK	Population pharmacokinetics
QOI	Questions of interest
QSP	Quantitative systems pharmacology
R&D	Research and development
RCT	Randomized clinical trial
RSV	Respiratory syncytial virus
RTI	Respiratory tract infection
SBGN	Systems biology graphical notation
SBML	Systems Biology Markup Language
SED-ML	Simulation Experiment Description Markup Language
SIRS	Susceptible, infectious, recovered, and susceptible
SoE	Strength of evidence
V&V	Verification and validation
Vpop	Virtual population

1 In Silico Trials Help Solve a Growing Drug Development Challenge

An apparent rise of costs and decline of drug research and development (R&D) efficiency (inverse to the lower cost and increased computer power) has been diagnosed [1]. This trajectory challenges the current process to bring new drugs to patients into question. Clinical trials are the most expensive part of development; they are also time-consuming and often burdensome for patients. That is why selecting the right protocol design, and the most appropriate patient groups in confirmatory trials, can make the difference between a drug reaching patients or not. Ideally, developers should test multiple doses, drug combinations, or regimens in all patients, but that is unfeasible for combinatorial reasons. Developers therefore urgently need new tools to help match drug, dose, and drug combinations to the patients most likely to benefit. Is it therefore not logical to consider in silico (digital) counterparts of entire clinical trials to provide digital evidence to drug developers' and regulators' and policy makers' decision-making?

In fact, in silico clinical trials—based on simulation of clinical trial outcomes for virtual patients—is not so far from reality today.

Modeling and simulation (M&S) has been considered for quite some time as a potential solution to boost drug development efficiency [2]. Drug developers and regulators increasingly see in silico approaches as part of a new era of a more ethical, cost-efficient, and evidence-based medicine. Ultimately, in silico approaches will constitute a third R&D pillar alongside in vitro and in vivo testing and would be applied as early as possible in development. In silico techniques, broadly defined, are already widely used within drug development. They are part of pharmacometrics—the quantification of drug, disease, and trial information to aid R&D and regulatory decision-making [3]. Pharmacokinetic (PK) models describe how a drug's concentration in the body changes over time [4], while pharmacodynamics (PD) models capture the body's observed physiological response. These two combined may be referred to as PK/PD models and are frequently used to uncover dose–exposure–response relationships (thereby constituting a core component of new drug development) [5].

Traditional PK/PD models are simple, using only a few variables. More complex quantitative models of drug–body interactions have emerged over the last two decades, thanks to advances in knowledge management, computer processing, data science, and data sources.

Physiologically based pharmacokinetic (PBPK) modeling captures drug distribution and metabolism effects within a more physiologically relevant context, incorporating blood flow and tissue composition, for instance, plus drug–drug interactions, bioequivalence, and cross-population effects [6].

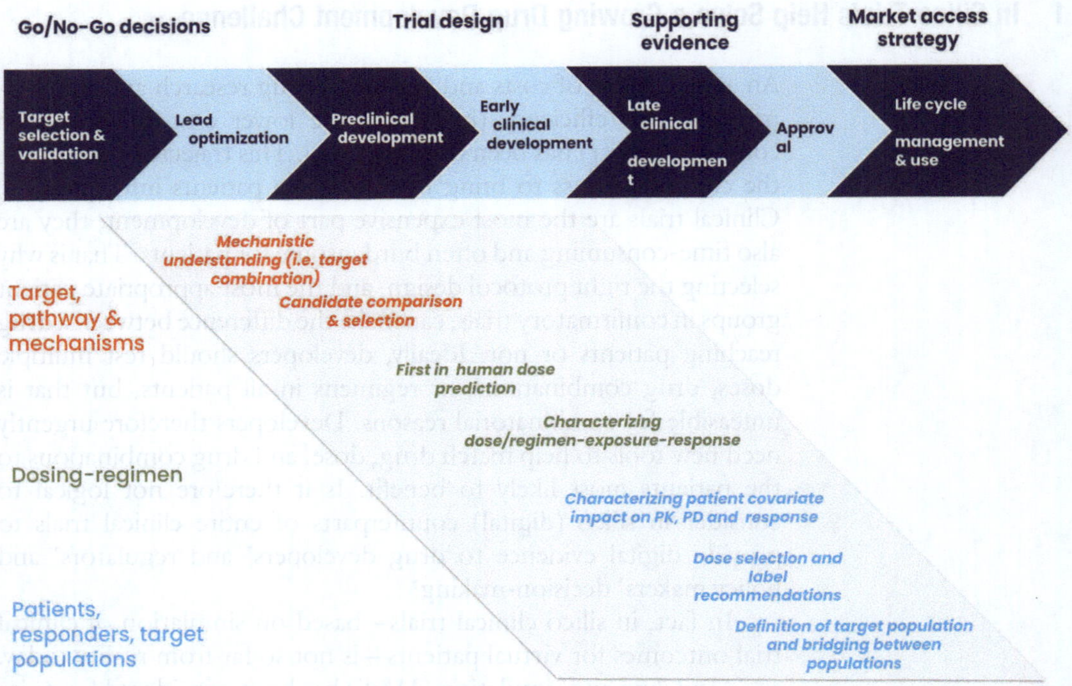

Fig. 1 Alignment of clinical development stages (top), central themes (left), and questions and decisions informed by MIDD

Quantitative systems pharmacology (QSP) approaches go even further, helping uncover and explore how drugs interact with a given target, biomarker, organ, or even organism to impact clinical outcomes [7–9]. Opposed to simpler models, QSP models promise to leverage a large body of preexisting knowledge, meaning they capture, describe, and may even help explain physiological behavior. They are ideal hypothesis testing tools and can be aligned with translational research (often in conjunction with PBPK) and thus have been most commonly applied early in the development of a drug.

Today, QSP approaches power applications across the drug value chain—from identifying biomarkers and informing trial design to supporting health economic assessment [10] (*see* Fig. 1). A recent survey reported that the top five decisions impacted by QSP are dosing/scheduling, drug combination choices, trial design, and biomarker-related, and better response and patient stratification [11].

As in silico tools have evolved to help fill data gaps in drug development, regulatory authorities have adapted to accommodate in silico evidence. The US Food and Drug Administration (FDA) uses the term "model-informed drug development" (MIDD) [12] to capture all drug exposure-based, (systems) biological, and statistical models derived from preclinical and clinical data sources to inform drug development and decision-making [13] (*see* Box 1 for more details).

Box 1: Model-Informed Drug Development (MIDD) Landscape

The most widespread fields of application within MIDD (or MIDDD including discovery [18]) with regulatory impact are population pharmacokinetics (popPK [19]) and pharmacodynamics (PK/PD [5]), exposure–response relationship modeling [20], and trial simulation, also termed disease–drug–trial models [21]. For the latter, data resembling a real clinical trial are generated and statistical analysis of that data can inform about protocol-dependent probability of achieving the target value, probability of success, and probability of a correct decision to support study design recommendations and quantitative decision-making [22–24]. Another class of models counting to the MIDD ecosystems are disease progression models (DPMs) that aim at integrating previously obtained disease knowledge to elucidate the impact of novel therapeutics or vaccines on disease course [25] (see an example projecting physiological knowledge, pivotal and follow-up trial data-based projection of hormone replacement in hypoparathyroidism on kidney disease progression [26, 27]). Modeling techniques that are applied in health technology assessment (HTA) can be found, e.g., in [28] and references within.

In 2017, recognizing modeling's potential to improve and accelerate drug R&D, the FDA launched an MIDD pilot program to explore how modeling can be used in drug development. The pilot aims to set out appropriate regulatory advice, including what kind of MIDD data should be reported in regulatory submissions [14]. Industry embraced the program, recognizing its role in encouraging implementation of MIDD strategies [15].

The European Medicines Agency (EMA)'s Committee for Medicinal Products for Human Use (CHMP) has pooled relevant expertise within a Methodology Working Party [16] which deals with all interactions involving in silico modeling and biostatistical approaches.

Efforts are underway to standardize the use of modeling in clinical development. In 2022, the International Council for Harmonisation of Technical Requirements for Pharmaceuticals for Human Use (ICH) released a timetable detailing its plans to issue general guidance on MIDD approaches in drug development [17].

What are then precisely in silico trials?

As per a common definition, in silico trials use computer M&S to evaluate the safety and efficacy of drugs, medical devices, or diagnostics [29]. For in silico trials dedicated to drugs, integration of disease and drug mechanisms are needed to predict clinical outcomes. Some approaches include virtual patients (digital counterparts of individual patient responses) to simulate drugs' effects across populations [21].

Traditionally, in silico trials, or trial simulation, has been understood as another (distinct) member among the other MIDD tools (next to PK/PD and later QSP). In most "traditional" in silico trials, simple statistical models describe the efficacy distribution (and factors impacting this distribution; *see*, e.g., [30]) and then Monte Carlo simulation of individual patients according to a predefined study protocol. In fact, the simple efficacy distribution can be replaced by a more complex model that employs mechanistic hypotheses to explain (based on the knowledge about interactions in biological systems, *see* [31, 32] as selected examples) rather than to infer the contributions to effect variability. In that way, several well-established or hypothesized impacts on the efficacy (age physiology, genetic polymorphism, drug regimen; see the variability of questions QSP can address) on top of trial designs (eligibility criteria, durations, endpoints, arms, etc.) at the group or even individual level can be obtained.

In silico trials could already today have many more applications beyond drug development. More complex, physiological models could be used to provide health economic data projections to support cost-effectiveness decisions [28], in the (a) synthesis and (b) cost-economic projection and (c) overall appraisal of clinical data, and therefore it seems also reasonable to assume that this technique can become mature enough to even impact clinical decision-making [33].

To reach this maturity, workflows still need to be harmonized, new tools need to be developed, and more stakeholders need to be integrated into the M&S life cycle for which the following sections make concrete propositions and highlight examples.

2 A Specific Workflow for In Silico Clinical Trials Powered by Knowledge-Based Modeling

Establishing in silico clinical trials is a scientifically and technically challenging task that needs a dedicated workflow, especially when a clinical trial simulation setup needs to be wrapped around a complex model representing the knowledge about physiology, disease (progression), and treatment effects as well as factors leading to between-patient variability.

Knowledge gaps in our understanding of biological processes impede the formulation of a wildcard, all-purpose multiscale physiological model—from pathways, cells, and organs up to the organism—covering all features of potential interest under every possible condition, for every disease, and for every type of drug. Therefore, current practice is to develop systems biology models (or QSP models—the more "pragmatic" version of the systems biology models dedicated to drug development and pharmacology questions) specific to diseases, features, and even (classes of) treatments. With this specialization comes a wide variety of modeling approaches: both in model nature (ordinary or partial differential equation, agent-based, logical, etc.) and in granularity (time and length scales, breakdown of the system in organs or compartments, restriction to pathway, cellular, organ, or system's level) and scope/breadth (selection of phenomena, pathways, organs). This makes best practices or a single gold standard model setup difficult to formulate.

Nevertheless, higher-level recommendations for workflows to build, assess, and apply such models have emerged [34] in the community. Model development, knowledge review, formalization of the model, and exploration and validation of its behavior followed by generation of simulations to answer certain questions of interest (QOIs) are the usual main stages proposed in workflows found in the literature [35–38]. The exact implementation of such workflows may still vary among different organizations and among their stage in their life cycle (developers will have different workflows than end users who will also probably have different uses among first application and reuse).

Novadiscovery (nova) uses mechanistic, knowledge-based modeling to run in silico trials [39]. The nova approach integrates physiological models of disease and of drug actions (built using ordinary differential equations (ODEs)) with virtual populations (Vpops), built using statistical models accounting for inter-patient variability. The definition of a dedicated in silico trial protocol, in line with real trial protocols, allows to mimic all aspects of real-life clinical trial results.

Typical deliverables of such workflow are the project or model development plan and the computational model (CM) and its documentation including the knowledge model (KM) detailing all physiological, pathophysiological, and treatment-related components; systems and processes considered; hypotheses, simplifications, and limitations (structured per submodel); the in silico study protocol; prospective validation protocol (or plan); a validation report and research report; and simulation datasets (and/or all constituents to reproduce simulations).

The proposed dedicated workflow (Fig. 2) consists of six stages, each summarized in Table 1 with general and specific steps to knowledge-based modeling combined with trial simulation and as detailed in the following.

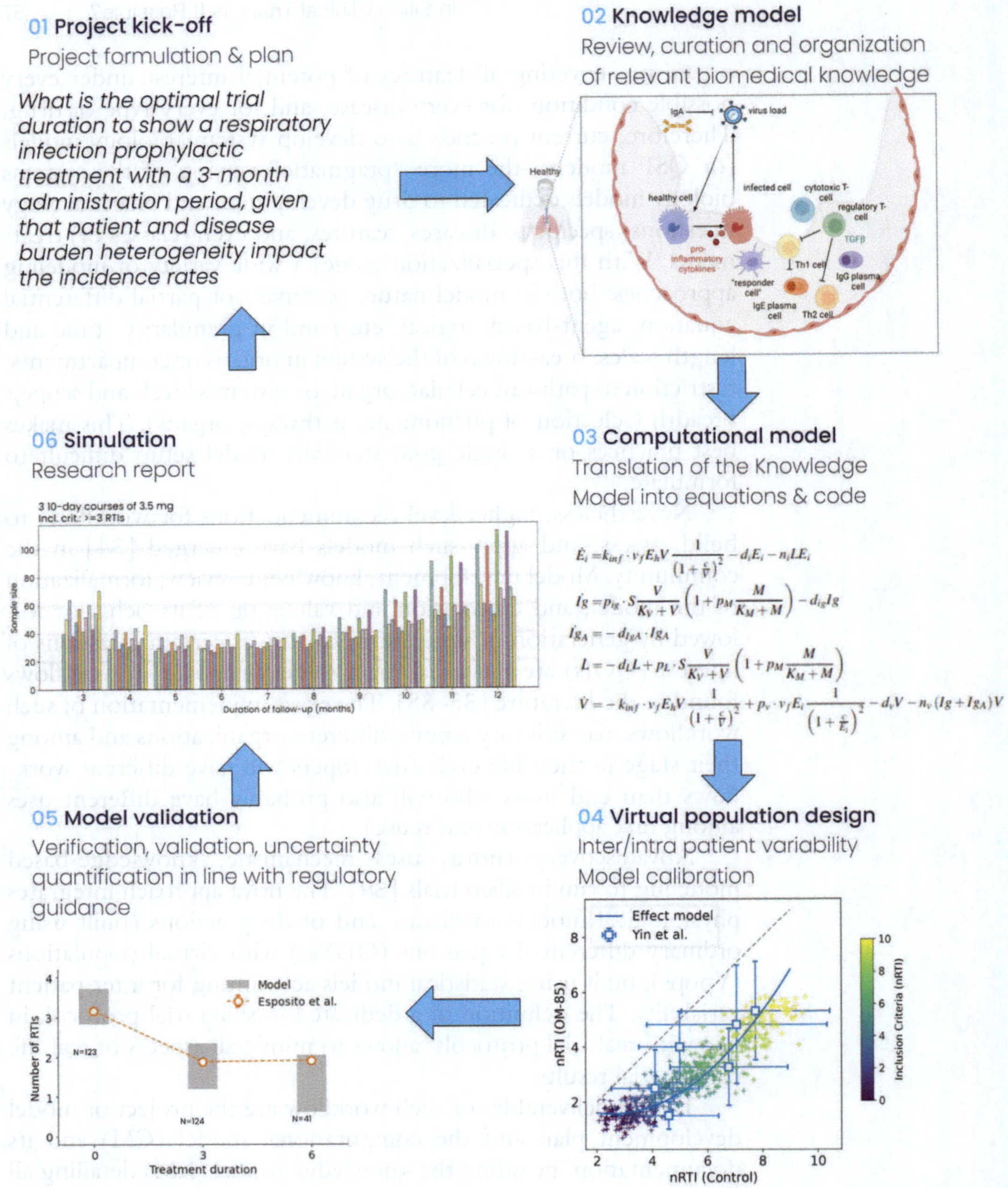

Fig. 2 In silico project workflow applied by nova to develop models dedicated to address drug development questions, with specific elements tied to the use case presented in 3: Step 1: specific research question defining physiology and pathophysiology, treatment, and trial settings for which the model and in silico trial has to be developed. Step 2: KM capturing mechanisms responsible for viral respiratory tract infection and immunological factors leading to infection rate variability including mechanism of action and pharmacology of prophylactic treatments. Step 3: ODE system implementing the regulation of the mechanisms described in the KMs translating patient characteristics and administration regimen into time-dependent modification of infection rate. Step 4: generation of a pediatric Vpop suffering from recurrent RTIs and in silico trial protocol to cover between-patient and between-trial variability and fine-tuning of model parameters so that meta-analyzed outcome distribution can be reproduced. Step 5: validation that the outcome of an administration, not included in the calibration set, is plausibly predicted. Step 6: simulation of clinical benefit metrics with in silico trials of varying follow-up period and reporting of the results. Step 6 to 1: feedback and continuous improvement following the "learn-and-confirm" paradigm [67]

Table 1
Generic and specific steps of the workflow to develop and conduct in silico clinical trials on the back of a knowledge-based model

Step	Steps typically performed in systems biology and quantitative systems pharmacology	Steps specific to knowledge-based modeling embedded in trial simulation
Model development and project planning	Formulating questions of interest Deriving anticipated context of use Requirements definition Project planning	Delineating the physiology, pathophysiology, and treatment mechanisms relevant Preliminary data availability check High-level model architecture, definition of submodels and connections High-level plan for calibration and validation
Systemic knowledge review and knowledge model (KM)	Review of the available knowledge (e.g., research and/or review articles, patents, books, reports)	Curation, annotation, and discussion of pieces of knowledge, needed to determine uncertainty in the knowledge to build the model Organization of the pieces of knowledge in textual and graphical form (knowledge model) in a white boxed manner
Disease and drug computational models (CM) and trial simulation setup	Formulation of a dynamic ODE model (with preference in portable format) Integration of phenomenological models for (clinical) outcomes Parameter identification and estimation from in vitro, in vivo, and clinical data defining the overall behavior of submodels and integrated model to be covered	Integration of knowledge model and automatically rendered model equations and PVC metadata into a dynamic, traceable model documentation
Model calibration and generation of the virtual population	Expansion of reference virtual individual models into a virtual and populated-based calibration against reference (clinical) data Goodness-of-fit characterization	Documentation of a reproducible calibration protocol with data collection, curation, flow, and calibration procedure Formulation of a simulation protocol resembling a clinical trial protocol defining the, e.g., study periods, interventions, population, and analyses in terms of Vpop design (patient-specific parameters) and protocol-specific parameters of model and solver as well as parameters used in the analysis

(continued)

Table 1
(continued)

Step	Steps typically performed in systems biology and quantitative systems pharmacology	Steps specific to knowledge-based modeling embedded in trial simulation
Establishing model credibility	Validation of the model is performed in order to assess the level of credibility attached to each answer to be obtained through modeling and simulation processes (M&S). In the absence of model validation guidelines for complex models	Verification, validation, and uncertainty characterization adapting relevant regulatory guidance on (computational) model assessment, evaluation, validation, and qualification (see main text) Validation of knowledge models by domain experts Assessment of a set of qualitative credibility factors (e.g., transparency checking, model reuse, protocol records) Continuous integration and deployment (CI/CD) unit (submodel), system (integrated model), and related regression tests Comparison of simulated data with validation data (not used for calibration and distinct from calibration data context), statistical analyses of prediction errors, assessment of fulfillment of predefined credibility goals, use of "model risk"-based approach and for a given context of use (CoU) of the model
Simulation and analysis	Running the simulations agreed upon in the in silico study protocol in order to answer the nonclinical and/or clinical questions	Project-specific analyses focusing on clinical efficacy or effectiveness, for example, using the effect model methodology Synthetic control arm simulations or long-term follow-up projection for single arm confirmatory trials or patient-specific outcome prediction

Step 1: Project scope definition

Depending on predefined QOIs (e.g., what are the optimal dose, optimal responder profile, optimal duration of treatment), the relevant high-level features of physiology are broken down into submodels. The type and number of physiological systems represented as submodels typically define the "breadth" of the overall model, while the level of detail a submodel implements (granularity, i.e., molecular, cellular, tissue level) defines the "depth" of the overall model. The QOIs also define the "user" and "mission" requirements for the future exact use of the model, which is termed its "context of use" (CoU) and which will guide the development planning. A disease drug and trial model development and project plan (already including a proposition for a model architecture, including high-level specifications for submodels and potential calibration and validation plans) summarizes these decisions and lays out how and when such a model can be realized and delivered. In very complex cases (unprecedented CoU, complex multiscale model development needed), this work represents in fact an entire feasibility assessment that will be concluded by the development and project planning.

Step 2: Knowledge model design

Model development starts with reviewing sources for biomedical knowledge and capturing pieces of knowledge about relevant biological entities and their functional relationships. While text-mining and natural language processing have been discussed as potential future technologies to automate biomedical knowledge extraction and creation of densely annotated biomedical knowledge graphs, still currently, human intervention is indispensable. Each piece of knowledge is based on an extract from a reviewed source reformulated into an assertion (Box 2) and that is humanly curated with an evaluation of its strength of evidence (SoE) (*see* definition in Box 3)—i.e., a critical appraisal using a scoring system. These assertions are then organized in graphical and textual form (a KM) capturing the available causality, sequence, and extent of phenomena to be captured quantitatively in the model. The KM adopts a physiology-centered paradigm centering around known physiological parameters, states, and behavior. Then, disruptions leading to the diseased state are introduced in the model. Often submodels contain both physiological and pathophysiological mechanisms but describe different physiological systems (e.g., hepatic metabolism, immune effectors at the organ level, etc.). Treatments in scope of the QOIs (comprising mechanisms determining the pharmacokinetics, targets and mode of action, and pharmacodynamics as well as downstream effects) are included in the model to rationalize and quantify the mode and overall mechanism of action as well as dose-exposure–response relationship and impact of relevant patient characteristics. A key element in the KM enabling simulation of realistic trial outcomes is the formulation of a dedicated so-called clinical submodel (e.g., establishing a link between objective biophysical measurements or physiological fluid markers and clinical outcomes of interest, often leveraging more a phenomenological model fitted to reference clinical data).

Box 2: Assertion

An arrangement of pieces of knowledge carrying scientific information in the form of plain language. An assertion in its simplest expression is composed of two biologic entities linked by a functional relationship.

The following assertion "kinase X is phosphorylated by interacting with Y" contains three entities: X, phosphorylated X and Y, and two relations—[kinase X → phosphorylated kinase X] and [X + Y → phosphorylated kinase X]. In addition to a textual statement in plain scientific language, an assertion is a knowledge graph which captures contextually each relation between entities referred to in the extract with a controlled vocabulary and relevant public identifiers (such as UniProt ID).

Entities may be molecules (e.g., protein, DNA codon), organelles (e.g., mitochondria, cell membrane), cells (e.g., T-cell, erythrocyte), aggregations of cells (e.g., tumor, vascular endothelium), organs (e.g., liver), and systems (e.g., immune system, arterial vascular tree).

Box 3: Strength of Evidence (SoE)

The SoE scoring system consists in the appraisal of confirmatory/contradictory research findings from the scientific literature for a particular piece of knowledge (better, an assertion, which is the more concise format of a piece of knowledge).

Another dimension of the SoE is the fitness for purpose of the study methodology used and presented by the authors of the research article which is used as a source of information.

Step 3: Computational model design

The heart of knowledge-based modeling is then to transform the KM into mathematical equations and computer code (a CM) so that the model captures fully and accurately the included pieces of knowledge in line with modeling approaches that have been successfully applied in the CoU or type of (sub)model. As for the exact mathematical formalism, there unfortunately still does not exist a unified systems biology approach (although progress has been made, e.g., in genome-wide mechanistic signaling network models [40] or to generalize and facilitate immune system effector function description in various contexts [41]). Therefore, in many cases, the equations (structural model) are typically problem-specific, which may introduce limitations as to the breadth of the contexts of use potentially covered and emphasize the need to validate the right choice of equations (see below).

Step 4: Calibration and virtual population design

In the model calibration phase, model parameters are informed and estimated using in vitro, in vivo, and human (clinical, marker) data obtained from the literature, dedicated biological and clinical databases, or preclinical and clinical study reports and databases. While several model parameters can be aligned with physiological characteristics, others can be derived under certain mathematical assumptions (steady state) or even must be entirely estimated from experimental data. For this estimation, scenarios with specific (experimental) protocol parameters are used to streamline the model to the conditions under which data have been obtained. However, to incorporate in vitro data or animal data, often only a discrete number of prototypical virtual individuals are needed, and human data often come with variability that requires the definition of a Vpop. This approach expands reference model behavior into known variability (e.g., age-dependent physiology) and unknown or assumed variability, all described as co-distributions of model parameters (patient descriptors) and dependent formulaic parameters that all define the Vpop. As an example, in nonalcoholic steatohepatitis (NASH) models, one of the patient descriptors is typically a known patient's body mass index (BMI) since lifestyle (and obesity) is a well-known risk factor for this disease and the disease mechanisms are well characterized. Here virtual patients, for example, would feature a BMI uniformly distributed between 25 and 40 kg/m^2; and formulaic (or correlated) parameters describe metabolic dysregulation mechanisms (and variability of the extent of this dysregulation), leading to BMI-dependent disease progression and outcomes.

Generally, population-level parameters can be estimated borrowing concepts from both nonlinear mixed effect modeling (NLMEM) like popPK and PBPK modeling, which makes neither parameter estimation method guaranteed to succeed and thus, calibration needs potentially stepwise, iterative, and specific workflows (see some examples from the QSP community in [42]). For dynamic systems biology models, a multitude of optimization algorithms exists that estimate parameters such as local gradient-based search algorithms [43] with Latin hypercube restarts (starting with wider coverage of parameter space and then performing efficient local optimization), but also stochastic and evolutionary algorithm frameworks (e.g., covariance matrix adaptation evolution strategy (CMAES); *see* [44]) or stochastic-deterministic algorithms based on scatter search (*see* [45]) that are global optimization techniques, meaning they share the intrinsic ability to escape local minima but may need more individual function evaluations to reach a minimum. It is nowadays acknowledged that the performance of global optimization algorithms largely depends on the data-simulation normalization technique [46, 47]. Palgen et al. have recently shed some light on how creation of normalized objective "scoring" functions with flexible context-dependent definition together with CMAES algorithm can be used to calibrate a multiscale lung adenocarcinoma model to a very heterogeneous set of data [48]. For NLMEM parameter estimation, other classes of approaches are adopted, which can be roughly classified according to four main categories: (a) methods based on individual estimates, (b) methods based on approximating the likelihood through linearization of the nonlinear function, (c) methods based on the exact likelihood which tackle the multidimensional integration directly, and (d) a Bayesian approach which uses both the likelihood based on the data and prior information about model parameters [49] (*see* [50] for a prominent implementation of an expectation maximization algorithm). It is to note that complex models may suffer from identifiability issues and that parameter uncertainty mixes into the overall unidentified variability [51]. Also, the latter methods are not yet computationally tractable for larger and complex models with potentially high number of parameters. Despite active development in this field (e.g., optimal control theory [52], hybrid algorithms [53]), more practical approaches [42]—especially via reweighting of individual virtual patient sampling within multivariate baseline and outcome distributions [54]—have gained attraction.

In conclusion of the above considerations, a combination of classical parameter estimation techniques, practical and pragmatic Vpop reweighting algorithms, and extensive characterization of the model behavior (notably including global sensitivity analysis [55, 56]) will constitute good practice in model calibration. Current unsolved questions in the calibration of complex models are the importance of parameter non-identifiability, handling of latent and hidden variables, and overall robustness with respect to operation outside of the calibration scope.

Step 5: Model validation

Before its use for simulations that may impact important decisions, and for each question to be answered in the nonclinical and/or clinical setting, a validation of the model should be performed to assess the level of model credibility, a process that is also central to acceptance of models by regulatory agencies. For some types of models, like popPK [57, 58] or PBPK [59] models, regulatory guidance on how to qualify, conduct, and report M&S for regulatory consideration does exist.

But what if no dedicated regulatory guidance is available, like for complex disease-trial models and QSP [60]? Generally, a validation protocol, written prior to any validation actions, helps to correctly define the scope and conditions of application and to communicate the anticipated validation strategy with the involved stakeholders and can be used to obtain formal feedback. It also serves as a record in a prospective setting, which should be regarded as the gold standard. A possibility for cases that models are not (yet) covered by any specific guideline is the alignment of the validation process with a consensus standard ("Assessing Credibility of Computational Modeling through Verification and Validation: Application to Medical Devices", V&V40 [61]) that has been first acknowledged by the FDA as draft guidance in medical device regulation [62], but which is formulated at such an abstract level that they apply to all kinds of modeling (e.g., PBPK models [63]). The definition of CoU and model credibility and from V&V40 may in fact apply in the drug development field without adaptation: "The CoU of the model is defined as the specific role and scope of the computational model used to address the question of interest. The CoU should include a detailed description of what will be modeled and how model outputs will be used to answer the question of interest" and "Model credibility refers to the trust in the predictive capability of a computational model for the Context of Use (CoU). Trust can be established through the collection of evidence from credibility activities. The process of establishing trust includes performing Verification and Validation (V&V) and then demonstrating the applicability of the V&V evidence to support the use of the computational model for the CoU" [61]. Once the CoU of the model is defined, the validation process starts with an estimation of the model risk which drives the selection of V&V activities and goals for the credibility factors:

- Model risk is "the possibility that the use of the CM leads to a decision that results in patient harm and/or other undesirable impacts." Model risk is evaluated as the combination of the model influence and the decision consequence.
- Model influence is the contribution of the CM relative to other contributing evidence in decision-making.
- Decision consequence is the significance of a harm to patient (s) resulting from an incorrect decision.

The adaptation of the V&V activities and their assessment, however, needs a more drug development-specific definition of credibility factors and the way to evaluate them, but it does not change the global framework of the V&V process. Model credibility is assessed with verification which ensures that the mathematical model is implemented correctly and then accurately solved, a validation assessing the degree to which the CM is an appropriate representation of the reality of interest, and applicability analysis evaluating the relevance of the validation activities to support the use of the CM for a CoU. Validation activities are concerned with showing the correctness of the underlying model assumptions and how much sensitivities and uncertainties of the CM are understood.

For knowledge-based models that are to be used in in silico clinical trial frameworks, first, a qualitative assessment of model quality is evaluated through several factors that can be summarized by the following questions:

- Are the KM perimeters relevant to the question(s) of interest?
- How are the KMs validated?
- Is the model granularity and complexity adapted to the question of interest?
- Is it possible to access the knowledge justifying the model form and the parameter values?
- Are the uncertainties associated with the model form and inputs and their impact on model predictions understood and controlled?
- Is the model sensitivity to input parameters and its impact on model predictions understood and controlled?
- Is the in silico experiment design relevant to address the question(s) of interest?
- Are the M&S results relevant for the clinical purpose?
- Are the M&S results relevant to the context of use?

These factors will give an insight to the entire model development and its use in resolving drug development issues.

Second, a quantitative assessment of the model quality is performed, where the model's predictions are quantitatively compared to sets of observations that were not used in the calibration process. The comparison of model predictions with observations from the validation dataset depends on the CoU and data availability. This comparison should be done with data observed in a context overlapping with the CoU. A validation strategy is agreed on during model development and fixed in the validation protocol. This strategy describes the validation dataset (not used for model calibration) that will be compared with model prediction. It also describes the comparison tests which will be conducted, and it predefines the associated acceptance criteria [64]. The quantitative evaluation is supported by an applicability analysis that evaluates if favorable validation leads to trustworthy predictions. Applicability analysis starts with a description of the aim of the computational modeling, the CoU, and the sources of validation evidence. The primary validation evidence is selected from the sources of validation evidence as the source of evidence that is the most relevant to the CoU [65]. The real-life and modeled components of the validation context are described in the validation protocol and the validation report. Following these descriptions, applicability analysis consists in comparing the contexts of validation and providing rationale regarding appropriateness of using the model, validated with a given methodology, to make predictions in the CoU. Other sources of evidence may be used to support this rationale. A written report of the validation should be filed and should contain the description of QOIs; the CoU(s) of the model; the definition of model acceptability criteria for the CoU(s); a discussion of the model risk analysis; a detailed description of the undertaken V&V activities including statistical analyses, applicability analysis, and uncertainty assessment; and finally the summary of the anticipated model-informed decision-making (see the explanation and examples in [66]).

Step 6: In silico trial simulations

In the last steps of the workflow, dedicated simulations and analyses are carried out to address the QOIs; these either are iterative (exploratory approach) or follow the predefined in silico trial protocol (confirmatory approach) to mimic the workflow of clinical research. Simulation of the drug effect on individual virtual patients has the benefit of connecting individual-, group-, and population-level metrics, and thus special consideration has been put to the presentation and analysis of in silico clinical trial data. Various statistical methods and metrics are available to characterize treatment efficacy. The following paragraph focuses on a method which is applicable to the context of in silico and individual predictions. One potential way to disentangle individual factors contributing to (nonconstant) efficacy is the application of the effect model methodology. The effect model approach [68–70] is a tool which relates the rates (or risks) of events without treatment (Rc) and with (Rt), as supported by empirical evidence, simulations, and theoretical considerations [71–74]. While simulations can be conducted for the same patient in different arms in in silico trials and yield paired observations, the effect model can also be reconciled with meta-analyses [71, 75]. The effect model can in fact be set up for a wide variety of risks (or rates or probabilities) of clinical−/disease-related outcomes such as binary events (tumor progression, side effects, death) or continuous markers (e.g., low-density lipoprotein cholesterol (LDLc), blood level). In brief, for rates, the effect model is constructed in a two-step process:

- First, the disease model is applied to each virtual patient to determine for patient (i) the probability of suffering from the clinical event of interest without treatment (Rc, i).
- Then, the pharmacological (or treatment) submodel is connected with the disease model and reapplied to the same virtual subject to generate the probability of exhibiting the same clinical event of interest with the therapeutic intervention (Rt, i).

A convenient metric for the efficacy deduced from Rc, Rt distributions is the absolute benefit (AB) being the difference between these two risks Rc−Rt, for a population, subgroups, or for an individual patient (Rc, i)−(Rt, i), the latter being available for virtual patients that can be allocated in different arms (including the control arm) at the same time. How Rc,i and Rt,i can be obtained in a workflow adopted by nova using drug (including target alteration), disease, and trial model (for a Vpop) is detailed in Fig. 3. On the whole population level, the sum of the predicted AB for each patient of the population or group gives the total number of prevented events (NPE) that—through prediction—can enable comparison and selection of available treatments for a group or a population, new targets, or drug under development with comparator(s). The effect model is both implicit and specific to each disease and treatment. In Fig. 4, the simulated effect model for everolimus to present allograft rejection after lung transplantation is indicated [76]. The cluster of patients at $Rc = 0.7$ is in a (Rc, Rt) region of overall high response (characterized by high AB) that can be enriched by, e.g., inclusion criteria defining patients at elevated but not high risk for allograft rejection. Further, analysis of (i.e., predicted) markers of patients within that risk group might give insight into further stratification options (subset of optimal responders). In another example, Arsène et al. have conducted trial simulation for respiratory disease prophylaxis under coronavirus disease (COVID-19) nonpharmaceutical intervention (NPI). Here the effect model type analysis could indicate the impact of NPI on different metrics tied to clinical efficacy and benefit [32].

3 A Collaborative Knowledge-Based Modeling and In Silico Trial Simulation Software Platform, *jinkō*

It is not only the know-how and processes that determine the success or failure of an in silico clinical trial. Despite efforts to increase synergy of different MIDD techniques (through parallel synchronized application, cross-informative use, and sequential integration), a true convergence will require new tools and methods that enable greater technical integration and more cross-discipline collaboration [77].

Model developers need tools for the conception, design, implementation, testing, and deployment of models and their documentation as well as constituents of the in silico trials. The users of such models need viewing, editing, and assessment capacities and will need to run models and conduct in silico clinical trials for several scenarios. Therefore, the available (software) tools (a non-exhaustive list below) to perform these tasks are of crucial importance to establish in silico trials in the mainstream of clinical research.

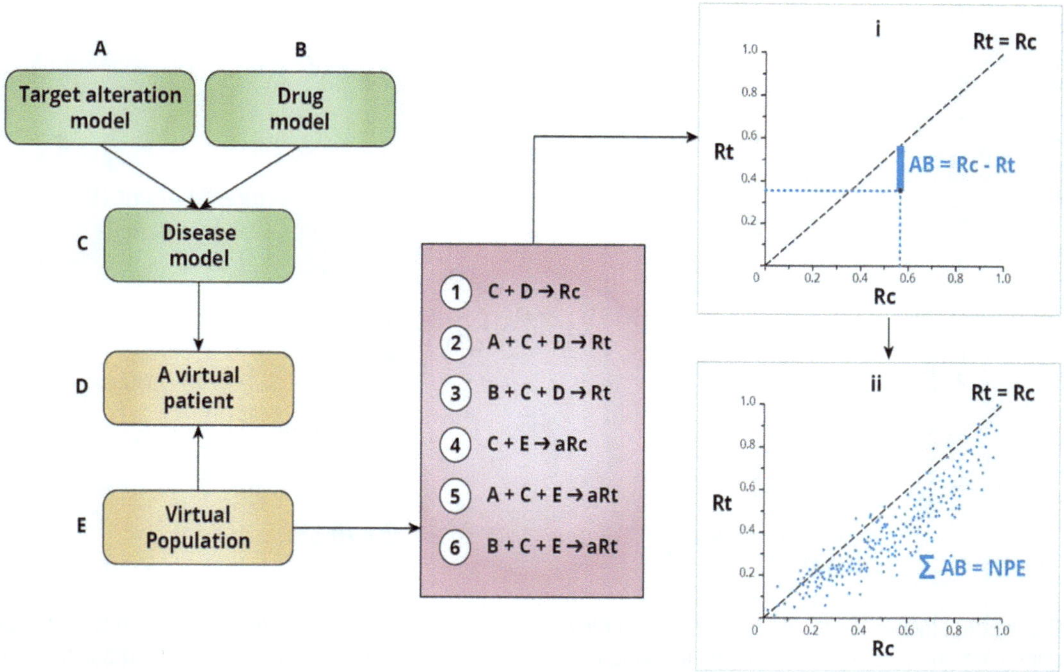

Fig. 3 Effect model building blocks for individual predictions: left flowchart: A: model of potential target(s) alteration; it integrates a target component of the disease model (C) and a target alteration profile. B: model of disease modifier(s); it can be a marketed drug, a compound under development, or a combination of either one or more, each altering a target. C: formal disease model, representing what is known about the disease and the affected physiological and biological systems. E: Vpop, a collection of virtual patients (D) characterized by values of a series of descriptors. Each descriptor represents one model parameter. Its distribution is derived from knowledge and/or data. Right graphs: i (single dot) and ii (multiple dots): each dot represents a virtual patient (or group of patients) in the Rc, Rt plane. The four components A–E enable to compute the risks (or rates, probabilities) indicated in the middle box: (1) When C is applied to E, simulation of the disease course on the virtual patient results in a value for Rc, the outcome probability in the control group (i.e., without treatment). (2) and (3) When A or B is applied to C, and A+C or B+C is applied to D, this yields Rt, the probability of outcome altered by the disease modifier(s), target alteration, or drug, respectively. The difference Rc – Rt is the predicted absolute benefit, i.e., the clinical event risk reduction the (virtual) patient is likely to get from the disease modifier. AB is an implicit function of two series of variables, patient descriptors which are relevant to A and/or B, and those relevant to C. Note that AB is the output of a perfect randomized trial since each patient is his/her own control. AB is the effect model of the potential target alteration (A) or of the treatment of interest. (4) When C is applied to each virtual patient in E, this results in all Rc for D (where "a" denotes average). Alternatively, if A or B is applied to all D from E, this yields the values of Rt for all virtual patients in D. By summing all the corresponding ABs, one derives the number of prevented outcomes (or NPEs). The processes can be represented graphically in the Rc, Rt plane, as shown in the accompanying visual: (i) for the AB for a virtual patient D; the dotted line represents the no-effect line, where Rt = Rc. (ii) for the NPE. Whether the patient is virtual, defined by an array of descriptor values, or real, defined by characteristics (genotype, phenotype, demographics, history), the process is the same. The disease model parameters are valued with the patient data and simulation is run to give the clinical outcome or the probability of clinical outcome when the disease is left evolving. This gives Rc. Then the treatment model is connected to the disease model to yield Rt

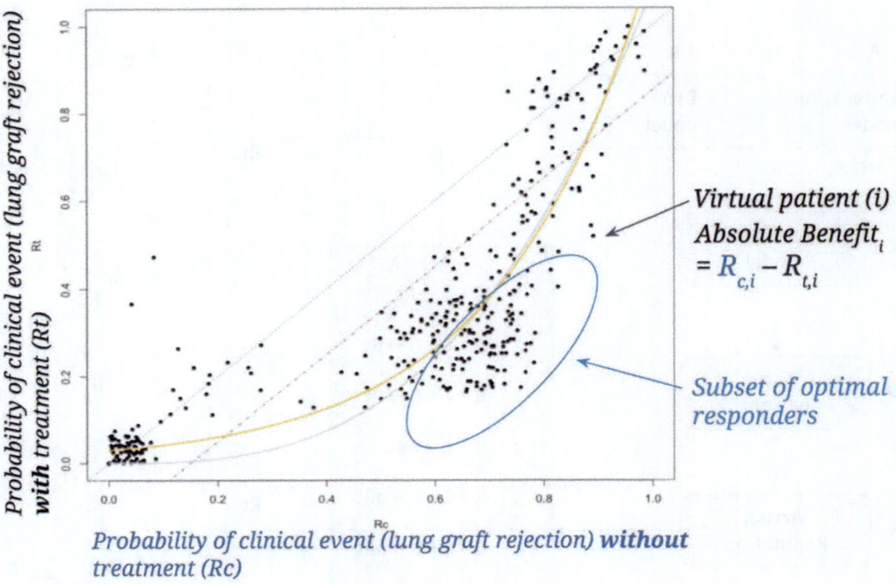

Fig. 4 Effect model of everolimus for the prevention of lung graft rejection generated with in silico clinical trials during the European sysCLAD project showing that the nonconstant efficacy is optimal for intermediate risk of rejection and that within a certain at-risk group, further patient profiling can lead to higher demonstrated size of the efficacy [76]

Software for multipurpose mathematical modeling is abundant (e.g., MATLAB [78] or R [79]), and systems biology and extensions for physiological modeling are available (e.g., SimBiology [80] and Mrgsolve [81], respectively). For solving modeling problems in systems biology, compiled (e.g., COPASI [82]) and scripted [83–85] packages have been proposed, and web-based simulator collections around standardized and transportable formats [86] aspire to solve the "reproducibility crisis" in this field of research [87]. There also exists more specialized commercial and open-source software for PK/PD (e.g., [88])/and PBPK [89–92]. With the pharmacometrics community, also specific software solutions have been ever evolving in dedicated functionality and maturity and are now naturally associated with pharmacometrics analysis (especially NLMEM), as implemented in NONMEM [93] or Monolix [94], but also here, multipurpose mathematical modeling tools like R are used in the community and industrial setting [95].

In silico clinical trials are somewhat blurred with trial simulation tools that are used for the statistical design of trials, with existing solutions reviewed by Meyer et al. [96]. Lately a few web-based specialized trial simulators have been put forward (see, e.g., the unified immune system simulator-tuberculosis (UISS-TB) [97, 98]).

Despite availability of many different software approaches, coherent workflows and multi-stakeholder engagement are still a problem in the community as the available solutions need in-depth technical expertise in the utilization, only target one facet of the model development, documentation, and reporting aspect—advocating for developing an integrated end-to-end solution. Especially with the aim enabling in silico clinical trials based on complex knowledge-based models, such a platform needs:

- Collaborative knowledge management facilitating human curation.
- Distributed model solving architecture allowing to run in silico trials with large size Vpop in "real time".
- Integrated and shared analytics and visualization enabling all stakeholders to take part in the research.

Under this premise, nova is currently developing "jinkō" (http://jinko.ai)—a unified in silico clinical trial simulation platform (example visuals in Fig. 5). With analytics tools, mechanistic modeling, and Vpops at the heart of the nova's core expertise, the platform's name was inspired by the fortunate homonymic collision

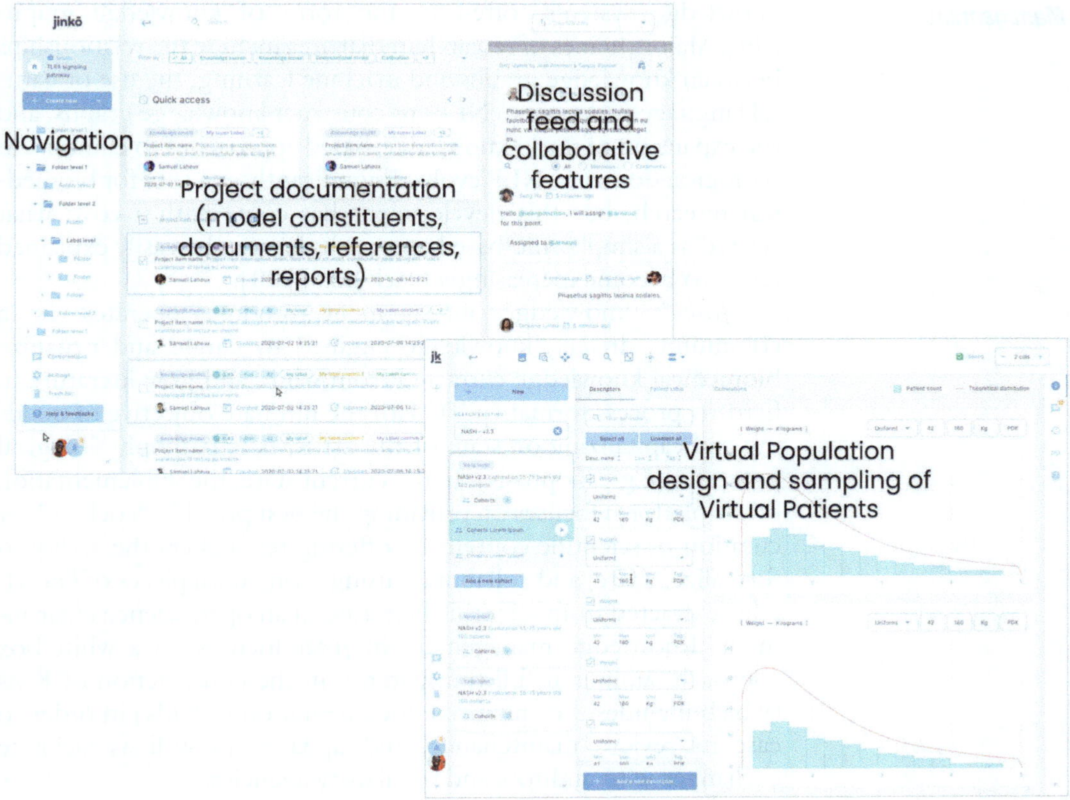

Fig. 5 *jinkō* overview: screenshots of the project-level documentation and the Vpop generation interface as examples for showcasing, collaborative editing, discussion, and facilitated in silico trial generation with respect to mathematical or scripting interfaces

of the Japanese words for "artificial" (人工, *jinkō*) and "population" (人口, *jinkō*).

From initial scientific knowledge curation to in silico trials and results exploration, *jinkō* embarks powerful collaborative features allowing for quick turnovers in communication between different stakeholders, including researchers, clinicians, and other experts. Building on human-readable KMs (*see* Subheading 3.1) to ultimately generate realistic in silico clinical trials allows both QSP experts and non-modeling experts to edit, share, and communicate under the same environment and thus speak the same language, promoting in silico practices. With this approach, non-modeling experts also become able to run their own simulations and bring their collaboration with modeling teams to another level. Using core libraries developed over the years, *jinkō* is equipped with powerful and intuitive simulation capabilities (*see* Subheading 3.2), from the generation of Vpops to the exploration of results, making use of easy-to-use data visualization packages (*see* Subheading 3.3).

3.1 Collaborative White-Box Knowledge Management

Knowledge-based modeling involves the design and implementation of computer systems that depend on formally represented knowledge systems, often in the form of knowledge graphs [99]. Major themes of research in that area include the relationships between knowledge graphs and machine learning, the use of natural language processing (NLP) to construct knowledge graphs, and the expansion of novel knowledge-based approaches to clinical and biological domains. NLP evolved significantly as a tool for biomedical research, but this development has come with a cost: what started as a simple rule-based approach that can be easily explained tends to become increasingly black box [100].

Jinkō's knowledge management module operates as a community-driven "knowledge engine" to curate and organize biomedical knowledge extracted from white and gray literature in hard-to-process format (pdf), with the ultimate objective of building and capitalizing on state-of-the-art KMs of pathophysiological processes (e.g., apoptosis). In the current state, the implementation of the platform focuses on providing the best possible "cockpit" for curation of scientific content by offering researchers the ability to formalize, share, and collaborate using biomedical pieces of knowledge extracted from scientific literature in an open science environment. Knowledge management in *jinkō* focuses on a white-box approach, aiming at full transparency in the construction of KMs (which includes the enhanced documentation of CMs) in order to ease its review, maintenance, and updates, as well as facilitate exchanges with auditors and regulatory agencies.

3.2 In Silico Clinical Trials Powered by a Distributed Solving Architecture

The approach chosen by *jinkō* is to very clearly separate the design of an in silico clinical trial and its simulation. That means that a trial is a resource, i.e., a type of document, in and of itself of the *jinkō* platform, with its own life cycle and versioning. The following elements are constituting an in silico clinical trial:

- The CM: a set of biochemical or biophysical equations, ODEs, and events through time, which describes the dynamics of the substances at play in a specific biological system, possibly under the action of a given treatment. This concept is already well formalized in the computational biology community, to the point where widespread standards like Systems Biology Markup Language (SBML) [101, 102] have been developed to encode, store, and share such models in a way that makes them program-readable and possible to simulate.
- The Vpop design, which defines the laws that underlie the composition of virtual patients in a Vpop. These are probability distributions and covariance of the vector of model parameters characterizing a virtual patient. Sampling this design yields the virtual cohort on which the trial is simulated. It is also possible to import pre-set Vpops, for instance, to create digital twins from a real-world cohort.
- The protocol which describes a set of simulation scenarios with parametric perturbations is used to define the arms of the trial. The important difference with a real-world trial is that with an in silico trial, patients can be duplicated at will. Thus, every patient can be in every arm and therefore act as its own control. Each arm generates a different version of the same patient, allowing unbiased comparison, with minimal noise, leading to a prediction of the "true" differences between arms.
- A set of measures which define several quantities of interest from simulation results and the rules to compute them. They can be, for example, usual longitudinal aggregates (average or area under the curve of a certain substance's quantity through time), values at specific times, or complex user-defined functions. It is also possible to define a measure across multiple protocol arms, for instance, the ratio of the effect of a drug on the treated arm over the protocol arm. This is how clinical endpoints can be declared.

Importantly, all these elements are also first-class resources of the platform. They are each editable separately, documentable separately, and reusable through various trials, and all have their own separate life cycles.

Once the elements described above, i.e., the CM, Vpop, protocol, and measures, are set up, the platform can be used to run the simulation of the in silico trial. At its core, *jinkō* is powered by the open-source SUNDIALS ODE solver [103] and was developed with the final objective of bestowing the end user with a high-level framework to design their models and trials and to simulate and then analyze in parallel entire virtual cohorts of possibly tens of thousands of virtual patients. The nature of the computations and the cloud environment of *jinkō* lend themselves well to distributed computing: patients on the various arms of a trial are independent from one another, so the simulations can be heavily parallelized. To do so, nova developed a distributed solving engine mostly from the ground up. Indeed, the development was facing a few well-known challenges from the fields of distributed build systems (such as Nix and Hydra [104]), classical distributed computation systems (such as Apache projects like Spark [105] or Storm [106]), and message queue systems (such as Apache Kafka [107] or ZeroMQ [108]). The following list summarizes the challenges and requirements for *jinkō's* distributed simulation engine:

- Deterministic and hierarchized computations that are idempotent and free of side effects, so the system is resilient to the failures and network shutdowns that are the bane of every distributed system.
- At-least-once semantics that guarantee (in the case of either of the aforementioned errors) that the loss of a computation cannot happen.
- Granular caching, so a trial or a calibration can as much as possible benefit from intermediary results obtained previously by another trial or calibration process, saving a lot of computation time.
- Fairness of access to the computational resources, so the time it takes for a simulation to complete is proportional to the complexity of the CM and to the size of the Vpop given by the user.
- Forward compatibility and migration of previously obtained simulation results, so simulation results can still be used months later, even if the platform underwent large changes.
- Coexistence of both long-running simulation tasks and short-running analytics tasks.
- Coexistence of both long-lived and short-lived results.
- Efficient indexation of simulations' results, for instance, to quickly find after a simulation all the patients which in each arm have some given value that satisfies a given criterion.
- Openness of the simulation process, so the platform can provide a good user experience by reporting and detailing live the current progress of their simulation, which patients are already simulated, what remains to do, etc.

- Resilience and adaptability to new types of analytics and new variants of trial and calibration workloads, so the platform can grow with the needs of its users.

The most basic simulation task, which is termed an "atomic solving job," consists in solving a model with a given (definitive) set of parameters, most often obtained from applying a scenario arm to a given patient. Each trial simulation process splits its workload into various atomic solving jobs, registers them in the system, waits for them to be completed, and then proceeds to post-processing steps. The trial simulation processes are themselves jobs of the platform: they obey the same rules, and they only work at a higher level of granularity. This hierarchy of jobs splitting their workload into subjobs and the deterministic nature of the models enables reusability. Reusability is technically realized through "granular caching," which means storing the results of an atomic task. In this way, single atomic jobs (using the exact same inputs) can be reused in complex optimization processes that, for example, iteratively repeat the trial process and/or apply alterations to the Vpop between each iteration instead of being recalculated. This caching mechanism is fully generic and works similarly in the case of analytics workloads too—where it is no longer about solving a model but about aggregating simulation results.

Another big challenge is how to schedule the different tasks in a way that ensures fairness between the users on the shared platform resources. The proposed approach defines a "fairness metric" for that purpose: the rate at which trials progress should be constant and equal for all users. It implies that a trial with 1000 patients should be ten times longer to complete than a trial with 100 patients (all other parameters being equal), regardless of which of these two trials was submitted first. To achieve load balancing under the use of this metric, a stochastic multi-queue system was devised. With that design, the simulation system distinguishes four entities: the core-webservice, the core-workers, the queue system, and the persistent cache. Concretely, Redis [109] is used to implement the queue system. The core-webservice is the entity that receives simulation submissions in the form of HTTP requests and will then translate them into job descriptions and push them in the queue system. Core-workers are a replicated service, which will then pick up jobs from the queue system, perform them (by spawning and waiting for subjobs on the same queue system), and write their results in the persistent cache. All the above—except for the fact that jobs can spawn and wait for subjobs—is a perfectly standard design for distributed systems [110]: the resources are not available to immediately start every simulation that the users may submit; therefore, an error-proof buffer space containing remaining jobs is needed. Most of the time, these queue systems have first-in-first-out semantics: jobs are performed in the order of their submission. This is not the ideal behavior for a platform such as *jinkō*, because it would entail that a user submitting a very big trial is potentially locking the whole platform. As a solution to this problem, a system has been designed that can dynamically create new queues, one for each new trial. The following question remains of how it can be ensured that, through time, each queue (and therefore each trial) is always given equal computational power. This can be considered as a simpler but more dynamic form of the "assignment problem" [111]. Assignment costs can be identical, which simplifies the problem as no worker is more costly to assign to a queue than another; however, the set of tasks (queues) to complete varies over time. The following implementation provides a potential solution to this bottleneck:

- First, the system creates a new queue on the fly for each new trial. This queue has an identifier obtained via a hash of the trial inputs, so resubmitting the same trial means reusing the same queue. Plus, the internal structure of a queue is not a list but is an ordered set. That means that it needs to be guaranteed that no job can ever be duplicated in a queue. That is what guarantees that submitting twice the same trial to the same queue is idempotent.
- Queues are ordered in the lexical order of their identifiers. This ordering is only used to assign a number to each queue, to uniformly sample them. There is no prioritization whatsoever.
- Workers randomly assign themselves to the queues, in a uniform manner, to make sure that these assignments remain both relevant and fair through time (i.e., even if new queues are created or consumed), followed by simple modular arithmetic. At startup, a worker is given a random positive integer number (uniformly), and it will use it to select which queue it will read, modulo the current number of queues. So, if a worker is assigned the number 262,144 at startup (for four queues), then that worker will pick up jobs in the first queue, because 262,144 modulo 4 = 0. That means each worker is loosely assigned to a queue: workers stick to their assigned queue while the number of queues does not change, but the creation or consumption of a queue will automatically shift those assignments. Thus, queues should always have on average the same number of workers assigned to them. And given our workers are all identical and run on machines with the same computational power, it ensures that the total computational power is always equally distributed between the queues.

The current implementation favors queues, i.e., structures that preserve some ordering notion, instead of, for instance, regular unordered sets. One may point out that the order in which patient simulations are done does not matter. But there are two reasons to maintain the notion of job ordering within a single queue:

- This allows workers to rotate the queues: when a worker picks a simulation job, it does not remove it; it moves it at the end of the queue. It will only delete the job when it is sure this job is complete, and its results are safely stored. Some locking mechanisms (such as distributed locks implemented in Redis [112]) can be added to ensure workers are aware of the jobs that are already taken care of and therefore that can just be skipped. This in return ensures that no job can be lost, without requiring an extra service that pushes back failed jobs into the queues.
- This system is not used only for trials, but it is also used for several types of workloads that require computation time and thus cannot yield an immediate answer to the end user. For instance, this system is used too by the analytics and visualization workloads, which need time to aggregate the results of large simulations. But given these workloads are much smaller than simulations, the end user will expect a synchronous answer. Therefore, in such cases, a first-in-first-out behavior is wanted if the objective is to give the end user the maximum responsiveness. Therefore, visualization jobs are pushed in the same queue; only trials have their dedicated queues. If needed, that dedicated visualization queue can be favorably biased, so workers sample it more often. Having a common scheduling system for all these different workloads alleviates the need for special casing, for instance, having a worker pool dedicated to simulations and another one dedicated to analytics and visualizations, with the risk that one remains idle while the other is overloaded. Here, any worker can perform either simulation or analytics workloads, depending on what is the most pressing, and no worker stays idle.

3.3 Barrierless Analytics and Visualization

An important part of the in silico trial workflow is the analysis and visualization of results, i.e., the simulation outputs of the Vpop. Various analytics modules have been developed and integrated in *jinkō* to assist the users in analyzing simulation results and producing reports in line with traditional clinical trial reports.

The most intuitive first step is data visualization. Data from in silico trial simulations typically come in two forms: longitudinal (the evolution of variables of interest over time) and cross-sectional (the aggregation of time-independent quantities of interest across the whole Vpop). Longitudinal data are best represented in the form of time series. While the time series associated with a single patient can be represented, it is often more informative to visualize an aggregation across the Vpop of it. One way to achieve this is to plot the quantiles (typically quartiles and 2.5% and 97.5% quantiles) of the variable of interest over time as shown in Fig. 6a. Cross-sectional data are best represented by histograms for continuous data and bin charts for discrete data as shown in Fig. 6b. The platform enables filtering and grouping capabilities (with both

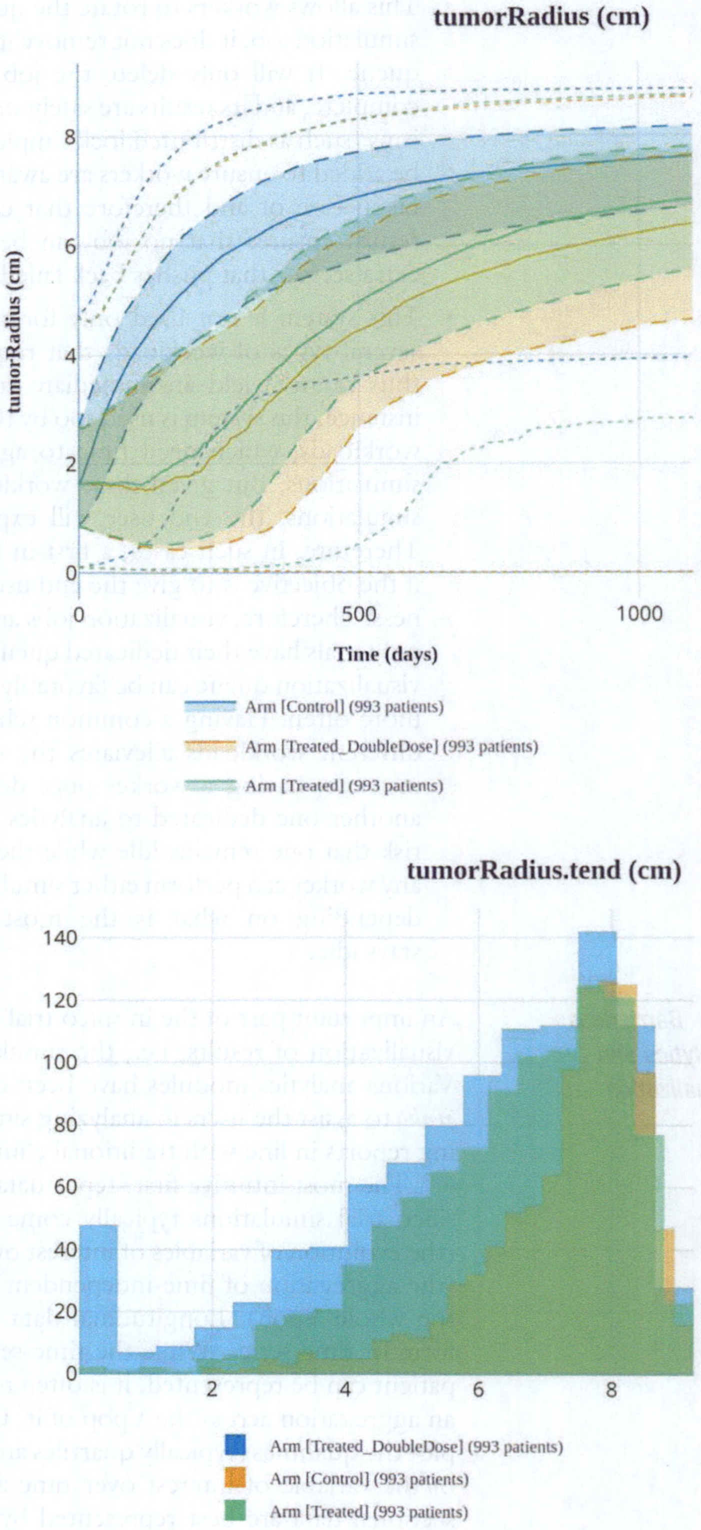

Fig. 6 Visualization of longitudinal (**a**) and cross-sectional (**b**) data from a simulated clinical trial

longitudinal and cross-sectional data) to visualize data (from multiple control arms), group the data by age categories, or filter patients that harbor a particular characteristic. The various filtering features also make it possible to investigate the impact of different sets of inclusion and exclusion criteria for the study population on the targeted clinical outcome.

State-of-the-art analytics is a crucial component of biomedical research in general. As such, *jinkō* integrates a library of analytics allowing users to design, run, and share the results of commonly used statistical methods. For example, in many fields such as oncology or cardiology, the analysis of time-to-event data has become essential. Construction of a table of survival analysis data has been implemented in the platform together with adequate empirical or parametric statistical methods such as Kaplan–Meier (K–M) analysis [113].

This allows the in silico assessment of treatment or intervention efficacy by comparing numbers of expected and observed events (tumor progression, etc.) for each group (Fig. 7).

Another example of an important analytical tool is global sensitivity analysis (GSA) which is used to identify the parameters of interest that contribute for most of the variability observed in quantities of interest [114]. Common GSA techniques include Sobol's method which relies on functional decomposition of the variance (or analysis of variance (ANOVA) based) [55, 56, 115] or the so-called tornado plot which consists in bucketizing parameter values and comparing the medians of the quantities of interest into each bucket with respect to the overall quantities of interest median (Fig. 8).

3.4 Model Editing and Calibration Tasks

One of *jinkō's* design principles is to allow full traceability and documentation of the simulation process. Hence, extensive effort is made to integrate **steps 2** through **6** of the workflow outlined in Subheading 2 including CM design and Vpop calibration as well as validation. As traceability, user-friendliness and collaborative aspects as well as the cloud architecture of the platform are current and urgent needs for the community, model editing, and calibration features of the platform (which are currently under development) will be outlined briefly in the following.

For model editing, established software or programmatic approaches from the systems biology space may be used (see the introductory paragraphs of Subheading 3) given that models can be exported in a serialized format (i.e., SBML [101, 102]). Nova uses an in-house simulation framework for a robust model development library, allowing flexibility, automation, and programmatic model editing and composition features, particularly models. Then created or edited models in serialized format may be uploaded to the platform, which also leverages existing models for use in *jinkō*, thereby augmenting reusability of models or parts of models.

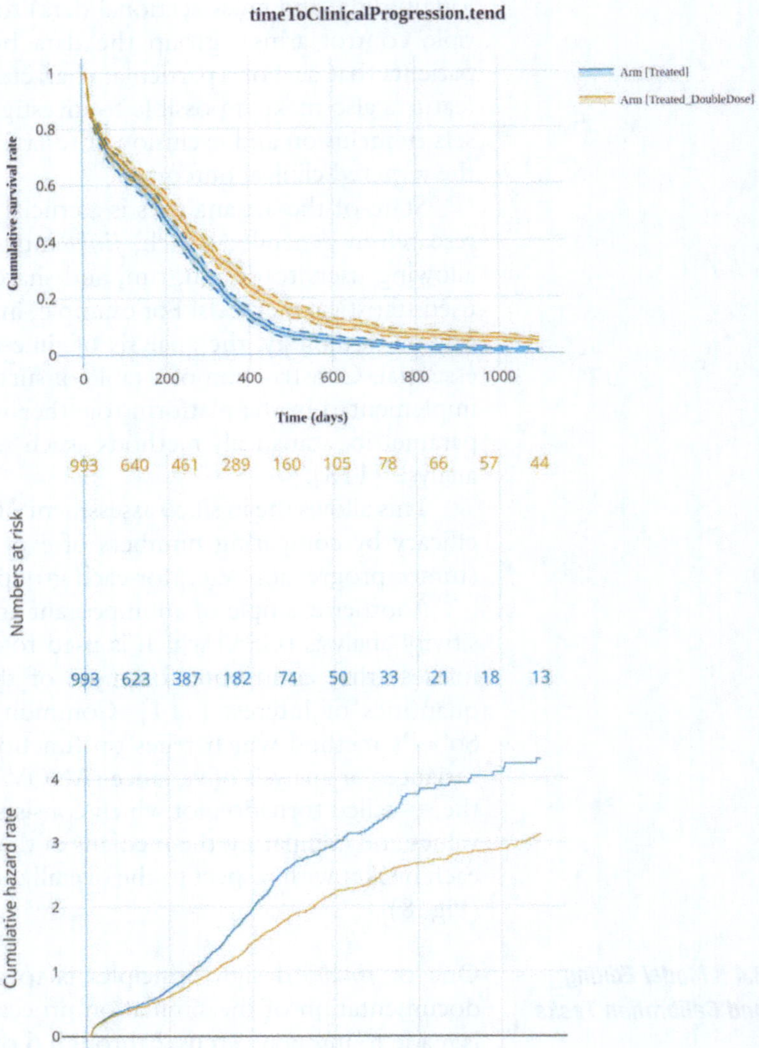

Fig. 7 Survival analysis visualization K–M curve (top), patient table (middle), and cumulative hazard rate (bottom) for two different protocol arms

Each model has its auto-generated LaTeX [116] documentation, with bioreactions, ODEs, and automated links to their dependencies, as well as tables of parameters and initial conditions. Editable annotations are also rendered alongside the element they refer to, allowing for full traceability of the primary source of information, that can stem from the literature review also performed in *jinkō*. Modelers and non-modelers, such as reviewers of the model, need a graphical overview of the model: *jinkō* leverages systems biology graphical notation (SBGN) [117, 118], the graphical "counterpart" of SBML, used to standardize graphical notations of biological models, to produce an extensive and comprehensible graph of the model, illustrating the links between the different

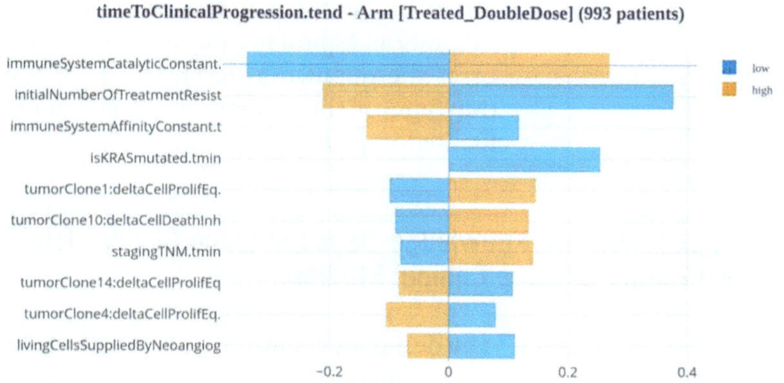

Fig. 8 Sensitivity analysis: tornado plot featuring the ten parameters that contribute the most to the observed variability of a key quantity of interest

variables. *Jinkō* will allow users to edit an existing model or create a new one, using scripting languages like Tellurium/Antimony [83, 84] (a simple yet powerful scripting language to build an SBML model, with only one line per bioreaction). The serialization of all additional constituents (Vpop design, protocol arm design, etc.) also allows for alignment with the Simulation Experiment Description Markup Language (SED-ML) encoding simulation setups, to ensure exchangeability and reproducibility of simulation experiments [119, 120].

Once the model structure is built, it needs to be parametrized to best fit the data. This step is usually one of the most difficult and most complex parts of the workflow and is furthermore difficult to document and to reproduce. Established systems biology software and multipurpose mathematical modeling tools provide model calibration algorithms and allow for allow for subsequent export and reimport of serialized models to enable interoperability (nova uses an in-house programmatic simulation framework for this purpose; *see* Subheading 2, **step 4**). To facilitate this process and enhance integration, calibration in *jinkō* is currently being developed. The architecture design includes definition of combined data-normalizing scalar and log-likelihood-based objective functions derived from heterogeneous data and/or qualitative behaviors of variables of the model. With these objective functions, optimization employs genetic algorithms (i.e., CMAES [44], as described in **step 4** of Subheading 2) to find the best parameter values. The versioning system implemented throughout *jinkō* will allow users to keep track of the successive versions of the models, how they have been obtained, and the different calibration constraints that have been used. Likewise, the Vpop can be refined using a subsampling algorithm (*see* [54] for a similar method working with this principle). In brief, the algorithm efficiently selects a subset of virtual patients that best fits a co-distribution of model inputs or outputs (*see* **step 4** of Subheading 2). During and after calibration, typical metrics of goodness of fit, whether at the patient or at the population level, will be made available to validate the outcome of the calibration process and will be traced and documented in *jinkō*.

4 Application of a Knowledge-Based In Silico Clinical Trial Approach for the Design of Respiratory Disease Clinical Studies

The design of clinical trials is a major factor of success for the entire development program, and in silico trials can be used to search for the optimal design. This search, however, suffers from the high number of dimensions and parameters involved in the optimization problem (trial design → probability of success) [23]—a limitation recognized across virtually all fields of medicine (e.g., atopic disorders [121, 122], rare disorders [123], cancer [124]) and all domains of trial design such as patient inclusion and exclusion (severity, concomitant disease and treatments, age, ethnicity, etc.), dosing regimen (frequency, duration, dose, route of administration, formulation, etc.), or other trial parameters (endpoint, statistical methods, seasonality, comparator, placebo, etc.).

Traditionally, to determine the optimal trial design, a multiactor iterative process is engaged which involves a panel of experts, historical data, statistical models, and hypotheses. In this context, computational exploration of design space is well adapted and is routinely performed to discriminate the best conditions and most sensitive parameters among this myriad of design choices [23]. However, these methods suffer from a major limitation because they are static with regard to predicted efficacy. Techniques that are able to capture and adjust for factors leading to effect size heterogeneity do exist (e.g., meta-regression, model-based meta-analysis, network meta-analysis [28, 125–130]), but their application is limited where clinical data is abundant. For exploratory and pivotal trial support where data are only being generated, these methods cannot thus be employed. Consequently, a large portion of the trial design space cannot be interrogated such as dosing regimen choices or extrinsic factors having an impact on efficacy as well as higher-order interactions between those factors.

To attempt filling this gap, the herein presented use case reports how a knowledge-based complex model can predict trial outcomes for multiple trial designs including combinations of treatment administration scheme and patients' characteristics. The following section reports an example use case with OM-85, a bacterial extract containing lyophilized fractions of 21 strains with immunomodulating properties, and which is used in prophylaxis of respiratory tract infections. Acute viral lower respiratory tract infections (RTIs) in children are responsible for many hospitalizations and are the main cause of wheezing in preschool children [131]. Viral upper RTIs are less severe but still cause significant burden for the patients and society (high prevalence, morbidity, complications, overuse of antibiotics, absenteeism) [132]. Efficient strategies to prevent or reduce the frequency of viral RTIs in patients at risk for recurrent infections is currently an unmet need, as vaccinations fail to target most respiratory viruses such as respiratory syncytial viruses [133] or rhinoviruses [134]. Efficient, non-targeted prophylaxis of respiratory tract infections could thus have a large impact. In preschool children with recurrent RTIs, OM-85 has proven to be efficacious as a prophylactic strategy in the recurrence of RTIs [135] and has been used in over 70 million patients. Nevertheless, bacterial lysates currently on the market need to provide new clinical efficacy evidence. It is therefore crucial to find optimal trial designs especially in such a space where patient heterogeneity is high and overall effect size remains modest, allowing little room for error.

In order to rationalize and inform trial design choices—especially in times where RTI disease burden is being perturbed by COVID-19 and countermeasures—a multiscale mechanistic model of the efficacy of OM-85 in RTI prophylaxis was developed [32, 136] that combines a PBPK and PD model of OM-85 oral administration, a within-host respiratory tract infection disease model, and a between-host epidemiological model of seasonal variations of viral disease burden. Knowledge from a large body of the published scientific literature was extracted to inform equation and parameter values as well as to calibrate parameters left undetermined. The model incorporates knowledge about the absorption, distribution, metabolism, and excretion (ADME) properties of OM-85 and its mechanism of action, the gut–lung axis, the human physiology involved in immune activation in mucosal tissues, and the resolution of viral respiratory tract infections by non-lytic and lytic immune effectors as well as data on disease burden of most prevalent viruses and RTIs in different populations. After calibration to reproduce key reference data, the setup has been used to interrogate a large ensemble of potential trial designs, informing the n-dimensional space of design choices with predicted trial power and effect size. Intersecting, these two constructed n-dimensional surfaces allowed us to identify optimal design choices in an otherwise conceptually intractable space.

The model (published in [32]) is composed of a system of ODEs coupled with a Vpop approach where parameters defining individual virtual patients are described by statistical distributions. The mechanistic model is composed of three components: a within-host mechanistic disease model, a between-host disease burden model, and a treatment model. Its development in light of the typical model development workflow (*see* Subheading 2) is sketched in Fig. 2.

The within-host model describes immune dynamics in response to respiratory virus exposure. It accounts for onset and resolution of infection, target cell infection and pathogen replication, preemptive clearance by the innate immune system and innate inflammatory response, activation of the adaptive response, neutralization of the virus by antibodies, and finally cytolysis of infected cells by cytotoxic lymphocytes. The model was calibrated to reproduce (i) the mean viral load during experimental human respiratory syncytial virus (RSV) infection and (ii) the RTI distribution reported in (prevalence distribution) reference datasets [137]. For this, the model translates stochastic processes determining viral exposure series and state of antiviral defenses into RTI occurrences by applying a threshold on the number of infected cells (20% on the proportion of infected cells).

The between-host model accounts for the RTI seasonality and is based on time-dependent transmission of RSV, rhinovirus, and influenza viruses, which are the most common respiratory viruses in children [138–140]. The model follows the classical formalism of susceptible, infectious, recovered, and susceptible (SIRS) epidemiological models [141]. Healthy susceptible individuals may become infected at a given infection rate. They can then recover after a given time (quantified by the recovery rate), leading them to become recovered individuals. Recovered individuals may lose their immunity after a given time (quantified by an immunity loss rate) and become as a consequence susceptible again to an infection. The model was calibrated against virus-specific incidence data [142–144] to reproduce virus-specific observed seasonality.

The treatment model describes the immunomodulating effect of OM-85 and efficacy on RTI prophylaxis. For this, a PBPK/PD model is linked to the immunological model through ingress in the respiratory tract of reprogrammed type-1 innate memory like cells [145], regulatory T-cells [146–148], and polyclonal immunoglobulin A—producing plasma cells [149, 150] originating from the intestinal Peye's patches according to the current understanding of OM-85's mechanism of action. The model drug-specific PK parameters were calibrated using rodent PK data of a similar product (OM-89) [151, 152] and were allometrically scaled to human physiology. Drug-specific PD parameters were calibrated against two sets of human PD response data under different treatment regimens [150, 153].

To calibrate the clinical efficacy of OM-85, the study of Razi et al. [154] was considered as reference. Following the real trial protocol, a Vpop of children aged 1 to 6 and with a history of three or more RTIs in the previous 12 months was generated. Virtual patients were allocated to receive either OM-85 or a placebo (3.5 mg per day for 10 consecutive days for 3 months) at the start of the trial in August. Model parameters were adjusted to reproduce the observed decrease in cumulative number of RTIs over

12 months between the placebo and the treatment group (observed, −2.18 with −3.22 to −1.13 for the 95% confidence interval, and simulations, −2.03 with −3.43 to −0.7 for the 95% confidence interval).

The setup then simulated a large ensemble of 4320 trial designs in order to identify optimal trial design choices [136]. To construct the trial design space, six trial parameters were varied, with various levels for each parameter: duration of treatment (1, 2, 3, 4, 6, or 12 months), number of days of administration at the beginning of each treatment month (1, 5, 10, 20, or 30 days), the trial starting month (January, April, July, or October), the duration of follow-up after treatment termination (0, 1, 3, or 6 months), the number of RTIs reported by the included patients the previous 12 months (0, 3, or 6 RTIs), and the age group of included patients (1–3, 4–6, or 1–6 years old). For each point in the trial design space, 1000 randomized placebo-controlled in silico trials with 20 patients per arm were conducted, sampled from a pre-run database of more than 100,000 simulated virtual patients.

The distribution and heatmap of monthly average absolute benefit for the 4320 trial designs is reported in Fig. 9. The monthly absolute benefit here is the difference between the mean number of RTIs in the placebo and the treatment group divided by the trial duration (including treatment and follow-up). It measures the number of prevented RTIs in the treatment group throughout the trial. The heavy right tail of this distribution shows that a fraction of trial designs (longer follow-up, more severe patients) is able to capture a larger clinical benefit. A threshold at 1 prevented RTI per year is defined as the minimal absolute benefit which can be considered clinically relevant. This value is chosen arbitrarily as a proof of context and can be adapted for the specific context of interest (regulatory, geographical).

The distribution and heatmap of empirical power as the second important measure of trial success is reported in Fig. 9b [155]. The distribution is almost uniformly distributed over [0, 1] showing that a large proportion of trial designs are not sufficiently powered with the chosen realistic sample size. A threshold of minimal power is set to 0.9 for this analysis, and here again the chosen value can be adapted to desired power.

An optimized trial design should satisfy both thresholds on monthly absolute benefit and on power. Intersection of the two selected ensembles results in the selection of 590 trial designs over the 4320 initial designs (Fig. 9c, top). A sample of eight selected trial designs is reported in Fig. 9c (bottom) which shows that optimal design can be obtained with various levels in each trial parameter category. This curated selection of trial designs can at this stage be explored by filtering out fewer practical options such as

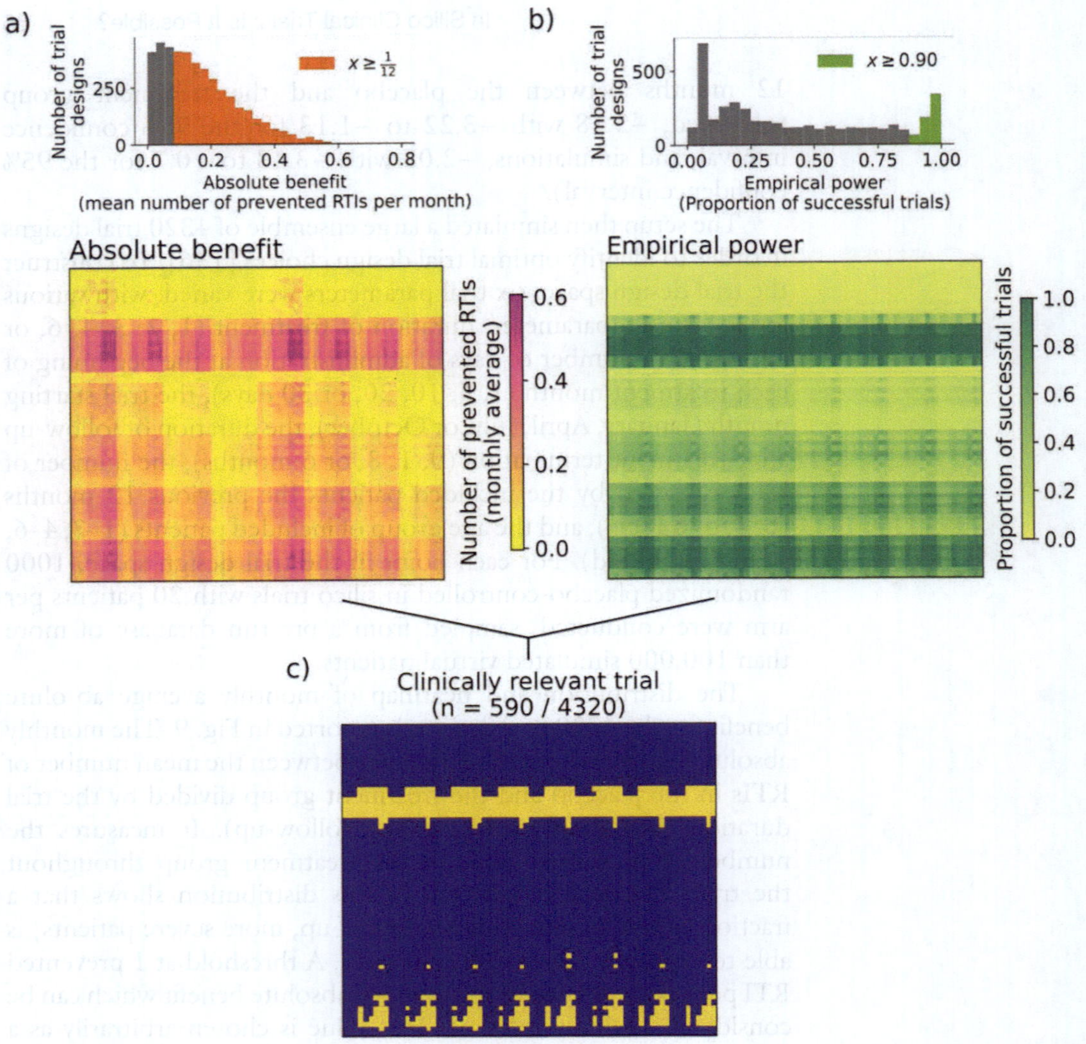

Trial number	4049	1226	1006	1192	1138	4098	1322	987
Age (years)	(1, 6)	(1, 6)	(4, 6)	(4, 6)	(4, 6)	(1, 3)	(1, 6)	(1, 3)
History of RTIs	6	0	6	3	3	3	6	6
Treatment duration (months)	6	12	12	12	12	6	12	12
Follow-up duration (months)	1	0	6	0	3	6	3	1
Number of doses per month	30	30	10	30	20	30	5	10
Treatment start	Jan	Jul	Oct	Apr	Oct	Apr	Jan	Oct
Absolute benefit (monthly, average)	0.42	0.28	0.31	0.37	0.3	0.2	0.36	0.46
Empirical power	0.987	0.944	0.994	0.991	0.978	0.933	0.906	0.922

Fig. 9 Application example with the immunomodulator OM-85 of an in silico clinical trial approach to support trial design. Trial design space is populated with 4320 designs by varying main parameters (e.g., age of included patients, history of events, or treatment duration). (**a**) Top: distribution of absolute benefit values for the full trial design space. Absolute benefit is defined as the difference of the mean monthly number of RTIs in the placebo group compared to the treatment group. Bottom: single absolute benefit values for each trial design are reported in a 72 × 60 matrix. (**b**) Top: distribution of empirical power values for the full trial design space. Bottom: single empirical power values for each trial design are reported in the similarly ordered 72 × 60 matrix as in (**a**). (**c**) Clinically relevant trial designs are reported (yellow) using the similarly ordered 72 × 60 matrix as in (**a**). Such designs are defined as having a yearly absolute benefit of at least 1 RTI and an empirical power of at least 0.8

longer trial durations, for example, or broader patient population for further narrowing the set of optimal trial designs. In a scenario where such a computational systematic exploration is performed, this list of optimal trial designs can then be presented to the various stakeholders and clinical and medical experts as a support tool for decision-making. In conclusion, here it was demonstrated how the combination of mechanistic disease and treatment modeling with high computing capacity can be used to exhaustively explore the high-dimensional space of trial parameters. Contrary to purely statistical models, here this search is not limited to trial parameters which do not impact treatment efficacy. For example, dosing regimen or patient characteristic considerations can be included. As a result, this represents a powerful tool for model-assisted drug development.

5 The Future of In Silico Clinical Trials

Can in silico trials—at some point—replace real clinical trials? Under certain conditions and for selected applications, in silico methods can already today be accepted as alternative to ethically questionable animal or human experimentation. In silico trials have been termed to potentially bring the "3R principle" into the human context which originally stands for "refine, reduce, and replace" animal experimentation by alternative or innovative methods. The 3Rs have been formulated by Russell and Burch [156] and are now used as an ethical framework for improving laboratory animal welfare throughout the world. Recently, FDA has even (partially, and under certain conditions) lifted the requirement to perform animal experimentation before human trials [157]. When projected into the human setting, the 3Rs would mean to refine so as to "reduce the risks [. . .] or improve the predictive accuracy" of human clinical trials, to "reduce the number of humans involved" in clinical trials, and to "replace human experiments in the prediction of the expected safety and/or efficacy for a new treatment" (see the suggested taxonomy in the consensus review by Vicceconti et al. [29]). Surely, in silico trials, as part of the MIDD paradigm, help with the optimal design of trials, which can be interpreted as the "refine" and "reduce" part of the 3R, but can in silico trials be used as "replacement" for human clinical trials? Clearly, such application would have high impact on regulatory decision-making and needs to be qualified accordingly. As all novel methodologies used in the regulatory evaluation of medical products, in silico clinical trials will have to be qualified by regulators such as FDA and EMA through formal processes. Therefore, the establishment of good simulation practice, collaboration with regulatory science experts, and establishing credibility in in silico clinical trials are pivotal for advancing into high-risk and high-impact applications that replace fully powered, randomized clinical trials (RCTs). Even so, one may guess that validation in patients will be required, at least in a small cohort of predicted responders [158]. Along these lines, several modeling-based alternative methods do find already application—even to replace clinical trials:

- Delays in cardiac repolarization can lead to life-threatening heart arrhythmias which sometimes needs to be tested in so-called QT trials, but an in vitro data-based in silico cardiomyocyte model can nowadays replace this evidence-based risk assessment [159].
- PBPK can be used to replace, e.g., drug–drug interaction studies based on in vitro data [160] as well as has an impact in study waivers for generic drug development [161].

Maybe because regulatory acceptance of in silico applications is a key enabling factor for uptake of a method in industry, more use cases with regulatory relevance and under the viewpoint of regulatory impact and assessment need to be established and discussed. Two topics of urgent unmet need in the regulatory science field open short- and medium-term opportunities for in silico trials (and maybe especially those based on complex physiological models): (a) to support synthetic control strategies and (b) to help determine individual absolute benefit on new medical products via "digital twins."

Single arm, uncontrolled studies are used in most rare disease trials, in a significant portion of pediatric drug development programs, and find increasing application in pivotal cancer trials [162] but are difficult to interpret in the approval (and even more so in HTA). Especially in accelerated approval, there is a risk for data gaps for real-world effectiveness assessment. Synthetic control methods, using external clinical [163] and real-world data [164], have been suggested to address this issue [165]. Synthetic control methods have been known in other fields (e.g., in epidemiology) for a while and now find more and more application in healthcare. Lambert et al. have summarized the process of synthetic controls in three basic steps [166]:

1. Definition of estimand, i.e., the comparison reflecting the clinical question.
2. Adequate selection of external controls, from previous RCTs or real-world data obtained from patient cohort, registries, or electronic patient files (and applying modeling techniques for the adjustment of individual- or group-level data).
3. Choosing the statistical approach depending on the available individual-level or aggregated external data and trial design.

There are already several examples for the utilization of synthetic control arms in drug development. Currently, different statistical frequentist inference methods and Bayesian techniques are being used to best match the interventional group with matched external control individuals [165]. All currently used synthetic control methods still face limitations, related to data (e.g., availability, quality, homogeneity, hidden variables, confounding variables, with the special issue of confounders that are not measured or cannot be measured) and statistical methodology (complexity, interpretability of findings). All statistical methods have in common that they rely on a pool of individual patient data records, meaning that these methods cannot interpolate or extrapolate on a single patient level. Mechanistic models are complementary to data-driven approaches as they promise to deterministically adjust the characteristics of individual patients by taking, e.g., an untested dose or another hypothetical age as input to the individual model and by then predicted the adjusted individual effect of that patient at a different age or under intervention with a different dose (*see* Fig. 10). This individual-level adjustment can in fact be added to the group adjustment of the statistical techniques which then could use an enlarged pool of real patients and synthetic, adjusted patients. In principle, then predictions of efficacy in the (synthetic) control group could be refined by adjusting for missing information (a mutation status, allergy, a missing set of blood markers, a dose not tested in the external database) or adjusting for confounders and biases that are hidden in the external data but not in the model. To this day, there seems very little precedence of the use of mechanistic models and in silico trials as synthetic controls (*see* [26, 27] for an example that comes close to such context of use). This seems especially critical since further advancement of this technique relies on illustrative use cases that can highlight the added value, opportunities, and challenges to stakeholders, including regulators.

As a second opportunity for in silico trial methodologies, and especially those implementing "digital twins," a breakdown of the traditional clinical development and benefit–risk assessment paradigm for personalized treatments can be mentioned. Although as ancient as Hippocrates, personalized medicine is being reinvented, thanks to the advent of precision medicine. It is becoming an innovative approach to tailoring disease prevention and treatment that takes into account differences in people's genes, environments, and lifestyles [168–170]. One of the main problems with the development of personalized medicine is—in a very simplified view—that most medical treatments are designed for the "average patient" as a one-size-fits-all approach, which may be successful for a majority of patients but not for all. In an RCT designed to show this average efficacy, the smaller the sample size of a subgroup of patients (with the extreme being 1 as the individual patient), the harder it is to show a benefit for this subgroup. For example, with "omics" becoming more widespread and affordable, individual therapeutic dosing or combination seems plausible but need not

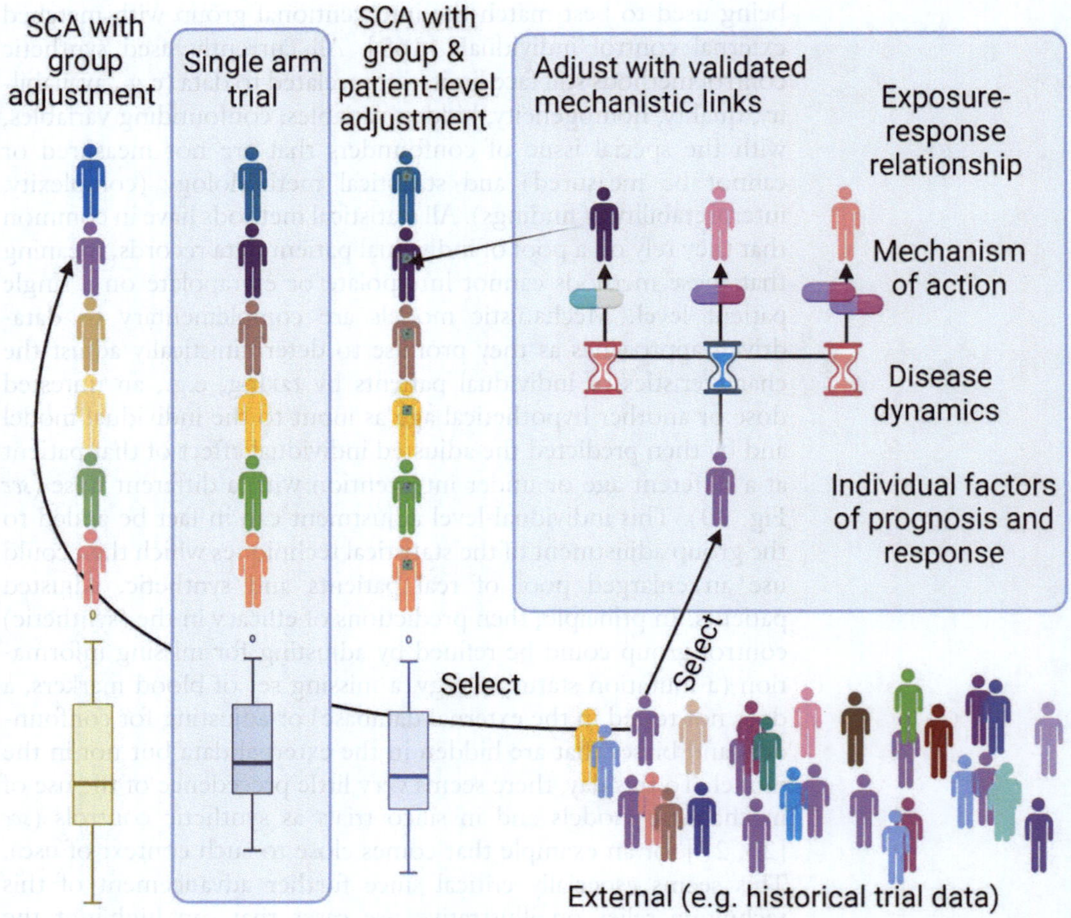

Fig. 10 Proposed workflow to augment synthetic control methods by mechanistic models: Group adjustment methods select individual patients from historical cohorts (e.g., using propensity score matching [167]) but cannot make up for missing data (time points, interventions, etc.). Mechanistic models can quantitatively link patient characteristics, exposure, and protocol changes with outcomes on the single patient level through mechanistic assumptions. These can be validated by sparse data and provide credible data extrapolation techniques, i.e., to adjust a single patient to the dose level or time point needed in the synthetic control. On top of that, the richer the mechanistic model, the better the synthetic control may account for hidden confounders that could reduce bias. (Figure created with BioRender.com)

only complex trial design (e.g., [171]) but also the qualification of all needed diagnostic tests [172] and still face uncertainty with respect to the regulatory approval requirements [168]. At the same time, data becomes more available and CMs evolve quickly and can be informed by personalized information. Recent advances have allowed for synthetic patient data to become statistically and clinically indistinguishable from real patient data. Bottom-up simulation and data-driven training of models are possible workflows; statistical models adding noise to real patient data and the use of machine learning and black box artificial intelligence techniques are the most prevalent methods to arrive at "high-fidelity" synthetic data [173, 174]. Still there are many hurdles for this kind of application of which integration of heterogeneous data from multiple sources and types needs to be tackled. In the long run, mechanistic computational disease and therapeutic models such as the ones used in in silico trials may generate digital twins of patients enrolled in clinical trials. Such development is of particular interest to capture the high degree of variability of immune-related disorders (immune digital twins [175]).

Overall, there are numerous promising applications in the field of in silico clinical trials. Mechanistic and physiologically and—more generally—knowledge-based models take a special position when embedded into a clinical trial simulation paradigm as they offer unique advantages (mainly leveraging preexisting information to lower the need for data) over purely statistical models but come with high complexity. The fact that they are knowledge and hypothesis driven can render them (if white-boxing/transparency guaranteed) ideal communication tools to convey mechanistic evidence and causality to stakeholders and decision-makers. Finally, integration of such transparent but highly complex and thus computationally expensive knowledge-based models into cloud-based simulation engines is the missing step to a computational platform that can be centric to clinical development, drug approval, and HTA and maybe even patient care.

Acknowledgments

Help with the editing by Melanie Senior is gratefully acknowledged.

Conflicts of Interest All authors are employees of Novadiscovery.

References

1. Scannell JW, Blanckley A, Boldon H, Warrington B (2012) Diagnosing the decline in pharmaceutical R&D efficiency. Nat Rev Drug Discov 11:191–200. https://doi.org/10.1038/nrd3681
2. Scannell JW, Bosley J, Hickman JA et al (2022) Predictive validity in drug discovery: what it is, why it matters and how to improve it. Nat Rev Drug Discov 21:915–931. https://doi.org/10.1038/s41573-022-00552-x
3. Standing JF (2017) Understanding and applying pharmacometric modelling and simulation in clinical practice and research. Br J Clin Pharmacol 83:247–254. https://doi.org/10.1111/bcp.13119
4. Williams PJ, Ette EI (2000) The role of population pharmacokinetics in drug development in light of the Food and Drug Administration's "Guidance for Industry: Population Pharmacokinetics". Clin Pharmacokinet 39:385–395. https://doi.org/10.2165/00003088-200039060-00001
5. Gobburu JVS, Marroum PJ (2001) Utilisation of pharmacokinetic-pharmacodynamic modelling and simulation in regulatory decision-making. Clin Pharmacokinet 40: 883–892. https://doi.org/10.2165/00003088-200140120-00001

6. Luzon E, Blake K, Cole S et al (2017) Physiologically based pharmacokinetic modeling in regulatory decision-making at the European Medicines Agency. Clin Pharmacol Ther 102: 98–105. https://doi.org/10.1002/cpt.539
7. Bai JPF, Earp JC, Pillai VC (2019) Translational quantitative systems pharmacology in drug development: from current landscape to good practices. AAPS J 21:72. https://doi.org/10.1208/s12248-019-0339-5
8. Knight-Schrijver VR, Chelliah V, Cucurull-Sanchez L, Le Novère N (2016) The promises of quantitative systems pharmacology modelling for drug development. Comput Struct Biotechnol J 14:363–370. https://doi.org/10.1016/j.csbj.2016.09.002
9. Azer K, Kaddi CD, Barrett JS et al (2021) History and future perspectives on the discipline of quantitative systems pharmacology modeling and its applications. Front Physiol 12:637999. https://doi.org/10.3389/fphys.2021.637999
10. Ermakov S, Schmidt BJ, Musante CJ, Thalhauser CJ (2019) A survey of software tool utilization and capabilities for quantitative systems pharmacology: what we have and what we need. CPT Pharmacometrics Syst Pharmacol 8:62–76. https://doi.org/10.1002/psp4.12373
11. Lemaire V, Bassen D, Reed M et al (2022) From cold to hot: changing perceptions and future opportunities for quantitative systems pharmacology modeling in cancer immunotherapy. Clin Pharmacol Ther. https://doi.org/10.1002/cpt.2770
12. Madabushi R, Seo P, Zhao L et al (2022) Review: role of model-informed drug development approaches in the lifecycle of drug development and regulatory decision-making. Pharm Res 39:1669–1680. https://doi.org/10.1007/s11095-022-03288-w
13. Model-informed drug development paired meeting program. https://www.fda.gov/drugs/development-resources/model-informed-drug-development-paired-meeting-program. Accessed 24 Feb 2023
14. Zineh I (2019) Quantitative systems pharmacology: a regulatory perspective on translation. CPT Pharmacometrics Syst Pharmacol 8:336–339. https://doi.org/10.1002/psp4.12403
15. Galluppi GR, Brar S, Caro L et al (2021) Industrial perspective on the benefits realized from the FDA's model-informed drug development paired meeting pilot program. Clin Pharmacol Ther 110:1172–1175. https://doi.org/10.1002/cpt.2265
16. European Medicines Agency—Methodology Working Party. https://www.ema.europa.eu/en/committees/working-parties-other-groups/chmp/methodology-working-party. Accessed 24 Feb 2023
17. International Council for Harmonisation—MIDD Discussion Group (2022) Considerations with respect to future MIDD related guidelines
18. Marshall S, Burghaus R, Cosson V et al (2016) Good practices in model-informed drug discovery and development: practice, application, and documentation. CPT Pharmacometrics Syst Pharmacol 5:93–122. https://doi.org/10.1002/psp4.12049
19. Hsu L-F (2022) A survey of population pharmacokinetic reports submitted to the USFDA: an analysis of common issues in NDA and BLA from 2012 to 2021. Clin Pharmacokinet 61:1697–1703. https://doi.org/10.1007/s40262-022-01182-7
20. Overgaard R, Ingwersen S, Tornøe C (2015) Establishing good practices for exposure–response analysis of clinical endpoints in drug development. CPT Pharmacometrics Syst Pharmacol 4:565–575. https://doi.org/10.1002/psp4.12015
21. Gobburu JVS, Lesko LJ (2009) Quantitative disease, drug, and trial models. Annu Rev Pharmacol Toxicol 49:291–301. https://doi.org/10.1146/annurev.pharmtox.011008.145613
22. Kimko HHC, Peck CC (2011) Clinical trial simulations. Springer, New York
23. Ankolekar S, Mehta C, Mukherjee R et al (2021) Monte Carlo simulation for trial design tool. In: Principles and practice of clinical trials. Springer, Cham, pp 1–23
24. Kowalski KG (2019) Integration of pharmacometric and statistical analyses using clinical trial simulations to enhance quantitative decision making in clinical drug development. Stat Biopharm Res 11:85–103. https://doi.org/10.1080/19466315.2018.1560361
25. Barrett JS, Nicholas T, Azer K, Corrigan BW (2022) Role of disease progression models in drug development. Pharm Res 39:1803–1815. https://doi.org/10.1007/s11095-022-03257-3
26. Etheve L, Courcelles E, Lefaudeux D et al (2022) Essais cliniques in silico, une approche innovante visant à compléter les essais cliniques dans le domaine des maladies rares: validation d'un modèle computationnel chez les patients atteints d'hypoparathyroïdie. Rev Epidemiol Sante Publique 70:S258–S259.

https://doi.org/10.1016/j.respe.2022.09.063

27. Bertocchio J-P, Gittoes N, Siebert U et al (2022) La modélisation in silico montre une réduction de la survenue de l'insuffisance rénale terminale après 20 ans de traitement par rhPTH (1-84) chez des patients atteints d'hypoparathyroïdie non adéquatement contrôlés par le traitement standard. Rev Epidemiol Sante Publique 70:S259–S260. https://doi.org/10.1016/j.respe.2022.09.064
28. Courcelles E, Boissel J-P, Massol J et al (2022) Solving the evidence interpretability crisis in health technology assessment: a role for mechanistic models? Front Med Technol 4:810315. https://doi.org/10.3389/fmedt.2022.810315
29. Viceconti M, Emili L, Afshari P et al (2021) Possible contexts of use for in silico trials methodologies: a consensus-based review. IEEE J Biomed Health Inform 25:3977–3982. https://doi.org/10.1109/JBHI.2021.3090469
30. Pappalardo F, Russo G, Tshinanu FM, Viceconti M (2019) In silico clinical trials: concepts and early adoptions. Brief Bioinform 20:1699–1708. https://doi.org/10.1093/bib/bby043
31. Gutiérrez-Casares JR, Quintero J, Jorba G et al (2021) Methods to develop an in silico clinical trial: computational head-to-head comparison of lisdexamfetamine and methylphenidate. Front Psychol 12:741170. https://doi.org/10.3389/fpsyt.2021.741170
32. Arsène S, Couty C, Faddeenkov I et al (2022) Modeling the disruption of respiratory disease clinical trials by non-pharmaceutical COVID-19 interventions. Nat Commun 13:1980. https://doi.org/10.1038/s41467-022-29534-8
33. Boissel J, Auffray C, Noble D et al (2015) Bridging systems medicine and patient needs. CPT Pharmacometrics Syst Pharmacol 4:135–145. https://doi.org/10.1002/psp4.26
34. Gadkar K, Kirouac D, Mager D et al (2016) A six-stage workflow for robust application of systems pharmacology. CPT Pharmacometrics Syst Pharmacol 5:235–249. https://doi.org/10.1002/psp4.12071
35. Visser SAG, de Alwis DP, Kerbusch T et al (2014) Implementation of quantitative and systems pharmacology in large pharma. CPT Pharmacometrics Syst Pharmacol 3:142. https://doi.org/10.1038/psp.2014.40
36. Friedrich C (2016) A model qualification method for mechanistic physiological QSP models to support model-informed drug development. CPT Pharmacometrics Syst Pharmacol 5:43–53. https://doi.org/10.1002/psp4.12056
37. Ghosh S, Matsuoka Y, Asai Y et al (2011) Software for systems biology: from tools to integrated platforms. Nat Rev Genet 12:821–832. https://doi.org/10.1038/nrg3096
38. Azeloglu EU, Iyengar R (2015) Good practices for building dynamical models in systems biology. Sci Signal 8:fs8. https://doi.org/10.1126/scisignal.aab0880
39. Novadiscovery SA homepage. https://www.novadiscovery.com/. Accessed 24 Feb 2023
40. Rian K, Hidalgo MR, Çubuk C et al (2021) Genome-scale mechanistic modeling of signaling pathways made easy: a bioconductor/cytoscape/web server framework for the analysis of omic data. Comput Struct Biotechnol J 19:2968–2978. https://doi.org/10.1016/j.csbj.2021.05.022
41. Palsson S, Hickling TP, Bradshaw-Pierce EL et al (2013) The development of a fully-integrated immune response model (FIRM) simulator of the immune response through integration of multiple subset models. BMC Syst Biol 7:95. https://doi.org/10.1186/1752-0509-7-95
42. Cheng Y, Straube R, Alnaif AE et al (2022) Virtual populations for quantitative systems pharmacology models. Methods Mol Biol 2486:129–179
43. Raue A, Schilling M, Bachmann J et al (2013) Lessons learned from quantitative dynamical modeling in systems biology. PLoS One 8:e74335. https://doi.org/10.1371/journal.pone.0074335
44. Hansen N (2007) The CMA evolution strategy: a comparing review. In: Towards a new evolutionary computation. Springer Berlin Heidelberg, Berlin/Heidelberg, pp 75–102
45. Rodriguez-Fernandez M, Egea JA, Banga JR (2006) Novel metaheuristic for parameter estimation in nonlinear dynamic biological systems. BMC Bioinformatics 7:483. https://doi.org/10.1186/1471-2105-7-483
46. Degasperi A, Fey D, Kholodenko BN (2017) Performance of objective functions and optimisation procedures for parameter estimation in system biology models. NPJ Syst Biol Appl 3:20. https://doi.org/10.1038/s41540-017-0023-2

47. Grodzevich O, Romanko O (2006) Normalization and other topics in multi-objective optimization. In: Proceedings of the fields–MITACS industrial problems workshop. Fabien
48. Palgen J-L, Perrillat-Mercerot A, Ceres N et al (2022) Integration of heterogeneous biological data in multiscale mechanistic model calibration: application to lung adenocarcinoma. Acta Biotheor 70:19. https://doi.org/10.1007/s10441-022-09445-3
49. Harring JR, Liu J (2016) A comparison of estimation methods for nonlinear mixed-effects models under model misspecification and data sparseness: a simulation study. J Mod Appl Stat Methods 15:539–569. https://doi.org/10.22237/jmasm/1462076760
50. Comets E, Lavenu A, Lavielle M (2017) Parameter estimation in nonlinear mixed effect models using saemix, an R implementation of the SAEM algorithm. J Stat Softw 80: i03. https://doi.org/10.18637/jss.v080.i03
51. Sher A, Niederer SA, Mirams GR et al (2022) A quantitative systems pharmacology perspective on the importance of parameter identifiability. Bull Math Biol 84:39. https://doi.org/10.1007/s11538-021-00982-5
52. Clairon Q, Pasin C, Balelli I et al (2021) Parameter estimation in nonlinear mixed effect models based on ordinary differential equations: an optimal control approach
53. Hu L, Jiang Y, Zhu J, Chen Y (2013) Hybrid of the scatter search, improved adaptive genetic, and expectation maximization algorithms for phase-type distribution fitting. Appl Math Comput 219:5495–5515. https://doi.org/10.1016/j.amc.2012.11.019
54. Allen R, Rieger T, Musante C (2016) Efficient generation and selection of virtual populations in quantitative systems pharmacology models. CPT Pharmacometrics Syst Pharmacol 5:140–146. https://doi.org/10.1002/psp4.12063
55. Ratto M, Tarantola S, Saltelli A (2001) Sensitivity analysis in model calibration: GSA-GLUE approach. Comput Phys Commun 136:212–224. https://doi.org/10.1016/S0010-4655(01)00159-X
56. Saltelli A, Annoni P, Azzini I et al (2010) Variance based sensitivity analysis of model output. Design and estimator for the total sensitivity index. Comput Phys Commun 181:259–270. https://doi.org/10.1016/j.cpc.2009.09.018
57. (2007) Guideline on reporting the results of population pharmacokinetic analyses. https://www.ema.europa.eu/documents/scientific-guideline/guideline-reporting-results-population-pharmacokinetic-analyses_en.pdf. Accessed 14 Mar 2023
58. (2022) Population pharmacokinetics guidance for industry. https://www.fda.gov/regulatory-information/search-fda-guidance-documents/population-pharmacokinetics. Accessed 15 Mar 2023
59. (2018) Guideline on the reporting of physiologically based pharmacokinetic (PBPK) modelling and simulation. https://www.ema.europa.eu/en/reporting-physiologically-based-pharmacokinetic-pbpk-modelling-simulation-scientific-guideline. Accessed 15 Mar 2023
60. Musuamba FT, Bursi R, Manolis E et al (2020) Verifying and validating quantitative systems pharmacology and in silico models in drug development: current needs, gaps, and challenges. CPT Pharmacometrics Syst Pharmacol 9:195–197. https://doi.org/10.1002/psp4.12504
61. American Society of Mechanical Engineers (2018) Assessing credibility of computational modeling through verification and validation: application to medical devices. https://www.asme.org/codes-standards/find-codes-standards/v-v-40-assessing-credibility-computational-modeling-verification-validation-application-medical-devices. Accessed 27 Feb 2023
62. U.S. Department of Health and Human Services, Food and Drug Administration, Center for Devices and Radiological Health (2021) Assessing the credibility of computational modeling and simulation in medical device submissions—draft guidance for industry and food and drug administration staff
63. Kuemmel C, Yang Y, Zhang X et al (2020) Consideration of a credibility assessment framework in model-informed drug development: potential application to physiologically-based pharmacokinetic modeling and simulation. CPT Pharmacometrics Syst Pharmacol 9:21–28. https://doi.org/10.1002/psp4.12479
64. Viceconti M, Pappalardo F, Rodriguez B et al (2021) In silico trials: verification, validation and uncertainty quantification of predictive models used in the regulatory evaluation of biomedical products. Methods 185:120–127. https://doi.org/10.1016/j.ymeth.2020.01.011
65. Pathmanathan P, Gray RA, Romero VJ, Morrison TM (2017) Applicability analysis of validation evidence for biomedical computational

models. J Verif Valid Uncertain Quantif 2: 021005. https://doi.org/10.1115/1.4037671

66. Musuamba FT, Skottheim Rusten I, Lesage R et al (2021) Scientific and regulatory evaluation of mechanistic in silico drug and disease models in drug development: building model credibility. CPT Pharmacometrics Syst Pharmacol 10:804–825. https://doi.org/10.1002/psp4.12669
67. Sheiner LB (1997) Learning versus confirming in clinical drug development. Clin Pharmacol Ther 61:275–291. https://doi.org/10.1016/S0009-9236(97)90160-0
68. Boissel J-P, Kahoul R, Marin D, Boissel F-H (2013) Effect model law: an approach for the implementation of personalized medicine. J Pers Med 3:177–190. https://doi.org/10.3390/jpm3030177
69. Kahoul R, Gueyffier F, Amsallem E et al (2014) Comparison of an effect-model-law-based method versus traditional clinical practice guidelines for optimal treatment decision-making: application to statin treatment in the French population. J R Soc Interface 11: 20140867. https://doi.org/10.1098/rsif.2014.0867
70. Boissel J-P, Kahoul R, Amsallem E et al (2011) Towards personalized medicine: exploring the consequences of the effect model-based approach. Perinat Med 8:581–586. https://doi.org/10.2217/pme.11.54
71. Boissel J-P, Collet J-P, Lievre M, Girard P (1993) An effect model for the assessment of drug benefit. J Cardiovasc Pharmacol 22: 356–363. https://doi.org/10.1097/00005344-199309000-00003
72. Boissel J-P (1998) Individualizing aspirin therapy for prevention of cardiovascular events. JAMA 280:1949. https://doi.org/10.1001/jama.280.22.1949
73. Glasziou PP, Irwig LM (1995) An evidence based approach to individualising treatment. BMJ 311:1356–1359. https://doi.org/10.1136/bmj.311.7016.1356
74. Wang H, Boissel J-P, Nony P (2009) Revisiting the relationship between baseline risk and risk under treatment. Emerg Themes Epidemiol 6:1. https://doi.org/10.1186/1742-7622-6-1
75. Boissel J-P, Cucherat M, Nony P et al (2008) New insights on the relation between untreated and treated outcomes for a given therapy effect model is not necessarily linear. J Clin Epidemiol 61:301–307. https://doi.org/10.1016/j.jclinepi.2007.07.007
76. Pison C, Magnan A, Botturi K et al (2014) Prediction of chronic lung allograft dysfunction: a systems medicine challenge. Eur Respir J 43:689–693. https://doi.org/10.1183/09031936.00161313
77. Joshi A, Ramanujan S, Jin JY (2023) The convergence of pharmacometrics and quantitative systems pharmacology in pharmaceutical research and development. Eur J Pharm Sci 182:106380. https://doi.org/10.1016/j.ejps.2023.106380
78. Matlab product homepage. https://fr.mathworks.com/products/matlab.html. Accessed 8 Mar 2023
79. R project homepage. https://www.r-project.org/. Accessed 8 Mar 2023
80. SimBiology homepage. https://uk.mathworks.com/products/simbiology.html. Accessed 8 Mar 2023
81. Mrgsolve homepage. https://mrgsolve.org/. Accessed 8 Mar 2023
82. Hoops S, Sahle S, Gauges R et al (2006) COPASI—a complex pathway simulator. Bioinformatics 22:3067–3074. https://doi.org/10.1093/bioinformatics/btl485
83. Tellurium Github repository. https://github.com/sys-bio/tellurium. Accessed 8 Mar 2023
84. Choi K, Medley JK, König M et al (2018) Tellurium: an extensible python-based modeling environment for systems and synthetic biology. Biosystems 171:74–79. https://doi.org/10.1016/j.biosystems.2018.07.006
85. Roadrunner Github repository. https://github.com/sys-bio/roadrunner. Accessed 8 Mar 2023
86. BioSimulators.org homepage. https://biosimulators.org/. Accessed 8 Mar 2023
87. Tiwari K, Kananathan S, Roberts MG et al (2021) Reproducibility in systems biology modelling. Mol Syst Biol 17:e9982. https://doi.org/10.15252/msb.20209982
88. Phoenix WinNonLin homepage. https://www.certara.com/software/phoenix-winnonlin/. Accessed 8 Mar 2023
89. El-Khateeb E, Burkhill S, Murby S et al (2021) Physiological-based pharmacokinetic modeling trends in pharmaceutical drug development over the last 20-years; in-depth analysis of applications, organizations, and platforms. Biopharm Drug Dispos 42:107–117. https://doi.org/10.1002/bdd.2257
90. SimCYP homepage. https://www.certara.com/software/simcyp-pbpk. Accessed 8 Mar 2023
91. GastroPlus homepage. https://www.simulations-plus.com/software/gastroplus/. Accessed 8 Mar 2023

92. PK-Sim Github repository. https://github.com/Open-Systems-Pharmacology/PK-Sim. Accessed 8 Mar 2023
93. NONMEM homepage. https://www.iconplc.com/innovation/nonmem/. Accessed 8 Mar 2023
94. Monolix homepage. https://lixoft.com/products/monolix/. Accessed 8 Mar 2023
95. Roche: shifting to an open-source backbone in clinical trials. https://posit.co/blog/roche-shifting-to-an-open-source-backbone-in-clinical-trials/. Accessed 8 Mar 2023
96. Meyer EL, Mesenbrink P, Mielke T et al (2021) Systematic review of available software for multi-arm multi-stage and platform clinical trial design. Trials 22:183. https://doi.org/10.1186/s13063-021-05130-x
97. Juárez MA, Pennisi M, Russo G et al (2020) Generation of digital patients for the simulation of tuberculosis with UISS-TB. BMC Bioinformatics 21:449. https://doi.org/10.1186/s12859-020-03776-z
98. UISS-TB simulator. https://www.strituvad.eu/uiss-tb-simulator. Accessed 8 Mar 2023
99. Callahan TJ, Tripodi IJ, Pielke-Lombardo H, Hunter LE (2020) Knowledge-based biomedical data science. Annu Rev Biomed Data Sci 3:23–41. https://doi.org/10.1146/annurev-biodatasci-010820-091627
100. Bhatnagar R, Sardar S, Beheshti M, Podichetty JT (2022) How can natural language processing help model informed drug development?: a review. JAMIA Open 5:ooac043. https://doi.org/10.1093/jamiaopen/ooac043
101. Keating SM, Waltemath D, König M et al (2020) SBML level 3: an extensible format for the exchange and reuse of biological models. Mol Syst Biol 16. https://doi.org/10.15252/msb.20199110
102. SBML homepage. https://sbml.org. Accessed 8 Mar 2023
103. SUNDIALS GitHub repository. https://github.com/LLNL/sundials. Accessed 8 Mar 2023
104. NixOS Hydra GitHub repository. https://github.com/NixOS/hydra. Accessed 8 Mar 2023
105. Apache Spark homepage. https://spark.apache.org/. Accessed 8 Mar 2023
106. Apache Storm GitHub repository. https://github.com/apache/storm. Accessed 8 Mar 2023
107. Apache Kafka GitHub repository. https://kafka.apache.org. Accessed 8 Mar 2023
108. ZeroMQ homepage. https://zeromq.org. Accessed 8 Mar 2023
109. Redis homepage. https://redis.io. Accessed 8 Mar 2023
110. Johnson T (1995) Designing a distributed queue. In: Proceedings. Seventh IEEE symposium on parallel and distributed processing. IEEE Computer Society Press, pp 304–311
111. Spivey MZ, Powell WB (2004) The dynamic assignment problem. Transp Sci 38:399–419. https://doi.org/10.1287/trsc.1030.0073
112. Redis "distributed locks" documentation. https://redis.io/docs/manual/patterns/distributed-locks. Accessed 8 Mar 2023
113. Kishore J, Goel M, Khanna P (2010) Understanding survival analysis: Kaplan-Meier estimate. Int J Ayurveda Res 1:274. https://doi.org/10.4103/0974-7788.76794
114. Iooss B, Lemaître P (2015) A review on global sensitivity analysis methods. In: Dellino G, Meloni C (eds) Uncertainty management in simulation-optimization of complex systems, Operations research/computer science interfaces series, vol 59. Springer, Boston, pp 101–122
115. Archer GEB, Saltelli A, Sobol IM (1997) Sensitivity measures, anova-like techniques and the use of bootstrap. J Stat Comput Simul 58:99–120. https://doi.org/10.1080/00949659708811825
116. LaTeX—a document preparation system. https://www.latex-project.org/. Accessed 16 Mar 2023
117. Schreiber F, Le Novère N (2013) SBGN. In: Encyclopedia of systems biology. Springer, New York, pp 1893–1895
118. SBGN GitHub repository. https://sbgn.github.io/. Accessed 8 Mar 2023
119. SED-ML homepage. https://sed-ml.org/. Accessed 16 Mar 2023
120. Smith LP, Bergmann FT, Garny A et al (2021) The simulation experiment description markup language (SED-ML): language specification for level 1 version 4. J Integr Bioinform 18:20210021. https://doi.org/10.1515/jib-2021-0021
121. Silverberg JI, Simpson EL, Armstrong AW et al (2022) Expert perspectives on key parameters that impact interpretation of randomized clinical trials in moderate-to-severe atopic dermatitis. Am J Clin Dermatol 23:1–11. https://doi.org/10.1007/s40257-021-00639-y
122. Knowles RG (2011) Challenges for the development of new treatments for severe asthma: a pharmaceutical perspective. Curr Pharm

Des 17:699–702. https://doi.org/10.2174/1381612117954290191

123. Kempf L, Goldsmith JC, Temple R (2018) Challenges of developing and conducting clinical trials in rare disorders. Am J Med Genet A 176:773–783. https://doi.org/10.1002/ajmg.a.38413
124. Loi S, Buyse M, Sotiriou C, Cardoso F (2004) Challenges in breast cancer clinical trial design in the postgenomic era. Curr Opin Oncol 16: 536–541. https://doi.org/10.1097/01.cco.0000142925.99075.a0
125. Mawdsley D, Bennetts M, Dias S et al (2016) Model-based network meta-analysis: a framework for evidence synthesis of clinical trial data. CPT Pharmacometrics Syst Pharmacol 5:393–401. https://doi.org/10.1002/psp4.12091
126. Mandema JW, Gibbs M, Boyd RA et al (2011) Model-based meta-analysis for comparative efficacy and safety: application in drug development and beyond. Clin Pharmacol Ther 90:766–769. https://doi.org/10.1038/clpt.2011.242
127. Milligan PA, Brown MJ, Marchant B et al (2013) Model-based drug development: a rational approach to efficiently accelerate drug development. Clin Pharmacol Ther 93: 502–514. https://doi.org/10.1038/clpt.2013.54
128. Mandema J, Cox E, Alderman J (2005) Therapeutic benefit of eletriptan compared to sumatriptan for the acute relief of migraine pain—results of a model-based meta-analysis that accounts for encapsulation. Cephalalgia 25:715–725. https://doi.org/10.1111/j.1468-2982.2004.00939.x
129. Li L, Ding J (2020) General considerations of model-based meta-analysis. Chinese J Clin Pharmacol Ther. https://doi.org/10.12092/j.issn.1009-2501.2020.11.006
130. Jansen JP, Naci H (2013) Is network meta-analysis as valid as standard pairwise meta-analysis? It all depends on the distribution of effect modifiers. BMC Med 11:159. https://doi.org/10.1186/1741-7015-11-159
131. Bonner K, Scotney E, Saglani S (2021) Factors and mechanisms contributing to the development of preschool wheezing disorders. Expert Rev Respir Med 15:745–760. https://doi.org/10.1080/17476348.2021.1913057
132. Niederman MS, Torres A (2022) Respiratory infections. Eur Respir Rev 31:220150. https://doi.org/10.1183/16000617.0150-2022
133. Green CA, Drysdale SB, Pollard AJ, Sande CJ (2020) Vaccination against respiratory syncytial virus. Interdiscip Top Gerontol Geriatr 43:182–192
134. Papi A, Contoli M (2011) Rhinovirus vaccination: the case against. Eur Respir J 37:5–7. https://doi.org/10.1183/09031936.00145710
135. Yin J, Xu B, Zeng X, Shen K (2018) Broncho-Vaxom in pediatric recurrent respiratory tract infections: a systematic review and meta-analysis. Int Immunopharmacol 54:198–209. https://doi.org/10.1016/j.intimp.2017.10.032
136. Arsène S, Chevalier A, Couty C et al (2021) Mechanistic model based meta-analysis for pediatric respiratory tract infection prophylaxis trial design. In: Pediatric respiratory infection and immun. European Respiratory Society, p PA3152
137. Carlsson CJ, Vissing NH, Sevelsted A et al (2015) Duration of wheezy episodes in early childhood is independent of the microbial trigger. J Allergy Clin Immunol 136:1208–1214.e5. https://doi.org/10.1016/j.jaci.2015.05.003
138. Acedo L, Díez-Domingo J, Moraño J-A, Villanueva R-J (2010) Mathematical modelling of respiratory syncytial virus (RSV): vaccination strategies and budget applications. Epidemiol Infect 138:853–860. https://doi.org/10.1017/S0950268809991373
139. Yu J, Xie Z, Zhang T et al (2018) Comparison of the prevalence of respiratory viruses in patients with acute respiratory infections at different hospital settings in North China, 2012–2015. BMC Infect Dis 18:72. https://doi.org/10.1186/s12879-018-2982-3
140. Pattemore PK, Jennings LC (2008) Epidemiology of respiratory infections. In: Pediatric respiratory medicine. Elsevier, pp 435–452
141. Kermack WO, McKendrick AG (1927) A contribution to the mathematical theory of epidemics. Proc R Soc Lond Ser A-Contain Pap Math Phys Character 115:700–721. https://doi.org/10.1098/rspa.1927.0118
142. White LJ, Mandl JN, Gomes MGM et al (2007) Understanding the transmission dynamics of respiratory syncytial virus using multiple time series and nested models. Math Biosci 209:222–239. https://doi.org/10.1016/j.mbs.2006.08.018
143. Zhang Y, Yuan L, Zhang Y et al (2015) Burden of respiratory syncytial virus infections in China: systematic review and meta–analysis. J Glob Health 5:020417. https://doi.org/10.7189/jogh.05.020417

144. Flahault A, Blanchon T, Dorléans Y et al (2006) Virtual surveillance of communicable diseases: a 20-year experience in France. Stat Methods Med Res 15:413–421. https://doi.org/10.1177/0962280206071639

145. Wang X, Peng H, Tian Z (2019) Innate lymphoid cell memory. Cell Mol Immunol 16: 423–429. https://doi.org/10.1038/s41423-019-0212-6

146. Navarro S, Cossalter G, Chiavaroli C et al (2011) The oral administration of bacterial extracts prevents asthma via the recruitment of regulatory T cells to the airways. Mucosal Immunol 4:53–65. https://doi.org/10.1038/mi.2010.51

147. Strickland DH, Judd S, Thomas JA et al (2011) Boosting airway T-regulatory cells by gastrointestinal stimulation as a strategy for asthma control. Mucosal Immunol 4:43–52. https://doi.org/10.1038/mi.2010.43

148. Fu R, Li J, Zhong H et al (2014) Broncho-Vaxom attenuates allergic airway inflammation by restoring GSK3β-related T regulatory cell insufficiency. PLoS One 9:e92912. https://doi.org/10.1371/journal.pone.0092912

149. Emmerich B, Pachmann K, Milatovic D, Emslander HP (1992) Influence of OM-85 BV on different humoral and cellular immune defense mechanisms of the respiratory tract. Respiration 59:19–23. https://doi.org/10.1159/000196126

150. Lusuardi M, Capelli A, Carli S et al (1993) Local airways immune modifications induced by oral bacterial extracts in chronic bronchitis. Chest 103:1783–1791. https://doi.org/10.1378/chest.103.6.1783

151. van Dijk A, Bauer J, Sedelmeier EA, Bessler WG (1997) Absorption, kinetics, antibody-bound and free serum determination of a 14C-labeled *Escherichia coli* extract after single oral administration in rats. Arzneimittelforschung 47:329–334

152. Burckhart MF, Mimouni J, Fontanges R (1997) Absorption kinetics of a 14C-labelled Escherichia coli extract after oral administration in mice. Arzneimittelforschung 47:325–328

153. Danek K, Felus E (1996) Influence of oral bacterial lysate stimulation on local humoral immunity on bronchial asthma patients. Int Rev Allergol Clin Immunol 2:42–45

154. Razi CH, Harmancı K, Abacı A et al (2010) The immunostimulant OM-85 BV prevents wheezing attacks in preschool children. J Allergy Clin Immunol 126:763–769. https://doi.org/10.1016/j.jaci.2010.07.038

155. Zhu H, Lakkis H (2014) Sample size calculation for comparing two negative binomial rates. Stat Med 33:376–387. https://doi.org/10.1002/sim.5947

156. Russell WMS, Burch RL (1960) The principles of humane experimental technique. Med J Aust 1:500–500. https://doi.org/10.5694/j.1326-5377.1960.tb73127.x

157. Wadman M (2023) FDA no longer needs to require animal tests before human drug trials. Science. https://doi.org/10.1126/science.adg6264

158. Boissel J-P, Pérol D, Décousus H et al (2021) Using numerical modeling and simulation to assess the ethical burden in clinical trials and how it relates to the proportion of responders in a trial sample. PLoS One 16:e0258093. https://doi.org/10.1371/journal.pone.0258093

159. Li Z, Ridder BJ, Han X, Wu WW, Sheng J, Tran PN, Wu M, Randolph A, Johnstone RH, Mirams GR, Kuryshev Y, Kramer J, Wu C, Crumb WJ Jr, Strauss DG (2019) Assessment of an in silico mechanistic model for proarrhythmia risk prediction under the CiPA initiative. Clin Pharmacol Ther 105(2): 466–475. https://doi.org/10.1002/cpt.1184

160. Hanke N, Frechen S, Moj D, Britz H, Eissing T, Wendl T, Lehr T (2018) PBPK models for CYP3A4 and P-gp DDI prediction: a modeling network of rifampicin, itraconazole, clarithromycin, midazolam, alfentanil, and digoxin. CPT Pharmacometrics Syst Pharmacol 7(10):647–659. https://doi.org/10.1002/psp4.12343

161. Yuvaneshwa K, Kollipara S, Ahmed T, Chachad S (2022) Applications of PBPK/PBBM modeling in generic product development: an industry perspective. J Drug Deliv Sci Technol 69:103152. https://doi.org/10.1016/j.jddst.2022.103152

162. Tenhunen O, Lasch F, Schiel A, Turpeinen M (2020) Single-arm clinical trials as pivotal evidence for cancer drug approval: a retrospective cohort study of centralized European marketing authorizations between 2010 and 2019. Clin Pharmacol Ther 108:653–660. https://doi.org/10.1002/cpt.1965

163. Hall KT, Vase L, Tobias DK et al (2021) Historical controls in randomized clinical trials: opportunities and challenges. Clin Pharmacol Ther 109:343–351. https://doi.org/10.1002/cpt.1970

164. U.S. Department of Health and Human Services, Food and Drug Administration (2023) Draft guidance: considerations for the design and conduct of externally controlled trials for drug and biological products guidance for industry. https://www.fda.gov/regulatory-information/search-fda-guidance-documents/considerations-design-and-conduct-externally-controlled-trials-drug-and-biological-products. Accessed 16 Mar 2023
165. Thorlund K, Dron L, Park JJ, Mills EJ (2020) Synthetic and external controls in clinical trials—a primer for researchers. Clin Epidemiol 12:457–467. https://doi.org/10.2147/CLEP.S242097
166. Lambert J, Lengliné E, Porcher R et al (2022) Enriching single-arm clinical trials with external controls: possibilities and pitfalls. Blood Adv. https://doi.org/10.1182/bloodadvances.2022009167
167. Reeve BB, Smith AW, Arora NK, Hays RD (2008) Reducing bias in cancer research: application of propensity score matching. Health Care Financ Rev 29:69–80. https://www.ncbi.nlm.nih.gov/pmc/articles/PMC4195028/
168. Knowles L, Luth W, Bubela T (2017) Paving the road to personalized medicine: recommendations on regulatory, intellectual property and reimbursement challenges. J Law Biosci 4:453–506. https://doi.org/10.1093/jlb/lsx030
169. Fournier V, Prebet T, Dormal A et al (2021) Definition of personalized medicine and targeted therapies: does medical familiarity matter? J Pers Med 11:26. https://doi.org/10.3390/jpm11010026
170. Carlsten C, Brauer M, Brinkman F et al (2014) Genes, the environment and personalized medicine. EMBO Rep 15:736–739. https://doi.org/10.15252/embr.201438480
171. Superchi C, Brion Bouvier F, Gerardi C et al (2022) Study designs for clinical trials applied to personalised medicine: a scoping review. BMJ Open 12:e052926. https://doi.org/10.1136/bmjopen-2021-052926
172. Chang L-C, Colonna TE (2018) Recent updates and challenges on the regulation of precision medicine: the United States in perspective. Regul Toxicol Pharmacol 96:41–47. https://doi.org/10.1016/j.yrtph.2018.04.021
173. Goncalves A, Ray P, Soper B et al (2020) Generation and evaluation of synthetic patient data. BMC Med Res Methodol 20:108. https://doi.org/10.1186/s12874-020-00977-1
174. Myles P, Ordish J, Tucker A (2023) The potential synergies between synthetic data and in silico trials in relation to generating representative virtual population cohorts. Prog Biomed Eng 5:013001. https://doi.org/10.1088/2516-1091/acafbf
175. Laubenbacher R, Niarakis A, Helikar T et al (2022) Building digital twins of the human immune system: toward a roadmap. NPJ Digit Med 5:64. https://doi.org/10.1038/s41746-022-00610-z

Chapter 5

Bayesian Optimization in Drug Discovery

Lionel Colliandre and Christophe Muller

Abstract

Drug discovery deals with the search for initial hits and their optimization toward a targeted clinical profile. Throughout the discovery pipeline, the candidate profile will evolve, but the optimization will mainly stay a trial-and-error approach. Tons of in silico methods have been developed to improve and fasten this pipeline. Bayesian optimization (BO) is a well-known method for the determination of the global optimum of a function. In the last decade, BO has gained popularity in the early drug design phase. This chapter starts with the concept of black box optimization applied to drug design and presents some approaches to tackle it. Then it focuses on BO and explains its principle and all the algorithmic building blocks needed to implement it. This explanation aims to be accessible to people involved in drug discovery projects. A strong emphasis is made on the solutions to deal with the specific constraints of drug discovery. Finally, a large set of practical applications of BO is highlighted.

Key words Bayesian optimization, Black box optimization, Global optimization, Active learning, Drug optimization, Drug discovery, Gaussian process, Acquisition function, Sequential design

1 Introduction

The drug discovery process ends with the development of a clinically effective therapy. It can be of multiple types (small molecules, peptides, proteins, antibodies, etc.). Also, depending on the disease, the stage of the disease, the biological target(s), the mode of action, the targeted administration route, and other constraints, each drug must exhibit different properties. For these reasons, the research of new drugs can be seen as the finding of a compromise between multiple properties. The difficulty lies in the fact that the compound profile satisfying these different criteria is poorly known a priori.

During the drug discovery process, the chemical structures of potential drug candidates are optimized to obtain the desired profile for compounds to become preclinical candidates. But despite the improvements in computational chemistry, the relationship between a chemical formula and its properties is still challenging.

Alexander Heifetz (ed.), *High Performance Computing for Drug Discovery and Biomedicine*, Methods in Molecular Biology, vol. 2716, https://doi.org/10.1007/978-1-0716-3449-3_5,

For this reason, the optimization process remains a trial-and-error path in which new molecules are designed (by humans or computers), produced, tested, and analyzed. After each round, experimental results are used to launch a new design cycle. These design cycles called DMTA (design, make, test, analyze) should allow the designer to drive the process across the unknown landscape of the relation between structures and properties.

Mathematically, the optimization process could be seen as a "black box" optimization (BBO) of an unknown function for which we are looking for its global extremum [1].

While a lot of BBO applications refer to the search for the maximum of the objective function, mathematics and programming favor minimization. Thus, the general form of the optimization problem is given by Eq. 1:

$$x^* = \underset{x \in \Omega}{\operatorname{argmin}} f(x) \tag{1}$$

where f is the objective function that must be optimized over the bounded hypervolume Ω. As explained by Alarie et al. [2], a BBO is characterized by an objective function and/or a parameter landscape that is unknown, inaccessible, unexploitable, or nonexistent. In other words, the analytical expression f is unknown, nor its derivative is known or accessible (e.g., too computationally expensive). Thus, evaluation of the function is restricted to sampling at a point x and getting a possibly noisy response.

In such a situation, one could think of using a grid search, random search [3], or numeric gradient estimation to evaluate f and find its global optimum. However, the objective function in a BBO process is expensive, particularly for drug discovery applications. Solutions are thus needed to minimize the number of evaluations of the black box function f.

In practice, the black box function is approximated by a machine learning (ML) model. This model is trained on the existing data assets and then applied to recommend new points for tests. These new data are fed back into the model to improve its prediction capacities. This cycle is repeated until the initial objective or the predetermined budget has been achieved.

In the big data regime, the optimization of a black box function relies on the availability or the capacity to acquire a lot of input data for the ML model. The quality and quantity of data are key for the training of an efficient ML model for the selection of the next point to evaluate.

A specificity of the drug discovery process is that it happens in a small data regime, where the design is driven by a limited availability of high-quality experimental data [4, 5]. A supplementary hurdle is data sparsity. While in a project, the activity on the main biological target and the physicochemical properties are measured for most of the compounds, preclinical and clinical data are obtained for a limited number of candidates.

Open and collaborative projects aim at a better and wider sharing of data in drug discovery. However, a large number of modalities, targets, and properties to measure, and the complexity and reproducibility problem of the experiments, don't allow to overcome the sparsity and small data regime barrier of drug design yet.

As explained by Terayama et al. [1], when the goal is to maximize an objective function, the simpler way is to take as a recommendation the highest predicted point by the ML model (see greedy acquisition function in Chap. 2). This method is suboptimal as it doesn't consider the uncertainty of the prediction. Thus, it tends to optimize more slowly toward a local minimum.

Various methods have been proposed to tackle a BBO problem [1, 2]. Among them, we want to highlight three widely used strategies that are active learning (AL), reinforcement learning (RL), and Bayesian optimization (BO) [6]. The three strategies are sequential, alternating predictions and measurements with the aim of optimizing an objective function. These strategies are similar and are regularly confounded and/or mixed.

Here are our definitions and the main characteristics that distinguish them:

- Reinforcement learning (RL)

  RL is a method that allows an agent to learn in an interactive environment by trial and error using feedback from its actions and experiences. Globally, RL aims at maximizing a reward. In each step of reinforcement, the agent decides on an action that will be applied to the environment. This action led to a new state of the environment allowing the calculation of a reward. The reward is feedback to the agent allowing its tuning.

  This strategy has the specificity that the agent never directly sees the data that are generated, but only the reward of its previous action.

  RL is a well-known approach in drug design, and it is generally used to optimize a generative deep learning model toward the generation of compounds with optimal chemical properties [7].
- Active learning (AL) and Bayesian optimization (BO): similarities

  AL and BO are strongly linked as both are meant to optimize objectives that are costly to evaluate. Unlike RL, AL and BO are part of the supervised learning approach where pairs of data label are available to train a central ML model equivalent to the RL's agent.

Rakhimbekova et al. proposed a definition for AL that is also applicable to BO: "iterative procedure in which an ML model proposes candidates for testing to the user, and the user then returns labelled candidates, which are then used to update the model" [8]. In other words, the ML model is trained with all the labeled data accessible, often in small quantities, whereas the number of potential candidates is huge and the test to label them is expensive.

Globally, both strategies aim at reducing at minimum the number of candidates tested by searching to reduce the uncertainty of the ML model.

– AL and BO: differences

AL and BO strategies differ by three aspects: their main objective, the source of the next selection candidate, and the policy used for this selection.

First, AL aims at improving the quality of the core ML model by increasing its applicability domain, reducing its global uncertainty, and thus increasing its accuracy.

BO aims at finding as fast as possible the extrema (minimum or maximum) of the objective. In a very simplistic view, at the end of the process, AL will have learned a global accurate model (but with no idea where the extreme in the applicability domain of the model is), whereas BO will return the position of the extrema of the objective (but the ML model will not be optimized).

Second, AL works for the selection of candidates in a finite set of possibilities. That's why AL is closely related to experimental design [9] and is widely used in the improvement of the detection of objects in images where a large number of images are accessible but unlabeled [10].

BO works generally in a continuous domain. Thus, the candidates are computed or generated coordinates in this multidimensional space.

Third, because of the global aim of improving the core ML model, the commonly used policy for AL is *uncertainty sampling* where the candidate of maximum uncertainty is selected [11].

For BO, both the predicted value and its uncertainty are considered for the selection. Thus, the candidates are the optimal points of the balanced function between their predicted values and the uncertainty of these predictions.

This review will describe the global process of BO and its components. It will also present examples of the multiple applications of BO in drug discovery.

2 Bayesian Optimization

This section presents the principle and the mathematical components behind Bayesian optimization (BO). Only the key elements are explained here, with the goal to be accessible to all drug hunter scientists from biologists to medicinal and computational chemists. For a more theoretical presentation of the BO options and mathematical concepts, we advise reading the review of Shahriari et al. [12], Brochu et al. [9], and Garnett [13].

2.1 Definition

Optimization is used everywhere in statistics and machine learning applications for decades. However, many applications have been settled for local refinement. On the contrary, BO belongs to the strategies that tend to do global optimization.

BO terminology refers to the sequential iteration of decision-based selection and model update. The heart of this strategy is the gathering of new information that will help improve the current knowledge. At each iteration, BO will propose new points that should complement the already known set of points. Each new point must maximize the added information. This information (known or added) is evaluated through the uncertainty of the ML model that will be reduced alongside the iterations.

2.2 BO Process

BO aims at finding, in a minimum number of iterations, the candidate that corresponds to the global minimum of an objective function. In practice, BO returns the candidate that corresponds to the observed minimum at the end of the iterations (*see* Algorithm 1). If the number of iterations is big enough, the convergence of the observed minimum can be studied. However, given the cost of an observation, the possible number of iterations is low. Thus, the accepted assumption is that the observed minimum is close or equal to the true minimum of the objective function.

To start its sequential process, BO needs a small number of observations of the objective function, and a way to acquire new observations. These observations are used for the training of an ML model that will serve as a surrogate of the objective function. In turn, this surrogate model is used as the base of an acquisition function that allows to balance exploration and exploitation during the BO loops. An optimization algorithm is applied to the acquisition function to determine its maximum. This found point is finally evaluated by the original objective function. If the initial budget (e.g., the number of loops) is not reached, the new observation is added to the initial dataset and a new BO loop is done. On the opposite, if the budget is consumed, the minimum of the observed point is considered as the minimum of the objective function.

Algorithm 1
Bayesian optimization

inputs:	
Objective function f	
Dataset D	⇒ set of observations of f
Budget B	⇒ e.g. number of iterations
while $B > 0$:	
$\mu(x), \sigma(x) = GP(D_i, f(x_i))$	⇒ build a surrogate model
$x' = argmin\ AF(\mu(x), \sigma^2(x))$	⇒ search the optimum of the acquisition function
$y = f(x')$	⇒ evaluate the optimum
$D[x'] = y$	⇒ augment the dataset
output:	
$x^* = argmin_{x \in D} f(x)$	⇒ return the global optimum observed

In consequence, the main considerations to apply BO are:

- The building of the surrogate model
- The selection of the acquisition function

In addition, the next points are of specific importance for BO in drug discovery:

- Multiobjective optimization
- Candidate generation

All these topics are discussed in the following paragraphs.

2.3 Surrogate Model

Like all Bayesian methods, BO depends on a prior distribution. The machine learning surrogate model uses it to approximate the true underlying landscape of the objective function [14]. In other words, it transforms a defined set of prior observations into a continuous landscape. For a lot of applications, and drug discovery in particular, this landscape is characterized by its non-convexity, meaning that it holds multiple local minima.

The key point is that the surrogate landscape uncertainties will be exploited by the acquisition functions in the sequential process (*see* next subheading). Thus, the goal is to learn a surrogate ML model that is not only accurate but also furnishes its predictions with a notion of uncertainty. The constraint is that the uncertainty must converge to zero (or a positive minimum value for noisy data) only when the distance to an observation converges to zero [9].

The uncertainty can be well captured by probabilistic models to produce accurate estimates of its confidence [14]. They manage noisy objective evaluation and they offer the means of uncertainty quantification (UQ), including the chance that local or global optima were missed [15].

Many approaches can be cited for building the surrogate model such as polynomial interpolation, neural networks, support vector machines, random forests, and Gaussian processes. Shahriari et al. describe in detail parametric and nonparametric models that can be used as surrogates [12]. We want to highlight their interesting analysis of the use of random forests (RF) for BO [16]. They make clear that if RF are good interpolators in the sense that they output good predictions in the neighborhood of training data, they are very poor extrapolators. More importantly, the predicted mean values and uncertainty of extrapolated points are constants, and the latter is also low. That means that for extrapolated points, the gradient of an acquisition function built upon RF will be zero [17]. In conclusion, despite RF method being widely used in cheminformatics, it is a poor choice to be used as a surrogate for BO.

In the end, two ML methods are currently considered of choice when dealing with uncertainty calibration [5, 14]: Bayesian neural networks (BNNs) [18–20] and Gaussian processes (GPs). Both methods have also been combined to improve the regression performances [21].

Recently, deep neural network ensemble has also been shown to be competitive to BNNs in the context of sequence optimization [22]. In particular, this method gives the advantage to use pre-trained models that are accurate with very few experiments and adapted to the high-dimensional space of sequence design [23].

2.4 Gaussian Process

For drug discovery and BO applications, considering the small data regime in which they operate, GPs appear to be preferred [24]. This is also reflected in Subheading 3 where most of the presented BO applications in drug discovery use GPs as the core surrogate model.

The Gaussian process (GP) method is named according to the notion of Gaussian distribution. A GP is a stochastic process (a collection of random variables), such that every finite collection of those random variables has a multivariate normal distribution, i.e., every finite linear combination of them is normally distributed [25]. GPs can be seen as an infinite-dimensional generalization of multivariate normal distributions. Thus, as with most of the ML methods, GPs are poor extrapolators (when used with local covariance functions), but they give high uncertainty predictions outside of the applicability domain (*see* Fig. 1). This corresponds to the desired behavior of a surrogate model for BO allowing to play with the exploration over exploitation trade-off.

Brochu et al. make a clever analogy between Gaussian distribution and GP [9]. The first one is a distribution over a random variable and is completely specified by its mean and covariance. By

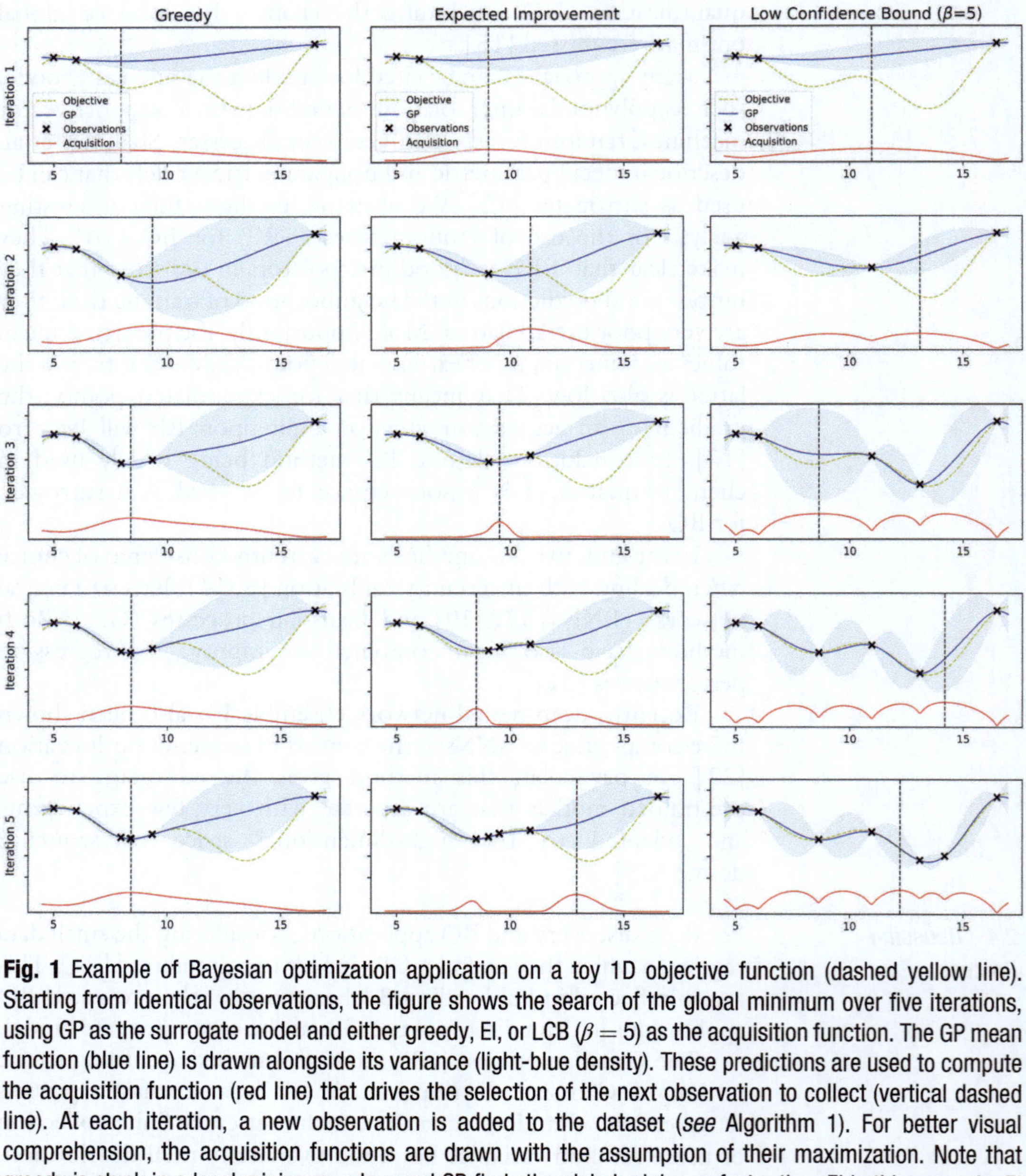

Fig. 1 Example of Bayesian optimization application on a toy 1D objective function (dashed yellow line). Starting from identical observations, the figure shows the search of the global minimum over five iterations, using GP as the surrogate model and either greedy, EI, or LCB ($\beta = 5$) as the acquisition function. The GP mean function (blue line) is drawn alongside its variance (light-blue density). These predictions are used to compute the acquisition function (red line) that drives the selection of the next observation to collect (vertical dashed line). At each iteration, a new observation is added to the dataset (*see* Algorithm 1). For better visual comprehension, the acquisition functions are drawn with the assumption of their maximization. Note that greedy is stuck in a local minimum, whereas LCB finds the global minimum faster than EI in this example. EI tends to select candidate points closer to the minimum of the predicted mean than LCB. This is largely due to the β value used for LCB that highly favors exploration of the landscape

analogy, a GP is a distribution over functions and is completely specified by its mean function m and covariance function k.

By notations, $X \in \mathbb{R}^{n*d}$ is a matrix of n points of dimension d. A row i of the matrix contains the representation x_i. The mean and covariance functions are defined by:

$$m(x) = \mathbb{E}[f(x)] \tag{2}$$

$$k(x, x') = \mathbb{E}[(f(x) - m(x))(f(x') - m(x))] \tag{3}$$

$K_\theta(X, X)$ is a kernel matrix where entries are computed by the kernel function:

$$[K]_{ij} = k(x_i, x_j) \tag{4}$$

where θ represents the set of kernel hyperparameters.

In this context, if a function $f(x)$ returns generally a scalar value, a GP returns the mean and variance of a normal distribution over the possible values of f at x:

$$f(x) \sim \mathrm{GP}(m(x), k(x, x')) \tag{5}$$

Figure 1 shows GP models trained at successive iterations of the BO process. At each iteration, the GP model (blue line) passes through all the observed points, and an uncertainty region (light blue) is derived. The width of this region is related to the distance to known observations.

Various methods derived from GP exist to tackle specific constraints of the datasets. For example, sparse GP methods deal with the computational cost of GPs for large datasets [12]. Interestingly, Cheng et al. applied RobustGP and Trust region Bayesian Optimization (TuRBO) to concentrate the GP training on the most interesting part of the search space [26]. RobustGP first detects outliers in the current BO iteration and then regresses without them using GP with Gaussian likelihood [27]. Thus, the potential bias of the surrogate model caused by the dataset outliers is reduced. TuRBO regresses local GP surrogates within a hyper-rectangle centered at the current best solution, known as trust region (TR) [28]. Doing this, it conducts BO locally, avoiding the exploration of highly uncertain regions in the search space. This is particularly useful for high-dimensional representations. The combination of both methods leads to a more robust and accurate surrogate model and a better evaluation and convergence of the acquisition functions.

2.5 Kernel and Input Representation

Whereas the GP prior mean function is generally assumed to be constant and equal to zero ($m = 0$), the covariance function encodes our belief about the objective we want to model [29]. Thus, the choice of the covariance function is an important inductive bias for the fit. For example, a response expected to be periodic can be fitted using a periodic function.

In the larger context of supervised learning, it is assumed that training points that are close to the point to predict will largely influence the prediction as it is assumed that they will have similar objective values. This behavior is a necessary condition for the convergence of the GP model optimization [30].

In GPs, the covariance function controls the similarity between two points, making the GP a distance-aware model [9, 29]. In most cases, covariance functions produce a higher correlation for closer points encoded [31].

By analogy in drug discovery, the kernels will specify an intuitive relation of similarity between two molecules. Their force is that they can be adapted to multiple molecular representations [32]. In the next paragraph, the most used covariance functions for drug discovery applications will be described and classified by the form of the used molecular representations. All described covariance functions are stationary, meaning they are a function of $x - x'$, thus invariant to translation in the input space (e.g., in opposition with periodic function). The function of two arguments mapping a pair of inputs is called a *kernel* [29].

- Continuous numerical values

Continuous numerical values are the most natural type of data for in silico optimization. In drug discovery, one domain that deals well with continuous numerical features is chemical reaction condition optimization [33]. On the opposite, it is not the case for molecular entities that cannot be directly described by numerical values. But a lot of physicochemical-based molecular descriptors can be computed to represent them such as clogP, molecular weights, number of atoms, number of H-bond donors and acceptors, etc. They are widely used as input for the training of predictive machine and deep learning models for biological properties [34].

In the last decade, big deep neural networks have been trained on various molecular representations, e.g., SMILES, molecular graphs, or protein sequences. New molecular representations, called embeddings, have been derived from these pre-trained models [35–41]. These embeddings are intended to capture in a continuous multidimensional space the vast and diverse chemical space accessible for drug design. Moreover, they provide the capacity to apply BO for molecular de novo generation, starting from a noncontinuous representation (*see* Subheading 2.8).

Outside drug discovery, GPs are mainly applied to continuous variables, e.g., for hyperparameter optimization of neural network training [42]. In this context, squared exponential (SE) (also called radial basis function (RBF)) and Matérn are the most common kernels for GPs. These are the same kernel used for GPs over continuous numerical values in drug discovery applications.

The SE function is defined by:

$$k_{\mathrm{SE}}(x, x') = \exp\left(-\frac{\|x - x'\|^2}{2l^2}\right) \tag{6}$$

The hyperparameter l, necessary for the generalization of the kernel, controls its length scale. This function respects the similarity assumption as its value approaches 1 when the points get closer and 0 when they go further apart.

All features of x are equally considered by this function, meaning they have the same influence on the fit. Thus, this function can be seen as naïve [9], but it applies well to drug discovery use cases where the influence of each individual molecular descriptor on the objective function is unknown.

Matérn kernels add a smoothness parameter ν to be more flexible in the fitting [43, 44]. ν is defined as $\nu > 0$ and such that when $\nu \to \infty$ the SE kernel is obtained. When $\nu = 1/2$ in the Matérn kernel gives the exponential covariance function:

$$k_{\text{MATERN } 1/2}(x, x') = \exp\left(-\frac{\|x - x'\|}{l}\right) \tag{7}$$

We refer the reader to Genton et al. [45] and Rasmussen and Williams [29] for a deeper mathematical presentation of the previous and of more kernels applicable to continuous variables.

- Fingerprints

Fingerprints (FPs) are another important molecular descriptor class used in QSAR modeling [34]. FPs are vectors of binary or integer values. Each value typically encodes the presence (or the number) of a given feature in a molecular entity. For small molecule fingerprints, the most popular molecular FPs is the Morgan fingerprint [46], also known as the extended-connectivity fingerprint ECFP4 [47]. Numerous other FPs have been described for small molecule representations such as ISIDA property-labeled fragment descriptors [48], atoms-pair FPs MAP 4 [49], or bioactivity profile-based fingerprints [50].

Pyzer-Knapp used Morgan FPs to train directly GPs associated with the previously presented SE kernel [51]. Scalar product or linear kernel can also be applied to molecular vectors x, x' [5].

If such a general kernel can be used on vectors, more specialized kernels have been created based on FP similarity measurements [52]. The most common similarity used in drug design is Tanimoto where two fingerprints are compared by the number of features present in both molecules normalized by the number of features occurring separately. Thus, it is not surprising that a Tanimoto kernel has been implemented [53] and recently applied to chemical FPs [14, 54]:

$$k_{\text{Tanimoto}}(x, x') = \sigma^2 \tilde{n} \frac{\langle x, x' \rangle}{\|x\|^2 + \|x\prime\|^2 - \langle x, x' \rangle} \tag{8}$$

where $\langle x, x' \rangle$ is the Euclidean inner product, $|| \cdot ||$ is the Euclidean norm, and σ is a hyperparameter. This kernel has a real chemical meaning when applied to molecular FPs as it drives the fitting of the GPs by the chemical feature similarity between two molecules.

- String

Before being represented by computed representations, molecular entities are represented by string data [32]. This is typical of the biological entities that are represented by chains of characters, each one coding for a pre-determined sub-entity in the chain sequence. The two main examples are DNA and proteins, both represented by sequences of nucleotides and amino acids, respectively. Each nucleotide and amino acid can be embodied by a one- or three-letter code. The full sequence is then the concatenation of all the single codes in the right successive order.

For small molecules, the primary meaningful string is their IUPAC representations [55]. However, the chemical names derived from this nomenclature are not efficient for use by a computer. That's why the simplified molecular input line system (SMILES) has been created [56]. SMILES is widely used in chemoinformatics as it is simple and traduces very well the molecular graph made by atoms and bonds of the molecules. Recently, SMILES deficiencies have been spotted, leading to the creation of more robust string notations like InChi [57] and SELFIES [36].

Derived from work on words and sentences, kernels dedicated to strings have been implemented [58, 59]. In these kernels, the similarity between two strings is evaluated based on the number of shared sub-strings. This can be efficiently applied to SMILES [54].

In their library for GAUssian processes in CHEmistry (GAUCHE), Griffiths et al. [5] implemented a string kernel developed for SMILES by Cao et al. [60]. This kernel computes an inner product between the occurrences of sub-strings, considering all contiguous sub-strings made from at most n characters. The sub-strings are obtained from a bag-of-character representation, and their counts ϕ are used in a Euclidean inner product between two strings S and S':

$$k_{\text{String}}(S, S') = \sigma^2 \tilde{n} \langle \phi(S), \phi(S') \rangle \quad (9)$$

These kernels consider sub-strings, whereas the SMILES grammar is of major importance. Interestingly, this grammar can be taken into account in the optimization process by the imposition of constraints over the desired string structure [61].

If the string kernels have been recently applied on protein sequences and SMILES strings, nothing prevents their usage on any other string representation.

- Graph

  Graph representation is considered the most natural way of representing chemical and biological entities and networks in drug discovery. They can adapt to various level of granularity allowing them to represent entities from small chemical compounds to biological systems.

  Atom-level graphs are used to represent small molecules and proteins in a consistent manner, where nodes encode individual atoms. Similarly, at the next granularity level, graphs can represent biological sequences where the nodes encode the elements, e.g., amino acids for proteins or nucleotides for DNA/RNA. In both types of graphs, the edges encode the relation between the nodes. They seize the special arrangement of the nodes by representing the chemical bonds and the intramolecular interactions between the nodes, e.g., 3D coordinates of the atoms, or interaction network of the residues inside the 3D structure of a protein.

  The granularity of a graph can go up to a multi-scale view of biological systems and functions as represented by knowledge graphs. In such graphs, each node encodes a pharmacologically meaningful object (compound, protein, DNA, RNA, cell, specie, tissue, experimental protocol, etc.), and the edges capture their multifaceted relations (interact, inhibit, activate, belong, etc.).

  Complex information can be included in a graph by assigning numerical features to corresponding nodes and edges as well as to the whole graph. These features represent properties of the nodes, edges, and graphs. For example, for nodes, the features could be atom or residue types; for edges, the features could be bond or interaction types, or distances; and for graphs, the features could be physicochemical properties of the compound described by the graph.

  Various software and Python libraries allow the representation of such graphs. We would like to cite Graphein which has been implemented in the specific context of deep learning calculations [62].

  Outside GP applications, graph kernels have been implemented to compare graphs based on the edges, random walks, subgraphs, or paths [63]. Graph kernels have also been applied to study molecular similarity [64–66]. These kernels have been applied later for GP modeling.

  In the GP context, Gao et al. used graph kernel on molecules represented as graph [67]. They used the modified marginalized graph kernel $k(G, G')$ described by Kashima et al. [68]. Starting from the molecular labeled graph, they compute the count of label paths appearing in the graph. A label path is produced by random walks on graphs and thus is regarded as a random variable. The kernel is then defined as the inner product of the count vectors averaged over all possible label paths. This kernel is also very similar to the Weisfeiler–Lehman kernel based on the inner products of label count vectors over λ iterations and is used in the GAUCHE library [5, 64].

2.6 Acquisition Function

The acquisition function is key in the BO process. It is computed over the predicted mean and variance of the GP surrogate model and is defined on the same domain. Its goal is to propose new points to evaluate for the next BO iteration (*see* Algorithm 1). Thus, the acquisition function is seen as the policy of BO [13], or equally, as the balance between the exploration and exploitation of the objective landscape [9, 12, 31, 51, 69]. This balance is typical of the human decision-making process where we have to deal between choices with immediate limited rewards and with less probable but higher rewards [70].

The acquisition function generalizes the objective landscape in a continuous surface. High acquisition values correspond to high potential for the BO process, because it has either a high predicted objective value (exploitation) or a high predicted uncertainty (exploration). In this sense, the acquisition function evaluates the informational benefit for the next cycle of the experimental test of a new point. Consequently, it guides the exploration of the objective function.

In BO, full exploitation will likely lead to local optimization, missing other more interesting local minima, and more importantly the global optimum. Adding exploration in the landscape span will prevent stagnation in a local optimum and encourages the testing of more diverse points. However, too much exploration will lead to an inefficient search [71]. In conclusion, a balance between exploration and exploitation is needed to converge upon an optimum and truly improve the objective function.

The surface of the acquisition function $AF(x)$ can be mathematically explored. The proposed point for the next BO cycle x' is then obtained by the search of the surface minimum:

$$x' = \text{argmin}\ AF\left(\mu(x), \sigma^2(x)\right) \tag{10}$$

This search can be done by conventional methods, one of the most used algorithms being L-BFGS-B which is applied inside a bounded hypervolume [72, 73].

Numerous acquisition functions have been implemented for BO applications. In the next part, we will cover the few main acquisition functions broadly used for BO in drug discovery: greedy, low confidence bound, probability of improvement, and expected improvement. We refer the reader to other reviews for a more comprehensive view of the existing acquisition functions [12, 13, 31], like Thompson sampling [74], entropy search [75, 76], and the knowledge gradient [77].

- Greedy

In a simple approach, the acquisition function simply equals the predictive mean of the surrogate model, regardless of the predicted uncertainty (*see* Fig. 1). This "function" is called greedy (G):

$$G(x) = \mu(x) \tag{11}$$

In drug discovery, this function is equivalent to the application of a QSAR model to a new compound. The driving hypothesis here is that the model is accurate for the evaluated compound. This can be true if at minimum the new compound belongs to the applicability domain of the model (condition not sufficient). Applied to BO, this means that greedy will allow local optimization around the already best-known region of the objective landscape but will lack the capability to leak out of this minimum. Figure 1 clearly illustrates this situation.

- Confidence bound

The greedy approach misses the consideration of the uncertainty of the predictions. A simple way to include it is through the lower confidence bound (LCB) heuristic [78, 79]. LCB is a linear combination between the predicted mean and standard deviation values:

$$LCB(x) = \mu(x) - \beta * \sigma(x) \tag{12}$$

We must note that the upper confidence bound (UCB) function is equivalent to LCB with the assumption of maximization of the acquisition and objectives functions.

LCB and UCB introduce a tuning parameter β to balance the exploration/exploitation ratio of the BO process. This β value is also known as *Jitter* [80]. Larger β pushes for high exploration, whereas smaller β leads to more exploitation with the limit $\mu(x)$ of the greedy acquisition. Figure 1 shows that tuning β parameter can allow a faster identification of the optimum. Whereas the choice of the balance (β) is generally left to the user, alternative implementations exist to automatically select the right value [79].

- Improvement

A family of acquisition functions is based on the notion of improvement. These functions drive the landscape exploration toward points that are likely to improve upon the best observation. The improvement $I(x)$, for the sake of minimizing the objective, can be defined by:

$$I(x) = \max(|f(x) - f(x^*)|, 0) \tag{13}$$

where $f(x^*)$ is the best solution measured so far. If the candidate point has a better predicted value than the best-known solution, the absolute difference is returned, else there is no improvement, and the formula returns 0.

An early function using $I(x)$ to measure the potential gain of a given point has been the probability of improvement (PI) [81]. Because the prediction distribution is Gaussian, PI can be analytically computed as follows [12]:

$$PI(x) = \Phi\left(\frac{|\mu(x) - f(x^*)|}{\sigma(x)}\right) \tag{14}$$

where Φ is the cumulative distribution function (CDF) of the standard normal cumulative distribution function. In this form, the next point will be selected by maximizing PI over the GP model. However, PI does not know how large the improvement will be. It will favor little improvements with high probability over larger but less probable ones. This corresponds to a high exploitation strategy [82], leading to small steps in the BO process.

Expected improvement (EI) takes better into account the amount of improvement we can expect from a new point over the current best observation [83]. EI is defined by:

$$\begin{aligned} EI(x) = (|\mu(x) - f(x^*)|)\Phi\left(\frac{|\mu(x) - f(x^*)|}{\sigma(x)}\right) \\ +\sigma(x)\phi\left(\frac{|\mu(x) - f(x^*)|}{\sigma(x)}\right) \end{aligned} \tag{15}$$

where ϕ is the probability density function (PDF) of the standard normal cumulative distribution function. EI(x) will be high when the prediction is better than the best-known observation, and when the uncertainty is high (*see* Fig. 1). Like PI, maximization of EI will allow the selection of the next point to test.

EI possess a strong theoretical asset to guarantee that it will never be stuck in a local minima [84]. It evaluates better than PI the risk–reward ratio of testing a new point by estimating the amount of expected improvement. Thus, it can select a point with a large but uncertain improvement, which is not the case for PI.

Nevertheless, EI is still seen as overly greedy as it explores the area around the best points without efficiently exploring farther regions even more uncertain. That's why a new hyper-parameter ξ can be included in the EI(x) formula:

$$EI(x) = (|\mu(x) - f(x^*) - \xi|)\Phi\left(\frac{|\mu(x) - f(x^*) - \xi|}{\sigma(x)}\right) + \sigma(x)\phi\left(\frac{|\mu(x) - f(x^*) - \xi|}{\sigma(x)}\right) \tag{16}$$

For LCB, β controls directly how much uncertainty we want to consider. Here, ξ can be seen as a modifier of the best-known observation to drive the selection toward more exploration [85]. This parameter is implemented in some BO libraries like GPyOpt [80].

2.7 Batch and Multiobjective Constraints

In all the previous sections, a single "simple" scenario has been considered: optimization of one objective by the training of a surrogate model, the application of an acquisition, and the sequential selection of points to evaluate one by one. More complex drug discovery use cases will be covered in this section.

- Batch selection

  Drug discovery is characterized by the complexity and the cost to develop and apply an experimental setting to obtain a reliable value. This difficulty has favored the implementation of high-throughput experiments that allow testing rapidly, in parallel or a row, batches of points [86]. Whereas BO aims at reducing the number of points tested, batch testing allows increasing the number of acquired observations in a short time frame. For the BO process, this is traduced by the selection of a batch of points to be tested. It is called batched BO [87].

  The batch selection brings up its questions. The main one is what is the amount of information that will bring the evaluation of the full batch. If we want to maximize this information, we must reduce the information overlap between all items of the batch. Interestingly, Englhardt et al. evaluate various batch selection methods using three criteria which are informativeness, diversity, and representativity [88].

  A simple batched BO method exists in the context of active learning optimization in which the selection is done inside a pre-defined set of unlabeled points (*see* Subheading 2.8). Once the surrogate model is trained on the known observations, all the candidates can be scored by the application of the acquisition function. The top x points can then be selected as items of the batch. This method is very naïve as the compounds in the batch are selected independently without considering the amount of information that should bring the other items of the same batch. However, it is very fast and cheap in computing as it required a single identical application of the acquisition function to all the candidates. This method has been successfully applied to high-throughput virtual screening [89] and on small molecule bioactivity optimization [90].

Without a pre-defined set of potential candidates, the BO process includes the optimum search of the acquisition function to determine the features of the next point to test. Given a fixed dataset, the derived acquisition surface is fixed; thus, multiple searches of its optimum will lead to the same or very close sampled point(s). To avoid this situation, when a batch is filled, the pendant points (previously added points in the batch but not yet evaluated) must be considered for the selection of the next point to add to the batch.

Multiple solutions have been proposed around BO parallelization. González et al. proposed a local penalization method [91], available in the GPyOpt package [80]. After each selection, the values of the acquisition function around the new pending points are reduced. Snoek et al. used a Monte Carlo approach to estimate the acquisition function under different possible results from pending evaluations [92]. Both methods need to select the batch points sequentially. Remarkably, Hernández-Lobato et al. overcome this constraint by implementing a selection based on Thompson sampling [93]. Because each batch item is selected independently, this method can be computationally parallelized. In addition, they retrospectively applied this method for the selection of active small molecules in the malaria dataset with success.

- Multitask optimization

In drug discovery, optimization is never done toward a single objective. In early drug discovery, people must deal with the therapeutic space (chemical or biological space), target and off-target activities, and physicochemical, ADME, in vitro, and in vivo PK properties. All these points must be considered on their own but also mainly regarding the other properties to find the best therapeutics. This multiobjective optimization is the heart of the drug design process.

A lot of papers for BO treat either multitasking or multiobjective BO optimization. However, we found that they tend to mix both notions. Roman Garnett describes the existing clear distinction [13]. Both optimizations address the sequential or simultaneous optimization of multiple objectives. Multitask optimization searches the optimum of each task individually. Nevertheless, the assumption is that all the tasks can provide information about the other tasks, hoping that each task optimization will be fastened by looking at the others. Without this consideration, multiple single-task optimizations would be done.

In drug discovery, the tasks are generally poorly correlated. Nevertheless, this could happen for the joint optimization of correlated activity measurements, e.g., in vitro and in vivo, or of correlated physicochemical properties and biological experiments, e.g., small molecule solubility and cell permeability.

If the tasks can be optimized sequentially, e.g., first solubility and then permeability that is more costly, the first task will be optimized alone. Then a joint GP model can be used to model both tasks and allow the faster optimization of the second [94]. In the inverse, if both tasks must be optimized simultaneously, an effective solution is the training of a surrogate model for each task independently and then realizing optimization of the weighted acquisition function.

- Multiobjective optimization

In multiobjective optimization, the optimum will be the point that jointly optimizes all the objectives. In drug discovery, it corresponds to the drug profile, by way of each intermediate key molecule from the hit to the lead and the preclinical candidate. Each compound has its own targeted profile starting from the simple activity profile of a hit to the very complex shape of a drug.

The global unequivocally optimum of multiobjective optimization is unreachable unless all the single objective optimums coincide [13]. Indeed, the more the objectives are anticorrelated, the more they compete with each other. Thus, a balanced profile between all the objectives is generally defined by applying minimal thresholds on some objectives and trade-offs between others.

In such a situation, multiple solutions can be found. Each solution is Pareto efficiency, meaning that no modification of the features of the point is available to improve one objective without making one of the others worse. The set of Pareto efficient points is called the Pareto front.

Several approaches exist to search the Pareto front. Fromer and Coley described them for molecular multiobjective optimization [69]. One major a posteriori method approximates the objectives either by independent or multitask surrogate models and then use the predictions to evaluate how a new point can expand the Pareto front. This expansion can be measured by the hypervolume improvement between the current Pareto front and the one with the evaluated point. The probability of hypervolume improvement (PHI) and the expected hypervolume improvement (EHI) are the two main acquisition functions derived from PI and EI (*see* Subheading 2.7) and based on hypervolumes for multiobjective optimization.

Another method for multiobjective optimization is to define a single multiparameter objective (MPO) by scalarization of the objectives' values. This method is widely used in drug discovery as it reduces the optimization to the search for the global optimum of a single objective [95]. The optimum of linear scalarization is guaranteed to lie on the Pareto frontier [13]. By doing multiple optimizations of multiple scalarizations in which the weight of each objective varies, multiple points of the Pareto front could be found. However, it may not allow determining the full front.

- Batch and multiple objectives optimization combination

If drug discovery is characterized by multiobjective optimization and batch selection, both constraints occur simultaneously. The diversity of the points in the selected batch is then extremely important to efficiently explore the Pareto front.

As described previously, hypervolume improvement is one of the main metrics for multiobjective optimization. Daulton et al. proposed a derived acquisition function called q-Expected HyperVolume Improvement (q-EHVI) that computes the hypervolume improvement over a batch of q points [96]. This formulation has been applied successfully to chemical reaction condition optimization over two objectives and by batches of three conditions [97].

Such measurement can potentially lead to the selection of close points, all providing a high EHVI. In a more complex approach, Konakovic Lukovic et al. also addressed this problem [98]. Based on hypervolume improvement, they introduced an explicit consideration of the diversity, in the feature space, of the selected batch's points. In brief, the Pareto front is approximated, and points are selected to maximize the hypervolume improvement while enforcing the samples to be taken from diverse regions of the design space. In drug design, this constraint would correspond to the selection of a structurally diverse batch of molecules [69].

2.8 Ranking or Sampling

As seen previously, the sampling of points in the search space is done by the acquisition function optimization. However, due to the input features used by the surrogate model, the sampled points may not have the right form to be decoded into the initial representation, e.g., a chemical structure cannot be rebuilt starting from a set of molecular descriptors. In this section, we will cover both situations where the input features can or cannot be decoded in a valid molecular structure. Figure 2 illustrates these situations inside the iterative process of BO.

- Library-based ranking

The decoding issue can be easily overcome if we accept to change the BO process toward active learning. To this end, the surrogate model and the acquisition function are used to score candidate points generated outside the "normal" acquisition function optimization process. The generation process of the

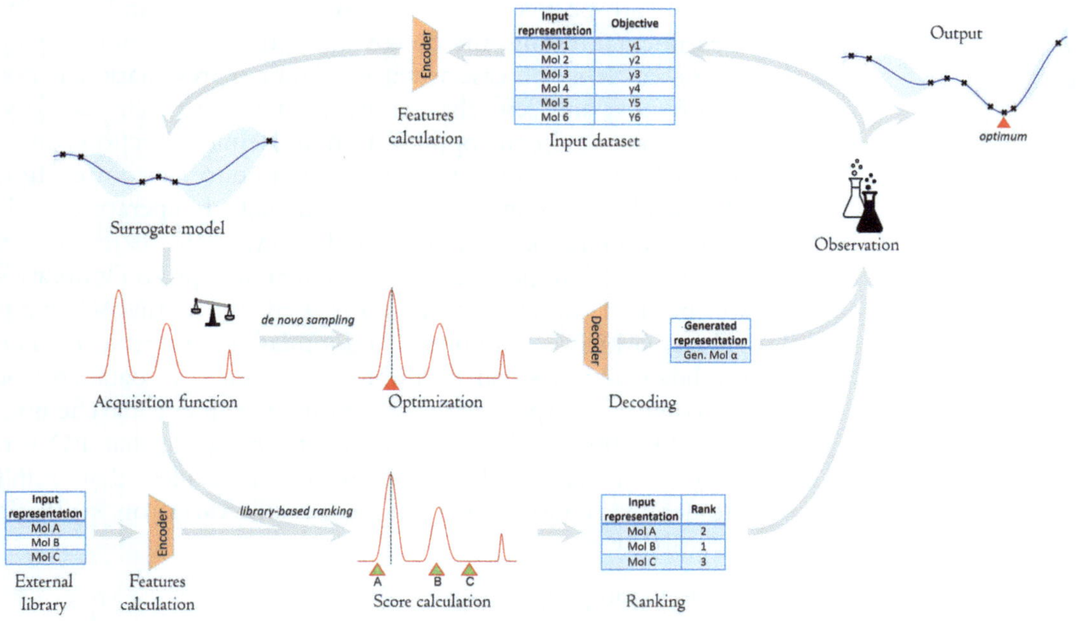

Fig. 2 Schematic representation of an LS-BO iterative process illustrating the two main strategies to select the candidate for the observation step. As described in Algorithm 1, the BO cycle starts with input data. Then, an encoder generates the features used to build a surrogate model. The predicted mean and variance values are then used to balance the exploration/exploitation ratio inside the acquisition function. For candidate selection, the first possibility is de novo sampling. The acquisition function is optimized and the decoder transforms the selected features into a usable representation. The second possibility relies on library-based ranking. The same encoder computes the input features and the acquisition function scores and ranks the candidates. In both cases, the objective value of the selected candidate is determined and becomes a new observation. If the budget is not reached, the observation is added to the input dataset. Otherwise, the best-acquired observation so far is considered the determined optimum of the objective function

candidates can be anything like a complete enumeration (when possible), a random sampling of the design space, or the use of pre-built libraries. Then, the candidates' features are calculated and used to compute an acquisition-based score. Finally, the candidates are ranked according to this score, and the top is selected for the observation step (*see* Fig. 2).

We want to mention that this situation is simulated in many papers that retrospectively apply BO to validate their method. The study is done on a fixed-length dataset with known objective values. A small part of the dataset will be used to initialize the first surrogate model, and the BO process will be used sequentially to select candidates in the second part of the dataset. If such an approach is reasonable, it biases the validation as the design space accessible initially is constrained by the dataset and is reduced alongside the retrospective analysis as fewer and fewer candidates are available for the selection. This in silico validation is regularly done to validate the BO process for drug discovery applications [51, 93].

In a prospective fashion, i.e., when we select candidates for experimental testing, the design space may be computationally accessible. It is the case when the input representation is not continuous, and few choices are possible for each category. That's the case for the optimization of chemical reaction conditions where the design space can be reduced to the combination of a small set of condition parameters, e.g., temperatures, solvents, reactants concentrations, and catalysts [97, 99].

If the design space cannot be enumerated, pre-built libraries will be used. In such cases, BO is used in a virtual screening process, which is common in molecular design projects. The candidate libraries can be public or commercial databases or can come from experimentalist ideation, e.g., medicinal chemists [89, 90, 100]. If this approach limits the space that BO can explore, it allows the selection of molecules that fulfill pre-defined constraints such as substructures or synthetic accessibility.

- De novo sampling

Library-based strategy is appealing but shares some drawbacks regarding the novelty and the diversity that can be accessed. De novo sampling allows exploring a wider region of the design space. The condition is that the optimum of the acquisition function can be decoded into the original representation (*see* Fig. 2). Deep learning-based embeddings enable this capacity.

Before going further, it is important to highlight that these embeddings are generally of high dimensionalities. However, GPs are limited to a small number of dimensions [31]. Moreover, finding the global optimum requires a good coverage of the design space, and the amount of data needed to cover the space increases exponentially with the number of dimensions [12]. The approaches presented in this section show how we can deal with this high-dimensionality curse.

In the first approach, a pre-trained model is used to convert the input representation into usable features. Starting from a protein sequence, Yang et al. extracted a continuous embedding from the UniRep pre-trained LSTM model [101]. It implicitly represents the probability of each character in the input sequence. Then, a surrogate model can be classically computed, and an optimum found. Finally, a sequence can be sampled indirectly from the optimized embedding [23].

A more straightforward process is obtained using the latent space (LS) coordinates of a variational auto-encoder (VAE) model. This approach is called LS-BO [102]. Through a first neural network called encoder, a VAE compresses an input representation into a lower-dimensional nonlinear representation named latent space. Then, a second neural network called decoder decompresses this manifold into the original representation. VAEs are widely used to convert chemical SMILES into a continuous vector [37]. In BO, the encoder is used to compute the features used to train the surrogate model, and the decoder allows easy conversion of the optimum into a functional molecular representation (*see* Fig. 2). As an example, Griffiths and Hernández-Lobato used the CDDD auto-encoder model in a BO strategy to successfully generate de novo small molecules with optimized properties [103].

The pre-trained latent space representation may not have the right inductive bias to correctly model the objective of interest. In other words, the assumption that close points have similar objective values in the trained surrogate model may not be true in the LS manifold [102]. Deshwal and Doppa overcame this weakness by applying multiple kernels for the surrogate model [104]. In a molecular design task, they combined kernels applied to the latent space vector and the corresponding molecular fingerprints. However, compared to surrogate models trained on the latent space features alone, their method performed only similarly.

Instead of using a pre-trained VAE model, recent papers propose to learn the LS representation alongside the training of the surrogate model. This method should allow the latent space representation to satisfy the assumption of the surrogate model and thus be more suitable for BO [102]. Grosnit et al. used this approach to jointly train a discriminative latent space and use it in a high-dimensional GP model [105]. Interestingly, they achieved SOTA performance on property-guided molecule generation, even if the quantity of data used to train the models (i.e., thousands of points) is higher than the standard of drug discovery.

3 Applications in Drug Discovery

With the rise of artificial intelligence, numerous recent applications are using machine/deep learning methods to optimize chemical entities or predict their properties. But this process can be long and costly. To fasten and reduce the experiments needed to achieve the holy grail, the use of Bayesian optimization can be of help. In the next section, we propose to review some examples of BO application in the context of drug discovery. The goal of this section is not to write a complete review but only to illustrate some applications of BO with recent papers.

3.1 Hyperparameter Optimization of Machine Learning Models

Unless a few exceptions, most of AI/ML methods need to be trained with optimal hyperparameters. Searching for these best hyperparameters can be time-consuming. Recently, a lot of papers [42, 71, 106] proposed the use of BO as a method to find optimal parameters in less time. Ahsan S. Alvi devoted his PhD thesis [107] to this subject. In 2020, a challenge [108] was put in place to demonstrate the superiority of Bayesian optimization over a random search for hyperparameter tuning. Sixty-five teams participated in it and had to optimize a collection of optimization problems. The top teams in the competition used a BO ensemble to reach the best performance. With this strategy, the best model was able to get more than 100-fold sample efficiency gain compared to random search. In the same year, Victoria et al. used BO for the tuning of hyperparameters of a convolutional neural network model. The goal of the model was to perform image classification with a minimum error rate calculated on a validation set. For this, the authors optimized the network depth, learning rate, momentum, and regularization hyperparameters. The objective was learned with a Gaussian process function and EI was used as the acquisition function. In only five iterations, the authors were able to obtain on both validation and test set an error rate of 0.212 and 0.216, respectively. In February 2022, Guan et al. published a paper about random forest hyperparameter tuning by including parameters for handling imbalance datasets (i.e., "class_weight" and "sampling_strategy"). The authors called their pipeline class imbalance learning with Bayesian optimization (CILBO). By using RDKit fingerprint [109] as the descriptor and ROC AUC as the objective, the tuning of the random forest model was performed using a stratified 5-CV procedure. The final model was tested and compared to the test set used in Stokes et al.'s paper [110]. The authors claim that, following this comparison, their random forest model performed similarly to Stokes' deep learning model.

3.2 Small Molecule Optimization

In drug discovery, a domain where BO is mostly adopted is small compound optimization [51]. Many different parameters have to be taken into account for a compound to be able to become a drug candidate. It is not an easy task to take into account all these different parameters, and sometimes even only one parameter can be very challenging to optimize. Furthermore, synthesizing and testing compounds are expensive, so methods that can lower the number of trials to get an optimal solution are warmly welcomed. Then, it's not surprising to see more and more papers [90, 103, 111, 112] dealing with this problem using BO.

In 2020, Korovina et al. published a paper about their ChemBO [113] tool. The framework goal is to generate and optimize compounds with desired properties and consider their synthesizability. To this end, the authors used a Gaussian process regression model associated with a graph kernel [66] and developed a new similarity measure. With this new similarity measure, the authors were able to see that when the measure between compounds was small, the difference in QED score was also small. Then, to consider the synthesizability of the compounds in the process, the Rexgen [114] library was used to randomly generate synthesizable compounds for given reagents and conditions. At each BO iteration, the compound having the highest acquisition value is recommended. The authors tested their method by optimizing the QED and a penalized logP score. They found out that using a linear combination of graph kernel and fingerprint provides the best trade-off compared to single kernel use. ChemBO was also compared to the work done in four other studies. It was found that ChemBO reached similar or better results compared to these studies but in lesser iterations. In 2022, Wand et al. published their 3D-based generative model called RELATION [115]. In this model, a deep learning model was trained to generate binding compounds for specific targets (i.e., AKT1 and CDK2) using as input 3D information of target–ligand complexes. The problem with such an approach is the validity of the generated compounds that satisfy the constraints. Indeed, only 30% of compounds generated were valid. To solve this problem, the authors introduced a conditional sampling of the chemical space driven by a BO framework with binding affinity as the objective. The activity was calculated with either docking or QSAR model using a Gaussian process surrogate model with EI as the acquisition function. The authors showed that by using BO, the number of valid molecules increased to 60%. Furthermore, using docking as an affinity estimator, compounds with better docking scores could be obtained, but also with similar binding modes compared to existing inhibitors. The same year, Mehta et al. shared their method called Multi-Objective Machine learning framework for Enhanced MolEcular Screening (MO-MEMES [116]). With their method, the authors claim to be able to retrieve more than 90% of the most interesting compounds while only docking 6% of an entire library. To do this, deep Gaussian process was used as the surrogate function, EI as the acquisition function, and CDDD embedding as the molecular descriptor. With this, the binding affinity (via docking), logP, and synthetic accessibility were simultaneously optimized. Then the complete framework was validated first on the ZINC-250 K library [117] and then on the Enamine HTS collection [118] containing 2 M compounds.

3.3 Peptide and Protein Sequence Optimization

In 2022, Yang et al. [23] showed how they modified pre-trained sequence models by deep ensembling to allow sequence design in Bayesian optimization for optimizing protein properties with only a few experiments. In more detail, the authors constructed a pre-trained sequence model using UniRep [101], which is a long short-term memory model (LSTM), to perform the next amino acid prediction. Then, to enable uncertainty predictions, an ensemble of multilayer perceptrons (MLP) was used to predict wanted properties. Finally, as the acquisition function for choosing sequences of interest, the upper confidence bound was selected. The approach was tested on three different tasks: design of hemolytic peptides, finding an unknown sequence target, and design of specific protein binding peptides. For the first task, the authors showed that on 9316 possible peptides, the algorithm was able to find a likely hemolytic peptide after only five iterations and a peptide that almost match the most hemolytic predicted peptide after 20 iterations. Here, the effect of using pre-trained models was also demonstrated. For the second task, using a similarity score to assess the similarity of the tested sequence with the unknown targeted sequence, it has been shown that the algorithm was able to find a closer sequence compared to random guesses. Also, with few data points, the use of pre-trained models showed better performances. For the third task, the goal was to optimize the binding of peptides to Ras GTPase. For this, the authors first predict the peptide/protein complex by using AlphaFold2-Multimer [119]. Then, they used as objectives for the BO the confidence score given by AlphaFold2-Multimer and the average distance of the peptide to the binding site. Unfortunately, no experimental validation of the peptides found was done to confirm these results.

Other papers also treat the problem of peptide/protein optimization. In the same year, Stanton et al. [120] presented their Latent Multi-Objective BayesOpt (LaMBO) approach. With their method, the authors proposed to optimize the sequence of proteins to get fluorescent proteins with improved brightness and thermostability. This framework consists of a combination of masked language models and deep kernel Gaussian processes. The authors showed that the mutations generated by the denoising autoencoder and the use of simulated stability and solvent-accessible surface area as objectives were able to outperform uniform random mutations. Cheng et al. [26] published a paper on directed protein evolution with closed-loop BO. In their approach, the authors proposed the use of a novel low-dimensional amino acid encoding strategy. Before using BO, they performed a search space prescreening to reduce the search space by removing outliers with an Extreme Gradient Boosting Outlier Detection (XGBOD) [121] model. Then, they built a surrogate model using RobustGP which consists of a GP model with an additional outlier filtering step and tested different acquisition functions (EI, UCB, PI, and TS). Finally, the authors also adapted and tested the Trust region Bayesian Optimization (TuRBO) [28] algorithm to build a collection of local GP models in the region showing the highest confidence. The authors showed on different tasks that their approach was able to find faster the highest fitness samples compared to random selection but also naïve BO. As a lot of recent publications can be found on peptide/protein design using BO frameworks [122, 123], it should be noted that we have also seen recently the use of BO for antibody design [124, 125].

3.4 Chemical Reaction Condition Optimization

Recently, there has been a lot of interest in using AI methods in synthetic organic chemistry [33, 126]. One of the ultimate goals is the automatization of chemical synthesis. Along these methods, Bayesian optimization showed some interest, especially for optimizing reactional conditions. In 2021, Shields et al. [127] showed on a benchmark dataset for a palladium-catalyzed direct arylation reaction that BO outperforms human decisions in terms of efficiency but also consistency. For this, each reaction is encoded with DFT (quantum mechanical) descriptors for each chemical component and raw value is considered for temperature and concentration. Then, reaction encoding is combined with a Gaussian surrogate model and expected improvement as the acquisition function. With this Bayesian optimization workflow, the authors showed its ability to surpass standard reaction conditions for a fluorination reaction. In fact, with a reaction space consisting of 312,500 possible configurations, the BO was able to find in ten rounds of experiments a reaction that led to a 69% yield versus a 36% yield for standard conditions. Following this work, in 2022, Kwon et al. [128] proposed a hybrid approach where they combined graph neural networks (GNN) models and BO. Using available reactions in Reaxys [129], GNN models were built to predict and prioritize reaction conditions for new chemical reactions when no data are available. Using these predictions for experimental validation as a cold start, and to reduce the space of possible reaction conditions, these experimental results were used to train a BO surrogate model. Then, using UCB as the acquisition function, the next experiments were prioritized by BO. Using this strategy, the authors showed their method led to better performance compared to baseline methods.

To turn BO frameworks for optimizing reaction conditions into processes usable for the vast community, more and more groups release toolkits or web applications. One of the most known is the GAUCHE library [5] where the authors proposed to describe the reactions with reaction fingerprints. Interestingly, their method works independently of the number of reactant and reagent categories unlike other approaches. Wang et al. [130] have proposed the NEXTorch library for BO. This library can be applied to various tasks, support both machine- and human-in-the-loop optimizations, and provide various visualization options and can be easily extended to other frameworks. Recently, Torres et al. [97] released their web application called EDBOApp for multiobjective BO by a cloud computing platform. This platform allows people without coding experience to easily use BO for reaction optimization, modification of conditions on the fly, and visualization of objective predictions and uncertainties.

Without going into too many details, there are more articles recently published that also tackle the reaction condition problem using Bayesian optimization. Okazawa et al. [131] made use of BO in combination with DFT calculations to predict the optimal alloy for nitrogen activation. For this, they used a Gaussian process regression and tested multiple acquisition functions to predict the heat of reaction of $N \equiv N$ bond cleavage. With such a strategy, the authors were able to find binary alloy catalysts more efficiently than through random search. In the publication of Kumar et al. [132], the authors used the BO strategy to do multiobjective optimization. By using Gaussian process regression (GPR), they optimized the condition reaction for the synthesis of methanol to obtain the best combination of conversion and selectivity. Rosa et al. [133] used BO to optimize in vitro transcription reaction conditions for mRNA vaccine production. By using this approach, the authors were able to reach a twofold production increase with 60 reactions tested.

3.5 Small Molecule 3D Conformation Minimization

There exist a lot of tools that aim to generate low-energy conformers, and very good results can be obtained with relatively small molecule (i.e., four or fewer rotatable bonds). The problem arises when molecules contain six or more rotatable bonds. The combinatorial explosion of possible conformers makes research for the lowest energy harder.

To solve this, Chan et al. [134] proposed using BO to reduce the sampling and so the time needed to explore the conformational space. They included in their method, through chosen kernels, prior knowledge about the torsional preferences of commonly occurring rotatable bonds. Then, they let the algorithm sampling freely the torsional bonds and learn from it. For compounds having a small number of rotatable bonds (e.g.,1–3), they saw that they could often obtain faster lower energy conformers compared to systematic search. In the following paper [135], the same authors proposed an optimization of their method. They included prior knowledge about correlated torsions. In fact, close rotatable bonds can be naturally constrained by each other, so considering this information could help avoid exploring high-energy regions. In this regard, the distribution of correlated torsions was used to constrain the search space of BO using a modified acquisition function. The authors compared the performance of their new method with their previous method on a validation set composed of 533 molecules ranging from 2 to 18 rotatable bonds. They showed that their improved method was able to find more frequently (>60%) lower-energy conformations than the previous one.

More recently, Fang et al. [136] made use of the BO to determine the lowest conformer energies for cysteine, serine, tryptophan, and aspartic acid. Their method (BOSS) uses a Gaussian process to model a potential surface area to the data points and refine it by minimizing a lower confidence bound acquisition function. In each iteration, the exploration is made possible by varying only the torsional angles present in the molecule and by calculating the energy of conformers with DFT. Finally, the most promising candidates representing potential local and global minima are refined by geometry optimization, entropy corrections, and coupled cluster calculations. The final list of conformers agreed with experimental measurements. The authors also compared their method with a genetic algorithm-based conformer search and showed their method being ten times more efficient in terms of computational cost.

3.6 Ternary Complex Structure Elucidation

There is an increasing interest in compounds having the capability of inducing protein degradation. We denote PROTACs and molecular glues. Both types of compounds are binding to an E3 ligase and a protein of interest (POI) which will be degraded. To rationally optimize the compounds, there is a need for structural information about these ternary complexes. Recently, Rao et al. [137] proposed a BO framework for the exploration of the space of possible ternary complex candidates. To do so, they first create a set of possible ternary complex candidates for each complex by taking independent PDB of the POI and the E3 ligase. Then, they docked from a known PROTAC the warhead in the POI and the E3 binder in the E3 ligase. The possible space of ternary complex candidates is then explored by applying relative rotation and translation of the POI with respect to the E3 ligase, and by testing multiple PROTAC conformations when the E3 binder and the warhead are not too far apart for the length of the linker. Having this space of possible conformations, a BO framework is used to recover the most promising ternary complex candidates. To do so, the BO objective function is optimized by considering a score related to the protein–protein interaction of the complex, and a score measuring the feasibility of finding a stable PROTAC conformer. The iteration process for finding valid candidates continues until some stopping criterion is reached. In the end, the final set of possible candidates undergoes local optimization with simulated annealing followed by clustering, filtering, and re-ranking. By applying such a framework, the authors were able to find near-native poses in the top 15 clusters for 13 out of 22 cases.

4 Conclusion

Bayesian optimization is a powerful method that has shown promising results in the last few years in drug discovery. It allows for a reduction in the number of experiments needed to achieve the desired goal, especially when the experiments are costly. This approach already successfully applies to the various steps faced along the drug discovery journey. We can denote among all the examples presented in this chapter optimization of AI/ML model hyperparameters, small molecule and peptide/protein properties and activities, chemical reaction conditions, small molecule 3D conformation, and ternary complex structures.

The BO framework comes as a combination of probabilistic modeling and the optimization of an acquisition function to navigate in the design space with the wanted exploration/exploitation balance. The goal is to select the most promising candidate at each iteration to fasten the research of an optimal solution. This optimal solution can come from single objective, multitasking, or multi-objective optimization. It can result from a library-based ranking or a de novo sampling. To this end, like other AI/ML methods, and depending on the modeling task, the studied objects (hyperparameters, small molecules, reactions, protein, etc.) may have to be transformed into convenient features. Different kernels already exist to deal with different types of representations, and new ones are developed. Also, frameworks are evolving and propose, for example, internal processes to discard outliers or use of BO ensemble. Still, many challenges exist and will have to be addressed soon. One example being how to efficiently control the exploration/exploitation balance in the successive BO iterations.

References

1. Terayama K, Sumita M, Tamura R, Tsuda K (2021) Black-box optimization for automated discovery. Acc Chem Res 54:1334–1346. https://doi.org/10.1021/acs.accounts.0c00713
2. Alarie S, Audet C, Gheribi AE, Kokkolaras M, Le Digabel S (2021) Two decades of blackbox optimization applications. EURO J Comput Optim 9:100011. https://doi.org/10.1016/j.ejco.2021.100011
3. Bergstra J, Bengio Y (2012) Random search for hyper-parameter optimization. J Mach Learn Res 13:281–305. https://doi.org/10.5555/2188385.2188395
4. Zhang Y, Ling C (2018) A strategy to apply machine learning to small datasets in materials science. Npj Comput Mater 4:25. https://doi.org/10.1038/s41524-018-0081-z
5. Griffiths R-R, Klarner L, Moss HB, Ravuri A, Truong S, Stanton S, Tom G, Rankovic B, Du Y, Jamasb A, Deshwal A, Schwartz J, Tripp A, Kell G, Frieder S, Bourached A, Chan A, Moss J, Guo C, Durholt J, Chaurasia S, Strieth-Kalthoff F, Lee AA, Cheng B, Aspuru-Guzik A, Schwaller P, Tang J (2022) GAUCHE: a library for Gaussian processes in chemistry. https://doi.org/10.48550/ARXIV.2212.04450
6. Mockus J, Tiesis V, Zilinskas A (1978) The application of Bayesian methods for seeking the extremum. In: Towards global optimization. Elsevier, Amsterdam, pp 117–129
7. Blaschke T, Arús-Pous J, Chen H, Margreitter C, Tyrchan C, Engkvist O, Papadopoulos K, Patronov A (2020) REINVENT 2.0: an AI tool for De Novo drug design. J Chem Inf Model 60:5918–5922. https://doi.org/10.1021/acs.jcim.0c00915
8. Rakhimbekova A, Lopukhov A, Klyachko N, Kabanov A, Madzhidov TI, Tropsha A (2023) Efficient design of peptide-binding polymers using active learning approaches. J Control Release 353:903–914. https://doi.org/10.1016/j.jconrel.2022.11.023
9. Brochu E, Cora VM, de Freitas N (2010) A tutorial on Bayesian optimization of expensive cost functions, with application to active user modeling and hierarchical reinforcement learning. https://doi.org/10.48550/arXiv.1012.2599
10. Stark F, Hazırbaş C, Triebel R, Cremers D (2015) CAPTCHA recognition with active deep learning. Aachen

11. Lewis DD, Gale WA (1994) A sequential algorithm for training text classifiers. In: Croft BW, van Rijsbergen CJ (eds) SIGIR '94. Springer London, London, pp 3–12
12. Shahriari B, Swersky K, Wang Z, Adams RP, de Freitas N (2016) Taking the human out of the loop: a review of Bayesian optimization. Proc IEEE 104:148–175. https://doi.org/10.1109/JPROC.2015.2494218
13. Garnett R (2023) Bayesian optimization. Cambridge University Press
14. Tom G, Hickman RJ, Zinzuwadia A, Mohajeri A, Sanchez-Lengeling B, Aspuru-Guzik A (2022) Calibration and generalizability of probabilistic models on low-data chemical datasets with DIONYSUS. https://doi.org/10.48550/arXiv.2212.01574
15. Gramacy RB (2021) Surrogates: Gaussian process modeling, design and optimization for the applied sciences. Chapman Hall/CRC, Boca Raton
16. Hutter F, Hoos HH, Leyton-Brown K (2011) Sequential model-based optimization for general algorithm configuration. In: Coello CAC (ed) Learning and intelligent optimization. Springer Berlin Heidelberg, Berlin/Heidelberg, pp 507–523
17. Zaytsev A. Acquisition function for Bayesian optimisation using random forests as surrogate model. In: StackExchange. https://stats.stackexchange.com/questions/455481/acquisition-function-for-bayesian-optimisation-using-random-forests-as-surrogate
18. Blundell C, Cornebise J, Kavukcuoglu K, Wierstra D (2015) Weight uncertainty in neural network. In: Bach F, Blei D (eds) Proceedings of the 32nd international conference on machine learning. PMLR, Lille, pp 1613–1622
19. Zhang Y, Lee AA (2019) Bayesian semi-supervised learning for uncertainty-calibrated prediction of molecular properties and active learning. Chem Sci 10:8154–8163. https://doi.org/10.1039/C9SC00616H
20. Ryu S, Kwon Y, Kim WY (2019) A Bayesian graph convolutional network for reliable prediction of molecular properties with uncertainty quantification. Chem Sci 10:8438–8446. https://doi.org/10.1039/C9SC01992H
21. Huang W, Zhao D, Sun F, Liu H, Chang EY (2015) Scalable Gaussian process regression using deep neural networks. In: International joint conference on artificial intelligence
22. Izmailov P, Vikram S, Hoffman MD, Wilson AG (2021) What are Bayesian neural network posteriors really like? In: International conference on machine learning
23. Yang Z, Milas KA, White AD (2022) Now what sequence? Pre-trained ensembles for Bayesian optimization of protein sequences. https://doi.org/10.1101/2022.08.05.502972
24. Bengio Y. What are some advantages of using Gaussian process models vs neural networks? In: Quora. https://www.quora.com/What-are-some-advantages-of-using-Gaussian-Process-Models-vs-Neural-Networks
25. Gaussian process. In: Wikipedia. https://en.wikipedia.org/wiki/Gaussian_process
26. Cheng L, Yang Z, Liao B, Hsieh C, Zhang S (2022) ODBO: Bayesian optimization with search space prescreening for directed protein evolution. https://doi.org/10.48550/arXiv.2205.09548
27. Martinez-Cantin R, Tee K, McCourt M (2018) Practical Bayesian optimization in the presence of outliers. In: Storkey A, Perez-Cruz F (eds) Proceedings of the twenty-first international conference on artificial intelligence and statistics. PMLR, pp 1722–1731
28. Eriksson D, Pearce M, Gardner J, Turner RD, Poloczek M (2019) Scalable global optimization via local Bayesian optimization. In: Wallach H, Larochelle H, Beygelzimer A, Alché-Buc FD, Fox E, Garnett R (eds) Advances in neural information processing systems. Curran Associates, Inc
29. Rasmussen CE, Williams CKI (2006) Gaussian processes for machine learning. MIT Press, Cambridge, MA
30. Mockus J (1994) Application of Bayesian approach to numerical methods of global and stochastic optimization. J Glob Optim 4:347–365. https://doi.org/10.1007/BF01099263
31. Frazier PI (2018) A tutorial on Bayesian optimization. https://doi.org/10.48550/arXiv.1807.02811
32. David L, Thakkar A, Mercado R, Engkvist O (2020) Molecular representations in AI-driven drug discovery: a review and practical guide. J Cheminform 12:56. https://doi.org/10.1186/s13321-020-00460-5
33. Hammer AJS, Leonov AI, Bell NL, Cronin L (2021) Chemputation and the standardization of chemical informatics. JACS Au 1:1572–1587. https://doi.org/10.1021/jacsau.1c00303
34. Orosz Á, Héberger K, Rácz A (2022) Comparison of descriptor- and fingerprint sets in machine learning models for ADME-Tox

targets. Front Chem 10:852893. https://doi.org/10.3389/fchem.2022.852893

35. Gómez-Bombarelli R, Wei JN, Duvenaud D, Hernández-Lobato JM, Sánchez-Lengeling B, Sheberla D, Aguilera-Iparraguirre J, Hirzel TD, Adams RP, Aspuru-Guzik A (2018) Automatic chemical design using a data-driven continuous representation of molecules. ACS Cent Sci 4:268–276. https://doi.org/10.1021/acscentsci.7b00572
36. Krenn M, Häse F, Nigam A, Friederich P, Aspuru-Guzik A (2020) Self-referencing embedded strings (SELFIES): a 100% robust molecular string representation. Mach Learn Sci Technol 1:045024. https://doi.org/10.1088/2632-2153/aba947
37. Winter R, Montanari F, Noé F, Clevert D-A (2019) Learning continuous and data-driven molecular descriptors by translating equivalent chemical representations. Chem Sci 10: 1692–1701. https://doi.org/10.1039/C8SC04175J
38. Ferruz N, Schmidt S, Höcker B (2022) ProtGPT2 is a deep unsupervised language model for protein design. Nat Commun 13: 4348. https://doi.org/10.1038/s41467-022-32007-7
39. Hu W, Liu B, Gomes J, Zitnik M, Liang P, Pande V, Leskovec J (2020) Strategies for pre-training graph neural networks. In: International conference on learning representations
40. Maziarz K, Jackson-Flux H, Cameron P, Sirockin F, Schneider N, Stiefl N, Segler M, Brockschmidt M (2021) Learning to extend molecular scaffolds with structural motifs. https://doi.org/10.48550/arXiv.2103.03864
41. Irwin R, Dimitriadis S, He J, Bjerrum E (2022) Chemformer: a pre-trained transformer for computational chemistry. Mach Learn Sci Technol 3:015022. https://doi.org/10.1088/2632-2153/ac3ffb
42. Nguyen V (2019) Bayesian optimization for accelerating hyper-parameter tuning. In: 2019 IEEE second international conference on artificial intelligence and knowledge engineering (AIKE). IEEE, Sardinia, pp 302–305
43. Matérn B (1986) Spatial variation, 2nd edn. Springer, Berlin/Heidelberg
44. Stein ML (1999) Interpolation of spatial data. Springer, New York
45. Genton MG (2001) Classes of kernels for machine learning: a statistics perspective. J Mach Learn Res 2:299–312
46. Morgan HL (1965) The generation of a unique machine description for chemical structures – a technique developed at chemical abstracts service. J Chem Doc 5:107–113. https://doi.org/10.1021/c160017a018
47. Rogers D, Hahn M (2010) Extended-connectivity fingerprints. J Chem Inf Model 50:742–754. https://doi.org/10.1021/ci100050t
48. Ruggiu F, Marcou G, Varnek A, Horvath D (2010) ISIDA property-labelled fragment descriptors. Mol Inform 29:855–868. https://doi.org/10.1002/minf.201000099
49. Capecchi A, Probst D, Reymond J-L (2020) One molecular fingerprint to rule them all: drugs, biomolecules, and the metabolome. J Cheminform 12:43. https://doi.org/10.1186/s13321-020-00445-4
50. Sturm N, Sun J, Vandriessche Y, Mayr A, Klambauer G, Carlsson L, Engkvist O, Chen H (2019) Application of bioactivity profile-based fingerprints for building machine learning models. J Chem Inf Model 59:962–972. https://doi.org/10.1021/acs.jcim.8b00550
51. Pyzer-Knapp EO (2018) Bayesian optimization for accelerated drug discovery. IBM J Res Dev 62:2:1–2:7. https://doi.org/10.1147/JRD.2018.2881731
52. Raymond JW, Willett P (2002) Effectiveness of graph-based and fingerprint-based similarity measures for virtual screening of 2D chemical structure databases. J Comput Aided Mol Des 16:59–71. https://doi.org/10.1023/A:1016387816342
53. Gower JC (1971) A general coefficient of similarity and some of its properties. Biometrics 27:857. https://doi.org/10.2307/2528823
54. Moss HB, Griffiths R-R (2020) Gaussian process molecule property prediction with FlowMO. https://doi.org/10.48550/arXiv.2010.01118
55. International Union of Pure and Applied Chemistry (1998) A guide to IUPAC nomenclature of organic compounds: recommendations 1993, Reprinted. Blackwell Science, Oxford
56. Weininger D (1988) SMILES, a chemical language and information system. 1. Introduction to methodology and encoding rules. J Chem Inf Model 28:31–36. https://doi.org/10.1021/ci00057a005
57. Heller SR, McNaught A, Pletnev I, Stein S, Tchekhovskoi D (2015) InChI, the IUPAC international chemical identifier. J

Cheminform 7:23. https://doi.org/10.1186/s13321-015-0068-4

58. Lodhi H, Shawe-Taylor J, Cristianini N, Watkins C (2000) Text classification using string kernels. In: Leen T, Dietterich T, Tresp V (eds) Advances in neural information processing systems. MIT Press
59. Cancedda N, Gaussier E, Goutte C, Renders JM (2003) Word sequence kernels. J Mach Learn Res 3:1059–1082. https://doi.org/10.5555/944919.944963
60. Cao D-S, Zhao J-C, Yang Y-N, Zhao C-X, Yan J, Liu S, Hu Q-N, Xu Q-S, Liang Y-Z (2012) In silico toxicity prediction by support vector machine and SMILES representation-based string kernel. SAR QSAR Environ Res 23:141–153. https://doi.org/10.1080/1062936X.2011.645874
61. Moss HB, Beck D, González J, Leslie DS, Rayson P (2020) BOSS: Bayesian optimization over string spaces. In: Proceedings of the 34th international conference on neural information processing systems. Curran Associates Inc, Red Hook
62. Jamasb AR, Viñas R, Ma EJ, Harris C, Huang K, Hall D, Lió P, Blundell TL (2020) Graphein – a Python library for geometric deep learning and network analysis on protein structures and interaction networks. https://doi.org/10.1101/2020.07.15.204701
63. Takimoto E, Warmuth MK (2002) Path kernels and multiplicative updates. In: Proceedings of the 15th annual conference on computational learning theory. Springer, Berlin/Heidelberg, pp 74–89
64. Shervashidze N, Schweitzer P, van Leeuwen EJ, Mehlhorn K, Borgwardt KM (2011) Weisfeiler-Lehman graph kernels. J Mach Learn Res 12:2539–2561
65. Rupp M, Schneider G (2010) Graph kernels for molecular similarity. Mol Inform 29:266–273. https://doi.org/10.1002/minf.200900080
66. Ralaivola L, Swamidass SJ, Saigo H, Baldi P (2005) Graph kernels for chemical informatics. Neural Netw 18:1093–1110. https://doi.org/10.1016/j.neunet.2005.07.009
67. Gao P, Yang X, Tang Y-H, Zheng M, Andersen A, Murugesan V, Hollas A, Wang W (2021) Graphical Gaussian process regression model for aqueous solvation free energy prediction of organic molecules in redox flow batteries. Phys Chem Chem Phys 23:24892–24904. https://doi.org/10.1039/D1CP04475C
68. Kashima H, Tsuda K, Inokuchi A (2003) Marginalized kernels between labeled graphs. In: Proceedings, twentieth international conference on machine learning. pp 321–328
69. Fromer JC, Coley CW (2022) Computer-aided multi-objective optimization in small molecule discovery. https://doi.org/10.48550/ARXIV.2210.07209
70. Whittle P (1983) Optimization over time: dynamic programming and stochastic control. Wiley, Chichester
71. Jasrasaria D, Pyzer-Knapp EO (2019) Dynamic control of explore/exploit trade-off in Bayesian optimization. In: Arai K, Kapoor S, Bhatia R (eds) Intelligent computing. Springer, Cham, pp 1–15
72. Byrd RH, Lu P, Nocedal J, Zhu C (1995) A limited memory algorithm for bound constrained optimization. SIAM J Sci Comput 16:1190–1208. https://doi.org/10.1137/0916069
73. Zhu C, Byrd RH, Lu P, Nocedal J (1997) Algorithm 778: L-BFGS-B: Fortran subroutines for large-scale bound-constrained optimization. ACM Trans Math Softw 23:550–560. https://doi.org/10.1145/279232.279236
74. Thompson WR (1933) On the likelihood that one unknown probability exceeds another in view of the evidence of two samples. Biometrika 25:285. https://doi.org/10.2307/2332286
75. Hennig P, Schuler CJ (2012) Entropy search for information-efficient global optimization. J Mach Learn Res 1809–1837. https://doi.org/10.5555/2188385.2343701
76. Villemonteix J, Vazquez E, Walter E (2009) An informational approach to the global optimization of expensive-to-evaluate functions. J Glob Optim 44:509–534. https://doi.org/10.1007/s10898-008-9354-2
77. Wu J, Poloczek M, Wilson AG, Frazier PI (2017) Bayesian optimization with gradients. https://doi.org/10.48550/ARXIV.1703.04389
78. Auer P (2003) Using confidence bounds for exploitation-exploration trade-offs. J Mach Learn Res 3:397–422. https://doi.org/10.5555/944919.944941
79. Srinivas N, Krause A, Kakade S, Seeger M (2010) Gaussian process optimization in the bandit setting: no regret and experimental design. In: Proceedings of the 27th international conference on international conference on machine learning. Omni Press, Madison, pp 1015–1022
80. (2016) GPyOpt: a Bayesian optimization framework in Python

81. Kushner HJ (1964) A new method of locating the maximum point of an arbitrary multipeak curve in the presence of noise. J Basic Eng 86: 97–106. https://doi.org/10.1115/1.3653121
82. Jones DR (2001) A taxonomy of global optimization methods based on response surfaces. J Glob Optim 21:345–383. https://doi.org/10.1023/A:1012771025575
83. Močkus J (1975) On Bayesian methods for seeking the extremum. In: Marchuk GI (ed) Optimization techniques IFIP technical conference Novosibirsk, July 1–7, 1974. Springer Berlin Heidelberg, Berlin/Heidelberg, pp 400–404
84. Vazquez E, Bect J (2010) Convergence properties of the expected improvement algorithm with fixed mean and covariance functions. J Stat Plan Inference 140:3088–3095. https://doi.org/10.1016/j.jspi.2010.04.018
85. Kamperis S (2021) Acquisition functions in Bayesian optimization. In: Lets Talk Sci. https://ekamperi.github.io/machine%20learning/2021/06/11/acquisition-functions.html
86. Mayr LM, Bojanic D (2009) Novel trends in high-throughput screening. Curr Opin Pharmacol 9:580–588. https://doi.org/10.1016/j.coph.2009.08.004
87. Azimi J, Fern A, Fern X (2010) Batch Bayesian optimization via simulation matching. In: Lafferty J, Williams C, Shawe-Taylor J, Zemel R, Culotta A (eds) Advances in neural information processing systems. Curran Associates, Inc
88. Englhardt A, Trittenbach H, Vetter D, Böhm K (2020) Finding the sweet spot: batch selection for one-class active learning. In: SDM
89. Graff DE, Shakhnovich EI, Coley CW (2021) Accelerating high-throughput virtual screening through molecular pool-based active learning. Chem Sci 12:7866–7881. https://doi.org/10.1039/D0SC06805E
90. Bellamy H, Rehim AA, Orhobor OI, King R (2022) Batched Bayesian optimization for drug design in noisy environments. J Chem Inf Model 62:3970–3981. https://doi.org/10.1021/acs.jcim.2c00602
91. González J, Dai Z, Hennig P, Lawrence N (2016) Batch Bayesian optimization via local penalization. In: Proceedings of the 19th international conference on artificial intelligence and statistics (AISTATS). pp 648–657
92. Snoek J, Larochelle H, Adams RP (2012) Practical Bayesian optimization of machine learning algorithms. In: Proceedings of the 25th international conference on neural information processing systems – volume 2. Curran Associates Inc, Red Hook, pp 2951–2959
93. Hernández-Lobato J, Gelbart M, Adams R, Hoffman M, Ghahramani Z (2016) A general framework for constrained Bayesian optimization using information-based search. https://doi.org/10.17863/CAM.6477
94. Swersky K, Snoek J, Adams RP (2013) Multi-task Bayesian optimization. In: Burges CJ, Bottou L, Welling M, Ghahramani Z, Weinberger KQ (eds) Advances in neural information processing systems. Curran Associates, Inc
95. Wager TT, Hou X, Verhoest PR, Villalobos A (2016) Central nervous system multiparameter optimization desirability: application in drug discovery. ACS Chem Neurosci 7:767–775. https://doi.org/10.1021/acschemneuro.6b00029
96. Daulton S, Balandat M, Bakshy E (2020) Differentiable expected hypervolume improvement for parallel multi-objective Bayesian optimization. In: Proceedings of the 34th international conference on neural information processing systems. Curran Associates Inc, Red Hook
97. Torres JAG, Lau SH, Anchuri P, Stevens JM, Tabora JE, Li J, Borovika A, Adams RP, Doyle AG (2022) A multi-objective active learning platform and web app for reaction optimization. J Am Chem Soc 144:19999–20007. https://doi.org/10.1021/jacs.2c08592
98. Konakovic Lukovic M, Tian Y, Matusik W (2020) Diversity-guided multi-objective Bayesian optimization with batch evaluations. In: Larochelle H, Ranzato M, Hadsell R, Balcan MF, Lin H (eds) Advances in neural information processing systems. Curran Associates, Inc, pp 17708–17720
99. Clayton AD, Pyzer-Knapp E, Purdie M, Jones M, Barthelme A, Pavey J, Kapur N, Chamberlain T, Blacker J, Bourne R (2022) Bayesian self-optimization for telescoped continuous flow synthesis. Angew Chem Int Ed 62:e202214511. https://doi.org/10.1002/anie.202214511
100. Agarwal G, Doan HA, Robertson LA, Zhang L, Assary RS (2021) Discovery of energy storage molecular materials using quantum chemistry-guided multiobjective Bayesian optimization. Chem Mater 33: 8133–8144. https://doi.org/10.1021/acs.chemmater.1c02040
101. Alley EC, Khimulya G, Biswas S, AlQuraishi M, Church GM (2019) Unified rational protein engineering with sequence-based deep representation learning. Nat

Methods 16:1315–1322. https://doi.org/10.1038/s41592-019-0598-1

102. Maus N, Jones HT, Moore J, Kusner M, Bradshaw J, Gardner JR (2022) Local latent space Bayesian optimization over structured inputs. In: Oh AH, Agarwal A, Belgrave D, Cho K (eds) Advances in neural information processing systems
103. Griffiths R-R, Hernández-Lobato JM (2020) Constrained Bayesian optimization for automatic chemical design using variational autoencoders. Chem Sci 11:577–586. https://doi.org/10.1039/C9SC04026A
104. Deshwal A, Doppa J (2021) Combining latent space and structured kernels for Bayesian optimization over combinatorial spaces. In: Beygelzimer A, Dauphin Y, Liang P, Vaughan JW (eds) Advances in neural information processing systems
105. Grosnit A, Tutunov R, Maraval AM, Griffiths R-R, Cowen-Rivers AI, Yang L, Zhu L, Lyu W, Chen Z, Wang J, Peters J, Bou-Ammar H (2021) High-dimensional Bayesian optimisation with variational autoencoders and deep metric learning. https://doi.org/10.48550/arXiv.2106.03609
106. Daulton S, Wan X, Eriksson D, Balandat M, Osborne MA, Bakshy E (2022) Bayesian optimization over discrete and mixed spaces via probabilistic reparameterization. https://doi.org/10.48550/arXiv.2210.10199
107. Alvi AS (2019) Practical Bayesian optimisation for hyperparameter tuning. University of Oxford
108. Turner R, Eriksson D, McCourt M, Kiili J, Laaksonen E, Xu Z, Guyon I (2021) Bayesian optimization is superior to random search for machine learning hyperparameter tuning: analysis of the black-box optimization challenge 2020. https://doi.org/10.48550/arXiv.2104.10201
109. Landrum G. RDKit: open-source cheminformatics
110. Stokes JM, Yang K, Swanson K, Jin W, Cubillos-Ruiz A, Donghia NM, MacNair CR, French S, Carfrae LA, Bloom-Ackermann Z, Tran VM, Chiappino-Pepe A, Badran AH, Andrews IW, Chory EJ, Church GM, Brown ED, Jaakkola TS, Barzilay R, Collins JJ (2020) A deep learning approach to antibiotic discovery. Cell 180:688–702.e13. https://doi.org/10.1016/j.cell.2020.01.021
111. Soleimany AP, Amini A, Goldman S, Rus D, Bhatia SN, Coley CW (2021) Evidential deep learning for guided molecular property prediction and discovery. ACS Cent Sci 7:1356–1367. https://doi.org/10.1021/acscentsci.1c00546
112. Graff DE, Aldeghi M, Morrone JA, Jordan KE, Pyzer-Knapp EO, Coley CW (2022) Self-focusing virtual screening with active design space pruning. J Chem Inf Model 62:3854–3862. https://doi.org/10.1021/acs.jcim.2c00554
113. Korovina K, Xu S, Kandasamy K, Neiswanger W, Poczos B, Schneider J, Xing E (2020) ChemBO: Bayesian optimization of small organic molecules with synthesizable recommendations. In: Chiappa S, Calandra R (eds) Proceedings of the twenty third international conference on artificial intelligence and statistics. PMLR, pp 3393–3403
114. Jin W, Coley CW, Barzilay R, Jaakkola T (2017) Predicting organic reaction outcomes with Weisfeiler-Lehman network. In: Proceedings of the 31st international conference on neural information processing systems. Curran Associates Inc, Red Hook, pp 2604–2613
115. Wang M, Hsieh C-Y, Wang J, Wang D, Weng G, Shen C, Yao X, Bing Z, Li H, Cao D, Hou T (2022) RELATION: a deep generative model for structure-based De Novo drug design. J Med Chem 65:9478–9492. https://doi.org/10.1021/acs.jmedchem.2c00732
116. Mehta S, Goel M, Priyakumar UD (2022) MO-MEMES: a method for accelerating virtual screening using multi-objective Bayesian optimization. Front Med 9. https://doi.org/10.3389/fmed.2022.916481
117. Sterling T, Irwin JJ (2015) ZINC 15 – ligand discovery for everyone. J Chem Inf Model 55:2324–2337. https://doi.org/10.1021/acs.jcim.5b00559
118. Enamine HTS Collection. https://enamine.net/compound-collections/screening-collection/hts-collection
119. Evans R, O'Neill M, Pritzel A, Antropova N, Senior A, Green T, Žídek A, Bates R, Blackwell S, Yim J, Ronneberger O, Bodenstein S, Zielinski M, Bridgland A, Potapenko A, Cowie A, Tunyasuvunakool K, Jain R, Clancy E, Kohli P, Jumper J, Hassabis D (2021) Protein complex prediction with AlphaFold-Multimer. https://doi.org/10.1101/2021.10.04.463034
120. Stanton S, Maddox W, Gruver N, Maffettone P, Delaney E, Greenside P, Wilson AG (2022) Accelerating Bayesian optimization for biological sequence design with denoising autoencoders. https://doi.org/10.48550/arXiv.2203.12742

121. Zhao Y, Hryniewicki MK (2019) XGBOD: improving supervised outlier detection with unsupervised representation learning. https://doi.org/10.48550/ARXIV.1912.00290
122. Hughes ZE, Nguyen MA, Wang J, Liu Y, Swihart MT, Poloczek M, Frazier PI, Knecht MR, Walsh TR (2021) Tuning materials-binding peptide sequences toward gold- and silver-binding selectivity with Bayesian optimization. ACS Nano 15:18260–18269. https://doi.org/10.1021/acsnano.1c07298
123. Hu R, Fu L, Chen Y, Chen J, Qiao Y, Si T (2022) Protein engineering via Bayesian optimization-guided evolutionary algorithm and robotic experiments. https://doi.org/10.1101/2022.08.11.503535
124. Park JW, Stanton S, Saremi S, Watkins A, Dwyer H, Gligorijevic V, Bonneau R, Ra S, Cho K (2022) PropertyDAG: multi-objective Bayesian optimization of partially ordered, mixed-variable properties for biological sequence design. https://doi.org/10.48550/arXiv.2210.04096
125. Khan A, Cowen-Rivers AI, Grosnit A, Deik D-G-X, Robert PA, Greiff V, Smorodina E, Rawat P, Akbar R, Dreczkowski K, Tutunov R, Bou-Ammar D, Wang J, Storkey A, Bou-Ammar H (2023) Toward real-world automated antibody design with combinatorial Bayesian optimization. Cell Rep Methods 3:100374. https://doi.org/10.1016/j.crmeth.2022.100374
126. de Almeida AF, Moreira R, Rodrigues T (2019) Synthetic organic chemistry driven by artificial intelligence. Nat Rev Chem 3: 589–604. https://doi.org/10.1038/s41570-019-0124-0
127. Shields BJ, Stevens J, Li J, Parasram M, Damani F, Alvarado JIM, Janey JM, Adams RP, Doyle AG (2021) Bayesian reaction optimization as a tool for chemical synthesis. Nature 590:89–96. https://doi.org/10.1038/s41586-021-03213-y
128. Kwon Y, Lee D, Kim JW, Choi Y-S, Kim S (2022) Exploring optimal reaction conditions guided by graph neural networks and Bayesian optimization. ACS Omega 7:44939–44950. https://doi.org/10.1021/acsomega.2c05165
129. Goodman J (2009) Computer software review: Reaxys. J Chem Inf Model 49:2897–2898. https://doi.org/10.1021/ci900437n
130. Wang Y, Chen T-Y, Vlachos DG (2021) NEXTorch: a design and Bayesian optimization toolkit for chemical sciences and engineering. J Chem Inf Model 61:5312–5319. https://doi.org/10.1021/acs.jcim.1c00637
131. Okazawa K, Tsuji Y, Kurino K, Yoshida M, Amamoto Y, Yoshizawa K (2022) Exploring the optimal alloy for nitrogen activation by combining Bayesian optimization with density functional theory calculations. ACS Omega 7:45403–45408. https://doi.org/10.1021/acsomega.2c05988
132. Kumar A, Pant KK, Upadhyayula S, Kodamana H (2023) Multiobjective Bayesian optimization framework for the synthesis of methanol from syngas using interpretable Gaussian process models. ACS Omega 8: 410–421. https://doi.org/10.1021/acsomega.2c04919
133. Rosa SS, Nunes D, Antunes L, Prazeres DMF, Marques MPC, Azevedo AM (2022) Maximizing mRNA vaccine production with Bayesian optimization. Biotechnol Bioeng 119:3127–3139. https://doi.org/10.1002/bit.28216
134. Chan L, Hutchison GR, Morris GM (2019) Bayesian optimization for conformer generation. J Cheminform 11:32. https://doi.org/10.1186/s13321-019-0354-7
135. Chan L, Hutchison GR, Morris GM (2020) BOKEI: Bayesian optimization using knowledge of correlated torsions and expected improvement for conformer generation. Phys Chem Chem Phys 22:5211–5219. https://doi.org/10.1039/C9CP06688H
136. Fang L, Makkonen E, Todorović M, Rinke P, Chen X (2021) Efficient amino acid conformer search with Bayesian optimization. J Chem Theory Comput 17:1955–1966. https://doi.org/10.1021/acs.jctc.0c00648
137. Rao A, Tunjic TM, Brunsteiner M, Müller M, Fooladi H, Weber N (2022) Bayesian optimization for ternary complex prediction (BOTCP). https://doi.org/10.1101/2022.06.03.494737

Chapter 6

Automated Virtual Screening

Vladimir Joseph Sykora

Abstract

Computational methods in modern drug discovery have become ubiquitous, with methods that cover most of the discovery stages: from hit finding and lead identification to lead optimization. The overall aim of these computational methods is to obtain a more efficient discovery process, by reducing the number of "wet" experiments required to produce therapeutics that have higher probability of succeeding in clinical development and subsequently benefitting end patients by developing highly effective therapeutics having minimal side effects. Virtual Screening is usually applied at the early stage of drug discovery, looking to find chemical matter having desired properties, such as molecular shape, electrostatics, and pharmacophores at desired three-dimensional positions. The aim of this stage is to search in a wide chemical space, including chemistry available from commercial suppliers and virtual databases of predicted reaction products, to identify molecules that would exert a particular biochemical response. This initial stage of the discovery process is very important since the subsequent stages will use the initial chemical motifs that have been found at the hit finding stage, and therefore the most suitable the compound is found, the more likely it is that subsequent stages will be successful and less time and resource consuming. This chapter provides a summary of various Virtual Screening methods, including shape match and molecular docking, and these methods are used in a Virtual Screening workflow that is provided as an example which is described to be run automatically in cloud resources. This automatic in-depth exploration of the chemical space using validated Virtual Screening methods can lead to a more streamlined and efficient discovery process, aiming to deliver chemical matter of high quality and maximizing the required biological effects while minimizing adverse effects. Surely, Virtual Screening pipelines of this nature will continue to play a central role in producing much needed therapeutics for the health challenges of the future.

Key words Virtual screening, Shape match, Shape similarity, Molecular docking

1 Introduction

Computational methods in modern drug discovery have become ubiquitous, with methods that cover most of the discovery stages: from hit finding and lead identification to lead optimization.

The overall aim of these computational methods is to obtain a more efficient discovery process, by reducing the number of "wet" experiments required to produce therapeutics that have the higher probability of succeeding in clinical development and subsequently

Alexander Heifetz (ed.), *High Performance Computing for Drug Discovery and Biomedicine*, Methods in Molecular Biology, vol. 2716, https://doi.org/10.1007/978-1-0716-3449-3_6,

benefitting end patients by developing highly effective therapeutics having minimal side effects.

With the advent of digitalization, the large collection of historical lab experiments covering both peer-reviewed scientific publications and patent applications are now available in computer format. This vast amount of scientific data, especially small molecule property data, can be used to develop computational systems that accurately predict physicochemical properties and likely biological activity.

Drug discovery projects require the physical measurement of the properties of the therapeutics being developed, being small molecules or biotherapeutics. Historically, large amounts of potential therapeutics (e.g. small molecules) were experimentally "screened" for activity, usually with the aid of robotic systems—in a process termed "high-throughput screening" (HTS). Computational methods have been developed with the aim of simulating this process in silico (i.e., on the computer). By analogy, the term "Virtual Screening" (VS) is used to refer to the use of computational methods to discriminate between "active" and "inactive" molecules.

In short, Virtual Screening (VS) can be seen as the ranking of molecules, irrespective of the method, for how likely they are to modulate a target of interest [1].

Virtual Screening is usually applied at the early stage of drug discovery, looking to find chemical matter having desired properties, such as molecular shape, electrostatics, and pharmacophores at desired three-dimensional positions. The aim of this stage is to search in a wide chemical space, including chemistry available from commercial suppliers and virtual databases of predicted reaction products, to identify molecules that would exert a particular biochemical response. This initial stage is important since the subsequent stages will use the initial chemical motifs that have been found at the hit finding stage, and therefore the most suitable the compound is found, the more likely it is that subsequent stages will be successful and less time and resource consuming.

With the industrialization of computational resources introduced by cloud computing, the availability of large CPU and GPU resources has become simpler and widely accessible. Platforms such as Amazon EC2 [2] and Microsoft Azure [3] can deploy hundreds of CPUs on demand, with little effort and up-front cost. The combination of this computing power with modern queuing and workflow systems offers the computational discovery scientist the basic components to build powerful Virtual Screening pipelines.

One particular feature of demands for Virtual Screening in drug discovery groups is that the process is required at specific time intervals: for example, when a new experimental crystal structure of the protein target has become available. At this particular point in time, a request to perform large Virtual Screening campaigns is demanded, which require large computational resources to process. These periods of burst peak demand when the Virtual Screening is needed is usually followed by result analysis and lab experimental work (chemical synthesis and assay testing), a period when large computing power is no longer needed. This alternation between peak demand and quiet periods of computing power is well suited for a cloud computing architecture [4], which can deploy computing power when required and scaled down when no longer needed.

As Virtual Screening workflows require the integration and processing of large datasets, in addition to the combination of multiple Virtual Screening software tools, keeping both the datasets and software methods up-to-date has been a traditionally time-consuming task for the computational discovery scientist. By automating Virtual Screening pipelines, which would run automatically as new experimental data becomes available, the early drug discovery process becomes more efficient as the scientist is not required to manually perform repetitive tasks (e.g., text file processing) each time there is a Virtual Screening request.

This chapter aims to summarize the concept of Virtual Screening and its automation. It starts by describing what Virtual Screening is and summarizing two popular methods: shape/pharmacophore similarity and molecular docking. Following, the chemical space to apply Virtual Screening is covered, including a summary of molecular standardization. It continues with a review of current workflow management systems, as these provide the tools to construct automated Virtual Screening pipelines. The chapter concludes by giving an example of an implementation of a Virtual Screening automated pipeline and summarizes technologies for its deployment.

2 Virtual Screening

As its name suggests, "Virtual Screening" is the in silico equivalent of screening a compound in the lab for a desired property or effect. Traditionally done in the lab, a compound is tested, for example, to identify if the compound binds or inhibits a particular protein target, and "screened" whether the compound has the desired effect or not. The advantages of doing this process fully in silico are obvious: cost per compound is negligible and speed is much accelerated.

Traditionally, Virtual Screening has been classified depending whether a reference ligand or a protein structure is available. Methods that require only a 3D structure of a reference compound (such as a natural ligand or known inhibitor) are called *ligand-based virtual screening*. Included in this group are methods for shape and electrostatic similarity (e.g., ROCS [5]) and pharmacophore search (e.g., CATS [6]). Methods that require the structure of the protein target are called ***structure-based virtual screening.*** In this group, methods such as molecular docking (e.g., Vina [7]) are included.

2.1 Ligand-Based Virtual Screening

This type of VS methods has as input a known inhibitor or natural ligand, which becomes a "template" (or reference), focusing on their structure to find molecules from a database that contain similar features, such as shape and/or pharmacophores. A molecular representation is computed for both the reference molecule and the database molecule, such as the Morgan fingerprint [8], and then applying a metric function (such as Tanimoto or Tversky similarity [9]) to calculate the similarity between the two representations.

2.2 Shape and Pharmacophore Similarity

A widely used method in this category is the calculation of the maximum volume overlay between two molecules. In Grant and Pickup [10], a Gaussian description of atomic volumes is used to calculate the volume intersection between two molecules. Using Gaussians as stand-alone atomic volumes, this representation gives a fast way to calculate the similarity between two molecular shapes. When pharmacophore types are included in the representation, such that only the same types of pharmacophores are overlaid on top of each other, the shape similarity will also include the relative positions in 3D where pharmacophores are located [5]. Usual pharmacophore types included are [11] hydrogen-bond donor, hydrogen-bond acceptor, hydrophobicity, logP, molar refractivity, and surface exposure.

2.3 Structure-Based Virtual Screening

When the structure of the protein target is available, methods have been developed that consider both the protein and the ligand structures to predict the orientation of binding of the ligand in the active site of the protein. The theoretical foundations for this model originated more than 100 years ago with the *lock and key* principle by Emil Fischer [12], in which a small molecule ligand (the "key") fits into the active site of a protein (the "lock") driven by steric and energy interactions. Modern methods also account for the flexibility of both the small molecule and the protein residues, accounting for the induced fit theory [13].

As the protein structure is more complex than that of the ligand, structure-based methods generally require considerably more numerical and therefore computational effort than ligand-based methods. However, one clear benefit of structure-based over ligand-based methods is that the structure of a known inhibitor or natural ligand is not required. As long as the three-dimensional structure of the protein is known, structure-based methods (such as docking summarized below) can be used to predict the binding orientation of a ligand in the context of a protein binding site and subsequently estimate the binding affinity between the ligand and the protein. The three-dimensional structure of the protein target can be determined experimentally by X-ray crystallography [14] or nuclear magnetic resonance [15], or predicted in silico from its sequence by using machine learning algorithms such as AlphaFold [16]. Once this structure is known, methods such as molecular docking can be applied in Virtual Screening to estimate the likelihood of binding of a database of small molecules.

2.4 Molecular Docking

Molecular docking is a computational procedure that attempts to predict the binding orientation between two molecules: a receptor and a ligand [17]. This procedure can be divided in two parts: correct positioning of the correct conformer of a ligand in the context of a binding site (positioning) and its successful recognition/scoring by a scoring function (scoring) [1]. Docking can then be applied to in silico screen a database of small molecules for their likelihood of binding to a known protein structure.

2.5 Benchmarking Virtual Screening Methods

There are a number of approaches that quantify the accuracy of VS tools. Usually, the VS tools produce a continuous numerical result that allows for the quantification of the probability of binding. For example, in ligand-based methods, a similarity metric (e.g., Tanimoto similarity [9]) can be calculated either on the structural fingerprint [18] (for the 2D structural similarity) or the volume overlay [10] (for the 3D structural similarity) of two molecules. The similarity metric score can then be used to rank unknown molecules for their similarity to the one which is known to be active. In a similar way, a structure-based method usually produces a "scoring function" result that can be used to rank the likelihood of binding [7]. In both cases, the continuous numerical value can be used to rank molecules based on their probability of binding to the target, from the most likely to the least. For these, an enrichment factor can be calculated which ranks the molecules in descending order of probability of binding and selects the molecules at desired percentages of the dataset.

If a classification method is employed, or a threshold is applied for a continuous numerical score, a receiver operating characteristic (ROC) curve can be calculated, and the area under the curve (AUC) of this line can be used to quantify the performance of the Virtual Screening tool. Following, these two methods are summarized.

2.6 Enrichment

A traditional method for benchmarking a Virtual Screening tool is to mix "active" molecules (i.e., molecules that exert a desired response in the biological target at hand) with "inactive" molecules, which are known not to affect a response in the target. In a process called enrichment, a specific VS tool is applied which ranks

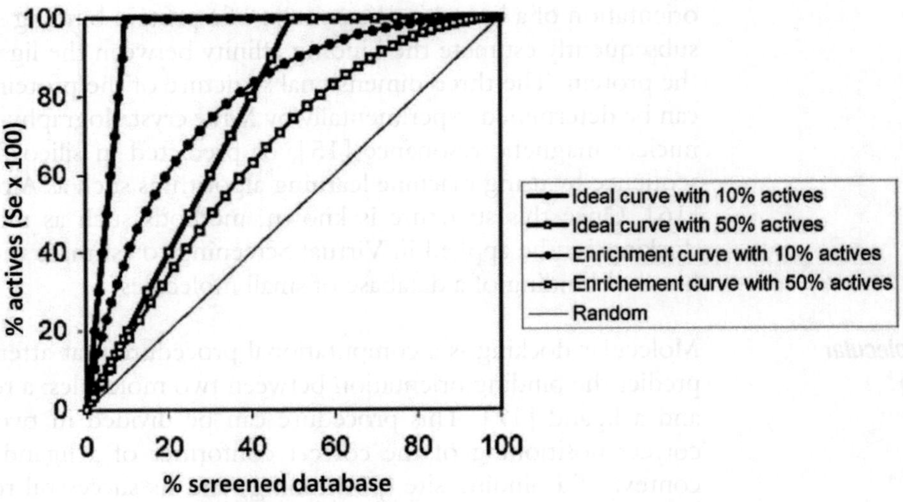

Fig. 1 Example of enrichment curves. (This plot was previously published by Triballeau et al. [19])

the molecules from the most likely active to the least likely active. The VS tool aims to rank the known "active" molecules in the top of the ranked list, sorted in descending order, from the most likely to the least, while the known inactive molecules would rank lower. The enrichment process is simple to calculate [1] and is defined as the sensitivity:

$$\mathrm{Se} = \frac{N_{\text{selected actives}}}{N_{\text{total actives}}}$$

Enrichment curves report the sensitivity as a function of the ranking (or as a percentage of the screened database) [19]. Enrichment curves rise from the lower left corner to the upper right, with the diagonal line being a random ranking. The further away the enrichment curve is from the diagonal line and toward the upper left corner, the better the discrimination between the two classes is.

As reported by Triballeau et al. in [19], enrichment curves suffer from two major drawbacks: one is that the enrichment is affected by the ratio of actives versus inactives, and the other is that enrichment captures only the sensitivity, while in reality the active/inactive classification problem should reflect both the ability to capture actives and the ability to discard inactives.

Figure 1 shows an example of typical enrichment curves as published in [19].

2.7 Receiver Operating Characteristic

A more robust metric for quantifying Virtual Screening is the calculation of receiver operating characteristic (ROC) curves. The ROC curve is a common method used in multiple fields [20] applied to evaluate the ability of a given test to discriminate between two populations [19].

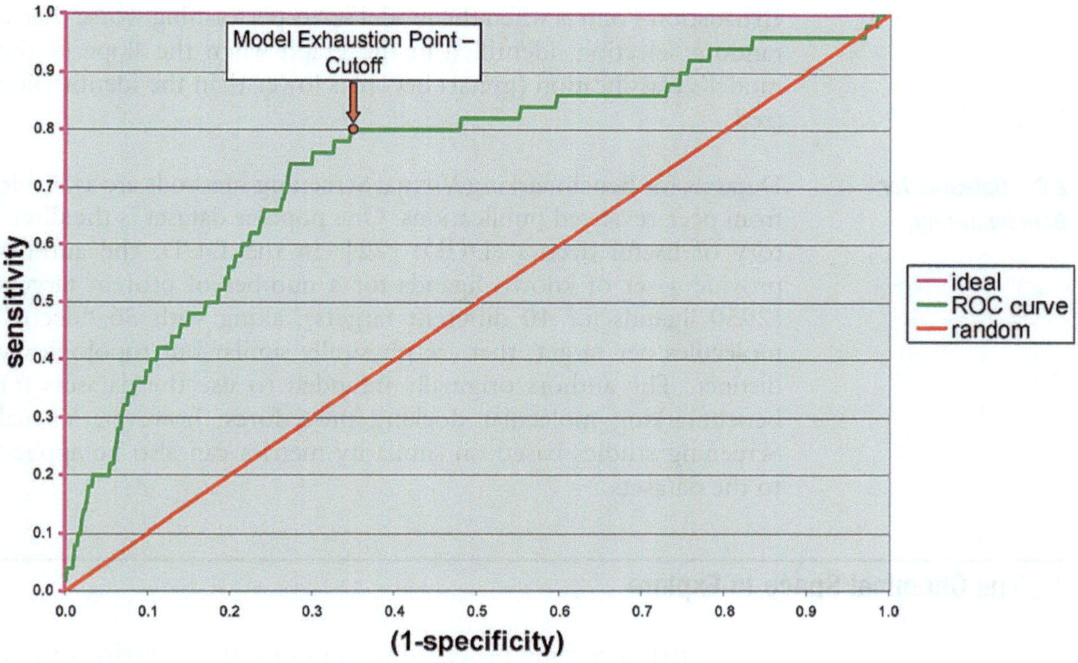

Fig. 2 Example of ROC curve. (This plot was previously published by Rizzi and Fioni [20])

In addition to the sensitivity, the specificity can also be calculated, giving an estimate of the ability to discard inactive molecules, as:

$$\mathrm{Sp} = \frac{N_{\text{discarded inactives}}}{N_{\text{total inactives}}}$$

The sensitivity is also known as the true positive rate (TPR), recall, or hit rate, and the specificity as the true negative rate (TNR) or selectivity.

The ROC curve can be created by plotting the true positive rate versus the false positive rate at different threshold settings. Therefore, the ROC is the curve created by plotting the sensitivity in respect to 1—specificity.

To quantify the ROC curve, the Area Under the Curve (AUC) of the ROC curve [21] can be used, which is a value that is bound between 0 and 1.

An example ROC curve is shown in Fig. 2, as described by Rizzi and Fioni [20].

As the sensitivity and specificity are bound between 0 and 1, the ROC curve rises from the bottom left to the top right. An ideal (perfect) classification would therefore follow the y axis to the top-left corner and then follow horizontally from the top-left to the top-right (purple line in Fig. 2). A random classifier would produce the identity line (red), while an average predictive classifier would produce a curve that resembles the green line. The Model

Exhaustion Point is when the model starts performing worse than a random selection, identified in the graph when the slope of the model's classification (green) becomes lower than the identity line (red).

2.8 Datasets for Benchmarking

Datasets for benchmarking Virtual Screening methods are available from peer-reviewed publications. One popular dataset is the directory of useful decoys (DUD) [22]. In the DUD, the authors provide a set of known ligands for a number of protein targets (2950 ligands for 40 different targets), along with 36 "decoy" molecules per target, that are physically similar but topologically distinct. The authors originally intended to use the datasets for benchmarking molecular docking procedures; however, Virtual Screening studies based on similarity metrics can also be applied to the datasets.

3 The Chemical Space to Explore

The Virtual Screening process gives an estimate of activity for one molecule. One important step in any robust Virtual Screening pipeline is to have well-curated compound libraries in which to search for molecules with desired properties. This part of the pipeline can be divided into two processes: first is to decide which chemical space to include in the search and then have a well-designed process for standardizing the molecules contained in this search space.

3.1 Search Space

The aim of the Virtual Screening process is to find compounds with desired properties and therefore provide the starting information in a drug discovery project. The ultimate goal of Virtual Screening is to have physical samples of the molecules that rank high in silico, for a subsequent test in "wet" lab experiments to confirm that their in silico prediction of the properties translate to the actual experimentally measured properties. If binding affinity toward a specific protein target is the property that is sought for, experimental techniques such as Surface Plasmon Resonance [23] or Isothermal Calorimetry [24] can be employed to validate that the physical sample of the compound has the desired binding affinity as predicted by the in silico method.

3.2 Catalogues of Chemical Suppliers

As this step requires having the physical sample of the compound, usually libraries that contain compound samples which have been already synthesized and purified are included. These include libraries from well-known compound and synthesis suppliers such as Merck [25]. If the Virtual Screening method identifies one of these, the compound can be readily sourced from the supplier and then experimentally tested.

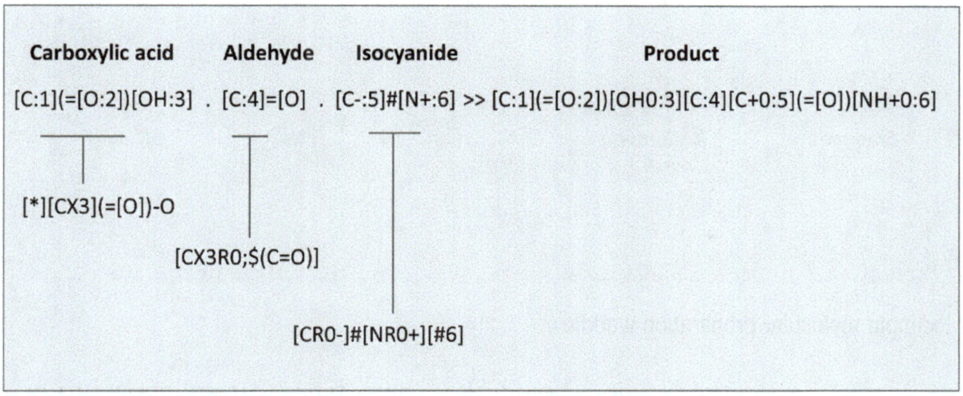

Fig. 3 An example SMIRK showing the Passerini reaction, and the SMARTS providing patterns for the selection of the necessary starting materials for the reaction

Irwing et al. have produced the ZINC database which compiles a number of catalogues from different chemical suppliers [26] that provide readily available compound samples for "wet" lab testing.

3.3 Virtual Compounds

To have a more thorough coverage of the chemical space, predicting chemical reaction products is a common way to enrich the search space. A traditional cheminformatics procedure to calculate theoretical reaction products has been using the SMIRKS reaction language [27, 28]. Developed by David Weininger, SMIRKS is an extension of the SMILES language and allows to define chemical reaction rules. For example, the Passerini multicomponent reaction [29] can be written as shown in Fig. 3. In this example, the reaction takes three starting materials: an aldehyde, a carboxylic acid, and an isocyanide. By defining a SMARTS [30] pattern for each starting material, a compound library that contains only one type of starting material can be produced. A programming routine can be written (e.g., using the RDKit software library [31]) that feeds each of the components from each starting material library into the reaction scheme, and therefore producing a combinatorial library for the Passerini reaction.

A more recent version of the ZINC database [32] contains a number of predicted "virtual" molecules that suppliers market as being possible to synthesize "on demand." This pre-built library provides an excellent starting chemical space to feed a Virtual Screening pipeline.

3.4 Compound Standardization Pipeline

One important component of the Virtual Screening process is to have the computational representation of the starting molecules of the searched chemical space in a form that is as close as possible to the actual physical form in the assay that the molecules will be subjected to. This includes the protonation state, tautomeric form, and 3D shape of the compounds in the library. In addition,

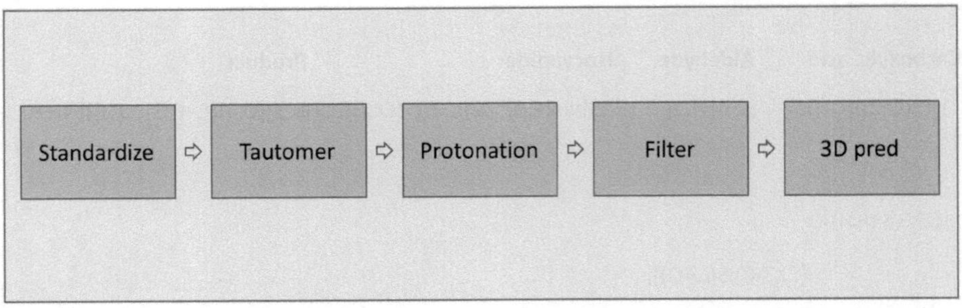

Fig. 4 Example molecular preparation workflow

the pipeline should filter chemotypes that are known to be problematic in medicinal chemistry programs, including reactive and nonspecific compounds. An example of a compound standardization pipeline is shown in Fig. 4.

The group behind the ChEMBL database [33] has published a molecular standardization pipeline [34] that can be readily used for molecular standardization. For tautomer enumeration and canonicalization, the RDKit can be used [31], as well as for the generation of 3D atomic coordinates [35]. For filtering, J. Baell and G. Holloway originally provided a set of substructures that are promiscuous across different bioassays, giving the Pan-Assay Interference Compounds (PAINS) [36], which provides a starting set of substructures to avoid. A further enhancement of this PAINS set was provided in Chakravorty et al. [37], giving a set of quantitative criteria to assess whether a compound is recommended to be excluded from HTS campaigns, not to remove it but to "proceed with caution," or one to proceed.

4 Workflow Systems

The workflow is the most obvious model for solving multistage problems and is well suited for automation. Traditional implementations of workflow systems have involved the sequential synchronous execution of each step of the workflow, whereby each block in the workflow executes sequentially as the previous step completes. Modern implementations use signals or message passing, on top of a queue manager, to orchestrate the asynchronous sequential execution of each step in the workflow. These systems have the benefit that once the request for the processing of a block is made, the block would only execute when it has the resources to do so. Examples of workflow management systems include KNIME [38, 39] and Apache Airflow [40].

The workflow provides multiple benefits for an automated Virtual Screening system, including the following:

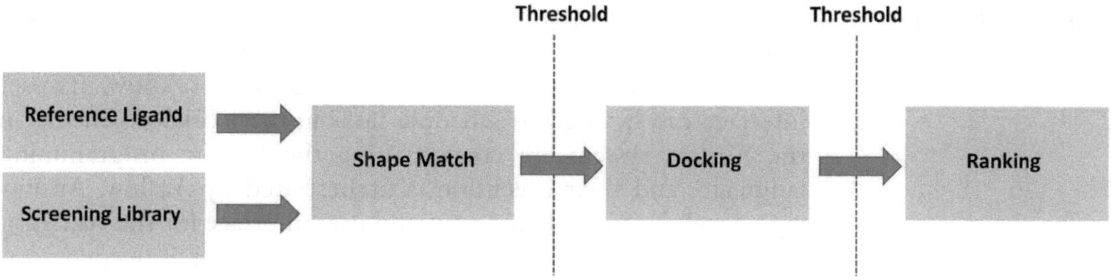

Fig. 5 Example Virtual Screening workflow

- Blocks can be implemented using a service-oriented architecture [41]. This gives multiple advantages to the application, including ease of updating and scaling.
- Each block is compartmentalized, and therefore if one block fails, previous blocks don't need to be re-executed. The failed block can be corrected and the process will restart from the repaired block.
- Each block can be deployed in different infrastructures (e.g., AWS [2]) and given specific hardware resources (e.g., custom CPUs, GPUs, and RAM resources)

4.1 Celery (Python)

In its basic format, Celery is an asynchronous task queue which is based on distributed message passing [42]. However, Celery provides a number of "primitives" which allow for the creation of groups of tasks. The primitives allow the user to create a number of tasks that execute either in sequence or in parallel, providing expressivity that allows for the passing of arguments between sequential tasks. Using this expressivity, the outputs of one function can become arguments of a sequential function. As such, Celery can provide the tools to execute complex workflows, combining sequential and parallel tasks.

Figure 5 shows an example diagram for a Virtual Screening pipeline that can be expressed using Celery.

4.2 Snakemake

Snakemake [43, 44] is a workflow management system that provides expressivity to define blocks and whole workflows in the Python programming language. One of the key features of Snakemake is that the workflow definition can contain a description of the required software. This concept, when compared to the Celery system, is useful for data pipelines that require a precise description of the software and data needed to be executed. This is critical when reproducibility is required, as the software and their versions required to reproduce a specific set of results are defined in the workflow itself. This is in contrast to a service architecture, where the actual details of the execution is left to the service, while the task manager is not responsible for these details; as long as the inputs and outputs are compliant with what is expected, the execution of the tasks in the workflow will proceed.

4.3 Apache Airflow

Apache Airflow [40] is an open-source workflow management system that includes a web user interface allowing the user to visualize and manage the different blocks of a workflow. Its user interface can be used to schedule tasks in the workflow. Blocks in the Airflow system are created using the Python programming language, and their execution is orchestrated by Airflow. Airflow is an excellent tool when a powerful user interface for the visualization of the workflow and management of each task is needed.

4.4 Microservice Architecture

There are two features of the Virtual Screening process that make a microservice architecture [45] well suited for its implementation: first is that Virtual Screening pipelines consist of multiple steps that combine CPU-/GPU-intensive steps with fairly small tasks (e.g., sorting a file). The second is that these two types of tasks scale differently: while a higher number of tasks are required for say shape match, less is required for other tasks. Using a microservice architecture, an application can customize both the type of computing resource (e.g., high CPU or RAM) and their number of instantiations, and with such the whole application gains in efficiency.

5 Django and Celery for Automating Virtual Screening

Powerful automated systems can be achieved when combining the Celery system with triggering signals—for example, the python Django web framework [46] provides a number of built-in signals that can be used to trigger the execution of a particular Celery workflow.

For example, assume a Celery workflow is available as shown in Fig. 6. The workflow is executed for a single compound library (the first step, at the leftmost of the diagram). The screening library is comprised of compound structure files that are split into chunks of equal number of molecules per chunk. This way, each chunk can be processed in a different server. This example system could store a number of different screening libraries (say one per compound supplier) and stored in a local database by a screening library data model which specifies the location of these structure chunks. Using Django signals, the application can detect when a new screening library object is added to the database. As this step is detected by a signal, the Django application could trigger the Celery workflow as shown in Fig. 6. By installing the Celery workers in different servers, each chunk of the screening library could be processed in a different server, as shown in Fig. 6.

The workflow showed in Fig. 6 combines both ligand-based and structure-based methods. Initially, a shape match procedure is applied, and subsequently a docking procedure is executed. In this

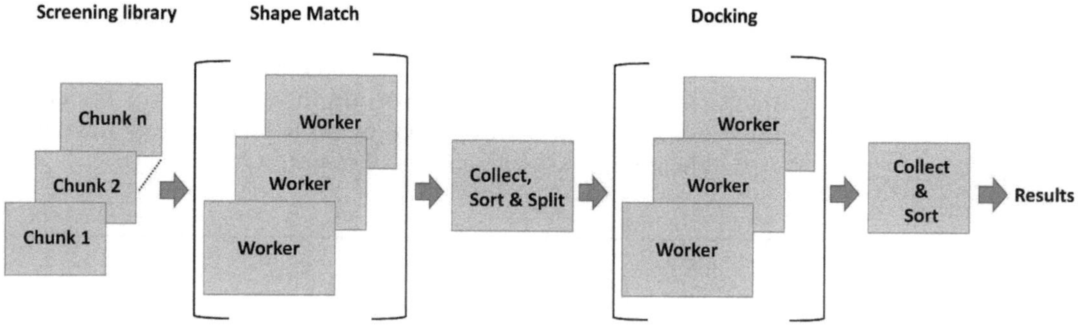

Fig. 6 Example implementation of a Virtual Screening workflow

example, a reference ligand is required for the shape match algorithm, and a protein structure is required for the docking procedure. Following, each step in the workflow is summarized.

Step 1: Shape match

The workflow starts by selecting the screening library the workflow is going to search in. For each structure file chunk, an instantiation of the shape match procedure is run. Each of these executions could be run in different servers.

Step 2: Collect, sort, apply threshold, and split

Once all instances of the shape match procedure have been applied, a single process can be executed which collects all resulting shape match files, each containing the structures in the chunk and the shape similarity score to the reference ligand. The process could sort all the structures by using the shape similarity score, and only proceed with the structures that have a similarity score greater than a pre-defined threshold. Finally, this stage of the workflow would produce the resulting structure files with a pre-determined number of compounds in each file—this in order for the following block in the workflow to be able to process a single file in a single computation thread and therefore increase the processing throughput.

Step 3: Docking

For each resulting shape match file, a docking procedure can be applied. This step requires a prepared protein structure file, and ideally the procedure will remove compounds that are likely to present steric clashes with the protein's residues. Each structure file can be processed in a single computation thread, and this in turn can be spread across a number of processing servers (or services).

Step 4: Final collect and sort

The final step in the workflow is the collection of all the resulting docking structure files, and sorting the structures by the docking score. The result of this procedure is the final result of the whole Virtual Screening workflow in this example.

6 Conclusion

A Virtual Screening system as described above, deployed using microservice architecture with sufficient cloud computing resources, could screen very large search spaces (i.e., over one billion chemical structures) in a matter of hours. The pipeline could be applied automatically when new data becomes available: for example, a new experimental protein crystal structure or the experimental three-dimensional structure of a ligand. Multiple implementations of the Virtual Screening methods could also be automatically applied (e.g., multiple implementations of molecular docking or shape/pharmacophore similarity). This automatic in-depth exploration of the chemical space using validated Virtual Screening methods can lead to a more streamlined and efficient discovery process, aiming to deliver chemical matter of high quality, maximizing the required biological effects while minimizing adverse effects. Surely, Virtual Screening pipelines of this nature will continue to play a central role in producing much needed therapeutics for the health challenges of the future.

References

1. Hawkins PCD, Skillman AG, Nicholls A (2007) Comparison of shape-matching and docking as virtual screening tools. J Med Chem 50(1):74–82
2. Amazon Web Services.: https://aws.amazon.com. Accessed 20 Mar 2023
3. Microsoft Azure.: https://azure.microsoft.com. Accessed 20 Mar 2023
4. Bicer DC, Agrawal G A framework for data-intensive computing with cloud bursting. IEEE international conference on cluster computing. Austin, TX, USA, 2011, pp 169–177
5. Nicholls A, MacCuish NE, MacCuish JD (2004) Variable selection and model validation of 2D and 3D molecular descriptors. J Comp-Aid Mol Des 18:451–474
6. Schneider G, Neidhart W, Giller T, Schmidt G (1999) Scaffold hopping by topological pharmacophore search: a contribution to virtual screening. Angew Chem Int Ed Eng 38:2894
7. Trott O, Olson AJ (2010) AutoDock Vina: improving the speed and accuracy of docking with a new scoring function, efficient optimization, and multithreading. J Comput Chem 31:455–461
8. Rogers D, Hahn M (2010) Extended-connectivity fingerprints. J Chem Inf Model 50:742–754
9. Willett P, Barnard J, Downs GM (1998) Chemical similarity searching. J Chem Inf Comput Sci 38:983–996
10. Grant JA, Pickup BT (1995) A Gaussian description of molecular shape. J Phys Chem 99:3503–3510
11. Kearsley SK, Smith GM (1990) An alternative method for the alignment of molecular structures: maximizing electrostatic and steric overlap. Tetrahedron Comput Method 3:615–663
12. Fischer E (1894) Einfluss der Configuration auf die Wirkung der Enzyme. Ber Dtsch Chem Ges 27:2985

13. Koshland DE (1994) The key-lock theory and the induced fit theory. Angew Chem Int Ed Eng 33:2375–2378
14. Galli S (2014) X-ray crystallography: one century of nobel prizes. JChem Ed 91(12): 2009–2012
15. Hu Y, Cheng K, He L, Zhang X, Jiang B, Jiang L, Li C, Wang G, Yang Y, Liu M (2021) NMR-based methods for protein analysis. Anal Chem 93(4):1866–1879
16. Jumper J, Evans R, Pritzel A et al (2021) Highly accurate protein structure prediction with AlphaFold. Nature 596:583–589
17. Halperin I, Ma B, Wolfson H, Nussinov R (2002) Principles of docking: an overview of search algorithms and a guide to scoring functions. Proteins Struct Funct Genet 47:409–443
18. Leach AR, Gillet VJ (2005) An Introduction to chemoinformatics. Springer, Dordrecht
19. Triballeau N, Acher F, Brabet I, Pin J, Bertrand H (2005) Virtual screening workflow development guided by the "receiver operating characteristic" curve approach. Application to high-throughput docking on metabotropic glutamate receptor subtype 4. J Med Chem 48(7): 2534–2547
20. Rizzi A, Fioni A (2008) Virtual screening using PLS discriminant analysis and ROC curve approach: an application study on PDE4 inhibitors. J Chem Inf Model 48(8):1686–1692
21. Hanley JA, McNeil BJ (1982) The meaning and use of the area under a receiver operating characteristic (ROC) curve. Radiology 143: 29–36
22. Huang N, Shoichet B, Irwin J (2006) Benchmarking sets for molecular docking. J Med Chem 49:6789–6801
23. Jönsson U, Fägerstam L, Ivarsson B, Johnsson B, Karlsson R, Lundh K, Löfås S, Persson B, Roos H, Rönnberg I, Sjölander S, Stenberg E, Ståhlberg R, Urbaniczky C, Östlin H, Malmqvist M (1991) Real-time biospecific interaction analysis using surface plasmon resonance and a sensor chip technology. Biotechniques 11(5):620
24. O'Neill M, Gaisford S (2011) Application and use of isothermal calorimetry in pharmaceutical development. Int J Pharm 417(1-2):83–93
25. Merck Screening Compounds.: https://www.sigmaaldrich.com/GB/en/technical-documents/technical-article/chemistry-and-synthesis/lead-discovery/screening-compounds. Accessed 20 Mar 2023
26. Irwin J.J.; Shoichet, B.K. ZINC—a free database of commercially available compounds for virtual screening. J Chem Inf Model 2005, 45(1), 177-182.
27. Gasteiger J, Martin Y, Nicholls A, Oprea T, Stouch T (2018) Leaving us with fond memories, smiles, SMILES and, alas, tears: a tribute to David Weininger, 1952–2016. J Comp-Aided Mol Design 32(2):313–319
28. Daylight Chemical Information Systems, SMIRKS: https://www.daylight.com/dayhtml/doc/theory/theory.smirks.html. Accessed 20 Mar 2023
29. Kazemizadeh A, Ramazani A (2012) Synthetic applications of Passerini reaction. Curr Org Chem 16(4):418–450
30. Daylight Chemical Information Systems, SMARTS: https://www.daylight.com/dayhtml/doc/theory/theory.smarts.html. Accessed 20 Mar 2023
31. RDKit: open-source cheminformatics software. https://www.rdkit.org. Accessed 20 Mar 2023
32. Irwin JJ, Tang KG, Young J, Dandarchuluun C, Wong BR, Khurelbaatar M, Moroz YS, Mayfield J, Sayle RA (2020) ZINC20—a free ultralarge-scale chemical database for ligand discovery. J Chem Inf Model 60(12):6065–6073
33. Gaulton A, Bellis LJ, Bento AP, Chambers J, Davies M, Hersey A, Light Y, McGlinchey S, Michalovich D, Al-Lazikani B, Overington JP (2012) ChEMBL: a large-scale bioactivity database for drug discovery. Nucleic Acids Res 40(D1):D1100–D1107
34. Bento AP, Hersey A, Félix E et al (2020) An open source chemical structure curation pipeline using RDKit. J Cheminf 12(51)
35. Riniker S, Landrum GA (2015) Better informed distance geometry: using what we know to improve conformation generation. J Chem Inf Comput Sci 55:2562–2574
36. Baell JB, Holloway GA (2010) New substructure filters for removal of pan assay interference compounds (PAINS) from screening libraries and for their exclusion in bioassays. J Med Chem 53(7):2719–2740
37. Chakravorty SJ, Chan J, Greenwood MN, Popa-Burke I, Remlinger KS, Pickett SD, Green DS, Fillmore MC, Dean TW, Luengo JI, Macarrón R (2018) Nuisance compounds, PAINS filters, and dark chemical matter in the GSK HTS collection. SLAS Discov 23(6): 532–544
38. Berthold MR, Cebron N, Dill F, Gabriel TR, Kötter T, Meinl T, Ohl P, Thiel K, Wiswedel B (2009) KNIME – the Konstanz information miner: version 2.0 and beyond. SIGKDD Explor Newsl 11(1):26–31

39. Knime workflow system.: https://www.knime.com/. Accessed 20 Mar 2023
40. Apache Airflow.: https://airflow.apache.org/. Accessed 20 Mar 2023
41. Laskey KB, Laskey K (2009) Service oriented architecture. WIREs Comp Stat 1:101–105
42. Python Celery system.: https://docs.celeryq.dev/en/stable/getting-started/introduction.html. Accessed 20 Mar 2023
43. Mölder F, Jablonski KP, Letcher B et al (2021) Sustainable data analysis with Snakemake. F1000Research 10:33
44. SnakeMake Workflow Management System.: https://snakemake.readthedocs.io/en/stable/. Accessed 20 Mar 2023
45. Thönes J (2015) Microservices. IEEE Softw 3(1):116–116
46. Django Python Framework.: https://www.djangoproject.com/. Accessed 20 Mar 2023

Chapter 7

The Future of Drug Development with Quantum Computing

Bhushan Bonde, Pratik Patil, and Bhaskar Choubey

Abstract

Novel medication development is a time-consuming and expensive multistage procedure. Recent technology developments have lowered timeframes, complexity, and cost dramatically. Current research projects are driven by AI and machine learning computational models. This chapter will introduce quantum computing (QC) to drug development issues and provide an in-depth discussion of how quantum computing may be used to solve various drug discovery problems. We will first discuss the fundamentals of QC, a review of known Hamiltonians, how to apply Hamiltonians to drug discovery challenges, and what the noisy intermediate-scale quantum (NISQ) era methods and their limitations are.

We will further discuss how these NISQ era techniques can aid with specific drug discovery challenges, including protein folding, molecular docking, AI–/ML-based optimization, and novel modalities for small molecules and RNA secondary structures. Consequently, we will discuss the latest QC landscape's opportunities and challenges.

Key words Quantum computing, Drug discovery, Drug development, Protein folding, Hybrid compute, Noisy intermediate-scale quantum (NISQ) era, Variational quantum eigensolver (VQE)

Acronyms

ADMET	Absorption, distribution, metabolism, excretion, and toxicity
DNA	Deoxyribonucleic acid
GAN	Generative adversarial networks
NISQ	Noisy intermediate-scale quantum
QA	Quantum annealing
QC	Quantum compute
QAOA	Quantum approximate optimization algorithm
QBM	Quantum Boltzmann machine
QGM	Quantum generative adversarial network model
QGA	Quantum genetic algorithms
QuANN	Quantum artificial neural network
QMCTS	Quantum Monte Carlo tree search
QMC	Quantum Monte Carlo
QMD	Quantum molecular dynamics
QML	Quantum machine learning

Alexander Heifetz (ed.), *High Performance Computing for Drug Discovery and Biomedicine*, Methods in Molecular Biology, vol. 2716, https://doi.org/10.1007/978-1-0716-3449-3_7,

QPCA	Quantum principle component analysis
QPE	Quantum phase estimation
QSVM	Quantum support vector machine
QVC	Quantum variational classifier
QW	Quantum walk
SNP	Single nucleotide polymorphism
TF	Transcription factor
VQE	Variational quantum eigensolver

1 Introduction

1.1 Computation

Several centuries ago, the abacus would have been the standard computing device, representing data in various numeric bases. Early versions of autonomous abstract machines like the Enigma and Turing machine used ciphers and other symbols to represent data. Nevertheless, the most common, if not the omnipresent, computing of today uses binary bits to represent data. Data processing by binary computers aims to alter bit sequences using logic gates such as *AND*, *OR*, and *NOT* gates and subsequent arithmetic and Boolean logic. Such binary machines have provided low-power, low-noise, and very cheap computing solutions. However, they often lead to very time and space complex solutions for complicated optimization problems. For example, optimizing energy systems is a complex task that involves identifying the optimal configuration or parameters for a system to function at peak efficiency. It often requires a large number of simulations or calculations to evaluate different scenarios and identify the best possible solution. While binary computers find these tasks challenging, nature has found optimal approaches to solve these. For example, photosynthesis in plants has evolved to optimize energy transfer by absorbing solar energy using chlorophyll molecules, which capture the solar energy as a photon. The resultant photoexcited electrons move through a series of molecules in a process called electron transport, ultimately producing ATP and NADPH, which are used in the production of glucose. Photosynthesis is incredibly efficient, with plants able to harvest as much as 95% of the solar energy that reaches them [1]. One reason for this efficiency, potentially, is the use of quantum coherence to transfer energy between molecules. Quantum coherence [2] is a phenomenon whereby particles can exist in multiple states simultaneously, allowing energy to move quickly and efficiently along a pathway of molecules without being lost to the environment. By using quantum coherence to transfer energy between molecules, plants can optimize the process of photosynthesis, maximizing the amount of energy they can extract from the sun [3]. One can argue that emerging quantum computation could provide a solution to some of the most challenging optimization tasks.

1.2 Quantum Computing

Quantum computer is a device that exploits quantum mechanical phenomena: primarily driven by two key phenomena—superposition and entanglement. For a thorough understanding of quantum computing fundamentals, see the standard textbook for the quantum information systems [4].

1.2.1 Superposition

To understand superposition, let us consider the notion of interference and imagine that there are two waves. These could be waves in water or electromagnetic waves. If we allow these waves to interfere, they will add up constructively if they are in phase. This means that if their peaks and troughs align, their amplitudes would add up, resulting in a larger wave with greater intensity. This constructive interference is analogous to a qubit being in a superposition of two states. On the other hand, if the two waves are out of phase, they would cancel each other out and result in no wave at all. This destructive interference is analogous to a qubit collapsing into a definite state rather than being in a superposition of states. In quantum computing, qubits can exist in a superposition of states, such as being in both 0 and 1 at the same time. This allows quantum algorithms to explore multiple possibilities simultaneously, leading to potentially faster solutions, compared to classical computing. The probability of measuring the qubit in a particular state depends on the amplitudes and phases of the superposition, similar to the probability distribution of a wave resulting from interference.

1.2.2 Entanglement

Let us now consider the second principle of entanglement. This occurs when two or more particles get coupled in such a way that the state of one particle cannot be described without also knowing the state of the others, regardless of their distance. When two particles are linked, a measurement on one particle immediately alters the state of the other particle. This can be explained using a simple example involving two entangled coins to illustrate the concept of entanglement. Let us imagine we have a magical pair of coins that, when flipped, always show the same side (both heads or both tails) no matter how far apart they are. Let us keep one coin and give the other to a friend who moves to another city. Now, we both agree to flip your coins at the exact same time. When we flip our coin and observe that it lands on heads, we instantly know that our friend's coin has also landed on heads, even though that friend is in another city. Similar conditions exist when we observe tails as the response. This "magical" correlation between the coins shows how they are entangled. Particles like electrons and photons can get entangled and exhibit this correlated behavior. When one measure the property of one entangled particle (such as its spin), one instantly knows the corresponding property of the other entangled particle, even if they are separated by vast distances. This instantaneous correlation goes against traditional ideas of locality, and causality, and has been described by Albert Einstein as "spooky action at a distance."

To understand how these two principles are employed in quantum computers, one can begin with initializing the quantum states (also known as data encoding from classical to quantum compute). Then quantum bits are fed to a quantum circuit, which selectively interferes with the components of the superposition and/or entanglement using a predefined mechanism, which is often an arrangement of gates on the quantum circuit. The solution to the quantum circuit's calculation is what remains after cancelling the relative amplitudes and phases of the input state [5]

1.2.3 Qubit/Quantum Bit

A qubit is the basic unit of quantum information. However, unlike traditional bits, which can be either 0 *or* 1, qubits can be in both states at the same time. Mathematically, a qubit can be represented as a linear combination of two states, usually denoted as $|0\rangle$ and $|1\rangle$. These states are also known as the computational basis states and can be written as:

$$|\psi\rangle = \alpha\,|\,0\rangle + \beta\,|\,1\rangle,$$

$$|0\rangle = \begin{bmatrix} 1 \\ 0 \end{bmatrix}$$

$$|1\rangle = \begin{bmatrix} 0 \\ 1 \end{bmatrix}$$

$$|\psi\rangle = \begin{bmatrix} \alpha \\ \beta \end{bmatrix}$$

where α and β are complex numbers, whose squared absolute values ($|\alpha|^2$ and $|\beta|^2$) represent the probabilities of measuring the qubit in the corresponding state at the time of measurement. The coefficients α and β must also satisfy the condition, $|\alpha|^2 + |\beta|^2 = 1$. This normalization condition ensures that the total probability of measuring the qubit in any possible state is always 1 [6].

1.2.4 Bloch Sphere

As an alternative, in the Bloch sphere representation of a qubit, its state is indicated by a point on the surface of a sphere (*see* Fig. 1). Here the north and south poles correspond to the states $|0\rangle$ and $|1\rangle$, respectively. The direction of the point on the sphere represents the superposition of the qubit in the $|0\rangle$ and $|1\rangle$ states. The values of α and β, which determine the probabilities of measuring the qubit in the $|0\rangle$ and $|1\rangle$ states, can be related to the coordinates of the point on the Bloch sphere [6, 7]. The values of α and β can be related to the coordinates of the point on the Bloch sphere as follows:

$$\alpha = \cos\theta/2; \beta = e^{i\varphi}\sin\theta/2$$

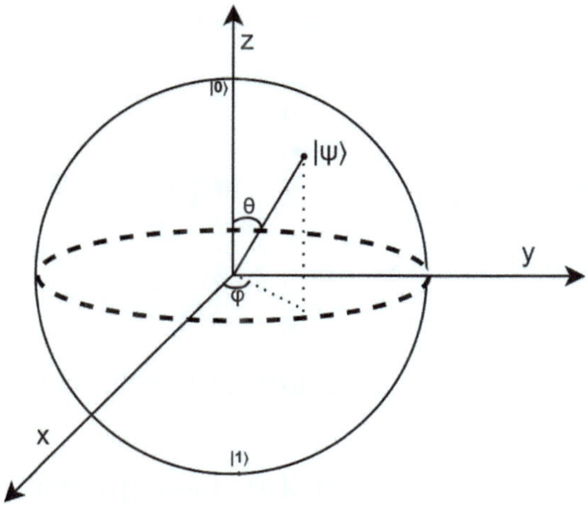

Fig. 1 Bloch sphere representation of the quantum bit (qubit)

where θ is the polar angle (measured from the north pole) and φ is the azimuthal angle (measured around the equator) [6, 8].

1.2.5 Quantum Circuit

A quantum circuit is a sequence of quantum operations, or gates, that manipulate the quantum states of qubits, which could be superposition, allowing them to represent multiple classical states simultaneously, and/or can be entangled, creating correlations between qubits that cannot be explained by classical physics alone. These quantum gates perform unitary transformations on the qubits' state vectors, enabling the manipulation of qubits in their superposition and entangled states [5, 9]. Some common quantum gates include the Pauli- X, Y, and Z gates, the Hadamard gate, and the *CNOT* gate. These gates are used in various combinations to create complex quantum algorithms that can solve problems more efficiently than classical algorithms [7].

The quantum circuit starts with an initial state, usually with all qubits in the $|0\rangle$ state, and then applies a series of quantum gates to transform the qubits into a final state. The final state can then be measured, which collapses the superposition of qubits into a single classical outcome, providing the result of the computation. It is important to note that quantum circuits are inherently probabilistic due to the nature of quantum mechanics. As a result, the outcomes of quantum computations are not always deterministic, and it may be necessary to repeat the quantum circuit multiple times to obtain a statistically significant result. Refer to Fig. 2 for details on quantum circuit operations and notations using Qiskit [5].

1.2.6 Quantum Gates

There are several quantum gates, each with specific functions and properties. Herein, we briefly review some of the commonly used quantum gates:

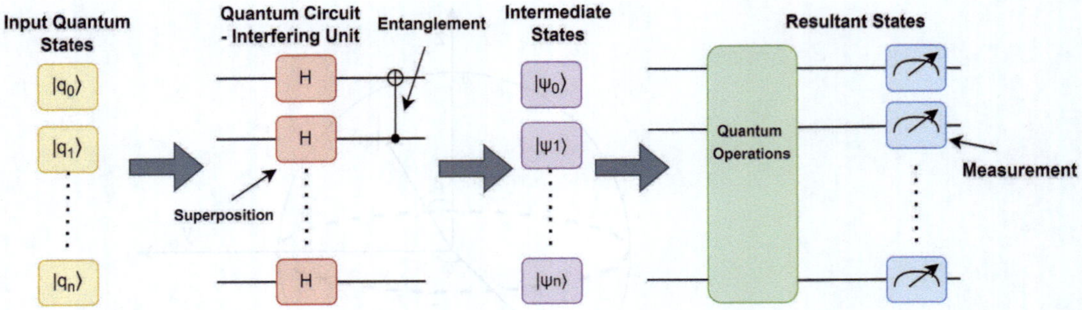

Fig. 2 Quantum computation is driven by quantum circuits as an interfering unit to produce a resultant state

Pauli Gates (*X*, *Y*, *Z*)

The simplest binary operator is the *NOT* gate. It's quantum equivalent, often referred to as the bit-flip gate or Pauli (X) gate, and it flips the state of a qubit, turning $|0\rangle$ to $|1\rangle$ and vice versa. On the other hand, Pauli- Z gate (Z), also known as the phase-flip gate, flips the phase of the $|1\rangle$ state by multiplying it by 1, leaving the $|0\rangle$ state unchanged. A combination of these two is the Pauli-Υ gate (Υ), with an added imaginary component. It applies both a bit-flip and a phase-flip to the qubit [5, 6].

Hadamard Gate (*H*)

The Hadamard gate is a fundamental single-qubit gate in quantum computing. It is used to create superposition and change the basis of a single qubit and plays a crucial role in many quantum algorithms like Grover's algorithm, the quantum Fourier transform, and the creation of bell states [5, 6]. The Hadamard gate can be represented as a 2 × 2 matrix operation as below:

$$H = \frac{1}{\sqrt{2}}\begin{bmatrix} 1 & 1 \\ 1 & -1 \end{bmatrix}$$

When applied to qubit, it performs the following transformations:

1. $$H\,|\,0\rangle = \frac{1}{\sqrt{2}}(|0\rangle + |1\rangle)$$

2. $$H\,|\,1\rangle = \frac{1}{\sqrt{2}}(|0\rangle - |1\rangle)$$

Phase Gates

These gates are different phases to the qubits. For example, the *S* gate, also called the phase gate, adds a phase of $\pi/2$ to the $|1\rangle$ state of a qubit, leaving the $|0\rangle$ state unchanged. Alternatively, the *T* gate, also known as the $\pi/8$ gate, adds a phase of $\pi/4$ to the $|1\rangle$ state, while the $|0\rangle$ state remains unchanged [7, 10].

Controlled Gates

These gates lead to a control operation between two qubits. For example, the controlled-NOT gate (*CNOT*) is a two-qubit gate where the first qubit (control) determines whether the second qubit (target) undergoes a bit-flip operation. If the control qubit is in the $|1\rangle$ state, the target qubit's state is flipped. This gate is crucial for creating entanglement between qubits. Controlled-U gate is a generalization of the *CNOT* gate, where the *U* gate is applied to the target qubit if the control qubit is in the $|1\rangle$ state. This *U*-gate can be any single-qubit unitary operation [6, 7].

Swap Gate (SWAP)

This two-qubit gate exchanges the quantum states of two qubits, effectively swapping their information [5].

Toffoli Gate (CCNOT)

Also known as the controlled-controlled-NOT gate, it is a three-qubit gate that applies an *X* gate to the target qubit only if both control qubits are in the $|1\rangle$ state. It can be used to implement classical reversible logic in quantum circuits [5].

1.2.7 Data Encoding Techniques

In e crucial for translating classical information into quantum states that can be processed by quantum circuits. Several encoding techniques have been presented, with some of the common ones being introduced below.

A. Amplitude encoding: As a first approach, the amplitudes of a quantum state can be used to encode classical information [5, 10]. For example, given a normalized classical data vector (c_1, c_2), it can be encoded into a single qubit as $c_1|0\rangle + c_2|1\rangle$. Amplitude encoding is often used in quantum machine learning algorithms such as the quantum support vector machine (QSVM).

B. Angle encoding: This technique encodes classical data as the relative angles between quantum states. For example, if we have a data point "x," we can use the angle $\theta = \pi x$ to create the quantum state $\cos\theta|0\rangle + \sin\theta|1\rangle$. This encoding is used in quantum algorithms like the quantum approximate optimization algorithm (QAOA) and variational quantum eigensolvers (VQE).

C. Basis encoding: Classical information is directly encoded into the computational basis states of qubits. For example, to encode the binary string "1101," we use the quantum state $|1101\rangle$. This encoding is straightforward and often used in quantum algorithms such as Grover's search algorithm, Shor's factoring algorithm, and the Bernstein–Vazirani algorithm.

1.2.8 Results' Interpretation

To interpret the results of a quantum computation, we perform measurements on the final state of the qubits. Since quantum computing is inherently probabilistic, the measurement collapses the quantum state into one of its basis states according to their corresponding probability amplitudes. The result is a classical bit string that represents the outcome of the computation [11]. For example, consider a qubit with state $|\psi\rangle = \alpha|0\rangle + \beta \mid 1\rangle$, where α and β are the amplitudes; when we measure the qubit in the computational basis (z-basis), it will collapse to either state $\mid 0\rangle$ or $\mid 1\rangle$. In practice, quantum algorithms need to be run multiple times to obtain a statistically significant result. For example, Grover's search algorithm requires multiple iterations to find the correct solution with high probability. The result can then be extracted by selecting the most frequent output among the measured outcomes. In some cases, additional classical post-processing steps may also be needed to extract useful information from the measured output. These may include error correction techniques, decoding the encoded data, or analyzing the probability distribution of the output states to find the optimal solution.

1.3 Quantum Annealing

Quantum annealing (QA) is a quantum optimization technique based on adiabatic theory used to find the global minimum of an objective function by exploiting quantum phenomena, like tunneling [12].

The adiabatic theorem states that if Hamiltonian of a quantum system changes slowly enough, the system will stay close to its instantaneous ground state [13]. Using this theorem, the quantum annealer starts with an initial Hamiltonian, whose ground state is easy to prepare. The Hamiltonian is then gradually evolved toward the problem Hamiltonian, which encodes the ground-state solution to the problem. The system is allowed to explore the solution space by exploiting quantum phenomena such as tunneling, which allows it to traverse energy barriers enabling it to find the global minimum faster and more accurately than classical methods [12, 14, 15].

1.4 Hamiltonians

The Hamiltonian is an operator representing the total energy of the system. It governs the time evolution of the quantum states according to Schrödinger's equation [16]. Hamiltonians play a central role in both quantum circuit-based algorithms and quantum annealing as they describe the energy landscape and the dynamics of quantum systems. In quantum computing, they represent the time evolution of a quantum system in quantum circuit-based algorithms. Finding the eigenvalues and eigenvectors of a given Hamiltonian, which reflect the energy levels and related quantum states of the system, is the objective of several algorithms including the quantum phase estimation (QPE) and the variational quantum eigensolver (VQE) [17, 18]. These methods have been used to solve a number range of problems, including combinatorial optimization and the simulation of quantum systems in chemistry and condensed matter physics [8, 19–21].

In quantum annealing, on the other hand, the objective is to identify the Hamiltonian's ground state, or lowest energy state, which corresponds to the best solution to the problem. The goal of these processes is to find the problem Hamiltonian's ground state by gradually switching from an initially prepared Hamiltonian [12, 14, 15, 22]. This allows the system to remain relatively close to its beginning state throughout evolution.

Hamiltonians are also critical for expressing and comprehending the underlying quantum systems. They bridge the gap between the physical implementations of quantum computing and their abstract mathematical formulations, directing the design and study of quantum algorithms for varied applications. There are some known Hamiltonians in quantum mechanics that describe the behavior of the physical system, with some commonly used ones shown in Table 1.

1.5 Physical Implementation

There are several physical systems that are currently being attempted to implement quantum computing. However, the optimal technology, if any, is yet to be identified. Here are a few examples:

- Photons: It's appealing to think of a qubit as the polarization state of a photon because photons do not decohere as much as other quantum systems do. "Waveplates" made of birefringent materials have been proposed for polarization rotations (one-qubit gates). Photons also enable encoding a qubit based on its location and timing. One can also encode quantum information in the continuous phase and amplitude variables of many-photon laser beams. However, getting photons to interact in the way that is needed for universal multi-qubit control is a significant problem.
- Trapped atoms: Using the right electric fields from nearby electrodes, individual ions of atoms can be held in place with nanometer accuracy in free space. Multiple trapped ion qubits can be linked together using a laser-induced coupling of the spins and a collective mode of harmonic motion in the trap. Trapped atoms have two key challenges: short life span at the initialization state and decoherence due to the noise interference leading to errors in the qubits.
- Quantum dots: Quantum computing with single atoms in vacuum requires cooling and entrapment; hence, large arrays of qubits may be easier to assemble and cool together if the

Table 1
Various Hamiltonians and their applications to quantum compute problems

Index	Description	Applications	References
1.	The ising model Hamiltonian is used to describe pairwise interactions between binary spins in a lattice. It is often applied to solve optimization problems and study phase transitions The formula is: $H = -J\sum_{i,j}\sigma_i\sigma_j - h\sum_{i}\sigma_i$ where σ_i are spins, J represents coupling strengths between spins, h is an external magnetic field, and the summations are over neighboring pairs (i, j) and all lattice sites i, respectively	Combinatorial optimization problems Quantum annealing Error-correcting codes Simulating quantum systems (to simulate the behavior of more complex quantum systems, such as interacting quantum particles) Machine learning (quantum Boltzmann machine) Quantum phase transitions	[77, 78]
2.	The Hubbard model Hamiltonian used to study strongly correlated electronic systems. It describes electrons hopping between lattice sites, with an energy cost for double occupancy. This model helps understand the behavior of electrons in materials, specifically in the context of high-temperature superconductivity and magnetism	From a drug discovery point of view, the Hubbard model may not be directly applicable, as it is primarily used to study strongly correlated electron systems in condensed matter physics. However, it has some potential applications Electronic structure calculations Efficiently simulate the quantum dynamics of biological systems such as protein–ligand interactions and protein folding The Hubbard model provides insights into the behavior of interacting particles, which can help researchers develop better optimization algorithms for drug discovery tasks such as optimal docking configuration	[79]
3.	The second-quantized Hamiltonian is a convenient formalism that allows for a more compact and efficient description of many-body systems, such as electrons in molecules or lattice models of interacting particles	Compute molecular properties, such as ground-state energies, excited state energies, and other properties like dipole moments and polarizabilities. Solving for drug candidates and their target proteins can provide insights into strength and nature of their interactions	[8]
4.	Transverse ising model used in quantum computing to describe the behavior of qubits. It consists of a sum of pairwise interactions between qubits, where each qubit can take on a value of 0 or 1 and there is a transverse magnetic field that acts on the qubits	Simulating complex quantum systems Combinatorial optimization problem Transverse ising model can be applied to quantum machine learning (QML)	[80]

(continued)

Table 1 (continued)

Index	Description	Applications	References
5.	In the coarse-grained Gō model, multiple atoms are grouped into single interaction sites, reducing the complexity of the model. This representation often uses a single interaction site per amino acid residue, typically representing the alpha-carbon (Cα) atom. The potential energy function considers both bonded and nonbonded interactions between native contacts. Bonded interactions include things like virtual bond stretching and virtual bond angle bending	The study of protein folding gives insight into the kinetics and thermodynamics of protein folding. Even though it is a simplified depiction, it can aid researchers in comprehending essential features of the folding process and in developing models that reflect the intricacy of protein folding in greater detail	[37]

"atoms" are integrated into a solid-state host [23]. Quantum dots offer a potential to do so, for example, electrostatically defined quantum dots and self-assembled quantum dots. Another proposed method is the use of arrays of quantum dots containing a single electron, whose two spin states provide a qubit [24]. Quantum logic is accomplished by changing voltages on the electrostatic gates to move electrons closer and farther from each other, activating and deactivating the exchange interaction.

1.6 General Applications

A number of application domains for quantum computing have been proposed [25]. Some of these include the following:

Cryptography: Quantum computers can potentially break many of today's encryption systems, which could have far-reaching ramifications for security in general and Internet-based encryption, in particular [26, 27]. Simultaneously, quantum computers can be utilized to develop novel encryption schemes that are resistant to traditional computer attacks [28].

Simulation: Quantum computers can simulate quantum processes efficiently as compared to conventional computers. Applications for these include chemical reactions or molecular behavior of proteins [29, 30]. This could result in advances in materials research and medication discovery [7, 30–35].

Optimization problems: Many real-world challenges involve determining the optimal answer from a large number of potential alternative solutions. Quantum computers can handle these optimization issues significantly quicker than traditional computers, resulting in more efficient solution to supply chains, logistics, finance, and scheduling applications [14, 36].

Machine learning: Quantum computers can accelerate various machine learning algorithms, which could have applications in fields such as image recognition and protein folding [37–39].

1.7 Limitations of Quantum Compute

The potential of quantum computing in addressing intricate problems that are beyond the capabilities of classical computers is highly significant. Nevertheless, various obstacles must be overcome before their timescale deployment.

The current quantum computing technology is constrained by a finite number of qubits, thereby imposing limitations on the scale and intricacy of computational tasks that can be accomplished. Increasing the quantity of qubits is a crucial factor in addressing intricate problems and attaining quantum advantage. However, the augmentation of qubit quantity also intensifies other concerns, such as qubit connectivity and error rates.

The precision of quantum gates is crucial in the manipulation of qubits to guarantee precise outcomes. Nevertheless, qubits are vulnerable to errors that arise from environmental factors, control imperfections, and various sources of noise [40]. Attaining quantum gates with high precision while simultaneously reducing error rates is a noteworthy obstacle.

Quantum error correction techniques are indispensable for alleviating the impact of errors in quantum operations. Notwithstanding, the utilization of these methodologies generally necessitates a substantial expense in the form of the quantity of qubits and the intricacy of the quantum circuits, thereby rendering their execution in devices of the immediate future a challenging task [40].

The development and upkeep of quantum computers incurs significant economic costs, primarily due to the advanced technology and infrastructure necessary to support them. This includes the provision of ultralow-temperature environments for superconducting qubits and intricate laser systems for trapped ion qubits. Minimizing the financial expenses linked with quantum computing is imperative to facilitate wider reach and acceptance.

Quantum computers are presently accessible exclusively to a restricted cohort of researchers and institutions, either through dedicated facilities or cloud-based access. The augmentation of quantum computing resources will be of paramount importance in promoting ingenuity and expediting the advancement of quantum algorithms and applications [40].

The development of efficient quantum algorithms that can optimally utilize the limited resources of near-term quantum devices is a persistent challenge in the field of software and algorithms. In addition, the development of software tools and programming languages that are user-friendly and specifically designed for quantum computing will be crucial in closing the divide between quantum hardware and pragmatic applications.

The issue of interoperability arises since quantum computers developed by various manufacturers employ unique qubit technologies, gate sets, and control systems. The standardization of quantum computing protocols and interfaces is a crucial factor for ensuring smooth integration and collaboration across various platforms and research groups.

The field of quantum computing is experiencing an increasing need for proficient individuals in various domains such as research, engineering, and practical implementation, resulting in a surge in demand for skilled professionals. The augmentation of education and training initiatives in the domain of quantum computing will be imperative in fulfilling the aforementioned demand and propelling the field toward progress.

The resolution of these obstacles necessitates sustained inquiry, originality, and cooperation among the academic community, private sector, and governmental entities. The potential of quantum computing to revolutionize various fields becomes increasingly tangible as advancements are made.

1.8 Hybrid Quantum Computing

Due to the above limitations, several studies have started shifting the focus on hybrid quantum computing, which combines the strengths of classical and quantum computers, to solve complex problems that neither system can handle alone. Quantum computers excel at optimization and simulation due to their ability to operate on many quantum states in parallel, but they face limitations in qubit count and coherence time. Classical computers are efficient in data handling, storage, and interfacing with other systems. By using classical computers to manage inputs and outputs and quantum computers for complex operations, hybrid quantum computing enables the solution of problems that would be challenging or impossible for classical computers alone [41]. This approach involves parameterized quantum circuits optimized by classical computers to tackle a wide range of quantum problems.

1.9 Parameterized Circuit

A parameterized circuit, also known as variational quantum circuit, has adjustable parameters, generally represented by set of angles [5, 39, 42]. A simple example of a parameterized gate (*see* Fig. 3) in the following single-qubit circuit:

By varying the value of θ, we can adjust the amount of superposition applied to the qubit and, therefore, change the output of the circuit. For example, if we set θ to 0, the output of the circuit will always be $|0\rangle$. If we set θ to $\pi/2$, the output will be equally likely to be $|0\rangle$ or $|1\rangle$. And if we set θ to π, the output will always be $|1\rangle$.

Parameterized circuits are well-suited for optimization problems in hybrid quantum computing. Variational quantum eigensolver (VQE), for example, employs a parameterized quantum circuit to solve for the ground state of a given Hamiltonian [39, 42–45].

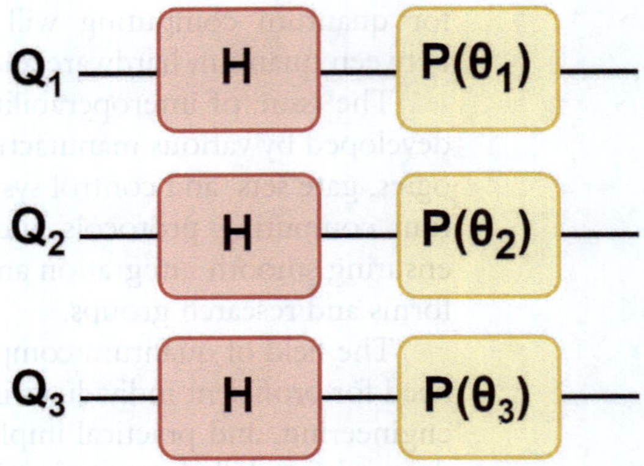

Fig. 3 An example of parameterized quantum circuit

1.10 Variational Quantum Eigensolver

The variational principle is a fundamental concept in physics that allows finding approximate solutions to complex problems. In quantum mechanics, it states that the expectation value of the energy for any trial wave function is always greater than or equal to the true ground-state energy of the system [42–44, 46].

In quantum computing, the variational principle is employed in the form of the variational quantum eigensolver (VQE) algorithm. VQE is a hybrid quantum-classical algorithm designed to find the ground-state energy of a quantum system, particularly useful for solving optimization problems and simulating quantum chemistry [8, 44, 47–49].

The algorithm uses a parametrized quantum circuit (ansatz) to prepare a trial wave function. The parameters of the ansatz are optimized classically to minimize the energy expectation value, thus converging to an approximation of the ground-state energy. VQE leverages the strengths of both quantum and classical computing, mitigating the limitations of current noisy intermediate-scale quantum (NISQ) devices. This makes it a promising technique for practical applications in the near term as choice for QC before the advent of fault-tolerant quantum computers.

VQE works by using a quantum computer to prepare a trial state of the system and then measuring the expectation value E_i of the Hamiltonian in that state. The trial state is prepared using a parameterized quantum circuit, known as an ansatz, which is designed to approximate the ground state of the system.

The goal of the algorithm is to find the ground-state energy of the system, which corresponds to the minimum possible value of the expectation value $\langle\psi|\mathrm{A}|\psi\rangle$ over all possible quantum states $|\psi\rangle$. The VQE algorithm uses a parameterized quantum circuit (ansatz) to prepare a trial quantum state $|\psi(\theta)\rangle$ that is close to the ground state of the Hamiltonian H. The expectation value E_i of the

Hamiltonian H is then calculated for the trial state $|\psi(\theta)\rangle$ using the expression:

$$E_i = \langle \psi(\theta) | H | \psi(\theta) \rangle$$

Here θ is the parameter for the trial quantum state.

Pseudocode

Input: Hamiltonian H, ansatz circuit U, classical optimizer O, number of iterations n

1. Initialize the parameters of the ansatz circuit (parameterized circuit) randomly
2. For $i = 1 \rightarrow n$:
   (a) Calculate the expectation value (E_i) of the Hamiltonian (H) using (U) and the current parameters
   (b) Calculate the gradients of E_i with respect to the parameters using backpropagation.
   (c) Update the parameters using the optimizer O based on the gradients and the current cost function value
3. Return the final parameters and the corresponding ground state energy

1.11 The Quantum Approximate Optimization Algorithm (QAOA)

The quantum approximate optimization algorithm (QAOA) is a quantum method that is used to find approximate solutions to NP-hard combinatorial optimization problems [19, 50, 51]. QAOA is a variational quantum algorithm that prepares an ansatz state using a parameterized quantum circuit and measures the predicted value of the objective function, which is subsequently optimized using classical (hybrid) optimization techniques.

The QAOA is inspired from quantum mechanics' adiabatic theorem, which claims that a slowly fluctuating Hamiltonian can be used to prepare the ground state of a problem Hamiltonian. To prepare the ansatz state, QAOA employs a hybrid quantum-classical technique with alternating unitary operators (*see* Fig. 4).

The QAOA works by approximating the ground state of the problem Hamiltonian using an alternative structure of two Hamiltonians: mixing Hamiltonian (H_M) and problem Hamiltonian (H_P)

This structure helps guide the quantum system toward the optimal solution. The mixing Hamiltonian is chosen such that its ground state is easy to prepare and is often a simple transverse field Hamiltonian; the main idea is to create superposition and promote exploration of the solution space by providing quantum tunneling between states.

Consider an undirected graph with vertices V and edges N as an example of the MaxCut [19, 52] issue. The purpose is to partition

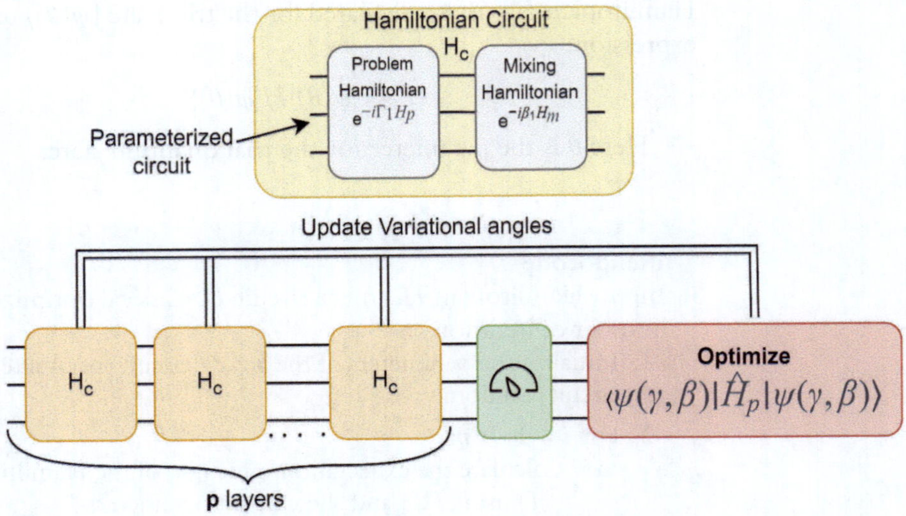

Fig. 4 The quantum approximate optimization algorithm (QAOA) implementation

the vertices into two sets with the least number of edges between them. This problem's cost function is:

$$C(z) = \frac{1}{2}\sum\nolimits_{(i,j)\epsilon N}(1 - z_i z_j)$$

where z_i is either +1 or −1, representing the two sets.

The QAOA circuit is constructed using two types of unitary operators:

The mixing Hamiltonian H_m: $\widehat{H_M} = \sum_i \sigma_x{}^{(i)}$

The problem Hamiltonian H_p: $\widehat{H_P} = \frac{1}{2}\sum_{i,j\,\epsilon\,N}\left(I - \sigma_z{}^{(i)}\sigma_z{}^{(j)}\right)$

The QAOA ansatz is prepared by applying the following unitary operations:

$$U(\gamma,\beta) = e^{-i\beta\widehat{H_M}}e^{-i\gamma\widehat{H_P}}$$

The ansatz state is then given by:

$$|\psi(\gamma,\beta)\rangle = U(\gamma_n,\beta_n)\,U(\gamma_{n-1},\beta_{n-1})\ldots\,U(\gamma_1,\beta_1)\mid s\rangle$$

where $|s\rangle$ is the initial state, usually an equal superposition of all possible bitstrings.

The objective is to find the optimal parameters β and γ that minimize the expected value of the cost function for MaxCut problem:

$$\left\langle \psi(\gamma,\beta) \middle| \widehat{H_P} \middle| \psi(\gamma,\beta) \right\rangle$$

Pseudocode for QAOA:

1. Initialize graph $G(V, E)$
2. Choose the number of QAOA layers n
3. Prepare the initial state $|s\rangle$
4. Initialize parameters γ and β randomly
5. Repeat until convergence or max number of iterations:
   (a) Prepare ansatz $|\psi(\gamma,\beta)\rangle$ using the QAOA circuit.
   (b) Estimate the expected value of the cost function $\left\langle \psi(\gamma,\beta) \middle| \widehat{H_P} \middle| \psi(\gamma,\beta.) \right\rangle$
   (c) Update the parameters γ and β using classical optimization techniques
6. Measure the final ansatz state to obtain the approximate solution

1.12 Quantum Machine Learning

Quantum machine learning (QML) unites quantum computing and machine learning principles to harness the power of quantum algorithms for data processing and analysis. The quantum approximate optimization algorithm for neural networks (QuANN) [53–56] is a well-known example of QML. QuANN adds quantum computing techniques to artificial neural network training, enabling faster optimization of model parameters. The core concept of QuANN is to express the optimization problem as a quantum ising model Hamiltonian that can be solved effectively using quantum algorithms.

The following equation shows how a classical neural network's cost function is turned into an ising Hamiltonian for a QuANN [54, 56]:

$$H(w) = \sum_{i=1}^{N} \left[L(x_i, y_i, w) + \frac{\lambda}{2} \sum_{j=1}^{M} w_j^2 \right] \rightarrow H_{\text{ising}}(\sigma) = -\sum_{i,j} J_{i,j}\sigma_i\sigma_j - \sum_i h_i\sigma_i$$

where:

- $H(w)$ is the classical cost function of the neural network
- L is a loss function
- λ is the regularization parameter
- w_j denotes model parameters
- H_{ising} is the ising model Hamiltonian

- $J_{i,j}$ represents the interaction strength between spins i and j
- h_i denotes the local field acting on spin i
- σ_i and σ_j are the spin variables
- N and M denote the number of training samples and model parameters, respectively.

This transformation of the cost function into the ising Hamiltonian allows the use of quantum algorithms, such as quantum approximate optimization algorithm (QAOA) or variational quantum eigensolver (VQE) [53, 54, 56–59], to find the ground state of the Hamiltonian, which corresponds to the optimal model parameters for the neural network. This process can result in significant speedup over classical optimization methods, especially for large and complex problems; however, there are many challenges to be addressed in this area, such as barren plateaus while training the quantum neural network [60].

2 Potential QC Applications to Drug Discovery

2.1 Drug Discovery

The drug discovery process is a complex, time-consuming, and resource-intensive endeavor, which involves multiple stages, such as target identification, target validation, hit identification, hit-to-lead optimization, lead optimization, and absorption, distribution, metabolism, excretion, and toxicity (ADMET) prediction. Quantum computing has the potential to significantly accelerate and optimize this process by providing more efficient algorithms and solving complex problems that are intractable for classical computers. In this section, we explore various quantum computing algorithms and their applications in the drug discovery process and provide the summary as a landscape in Fig. 5.

2.2 Target Identification

Target identification involves finding suitable biological targets, such as proteins or nucleic acids, for therapeutic intervention. Quantum computing can aid in this process through the following approaches.

2.2.1 Protein Structure Prediction

Quantum algorithms, such as the quantum approximate optimization algorithm (QAOA) [61], variational quantum eigensolver (VQE) [17, 47], and quantum phase estimation (QPE) [18], can be used to predict protein structures more efficiently with significant advantages in computational resources and speed [37, 38, 62].

2.2.2 Biomarker Identification

Quantum machine learning techniques, such as quantum support vector machines (QSVM) [39], quantum Boltzmann machine (QBM), and quantum artificial neural networks (QuANN), can help identify biomarkers more effectively by exploiting the inherent parallelism and unique properties of quantum computing [7].

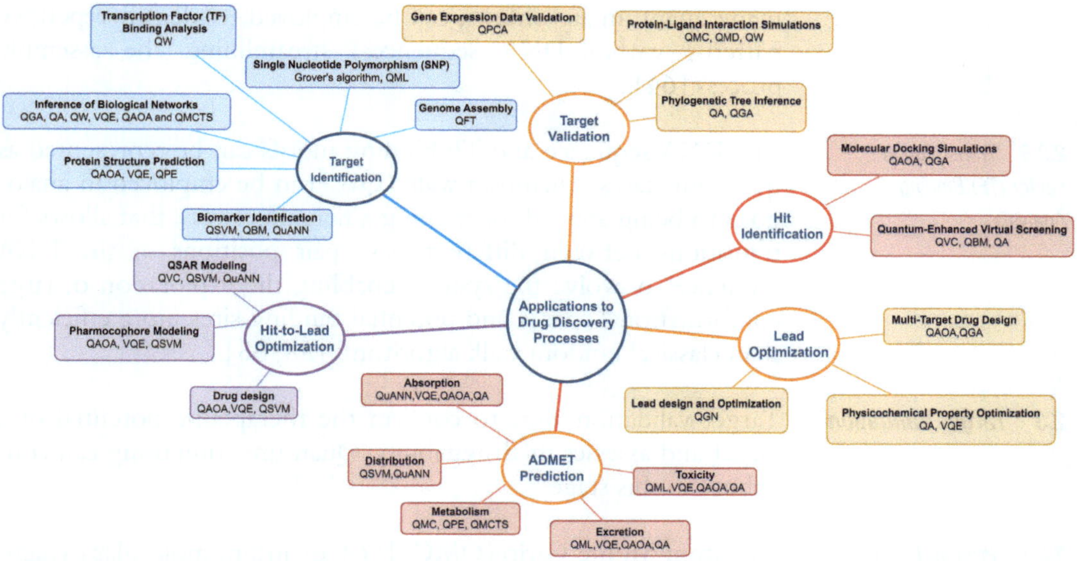

Fig. 5 The quantum computing landscape for applications to drug discovery

2.2.3 Inference of Biological Networks

The biological data, such as gene expression levels, protein interactions, or metabolic reactions, can be encoded into quantum states, typically using qubits. For example, in a protein interaction network, proteins can be encoded as nodes and their interactions as edges in a graph. This encoding allows the quantum algorithms to work directly with the biological data, taking advantage of quantum parallelism and superposition to explore multiple network structures simultaneously. By using QC algorithms like quantum genetic algorithms (QGA), QA, QW, VQE, QAOA, and QMCTS to optimize the objective function, inferences in biological networks can be made such as gene regulatory networks or protein–protein interaction networks [34, 63].

2.2.4 Single Nucleotide Polymorphism (SNP)

Quantum algorithms can potentially accelerate the process of SNP analysis by handling the large-scale computations more efficiently. For instance, quantum machine learning algorithms can be employed to identify patterns and make predictions about the effects of SNPs on phenotypes. Furthermore, QC algorithms like Grover's search algorithm could be used to search through large databases of genetic information to identify relevant SNPs.

2.2.5 Genome Assembly

Quantum algorithms have the potential to process complex data more efficiently by leveraging the principles of superposition and entanglement. Superposition allows for concurrent handling of multiple possible combinations, while entanglement aids in managing interdependencies between different sequence sections. Notably, quantum Fourier transform (QFT), a key component in many quantum algorithms, can be employed to identify repetitive patterns within DNA sequences, streamlining the assembly process [64].

2.2.6 Transcription Factor (TF) Binding Analysis

The DNA sequence and TF binding motifs can be represented as quantum states. Quantum walk (QW) can be employed in analogous to being applied for training a neural network that allows for transitions between different base pair positions in the DNA sequence to evolve the system, enabling the exploration of large conformational spaces and potential binding sites more efficiently than classical random walk algorithms [38, 65].

2.3 Target Validation

Target validation aims to confirm the therapeutic potential of a target and assesses its druggability. Quantum computing can contribute to this stage.

2.3.1 Protein–Ligand Interaction Simulations

Quantum Monte Carlo (QMC) [66], quantum molecular dynamics (QMD) [67–69], and quantum walk (QW) [65] can be employed by using random sampling techniques like Metropolis–Hastings or importance sampling to generate configurations (particle positions) in the quantum system to simulate protein–ligand interactions more accurately and efficiently, providing insights into the binding mechanisms and energetics of potential drug candidates [30, 70, 71]. Antibody modeling can be done using QMC on NISQ-based quantum computing [72].

2.3.2 Gene Expression Data Validation

Quantum principal component analysis (QPCA) can be applied to validate gene expression data, offering a quantum-enhanced approach to dimensionality reduction and feature extraction in high-dimensional gene expression datasets.

2.3.3 Phylogenetic Tree Inference

The quantum system in a superposition of all possible phylogenetic trees, each associated with a particular energy level (corresponding to the objective function value), can be optimized by using time evolutionary algorithms like quantum annealing (QA) and quantum genetic algorithms (QGA), potentially leading to improved understanding of the evolutionary relationships between biological targets and the identification of novel therapeutic targets.

2.4 Hit Identification

Hit identification involves the discovery of compounds that interact with the target of interest. Quantum computing can help accelerate this process.

2.4.1 Quantum-Enhanced Virtual Screening

Quantum variational classifier (QVC), QBM, and QA can be employed for quantum-enhanced virtual screening, providing more efficient and accurate predictions of compound–target interactions than classical approaches.

2.4.2 Molecular Docking Simulations

QAOA and QGA can be used for molecular docking simulations, exploring the vast conformational and chemical spaces of ligands and targets more efficiently and potentially identifying better hit compounds [73].

2.5 Hit-to-Lead Optimization

The hit-to-lead optimization stage aims to improve the properties of hit compounds, making them more suitable for further development. Quantum computing can contribute to this stage.

2.5.1 Quantitative Structure–Activity Relationship (QSAR) Modeling

Quantum-based structure–activity relationship (QSAR) modeling can be achieved using QVC, QSVM [39], and QuANN, providing a more efficient way to predict and optimize the relationships between the chemical structures of compounds and their biological activities.

2.5.2 Pharmocophore Modeling

Quantum computing algorithms have the potential to revolutionize pharmacophore modeling in drug discovery by offering more accurate and efficient solutions. Quantum algorithms such as the quantum approximate optimization algorithm (QAOA) can optimize molecular alignments and identify optimal pharmacophore hypotheses, while the variational quantum eigensolver (VQE) can accurately calculate binding energies between target proteins and ligands. Quantum support vector machines (QSVM) can classify compounds based on their pharmacophore features, and Grover's algorithm can expedite the search for matching compounds in large chemical search space.

2.5.3 Drug Design

Quantum computing algorithms have the potential to revolutionize drug design by providing accurate and efficient solutions. For example, in retrosynthesis, the process of breaking down complex target molecules into simpler precursors to design synthetic routes, quantum algorithms can play a crucial role. Quantum approximate optimization algorithm (QAOA) can optimize reaction pathways to identify the most efficient synthetic routes, while variational quantum eigensolver (VQE) can predict reaction energies accurately and help determine the feasibility of proposed reactions. Quantum support vector machines (QSVM) can be used to classify reaction types or predict reaction outcomes based on molecular features. Grover's algorithm can also expedite the search for suitable precursors or catalysts in large chemical databases.

2.6 Lead Optimization

Lead optimization aims to further refine the lead compounds to maximize their therapeutic potential. Quantum computing can help in this process.

2.6.1 Multi-target Drug Design

QAOA and QGA can be employed to design multi-target drugs, optimizing the activity profiles of compounds against multiple targets and potentially improving their therapeutic efficacy.

2.6.2 Physicochemical Property Optimization

The molecular structure, electronic configurations, and other essential features that influence the physicochemical properties can be encoded on qubits. Quantum algorithms like VQE and QA can be applied to evaluate or optimize the physicochemical properties of lead compounds, such as solubility, lipophilicity, and stability, enhancing their overall drug-like characteristics.

2.6.3 Lead Design and Optimization

Classical generative adversarial networks (GAN) is used for generating superior lead molecules with desired chemical and physical properties and higher affinity toward binding to the receptor for a target disease; GAN has certain limitation in exploring regions of the chemical space due to the high-dimensional search space. Quantum GAN model (QGM) can use larger (~100 qubits) to explore efficient and richer representation of high-dimensional chemical space, providing a significant advantage over the classical GAN [74, 75].

2.7 ADMET (Absorption, Distribution, Metabolism, Excretion, Toxicity) Prediction

In the context of ADMET properties, quantum computing can be applied to various aspects of absorption, distribution, metabolism, excretion, and toxicity.

2.7.1 ADMET Modeling

Absorption

Molecular structure and properties affecting absorption can be encoded on a quantum circuit. Quantum machine learning algorithms, such as QuANN, can be employed for predicting membrane permeability or oral bioavailability. Quantum optimization algorithms like QAOA, VQE, or QA can be used to optimize molecular properties related to absorption.

Distribution

Quantum machine learning (QML) algorithms like QSVM [39] or QuANN can be applied for predicting volume of distribution or tissue-specific partition coefficients. Quantum optimization algorithms can be used to optimize properties related to plasma protein binding or tissue penetration.

Metabolism

The metabolite profiles can be predicted by quantum Monte Carlo, or quantum phase estimation [76] can be applied for evaluating enzyme–substrate interactions and guiding the design of compounds with improved metabolic profiles.

Excretion

Molecular structure and properties affecting excretion can be encoded on a quantum circuit. Quantum machine learning algorithms like QSVM or QuANN can be employed for predicting renal or biliary clearance. Quantum optimization algorithms like QAOA,

VQE, or QA can be used to optimize molecular properties related to excretion, guiding the design of compounds with favorable elimination profiles.

Toxicity

Molecular structure and properties affecting toxicity can be encoded on a quantum circuit. Quantum machine learning algorithms like QSVM or QuANN can be applied for predicting cytotoxicity, genotoxicity, or organ-specific toxicities. Quantum optimization algorithms like QAOA, VQE, or QA can be employed to optimize molecular properties to minimize off-target interactions and adverse effects, guiding the design of compounds with improved safety profiles. QAOA, QSVM, and QW can be used for quantum-enhanced prediction of drug–drug interactions, helping to identify and mitigate potential safety risks associated with drug combinations.

3 Summary

In this review, we have provided an extensive review of how quantum computing can be applied to early drug discovery. Despite the promises, the actual examples of using quantum computing for drug discovery is sparse which highlights the key hurdles or challenges. One of the key challenges in quantum computing is the data encoding for a given biological or scientific problem, which could also be due to the practically limited number of quantum bits (in the range of 50–100) available as of today [40].

The second major issue is the design of algorithms, selection of precise logical quantum gates, and well-educated initial ansatz. There is a potential to use AI-trained approaches that can aid in designing of quantum circuits and selection of gates and ansatz to eliminate the current bottleneck.

Finally, the lack of head-to-head benchmark studies which can show benefits of using conventional vs quantum computers is really limiting the potential users who would like to explore the quantum computing-based solution space for their problems [24]

In summary, the recent development of quantum gate-based hardware and algorithms together has potential to tackle the computationally hard problem via hybrid quantum computation.

References

1. Sension RJ (2007) Quantum path to photosynthesis. Nature 446:740–741. https://doi.org/10.1038/446740a
2. Panitchayangkoon G, Hayes D, Fransted KA et al (2010) Long-lived quantum coherence in photosynthetic complexes at physiological temperature. Proc Natl Acad Sci U S A 107: 12766–12770. https://doi.org/10.1073/PNAS.1005484107
3. Higgins JS, Lloyd LT, Sohail SH et al (2021) Photosynthesis tunes quantum-mechanical mixing of electronic and vibrational states to steer exciton energy transfer. Proc Natl Acad Sci U S A 118:e2018240118. https://doi.org/10.1073/PNAS.2018240118

4. Nielsen MA, Chuang IL (2010) Fundamental concepts. In: Quantum computation and quantum information, 10th Ann edn. Cambridge University Press, pp 1–58
5. Treinish M, Gambetta J, Thomas S, et al (2023) Qiskit/qiskit: Qiskit 0.42.1. In: https://github.com/Qiskit. https://zenodo.org/record/7757946. Accessed 25 Apr 2023
6. Claudino D (2022) The basics of quantum computing for chemists. Int J Quantum Chem 122:e26990. https://doi.org/10.1002/qua.26990
7. Cordier BA, Sawaya NPD, Guerreschi GG, Mcweeney SK (2022) Biology and medicine in the landscape of quantum advantages. J R Soc Interface 19. https://doi.org/10.1098/RSIF.2022.0541
8. Lanyon BP, Whitfield JD, Gillett GG et al (2010) Towards quantum chemistry on a quantum computer. Nat Chem 2:106–111. https://doi.org/10.1038/nchem.483
9. Arrazola JM, Bergholm V, Brádler K et al (2021) Quantum circuits with many photons on a programmable nanophotonic chip. Nature 591:54–60. https://doi.org/10.1038/S41586-021-03202-1
10. Weigold M, Barzen J, Leymann F, Salm M (2021) Encoding patterns for quantum algorithms. IET Quantum Commun 2:141–152. https://doi.org/10.1049/QTC2.12032
11. Djordjevic I (2012) Quantum circuits and quantum information processing fundamentals. In: Quantum information processing and quantum error correction: an engineering approach. Academic, Amsterdam, pp 91–117
12. Rajak A, Suzuki S, Dutta A, Chakrabarti BK (2023) Quantum annealing: an overview. Phil Trans R Soc A 381. https://doi.org/10.1098/RSTA.2021.0417
13. Born M, Fock V (1928) Beweis des Adiabatensatzes. Z Phys 51:165–180. https://doi.org/10.1007/BF01343193/METRICS
14. Domino K, Koniorczyk M, Krawiec K et al (2023) Quantum annealing in the NISQ era: railway conflict management. Entropy 25: e25020191. https://doi.org/10.3390/E25020191
15. Lechner W, Hauke P, Zoller P (2015) A quantum annealing architecture with all-to-all connectivity from local interactions. Sci Adv 1: e150083. https://doi.org/10.1126/SCIADV.1500838
16. Prokhorov LV (2008) Hamiltonian mechanics and its generalizations. Phys Part Nucl 39:810–833. https://doi.org/10.1134/S1063779608050055
17. Fedorov DA, Peng B, Govind N, Alexeev Y (2022) VQE method: a short survey and recent developments. Mater Theory 6:1–21. https://doi.org/10.1186/S41313-021-00032-6
18. Cruz PMQ, Catarina G, Gautier R, Fernandez-Rossier J (2020) Optimizing quantum phase estimation for the simulation of Hamiltonian eigenstates. Quantum Sci Technol 5:044005. https://doi.org/10.1088/2058-9565/ABAA2C
19. Wang Z, Hadfield S, Jiang Z, Rieffel EG (2018) Quantum approximate optimization algorithm for MaxCut: a fermionic view. Phys Rev A (Coll Park) 97:022304. https://doi.org/10.1103/PHYSREVA.97.022304
20. Zhou L, Wang ST, Choi S et al (2020) Quantum approximate optimization algorithm: performance, mechanism, and implementation on near-term devices. Phys Rev X 10:021067. https://doi.org/10.1103/PHYSREVX.10.021067
21. Low GH, Chuang IL (2019) Hamiltonian simulation by Qubitization. Quantum 3:163. https://doi.org/10.22331/q-2019-07-12-163
22. Boixo S, Rønnow TF, Isakov SV et al (2014) Evidence for quantum annealing with more than one hundred qubits. Nat Phys 10:218–224. https://doi.org/10.1038/nphys2900
23. Ladd TD, Jelezko F, Laflamme R et al (2010) Quantum computers. Nature 464:45–53. https://doi.org/10.1038/nature08812
24. Loss D, DiVincenzo DP (1998) Quantum computation with quantum dots. Phys Rev A (Coll Park) 57:120. https://doi.org/10.1103/PhysRevA.57.120
25. Bayerstadler A, Becquin G, Binder J et al (2021) Industry quantum computing applications. EPJ Quantum Technol 8:25. https://doi.org/10.1140/EPJQT/S40507-021-00114-X
26. Castelvecchi D (2022) The race to save the internet from quantum hackers. Nature 602: 198–201. https://doi.org/10.1038/D41586-022-00339-5
27. Castelvecchi D (2023) Are quantum computers about to break online privacy? Nature 613: 221–222. https://doi.org/10.1038/D41586-023-00017-0
28. Yin J, Li YH, Liao SK et al (2020) Entanglement-based secure quantum cryptography over 1,120 kilometres. Nature 582:501–505. https://doi.org/10.1038/s41586-020-2401-y
29. Ma H, Govoni M, Galli G (2020) Quantum simulations of materials on near-term quantum computers. NPJ Comput Mater 6:1–8.

https://doi.org/10.1038/s41524-020-00353-z

30. Liu H, Elstner M, Kaxiras E et al (2001) Quantum mechanics simulation of protein dynamics on long timescale. Proteins Struct Funct Bioinform 44:484–489. https://doi.org/10.1002/PROT.1114
31. Cheng HP, Deumens E, Freericks JK et al (2020) Application of quantum computing to biochemical systems: a look to the future. Front Chem 8:1066. https://doi.org/10.3389/FCHEM.2020.587143
32. Maheshwari D, Garcia-Zapirain B, Sierra-Sosa D (2022) Quantum machine learning applications in the biomedical domain: a systematic review. IEEE Access 10:80463–80484. https://doi.org/10.1109/ACCESS.2022.3195044
33. Outeiral C, Strahm M, Shi J et al (2021) The prospects of quantum computing in computational molecular biology. Wiley Interdiscip Rev Comput Mol Sci 11:e1481. https://doi.org/10.1002/WCMS.1481
34. Weidner FM, Schwab JD, Wölk S et al (2023) Leveraging quantum computing for dynamic analyses of logical networks in systems biology. Patterns 4:100705. https://doi.org/10.1016/J.PATTER.2023.100705
35. Lau B, Emani PS, Chapman J et al (2023) Insights from incorporating quantum computing into drug design workflows. Bioinformatics 39:btac789. https://doi.org/10.1093/bioinformatics/btac789
36. Aboussalah AM, Chi C, Lee CG (2023) Quantum computing reduces systemic risk in financial networks. Sci Rep 13:3990. https://doi.org/10.1038/s41598-023-30710-z
37. Robert A, Barkoutsos PK, Woerner S, Tavernelli I (2021) Resource-efficient quantum algorithm for protein folding. npj Quantum Inf 7:38. https://doi.org/10.1038/s41534-021-00368-4
38. Casares PAM, Campos R, Martin-Delgado MA (2022) QFold: quantum walks and deep learning to solve protein folding. Quantum Sci Technol 7:025013. https://doi.org/10.1088/2058-9565/AC4F2F
39. Ezawa M (2022) Variational quantum support vector machine based on Gamma matrix expansion and variational universal-quantum-state generator. Sci Rep 12:6758. https://doi.org/10.1038/s41598-022-10677-z
40. Fellous-Asiani M, Chai JH, Whitney RS et al (2021) Limitations in quantum computing from resource constraints. PRX Quantum 2:040335. https://doi.org/10.1103/PRXQUANTUM.2.040335
41. Ge X, Wu RB, Rabitz H (2022) The optimization landscape of hybrid quantum–classical algorithms: from quantum control to NISQ applications. Annu Rev Control 54:314–323. https://doi.org/10.1016/J.ARCONTROL.2022.06.001
42. Tilly J, Chen H, Cao S et al (2022) The variational quantum eigensolver: a review of methods and best practices. Phys Rep 986:1–128. https://doi.org/10.1016/J.PHYSREP.2022.08.003
43. Peruzzo A, McClean J, Shadbolt P et al (2014) A variational eigenvalue solver on a photonic quantum processor. Nat Commun 5:4213–4220. https://doi.org/10.1038/NCOMMS5213
44. Parrish RM, Hohenstein EG, McMahon PL, Martínez TJ (2019) Quantum computation of electronic transitions using a variational quantum eigensolver. Phys Rev Lett 122:401. https://doi.org/10.1103/PhysRevLett.122.230401
45. Mansuroglu R, Eckstein T, Nützel L et al (2023) Variational Hamiltonian simulation for translational invariant systems via classical pre-processing. Quantum Sci Technol 8. https://doi.org/10.1088/2058-9565/ACB1D0
46. Rosenbrock HH (1985) A variational principle for quantum mechanics. Phys Lett A 110:343–346. https://doi.org/10.1016/0375-9601(85)90050-7
47. Grimsley HR, Economou SE, Barnes E, Mayhall NJ (2019) An adaptive variational algorithm for exact molecular simulations on a quantum computer. Nat Commun 10:1–9. https://doi.org/10.1038/s41467-019-10988-2
48. Kandala A, Mezzacapo A, Temme K et al (2017) Hardware-efficient variational quantum eigensolver for small molecules and quantum magnets. Nature 549:242–246. https://doi.org/10.1038/NATURE23879
49. Ma H, Fan Y, Liu J, et al (2022) Divide-and-conquer variational quantum algorithms for large-scale electronic structure simulations. arXiv paper. https://doi.org/10.48550/arXiv.2208.14789
50. Sarkar A, Al-Ars Z, Almudever CG, Bertels KLM (2021) QiBAM: approximate sub-string index search on quantum accelerators applied to DNA read alignment. Electronics (Basel) 10:2433. https://doi.org/10.3390/ELECTRONICS10192433
51. Farhi E, Goldstone J, Gutmann S, Zhou L (2022) The quantum approximate optimization algorithm and the Sherrington-

Kirkpatrick model at infinite size. Quantum 6: 759. https://doi.org/10.22331/q-2022-07-07-759

52. Guerreschi GG, Matsuura AY (2019) QAOA for Max-Cut requires hundreds of qubits for quantum speed-up. Sci Rep 9:6903. https://doi.org/10.1038/s41598-019-43176-9
53. Jain N, Coyle B, Kashefi E, Kumar N (2022) Graph neural network initialisation of quantum approximate optimisation. Quantum 6:861. 10.22331/q-2022-11-17-861
54. Narayanan A, Menneer T (2000) Quantum artificial neural network architectures and components. Inf Sci (N Y) 128:231–255. https://doi.org/10.1016/S0020-0255(00)00055-4
55. Acampora G, Schiattarella R (2021) Deep neural networks for quantum circuit mapping. Neural Comput Appl 33:13723–13743. https://doi.org/10.1007/S00521-021-06009-3
56. Sagingalieva A, Kordzanganeh M, Kenbayev N, et al (2022) Hybrid quantum neural network for drug response prediction
57. Broughton M, Verdon G, McCourt T, et al (2020) TensorFlow quantum: a software framework for quantum machine learning. Arxiv paper 12:23. https://doi.org/10.48550/arXiv.2003.02989
58. Biamonte J, Wittek P, Pancotti N et al (2017) Quantum machine learning. Nature 549:195–202. https://doi.org/10.1038/nature23474
59. Banchi L, Pereira J, Pirandola S (2021) Generalization in quantum machine learning: a quantum information standpoint. PRX Quantum 2: 040321. https://doi.org/10.1103/PRXQUANTUM.2.040321/FIGURES/6/MEDIUM
60. McClean JR, Boixo S, Smelyanskiy VN et al (2018) Barren plateaus in quantum neural network training landscapes. Nat Commun 9: 4812. https://doi.org/10.1038/s41467-018-07090-4
61. Farhi E, Harrow AW (2019) Quantum supremacy through the quantum approximate optimization algorithm. Arxiv paper. https://doi.org/10.48550/arXiv.1602.07674
62. Fingerhuth M, Babej T, Ing C (2018) A quantum alternating operator ansatz with hard and soft constraints for lattice protein folding. Arxiv paper. https://doi.org/10.48550/arXiv.1810.13411
63. Barrett J, Lorenz R, Oreshkov O (2021) Cyclic quantum causal models. Nat Commun 12:1–15. https://doi.org/10.1038/s41467-020-20456-x
64. Boev AS, Rakitko AS, Usmanov SR et al (2021) Genome assembly using quantum and quantum-inspired annealing. Sci Rep 11: 13183. https://doi.org/10.1038/s41598-021-88321-5
65. de Souza LS, de Carvalho JHA, Ferreira TAE (2022) Classical artificial neural network training using quantum walks as a search procedure. IEEE Trans Comput 71:378–389. https://doi.org/10.1109/TC.2021.3051559
66. Montanaro A (2015) Quantum speedup of Monte Carlo methods. Proc R Soc A Math Phys Eng Sci 471. https://doi.org/10.1098/RSPA.2015.0301
67. Gaidai I, Babikov D, Teplukhin A, et al (2022) Molecular dynamics on quantum annealers. Sci Rep 2022 12:112:16824. https://doi.org/10.1038/s41598-022-21163-x
68. Miessen A, Ollitrault PJ, Tavernelli I (2021) Quantum algorithms for quantum dynamics: a performance study on the spin-boson model. Phys Rev Res 3:4229–4238. https://doi.org/10.1103/PhysRevResearch.3.043212
69. Fedorov DA, Otten MJ, Gray SK, Alexeev Y (2021) Ab initio molecular dynamics on quantum computers. J Chem Phys 154:164103. https://doi.org/10.1063/5.0046930/13975532
70. Kirsopp JJM, Di Paola C, Manrique DZ et al (2022) Quantum computational quantification of protein–ligand interactions. Int J Quantum Chem 122:e26975. https://doi.org/10.1002/QUA.26975
71. King AD, Raymond J, Lanting T et al (2023) Quantum critical dynamics in a 5,000-qubit programmable spin glass. Nature 2023:1–6. https://doi.org/10.1038/s41586-023-05867-2
72. Allcock J, Vangone A, Meyder A et al (2022) The prospects of Monte Carlo antibody loop modelling on a fault-tolerant quantum computer. Front Drug Discov 2:13. https://doi.org/10.3389/FDDSV.2022.908870
73. Ghamari D, Hauke P, Covino R, Faccioli P (2022) Sampling rare conformational transitions with a quantum computer. Sci Rep 12: 16336. https://doi.org/10.1038/s41598-022-20032-x
74. Li J, Topaloglu RO, Ghosh S (2021) Quantum generative models for small molecule drug discovery. IEEE Trans Quantum Eng 2:1. https://doi.org/10.1109/TQE.2021.3104804
75. Moussa C, Wang H, Araya-Polo M, et al (2023) Application of quantum-inspired generative models to small molecular datasets. Arxiv paper. https://doi.org/10.48550/arXiv.2304.10867

76. Andersson MP, Jones MN, Mikkelsen KV et al (2022) Quantum computing for chemical and biomolecular product design. Curr Opin Chem Eng 36:100754. https://doi.org/10.1016/J.COCHE.2021.100754
77. The Ising Model. https://web.stanford.edu/~jeffjar/statmech/intro4.html. Accessed 26 Apr 2023
78. Verresen R (2023) Everything is a quantum Ising model. https://doi.org/10.48550/arXiv.2301.11917
79. Arovas DP, Berg E, Kivelson SA, Raghu S (2022) The Hubbard model. Annu Rev Condens Matter Phys 13:239–274. https://doi.org/10.1146/ANNUREV-CONMATPHYS-031620-102024
80. Aksela SS, Turunen P, Kantia T et al (2011) Quantum simulation of the transverse Ising model with trapped ions. New J Phys 13:105003. https://doi.org/10.1088/1367-2630/13/10/105003

Chapter 8

Edge, Fog, and Cloud Against Disease: The Potential of High-Performance Cloud Computing for Pharma Drug Discovery

Bhushan Bonde

Abstract

The high-performance computing (HPC) platform for large-scale drug discovery simulation demands significant investment in speciality hardware, maintenance, resource management, and running costs. The rapid growth in computing hardware has made it possible to provide cost-effective, robust, secure, and scalable alternatives to the on-premise (on-prem) HPC via Cloud, Fog, and Edge computing. It has enabled recent state-of-the-art machine learning (ML) and artificial intelligence (AI)-based tools for drug discovery, such as BERT, BARD, AlphaFold2, and GPT. This chapter attempts to overview types of software architectures for developing scientific software or application with deployment agnostic (on-prem to cloud and hybrid) use cases. Furthermore, the chapter aims to outline how the innovation is disrupting the orthodox mindset of monolithic software running on on-prem HPC and provide the paradigm shift landscape to microservices driven application programming (API) and message parsing interface (MPI)-based scientific computing across the distributed, high-available infrastructure. This is coupled with agile DevOps, and good coding practices, low code and no-code application development frameworks for cost-efficient, secure, automated, and robust scientific application life cycle management.

Key words Cloud computing, Edge computing, Fog computing, Scientific applications, High-performance computing (HPC), DevOps, Scientific DevOps, Scalable computing, Cloud-batch computing, Grid and cluster computing, Containerization, Docker, Podman, Singularity, Swarm and Kubernetes for Scientific computing applications

1 Introduction

High-performance computing in drug discovery dates back to the 1980s, when the race to generate the best-in-class novel small molecules on mainframe computers began [1]. Since the mid-twentieth century, software development has been driven mainly by choice of hardware and programming languages suitable to run on those resources and in line with Moore's law [2] which states "the number of processors (and hence the speed of processing the floating points) doubles almost every two years in the

Alexander Heifetz (ed.), *High Performance Computing for Drug Discovery and Biomedicine*, Methods in Molecular Biology, vol. 2716, https://doi.org/10.1007/978-1-0716-3449-3_8,

integrated circuit," giving a significant boost to use high-performance computing in the race to get better novel drug candidates to the market. The plethora of available software used in drug discovery is beyond the scope of this chapter. Therefore, I will limit the content to the high-performance computing (HPC) based scientific software landscape in the early and late phases of drug discovery. The aim is to systematically review the software applications that could benefit from the next-generation high-performance computing infrastructures such as Cloud, Fog, and Edge computing to accelerate drug discovery [3].

1.1 Scientific Application User Persona in Drug Discovery

Let us first understand the high-performance computing (HPC)-based end-user personas or key stakeholders of the scientific software users across drug discovery.

Table 1 provides a detailed overview of those personas from the early to late phase stages of drug discovery. The use of scientific software in medicinal chemistry methods such as molecular modeling, structure-based drug design, structure-based virtual screening, ligand-based modeling and molecular dynamics (MD), fragment molecular orbital (FMO), and free energy perturbation (FEP) [4] are gaining more popularity due to reduction in the cost of computer hardware, while other areas such as pharmacokinetic, pharmacodynamic, and structural activity relationship of ligands with its target-binding activity and toxicity prediction are also gaining popularity due to approval by regulatory authorities [5].

The key three categories of scientific software include the following:

1. Commercially available tools/platforms that could be purchased commercially from various software vendors, e.g., MATLAB and Schrodinger docking suite.
2. Open-source platform or software, e.g., Python, R (Bioconductor), Perl, Java (OpenJDK), GNU C/C++, Fortran, Go, shell scripts (Bash, csh), and standalone binary executable programs, e.g., BLAST Sequence Alignment tool.
3. Closed source or internally developed platforms or software comprising a combination of scripts or binary executables. This mostly includes use of the abovementioned two types to derive a new software tool or library.

1.2 Common Types of Scientific App

Table 2 provides the classification of the scientific applications:

- Command line application, e.g., Sequence Alignment tools (BLAST/CLUSTALW2)
- Standalone graphical user applications, e.g., PyMOL
- Web application:

Table 1
Key personas/end users of HPC in drug discovery

	Persona/end user	Drug discovery lab	HPC use
1	Lab/bench scientist smart lab/robotic platform scientist	Biology/chemistry/ pharmacology lab scientist	The lab scientist uses software for their day-to-day research and analysis. This includes histopathology assay imaging technologies for cytometry and binding assay kinetics.
2	x-Informatics	Bioinformatics, cheminformatics, structural biology	Bioinformatics, cheminformatics, and structural biologist use specialty tools for their day-to-day analysis, e.g., DNA/RNA sequence alignment. Solve the structure of a crystal structure of a protein. Modeling QSAR/QSPR.
3	Mathematical modeller	Computational biology, systems biology, statistical genetics	Computational biology deals with use of various machine learning tools and novel algorithms for analyzing complex biological data and statistical and meta-analysis. Systems biology involves large simulation of complex biological systems.
4	Bio/clinical statisticians	Systems pharmacology, DMPK, clinical (trial) biostatistician	The drug discovery involves design of clinical analysis DMPK analysis to select drug dosage prediction and pharmacology/ toxicity prediction. Simulations of kinetic and dynamic systems, which are represented by ordinary differential equations (ODE) and partial differential equations (PDE), are computationally demanding due to their complex nature need HPC.
5	Translational data scientist/precision medicine	Translational data scientist/business analysis	Use of machine learning/data analytics solutions. Identify insights for data-driven advanced analytics.
6	AI practitioner	Deep learning model building	Building machine learning and deep learning models from the biological/imaging/ chemistry data.
7	HPC DevOps engineer	Deployment of apps at scale	UI developer for apps, deployment of apps with DevOps, AutoML, etc. Need to work with Information Technology (Infrastructure, Network, Security, Compliance)

  - Client-Server web frameworks (e.g., Flask/Django or nodejs)
  - Micro applications (e.g., R-shiny or Dash/Streamlit)
- Web Interactive application: Jupyter, RStudio-Server

Table 2
Classification of the scientific applications

	Type	Description	Pro	Cons
1	Command-line interface (CLI)	Command-line interface (CLI)-based scientific tools are applications that are executed using text-based commands entered into a terminal window or shell. These tools have been used extensively in scientific research and are often preferred by experienced researchers due to their flexibility and efficiency	Flexibility, often have a small footprint, easier to reproduce research results, highly customizable, very easy maintainability	Learn complex command line syntax and workflows, lack of interactivity, difficult to visually explore data, less user-friendly, sometimes follow tedious installation steps
2	Standalone graphical user interface (GUI)	Standalone GUI applications are self-contained and can be run independently on a computer without requiring additional software or libraries. They are often used for scientific applications such as data analysis and visualization, as they can provide a more intuitive and user-friendly experience than CLI-based tools	User-friendly, interactive visualization, error prevention by providing input validation, easy to install	Limited flexibility, large footprint, limited reproducibility, automation challenges, development complexity, expensive solution
3	Client-server application	A software program is divided into two parts: a client and a server. The client is the user interface that interacts with the user, while the server provides the underlying functionality and data processing capabilities	Applications can be scaled easily, centralize data management, maintained centrally on the server	Applications can be slower than standalone applications, require an active network connection, more complex to develop and maintain, single point of failure

(continued)

Table 2 (continued)

Type	Description	Pro	Cons
4 Web-based interactive computing platform	A web-based interactive computing platform, such as Jupyter Notebook, Rshiny, is a type of software application that allows users to create, share, and collaborate on documents that combine code, text, and visualizations. These documents are often referred to as notebooks, and they are created and edited through a web browser	High accessibility, easy to reproduce research, allow multiple users to collaborate, interactive visualization tools	Limited offline capability, can be vulnerable to security threats, can be less responsible at times depending on the network connectivity
5 Integrated development/ platform	Software application that provides a comprehensive set of tools and features to support software development, testing, and debugging. Simplifies development process by providing a single interface for managing code, compiling programs, and debugging errors	Improve productivity, more flexible, easy integration, highly customizable	Steep learning curve, resource-intensive, expensive
6 Microservices-driven full stack application	Client-server applications that are built using a combination of microservices. The application is broken down into smaller, independently deployable services that can be developed, tested, and deployed separately from each other	Applications are modular, highly scalable, more resilient, gives the flexibility to choose the best tools for a particular job, seamless integration with other services/ platforms	Require an active network connection, more complex to develop and maintain

- Integrated Development/platform: Rstudio/MATLAB/IDLE
- Full Stack development app: Full UI/UX frontend to backend with/without API
- Microservices-based full stack (with API)
- Cloud versus Fog/Edge apps: runs on the devices at the Edge or Cloud (e.g., Google Colaboratory)

2 High-Performance Computing Overview

2.1 Best Practices for HPC Scientific Applications

The drug discovery and development process is a highly regulated process with quality control (QC) and quality assurance (QA). When software is used for drug discovery and development, the same can be applied, and there is a need for good coding practices (GCP) for scientific software development:

- Use a version control system for everything from code to container images to (automate) deployment (e.g., Git with GitHub, GitLab, Azure DevOps [6]).
- Let the computer do the work and write code with meta-programming and meta-classes.
- Use a build tool to automate workflows, e.g., continuous integration (CI) and continuous deployment (CD).
- Modularize code and code repurpose, do not repeat yourself or others.
- Test, Test...Testing with test-driven development and use assertions to check the program operations and turn bugs into test cases.
- Document design and purpose (plans but not mechanics), use document automation tools and markdowns, and write "how-to" user guide(s) for end users, document the Application-Programming Interfaces (API) for other developers.
- Refactor code in preference to explaining how it works.
- Collaborate, write programs for people and for not computers, and do periodic code reviews for improvement of the complex codebase.
- "Security first, Optimization and Scaling second" considerations. (There should not be any exception or compromise for security concerns even for on-prem software development.)
- Another good coding guideline practice is called Zen of Python [7], which provides 19 excellent guidelines for pythonic software development but can be extended to most of the high-level programming languages.

To enable the scientific software on HPC, one needs to understand the complexity of dependency in such software. Many software needs other third-party libraries for their runtime. Managing the libraries with compatibility matrix of, e.g., Python, GPU drivers, TensorFlow, NVidia CUDA, mathematical GNU C/C++ based libraries and lifecycle management of those could be one of the biggest issues in reproducing the results in HPC. The following tools help solve these challenges and aid in development and deployment of platform agnostics scientific software.

- A *kernel module system* is a code that can be loaded (and unloaded) into the kernel on-demand, simplifying the use of different software (versions) in a precise and controlled manner [8]. The modules get loaded without the need for a system reboot. One downside of the kernel module is that they need admin access for installation on the HPC nodes.
- *Spack* is a better kernel module package management tool designed to support multiple versions and configurations of software on a wide variety of platforms and environments enable versioned libs and reproducibility across the HPC nodes [9]. The key advantage of Spack over the native kernel modules is that a user can develop, install, and load/unload kernel modules eliminating the admin dependency.
- *Containers*

  Containers as the name suggests are analogous to the shipment of goods (e.g., a piece of code), and in software development, a container packs all the code and its dependencies and provides an isolated virtual runtime environment to run that code.

  The three most common containers used in HPC are *Docker, Podman, and Singularity.*

  Docker is a platform to build, share, ship, and run modern applications and is popular across the wider software development including drug discovery [10–13].

  Podman is the in-place replacement of Docker developed by RedHat Inc. as an open-source project, with native support for Kubernetes (*see* Subheading 2.2) [14, 15].

  Singularity was specifically created as open-source project at Lawrence Berkeley National Laboratory (LBNL) to run complex applications on HPC clusters with security, code isolation, and reproducibility as key objectives [16].

  Podman and Singularity natively allow running rootless containers making the container runtime more secure in the standalone server to Cloud node deployment.

  Docker Compose is a tool for defining and running multi-container Docker applications. It is used to orchestrate and manage Docker containers that work together as a single application [11, 17, 18].

Docker persistent storage [18] refers to the ability to store and retrieve data from a container beyond the lifetime of that container. By default, when a Docker container is stopped or deleted, any data stored within it is also deleted. However, in some cases, it is necessary to persist data across multiple container instances or even across multiple hosts. Docker can mount folders as volume and other file storage, such as support for distributed storage systems or cloud storage providers and enables scalable and fault-tolerant storage solutions for Docker applications.

2.2 Container Orchestration

The most common and popular platforms are Docker Compose, Docker Swarm, and Kubernetes.

Docker Swarm is a Docker native clustering and scheduling tool for Docker containers. Swarm mode allows the use of multiple Docker hosts to act as a swarm cluster, assign certain node (s) as manager and worker node; the manager node then manages membership and delegation, and worker nodes run the swarm services [19].

Docker Compose is a lightweight platform of choice for deploying containerized microservices [20–22] for smaller projects with few microservices or proof of concept work.

Kubernetes (Greek meaning: pilot or helmsman) is an open-source, portable, extensible container orchestration platform for managing containerized workloads and microservices. The platform is highly scalable with features such as autoscaling that facilitates both declarative configuration and automation. Kubernetes services, support, and tools are commonly available from various cloud providers as well as it is suitable for on-prem deployment [23].

2.3 Type of Infrastructure Based on Scientific App Deployment Paradigms

There are several paradigms for deploying the scientific applications. The best deployment strategy depends on several factors, such as the size and complexity of the application, the availability of resources, and the budget. Here are some examples of how these paradigms can be used in drug discovery applications.

2.3.1 Standalone Applications

These applications are installed locally on a user's computer, run independently of any network or server, and typically require fewer resources. For example, PyMOL is a molecular visualization tool that allows users to visualize and analyze 3D protein structures.

2.3.2 HPC Computing Involves Cluster and Grid Computing

- Cluster Computing: Custer is a local network of two or more homogeneous computers. A computation process on such a computer network, i.e., cluster is called cluster computing.
- Grid Computing: Grid computing is a network of homogeneous or heterogeneous computers working together over a long distance to perform a task that would rather be difficult for a single machine.

This is a good option for large and complex applications that require considerable resources. High-performance computing (HPC) clusters comprise many interconnected nodes, each with multiple processors and a large amount of memory. This makes HPC clusters highly efficient and scalable, ideal for tasks such as molecular dynamics simulations and virtual screening of large compound libraries. Some examples of software that can be used in a HPC environment include GROMACS for molecular dynamics simulations and Autodock for virtual screening.

2.3.3 Fat Node High-Performance Computing

Fat node is a hypervisor-based isolation computing with high-end, scalable CPU-GPU-RAM, making it highly efficient in processing large volume of numerical data. For example, HP Superdome X blade/server provides massively large (scalable processors with up to 892 cores, up to 48 TB of shared memory), highly flexible, and unbounded I/O file handling [24] making it attractive fat node-based supercomputing server node.

Fat nodes are used in the following:

- Big-data analytics, the single-system compute power and shared memory of the platform, along with the system's capacity to accelerate data ingest, enable large (~GB to TB) data to be loaded to GPU/RAM to unlock deeper insights without much code optimization.
- Computer-aided engineering, with advanced simulation technology, HPE Superdome Flex has been instrumental in improving the design from engineering designs of turbines of aircraft components to drug discovery simulations.
- Genomics and -omics data sets, such as Genome alignment, RNA-Seq/ChIP-Seq, and gene/protein expression, help predict, diagnose, and treat disease.

2.3.4 Cloud Computing

The cloud computing paradigm refers to delivering computing services over the internet and providing users with an on-demand access to scalable computing resources. Cloud computing is well suited for applications that do not require real-time data processing, such as batch processing or data analysis [25]. These applications can be run on the cloud at any time, taking advantage of the computing resources that are available. Another benefit of cloud computing is that it provides users with the flexibility to increase or decrease the amount of computing resources they use as their workload changes [26]. This allows users to pay only for the resources they need and avoid the costs associated with maintaining and upgrading their own infrastructure. However, a reliable internet connection is a crucial requirement for cloud computing, as applications are typically accessed over the internet. Examples of cloud-based drug discovery tools include ChemAxon's JChem for chemical structure handling and KNIME for data analysis and workflow management and COVID-19 Moonshot project [27].

2.3.5 *Fog Computing*

Fog devices are typically located at the edge of the network, which makes them well suited for applications that require low latency. This can be useful for applications that require real-time analysis of data, such as drug discovery workflows that involve high-throughput screening. Fog computing could be used to pre-process the data generated by tools such as Plate Mapper, allowing for real-time analysis and visualization of the results. Fog devices, such as edge servers or IoT gateways, could be placed in the lab near the screening equipment to perform data preprocessing and analysis, reducing the need for data to be sent back to a central server for processing [28–30].

2.3.6 *Edge Computing*

Edge computing is a distributed computing paradigm in which data processing and analysis is performed at the edge of the network, closer to where data is generated. By performing data processing and analysis at the edge, it significantly reduces the time required to process and act on data, making it well suited for applications that require real-time responses, such as industrial control systems or autonomous vehicles. Another advantage of edge computing is its ability to improve the reliability and accuracy of data readings. By performing error correction and filtering at the edge, it can help to identify and correct errors in data readings, ensuring that data is accurate and reliable. In the remote locations with limited or no internet connectivity, edge computing can be useful, enabling real-time data processing and analysis at the edge device. By performing data processing and analysis at the edge of the network, edge computing can reduce the need for data to be transmitted to a central server, reducing bandwidth requirements, and enabling applications to operate in low or no connectivity environments [25, 29, 31].

Figure 1 illustrates the integration of fog nodes, edge devices, cloud nodes, and on-prem servers in a unified network to enable efficient drug discovery applications. This interconnected system leverages the advantages of each component to optimize data processing, storage, and analysis [32].

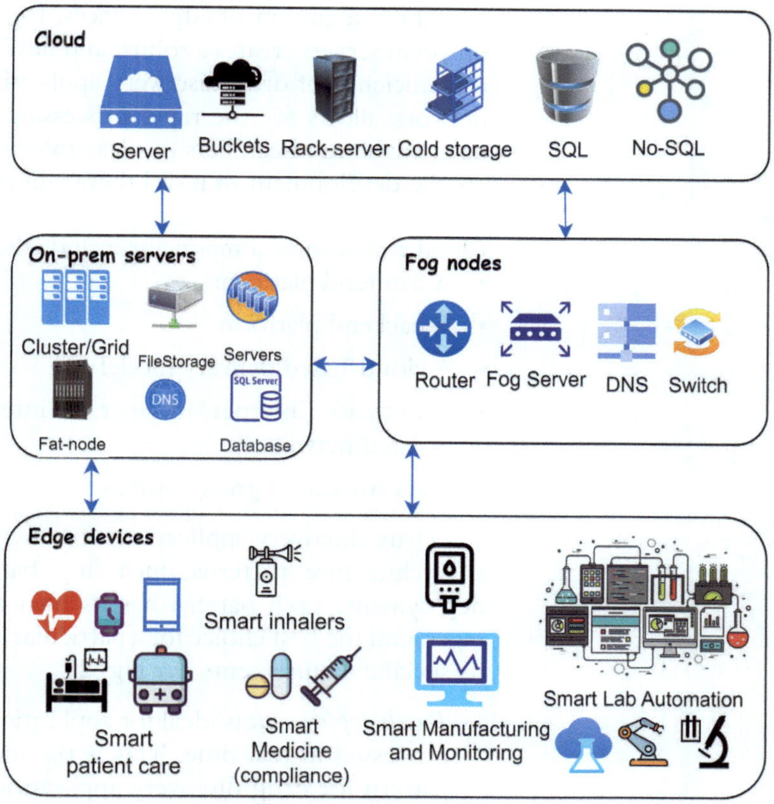

Fig. 1 On-Premise-Cloud-Fog-Edge overview

Edge devices, such as IoT sensors and portable equipment, play a crucial role in the initial data collection process. These devices generate real-time data from drug discovery experiments and transmit this information to the nearest fog nodes. The fog nodes, located at an intermediate level between edge devices and cloud nodes, perform low-latency processing, filtering, and preliminary analysis of the data. This local processing capability allows for faster response times, decreased network congestion, and improved data privacy [33].

Cloud nodes provide scalable computational resources and storage for more complex and resource-intensive tasks, such as molecular modeling, machine learning, and data analysis. They offer virtually unlimited capacity to handle the vast amounts of data generated during drug discovery processes and allow for collaboration between researchers across different geographical locations.

On-prem servers play a vital role in secure data storage, processing, and management. These servers are often used to store sensitive information, such as intellectual property and proprietary algorithms, while ensuring compliance with data protection regulations.

The integration of edge devices, fog nodes, cloud nodes, and on-prem servers creates a robust and flexible system that maximizes the efficiency of drug discovery applications. This interconnected network allows for the rapid processing, storage, and analysis of data, enabling researchers to make informed decisions and accelerate the development of novel therapeutics [34, 35].

Cloud architecture components include the following:

- A frontend platform
- A backend platform
- A cloud-based delivery model
- A network (internet, intranet, or intercloud) or virtual private isolated network
- Data storage (ingress/egress)

Drug discovery applications are often deployed using a variety of architecture patterns, including batch, real-time, and edge deployment. Each pattern has its own advantages and disadvantages, and the best choice for a particular application will depend on its specific requirements (*see* Fig. 2).

– *Batch deployment* is ideal for applications that do not need to be processed in real time. This is the most common deployment pattern for drug discovery applications, as it is the most cost-effective and scalable.

– *Real-time deployment* architecture is well suited for applications that require immediate responses, such as molecular dynamics simulations or real-time data analysis. By analyzing data in the real time, researchers can quickly identify and respond to changes, making it easier to optimize drug candidates or make critical decisions.

– *Edge deployment* is useful for applications that need to be processed close to the data source especially in distributed devices (e.g., smart devices, Lab of Future). This is a more recent deployment pattern, but it is becoming increasingly popular as the cost of edge devices decreases.

2.4 API Types for Communication in Scientific Apps

An API (Application Programming Interface) is a set of protocols, routines, and tools that allow software programs to communicate with each other. APIs share data and functionality to extend or build new applications.

APIs work by defining a set of methods or functions that can be called by one application from another. Each method has a specific name and parameters and returns the output.

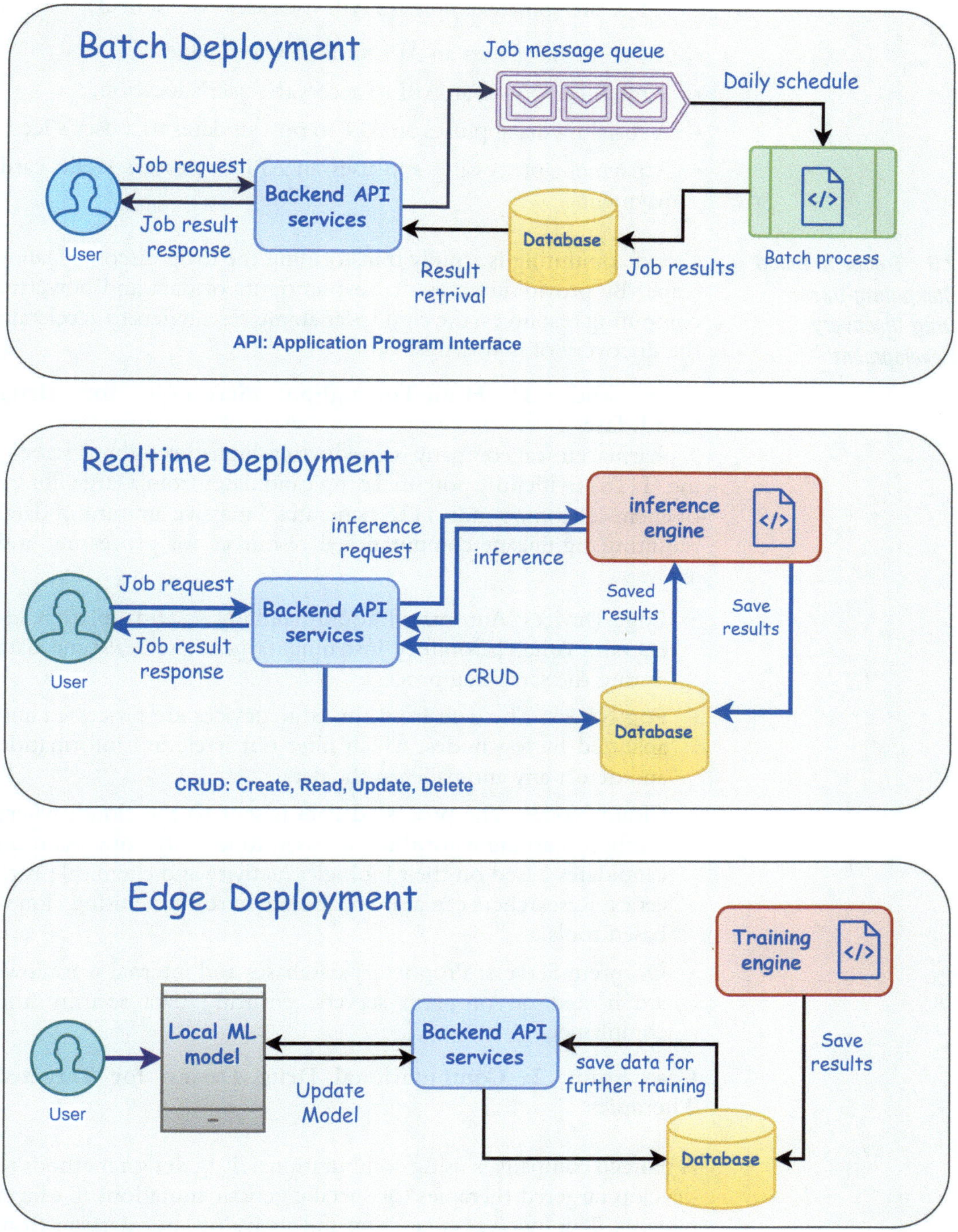

Fig. 2 Scientific application deployment architecture patterns

APIs allow developers to create applications abstracting functionalities from the existing application without plunging into the inner working of other applications.

Here are some examples of API use cases (*see* Table 3):

- A web browser uses an API to display web pages.
- A mobile app uses an API to access the user's location.
- A social media app uses an API to post updates to a user's feed.
- A payment processing app uses an API to process credit card payments.

2.5 Trends in Cloud Computing-Based Drug Discovery Development

Cloud computing is rapidly transforming the drug discovery landscape. By providing access to vast amounts of data and powerful computing resources, the cloud is enabling researchers to accelerate the discovery of new drugs.

Case Study 1: High-Throughput Screening for Drug Candidates

A pharmaceutical company is conducting high-throughput screening (HTS) to identify potential drug candidates from a large library of chemical compounds. HTS generates a massive amount of data, requiring significant computational resources for processing and analysis.

- Edge Devices: Automated liquid handling systems, plate readers, and other laboratory instruments generate real-time data during the screening process.
- Fog Nodes: The data from the edge devices are processed and analyzed by fog nodes, which filter out irrelevant information and detect any anomalies in the data.
- Cloud Nodes: The processed data is sent to the cloud, where machine learning algorithms are used to identify potential drug candidates based on their biological activity and chemical properties. Researchers can also collaborate in real time using cloud-based tools.
- On-prem Servers: Proprietary databases and internal workflows are hosted on on-prem servers, ensuring data security and compliance.

Case Study 2: Computational Drug Design for Targeted Therapies

A biotech company is using computational drug design methods to develop targeted therapies for specific genetic mutations in cancer patients. This involves generating and analyzing large datasets from genomic data, molecular simulations, and machine learning models.

- Edge Devices: Sequencing machines and other laboratory equipment generate genomic data and structural information for target proteins and candidate molecules.

Table 3
Overview of Application Programming Interface (API) used in scientific computing

	API	Definition	Pros	Cons
1	SOAP	Simple Objects Access Protocol	Using http-based XML based communication Well adapted to enterprise-wide applications	Network bandwidth usage is higher due to XML complicated to use compared to the other protocols
2	RESTful	Representational State Transfer (REST)	Support all data types, faster, works on all common web browsers Simple and manage load efficiently, easy to deploy and discover Server-driven architecture	Lack of statefulness: the request does not keep record in memory from previous request Lack security and needs added layer to protect
3	GraphQL	GraphQL is a query language, uses server-side runtime and it provides exact data requested by client	Unlike REST, its client-driven architecture Network performance (faster than REST), suitable for hierarchical data Allows loose coupling between client and server, strong typing and developer tooling	GraphQL query could be complex Difficult to use caching Rate limiting as one can does not define the limit on number of records to return per query
4	gRPC	Google Remote Procedure Call (gRPC) is a remote procedure call (RPC) framework, brings performance benefits and modern features to client-server applications	Lightweight, event-driven architecture High-performance, built-in code generation allows data streaming Uses HTTP/2 protocol and allows multiple parallel requests	Limited browser support Lack of maturity Steep learning curve Non-human readable (binary) data format
5	Web-Socket	Web-Socket enables two-way communication between a client and a server	Web-Socket allows two-way communication, sends faster, and receives than http/AJAX, communication cross platform (mobile, web, server)	Only compatible with HTML 5 compliant web browsers. No intermediate caching like
6	Webhook	A webhook is an event driven (also known as "reverse APIs" which are request driven)	Asynchronous, real time data transfer in response to event Suitable for automation processes	Limited to only receiving data, and not for bi-directional data transfer use Unreliable and insecure due to dependency on availability and performance of the publisher and the subscriber

- Fog Nodes: Fog nodes pre-process the data, including alignment, quality control, and feature extraction, before sending it to the cloud or on-prem servers for further analysis.
- Cloud Nodes: Advanced computational resources in the cloud are used to run molecular dynamics simulations, virtual screening, and machine learning models to predict drug-target interactions and optimize candidate molecules.
- On-prem Servers: Sensitive patient data and proprietary algorithms are hosted on on-prem servers, ensuring data security and compliance with regulations.

Here are some of the recent trends in cloud computing for drug discovery (*See* Table 4).

1. Increased use of artificial intelligence (AI) and machine learning (ML): AI and machine learning are already being used to significant effect in drug discovery, and their use is only going to grow in the future. These technologies can be used to analyze large genomic, proteomic, and other "omics" datasets to identify potential drug targets and design new drugs.
2. Enabling collaborations across academia and industry: Cloud computing is making it easier for researchers from academia and industry to collaborate on drug discovery projects. This is leading to a more efficient and effective drug discovery process [36].
3. Democratization of HPC: In the past, accessing high-performance computing involved investing in costly hardware, but with cloud computing, even startups can access the cutting-edge technologies in a cost-efficient way. Google cloud platform provides free Google Colaboratory, a decent compute node with Jupyter interactive programming and access to 12 GB of RAM with 100 GB fast I/O storage disk, and allows selection of CPU (4 core), one GPU (k80) or 8 TPUs, with amazing 180 teraflops of computational power.
4. Ultra large virtual screening: Cloud computing allows pharmaceutical companies to perform the virtual screening of millions of compounds quickly and efficiently [37]. By using virtual screening, researchers can identify potential drug candidates more quickly and accurately, reducing the time and cost of drug discovery [38, 39].
5. Cloud-based drug development platforms: Cloud-based drug development platforms have emerged, providing end-to-end drug discovery and development solutions. These platforms provide various services, including data analysis, virtual screening [37], and drug design, all in a cloud-based environment [40].

Table 4
Comparison of a computing platform: HPC-driven drug discovery

Characteristic	Cluster computing	Grid computing	Cloud computing	Fog computing	Edge computing
Definition	Group of computers working together to solve a problem or perform a task	Collection of distributed computing resources connected via a network to share resources	On-demand delivery of computing resources, services, and applications via the internet	Extension of cloud computing to the edge of a network, providing localized processing and storage	Decentralized computing infrastructure where data processing happens at or near the data source
Topology	Tightly coupled	Loosely coupled	Centralized	Distributed	Decentralized
Location	Commonly, in a single location	Commonly, in multiple locations	Commonly, in the cloud	Commonly, at the edge of the network	Commonly, at the edge of the network
Scale	Small to medium	Large	Large	Small to medium	Small
Data processing & latency	Process data within the cluster	Process data across multiple sites and hence greater latency	Process data in the cloud data centers	Process data close to the data source	Process data at or near the data source
Ownership & accountability	Single organization	Multiple organizations	Cloud service providers	Mix of cloud providers, private owners, and third-party providers	Device owners or IoT ecosystem
Cost & pricing	Fixed cost based on hardware and software	Variable cost based on resource usage	Pay-as-you-go or subscription based	Variable cost based on resource usage	Variable cost based on resource usage
Use cases	High-performance computing, parallel processing, molecular dynamics simulations	Collaborative research, large-scale data processing, sharing drug discovery resources	Web applications, data storage, data analytics, virtual screening, machine learning for drug discovery	Real-time analytics and monitoring of clinical trials	Remote monitoring of clinical trials
Security risks	Limited exposure to external threats	Possible exposure to external threats due to multiple sites	Depends on the cloud security configurations	Enhanced security due to localized processing	Enhanced security due to localized processing

6. Increased use of cloud computing in early-stage drug discovery: Cloud computing is being increasingly used in early-stage drug discovery. This is because it can help researchers quickly and efficiently screen large numbers of compounds for potential therapeutic activity [41].

2.6 Ethical Issues in Drug Discovery Using Cloud Computing

Several ethical issues may arise in drug discovery using cloud computing. Some of the most significant ethical concerns are as follows:

1. Data Privacy and Security: Since cloud computing involves storing and processing large amounts of data in remote servers, pharmaceutical companies must ensure that they use secure cloud infrastructure and have appropriate data access and usage policies to protect patient privacy [42].
2. Access to Data: Cloud computing can facilitate access to large amounts of data, but there may be ethical concerns about how that data is obtained and processed. For example, pharmaceutical companies may obtain data from patient electronic health records (EHRs) or social media platforms, raising questions about informed consent and privacy.
3. Intellectual Property: When data is shared among multiple parties through cloud computing, there may be concerns about protecting intellectual property. Pharmaceutical companies must have formal agreements to safeguard their intellectual property rights and prevent unauthorized use of their data.
4. Transparency and Accountability: Using cloud computing in drug discovery may raise concerns about transparency and accountability. Pharmaceutical companies must ensure that they are transparent about their data usage policies and accountable for their research's ethical implications.

3 Summary

Scientific software could be complex, may use dependency on other libraries, and often contains thousands of lines of code. Management of the large code base is needed for bug fix, collaborated development, and reproducibility (the capability to produce exactly the same inference or result when the code is executed repeatedly) in a heterogeneous environment [43]. Using the source version control (e.g., Git, GitHub, GitLab [44], and Azure DevOps [6]) solve this issue, and in addition help maintain project documentation, execution manual, code documentation, developer guidelines for API and services examples for large microservice-based projects [44].

The large complex scientific software uses collaborative software development, multiple iteration, code testing, and automated deployment. Maintainability of scientific software is also very important in the software lifecycle management and can consume 50–75% of project's cost and time. To keep the maintenance costs as low as possible, good coding practices, version control, and automated deployment should be used. In practice, a multidisciplinary team is required to take input from scientists with different expertise (e.g., mathematicians, biologists, chemists, scientists) to develop drug discovery software [45].

Scientific software often needs numerical and mathematical routines to crunch numbers. The rule of thumb is "do not repeat yourself but reuse" well-tested mathematical and statistical standard libraries to increase productivity [46].

Finally, the most obvious question is: how expensive is it for one to migrate from on-prem to cloud? This really depends on type of data transfer, e.g., data ingress cost which accounts for data upload to cloud and egress cost where large data is get downloaded from cloud storage; type of computing (CPU, GPU, TPU, etc); and other services needed to run the project in the cloud with security, e.g., logically isolated virtual private cloud (VPC) network to allow the secure deployment of services in the cloud.

The cloud providers help provide an estimated cost for the use of cloud components; however, careful architecture design of the implementation could provide better cost optimization [47]. The cost benefit of cloud use is a difficult problem which could be subjective to individual use cases, however, with careful deployment consideration, DevOps optimization and automation, and security considerations, one can achieve an optimal balance. Most of the scientific applications are either suitable for cloud-batch computing or deployment as microservices and hence should be easy to migrate with minimal intervention on the cloud [6].

The final and most critical question is: how can we deploy services most safely and securely on the cloud? The security implications should not be ignored even for the on-prem deployment of scientific applications. Security first mindset [48] is a must for next-generation software development and deployment, and it should be made compulsory for all software projects in the organization.

Acknowledgments

I am grateful to my PhD student, Pratik Patil, for discussion on cloud deployment architecture, good coding practice, and API development, Vedant Bonde for assistance with proofreading and improving the readability of the manuscript significantly.

References

1. Saharan VA, Banerjee S, Penuli S, Dobhal S (2022) History and present scenario of computers in pharmaceutical research and development. Comput Aided Pharm Drug Deliv 1–38. https://doi.org/10.1007/978-981-16-5180-9_1
2. Schaller RR (1997) Moore's law: past, present, and future. IEEE Spectr 34(52–55):57. https://doi.org/10.1109/6.591665
3. Grannan A, Sood K, Norris B, Dubey A (2020) Understanding the landscape of scientific software used on high-performance computing platforms. Int J High Perform Comput Appl 34:465–477. https://doi.org/10.1177/1094342019899451/
4. Jamkhande PG, Ghante MH, Ajgunde BR (2017) Software based approaches for drug designing and development: a systematic review on commonly used software and its applications. Bull Fac Pharm Cairo Univ 55: 203–210. https://doi.org/10.1016/J.BFOPCU.2017.10.001
5. Badwan BA, Liaropoulos G, Kyrodimos E et al (2023) Machine learning approaches to predict drug efficacy and toxicity in oncology. Cell Rep Methods 3:100143. https://doi.org/10.1016/J.CRMETH.2023.100413
6. Pauli W (2019) Breaking the wall between data scientists and app developers with Azure DevOps | Azure Blog and Updates | Microsoft Azure. https://azure.microsoft.com/en-gb/blog/breaking-the-wall-between-data-scientists-and-app-developers-with-azure-devops/. Accessed 28 Apr 2023
7. Peters T (2022) PEP 20 – the Zen of python. https://peps.python.org/pep-0020/. Accessed 3 May 2023
8. Salzman PJ, Burian M, Pomerantz O et al (2023) The Linux kernel module programming guide. https://sysprog21.github.io/lkmpg/. Accessed 28 Apr 2023
9. Gamblin T, Legendre M, Collette MR et al (2015) The Spack package manager: bringing order to HPC software chaos. In: International conference for high performance computing, networking, storage and analysis, SC, 15–20, November 2015. https://doi.org/10.1145/2807591.2807623
10. Saha P, Uminski P, Beltre A, Govindaraju M (2018) Evaluation of Docker containers for scientific workloads in the cloud. ACM Int Conf Proc Ser. https://doi.org/10.1145/3219104.3229280
11. List M (2017) Using Docker compose for the simple deployment of an integrated drug target screening platform. J Integr Bioinform 14: 20170016. https://doi.org/10.1515/JIB-2017-0016
12. Kononowicz T, Czarnul P (2022) Performance assessment of using Docker for selected MPI applications in a parallel environment based on commodity hardware. Appl Sci 12:8305. https://doi.org/10.3390/APP12168305
13. Novella JA, Khoonsari PE, Herman S et al (2019) Container-based bioinformatics with pachyderm. Bioinformatics 35:839–846. https://doi.org/10.1093/BIOINFORMATICS/BTY699
14. Stephey L, Younge A, Fulton D et al (2023) HPC containers at scale using Podman. https://opensource.com/article/23/1/hpc-containers-scale-using-podman. Accessed 2 May 2023
15. Gantikow H, Walter S, Reich C (2020) Rootless containers with Podman for HPC, Lecture notes in computer science (including subseries lecture notes in artificial intelligence and lecture notes in bioinformatics) LNCS 12321, pp 343–354. https://doi.org/10.1007/978-3-030-59851-8_23/COVER
16. Kurtzer GM, Sochat V, Bauer MW (2017) Singularity: scientific containers for mobility of compute. PLoS One 12:e0177459. https://doi.org/10.1371/JOURNAL.PONE.0177459
17. Gupta A (2020) Reverie labs: scaling drug development with containerized machine learning | AWS Startups Blog. https://aws.amazon.com/blogs/startups/reverie-labs-scaling-drug-development-with-containerized-machine-learning/. Accessed 28 Apr 2023
18. Tarasov V, Rupprecht L, Skourtis D et al (2019) Evaluating Docker storage performance: from workloads to graph drivers. Clust Comput 22:1159–1172. https://doi.org/10.1007/S10586-018-02893-Y/METRICS
19. Farshteindiker A, Puzis R (2021) Leadership hijacking in Docker swarm and its consequences. Entropy 23. https://doi.org/10.3390/E23070914
20. Saboor A, Hassan MF, Akbar R et al (2022) Containerized microservices orchestration and provisioning in cloud computing: a conceptual framework and future perspectives. Appl Sci 12:5793. https://doi.org/10.3390/APP12125793
21. Jha DN, Garg S, Jayaraman PP et al (2021) A study on the evaluation of HPC microservices in containerized environment. Concurr

Comput 33:1–1. https://doi.org/10.1002/CPE.5323

22. Nadeem A, Malik MZ (2022) Case for microservices orchestration using workflow engines. In: IEEE/ACM 44th international conference on software engineering: new ideas and emerging results (ICSE-NIER), pp 6–10. https://doi.org/10.1109/ICSE-NIER55298.2022.9793520
23. Poniszewska-Marańda A, Czechowska E (2021) Kubernetes cluster for automating software production environment. Sensors 21: 1910. https://doi.org/10.3390/S21051910
24. HPE superdome flex server architecture and RAS technical white paper. https://www.hpe.com/psnow/doc/A00036491ENW.pdf. Accessed 2 May 2023
25. Yeung T (2022) What's the difference: edge computing vs cloud computing. https://blogs.nvidia.com/blog/2022/01/05/difference-between-cloud-and-edge-computing/. Accessed 28 Apr 2023
26. Golightly L, Chang V, Xu QA et al (2022) Adoption of cloud computing as innovation in the organization. Int J Eng Bus Manag 14. https://doi.org/10.1177/18479790221093992
27. Puntel E (2020) COVID-19 how AI partnership is helping UCB search for new therapies. In: UCB Science News. https://www.ucb.com/our-science/magazine/detail/article/COVID-19-How-AI-partnership-is-helping-our-search-for-new-therapies. Accessed 28 Apr 2023
28. Kraemer FA, Braten AE, Tamkittikhun N, Palma D (2017) Fog computing in healthcare – a review and discussion. IEEE Access 5:9206–9222. https://doi.org/10.1109/ACCESS.2017.2704100
29. Earney S (2022) Edge computing vs fog computing: a comprehensive guide. https://xailient.com/blog/edge-computing-vs-fog-computing-a-comprehensive-guide/. Accessed 28 Apr 2023
30. Bukhari A, Hussain FK, Hussain OK (2022) Fog node discovery and selection: a systematic literature review. Futur Gener Comput Syst 135:114–128. https://doi.org/10.1016/J.FUTURE.2022.04.034
31. Jamshidi M (Behdad), Moztarzadeh O, Jamshidi A et al (2023) Future of drug discovery: the synergy of edge computing, internet of medical things, and deep learning. Future Internet 15:142. https://doi.org/10.3390/FI15040142
32. Daraghmi YA, Daraghmi EY, Daraghma R et al (2022) Edge–fog–cloud computing hierarchy for improving performance and security of NB-IoT-based health monitoring systems. Sensors 22. https://doi.org/10.3390/S22228646
33. Kunal S, Saha A, Amin R (2019) An overview of cloud-fog computing: architectures, applications with security challenges. Secur Priv 2:e72. https://doi.org/10.1002/SPY2.72
34. Younge AJ, Pedretti K, Grant RE, Brightwell R (2017) A tale of two systems: using containers to deploy HPC applications on supercomputers and clouds. In: Proceedings of the international conference on cloud computing technology and science, CloudCom 2017-December, pp 74–81. https://doi.org/10.1109/CLOUDCOM.2017.40
35. Spjuth O, Frid J, Hellander A (2021) The machine learning life cycle and the cloud: implications for drug discovery. Expert Opin Drug Discovery 16:1071–1079. https://doi.org/10.1080/17460441.2021.1932812
36. Puertas-Martín S, Banegas-Luna AJ, Paredes-Ramos M et al (2020) Is high performance computing a requirement for novel drug discovery and how will this impact academic efforts? Expert Opin Drug Discovery 15:981–986. https://doi.org/10.1080/17460441.2020.1758664
37. Guerrero GD, Pérez-Sánchez HE, Cecilia JM, García JM (2012) Parallelization of virtual screening in drug discovery on massively parallel architectures. In: Proceedings – 20th Euromicro international conference on parallel, distributed and network-based processing, PDP 2012, pp 588–595. https://doi.org/10.1109/PDP.2012.26
38. Gorgulla C, Boeszoermenyi A, Wang ZF et al (2020) An open-source drug discovery platform enables ultra-large virtual screens. Nature 580:663. https://doi.org/10.1038/S41586-020-2117-Z
39. Gentile F, Yaacoub JC, Gleave J et al (2022) Artificial intelligence–enabled virtual screening of ultra-large chemical libraries with deep docking. Nat Protoc 17:672–697. https://doi.org/10.1038/s41596-021-00659-2
40. Pun FW, Liu BHM, Long X et al (2022) Identification of therapeutic targets for amyotrophic lateral sclerosis using PandaOmics – an AI-enabled biological target discovery platform. Front Aging Neurosci 14:638. https://doi.org/10.3389/FNAGI.2022.914017
41. Boogaard P (2011) The potential of cloud computing for drug discovery & development. In: Drug discovery world. https://www.ddw-online.com/the-potential-of-cloud-computing-for-drug-discovery-development-1070-201110/. Accessed 28 Apr 2023

42. Faragardi HR (2017) Ethical considerations in cloud computing systems. PRO 1:166. https://doi.org/10.3390/IS4SI-2017-04016
43. Schaduangrat N, Lampa S, Simeon S et al (2020) Towards reproducible computational drug discovery. J Cheminform 12:1–30. https://doi.org/10.1186/S13321-020-0408-X
44. Hupy C (2022) DevOps and the scientific process: a perfect pairing | GitLab. https://about.gitlab.com/blog/2022/02/15/devops-and-the-scientific-process-a-perfect-pairing/. Accessed 28 Apr 2023
45. Leroy D, Sallou J, Bourcier J, Combemale B (2021) When scientific software meets software engineering. Computer (Long Beach Calif) 54: 60–71. https://doi.org/10.1109/MC.2021.3102299
46. Arvanitou EM, Ampatzoglou A, Chatzigeorgiou A, Carver JC (2021) Software engineering practices for scientific software development: a systematic mapping study. J Syst Softw 172:110848. https://doi.org/10.1016/J.JSS.2020.110848
47. Kumar S, Chander S (2020) Cost optimization techniques in cloud computing: review, suggestions and future scope. In: Proceedings of the international conference on innovative computing & communications. https://doi.org/10.2139/SSRN.3562980
48. Seven imperatives to build a "security-first" mindset. https://www.linkedin.com/pulse/seven-imperatives-build-security-first-mindset-kumar-mssrrm/?trk=articles_directory. Accessed 3 May 2023

Chapter 9

Knowledge Graphs and Their Applications in Drug Discovery

Tim James and Holger Hennig

Abstract

Knowledge graphs represent information in the form of entities and relationships between those entities. Such a representation has multiple potential applications in drug discovery, including democratizing access to biomedical data, contextualizing or visualizing that data, and generating novel insights through the application of machine learning approaches. Knowledge graphs put data into context and therefore offer the opportunity to generate explainable predictions, which is a key topic in contemporary artificial intelligence. In this chapter, we outline some of the factors that need to be considered when constructing biomedical knowledge graphs, examine recent advances in mining such systems to gain insights for drug discovery, and identify potential future areas for further development.

Key words Artificial intelligence, Drug discovery, Explainable AI, Graph embedding, Graph convolution, Knowledge graph, Machine learning, Natural language processing, Transformers

1 Introduction

Contemporary drug discovery is characterized by a diverse array of data types. Intervention strategies include small molecules, antisense oligonucleotides, proteins, gene and cell therapies, among others, both individually and in combination. During the discovery and development process, these agents are characterized by a wide array of technologies and at scales ranging from isolated molecules to human clinical trials. Performing analyses across all of this data is a considerable challenge, but connecting apparently disparate information can yield valuable insights. For example, the repurposing of candidate drugs for additional or alternative indications can be driven by observations of clinical side effects [1].

Knowledge graphs are one approach to data integration and inference that have recently attracted a lot of interest in the biomedical domain. In this approach, information is represented as nodes or entities connected by edges or relationships. Most commonly, edges are used to connect pairs of nodes, but *hyperedges* connecting more than two nodes are also possible. The decision

Alexander Heifetz (ed.), *High Performance Computing for Drug Discovery and Biomedicine*, Methods in Molecular Biology, vol. 2716, https://doi.org/10.1007/978-1-0716-3449-3_9,

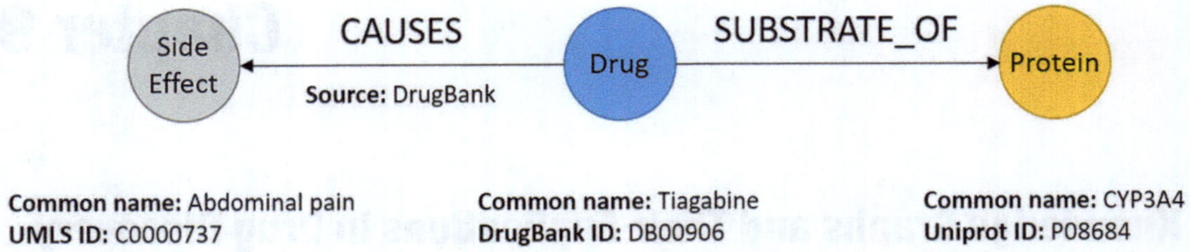

Fig. 1 An example of a simple, heterogeneous knowledge graph. Each circle represents a node of the specified type, with properties listed below. The lines represent relationships between the entities, again with a specified type and, optionally, additional properties. In this case, the arrows indicate that the relationships are directed, such that Tiagabine is a SUBSTRATE_OF CYP3A4, rather than the opposite way around. Unified Medical Language System (UMLS) [2]

about which representation to use is, at least in part, driven by the technology used to implement the graph (*vida infra*). Knowledge graphs can be homogeneous, with a single node type and a single edge type, or heterogeneous, with multiple types for each. As illustrated in Fig. 1, in a heterogeneous graph, one node could represent a particular drug, while a second could represent a specific protein. An edge between those nodes could then be used to indicate that a binding interaction between the two has been experimentally verified. Both nodes and edges can have additional properties, such as molecular structures or external identifiers. Edges can also have a direction. For example, an edge representing a connection between nodes representing a drug and an observed side effect would potentially be directed to show that the drug has the side effect, but not vice versa. On the other hand, an edge representing an interaction between two proteins nodes denoting that they bind to form a complex would typically not be directed. It is also possible to have multiple edges of the same type connecting the same nodes. For example, this could represent the situation where multiple measurements have been reported characterizing the interaction between the same drug:protein pair.

A number of knowledge graphs with the potential for drug discovery applications have been published in recent years. Some examples are listed in Table 1. The data included in these graphs is typically drawn from two different types of source. The first type is structured data sources specific to particular domains [14]. Examples of this type include Uniprot [15] for proteins, ChEMBL [16] for small molecule-protein interactions, Signor [17] for protein-protein interactions, and SIDER [18] for side effects. An advantage of these data sources is that they are often highly curated and therefore the information they contribute to the graph is, relatively speaking, robust. A disadvantage is that such curation typically limits the coverage that they are able to achieve, so that there are many potentially valuable pieces of information that are not captured. The second type of data source—natural language

Table 1
Examples of public domain knowledge graphs and related resources designed or potentially useful for drug discovery

Resource	URL	Node count	Edge count	Node types	Edge types	Last update
BioKG [3]	https://github.com/dsi-bdi/biokg	105 K	2 M	10	17	2020
Bioteque [4]	https://bioteque.irbbarcelona.org/	448 K	30 M	12	67	2022
Clinical Knowledge Graph [5]	https://data.mendeley.com/datasets/mrcf7f4tc2/1	16 M	220 M	35	57	2021
DRKG [6]	https://github.com/gnn4dr/DRKG	97 K	5.9 M	13	107	2020
GNBR [7]	https://zenodo.org/record/3459420#.Y7654HbP2Uk			3	36	2019
Hetionet [8]	https://het.io	47 K	2.3 M	11	24	2017
INDRA [9]	http://www.indra.bio					2022
OpenBioLink	https://zenodo.org/record/3834052#.Y76U1nbP2Uk	184 K	4.7 M	7	30	2020
PharmKG [10]	https://github.com/MindRank-Biotech/PharmKG	7.6 K	500 K	3	29	2020
PrimeKG [11]	https://zitniklab.hms.harvard.edu/projects/PrimeKG/	129 K	4.1 M	10	30	2023
SPOKE [12]	https://spoke.ucsf.edu/	47 K	2.3 M	11	24	2019[a]
TargetMine [13]	https://targetmine.mizuguchilab.org/	102 M[b]	28 M[b]	70[b]	14[b]	2023

[a] SPOKE was developed from Hetionet. The SPOKE website states that data sources are updated on a rotating weekly basis, but no statistics are provided in terms of current counts

[b] TargetMine is structured as a data warehouse rather than directly as a graph, so the distinction between nodes and edges is based on the authors' assessment of the data type descriptions

processing (NLP) [7, 19]—has the opposite characteristics. This type of source scales relatively well, for example, to the entire corpus of electronically published literature, but the quality of the data that is so derived can be variable. One key step in any NLP pipeline is the mapping of the extracted terms to relevant ontologies to enable normalization. For example, in scientific papers, entities such as proteins can be referred to using a wide range of synonyms. Mapping these to a unique set of identifiers such as that provided by Uniprot or Ensembl is a pre-requisite for integrating such information into a knowledge graph. Even with structured data, combining information from sources that use different identification systems can involve a non-trivial mapping task. In some cases, no simple correspondence exists between different systems even when they ostensibly pertain to the same scientific domain.

Although larger, proprietary drug discovery resources have been published [20], knowledge graphs in this domain generally remain relatively modest in size compared to those in some other areas. For example, the Google and Bing knowledge graphs contain billions of entities and tens of billions of relationships, while the state of the art is currently around the trillion relationship level [21]. An important practical factor that can constrain the size of knowledge graphs is the underlying technology—both hardware and software—that is available to implement them. The most important performance aspect of knowledge graphs is the ability to traverse and extract information from them. Updating the existing data or inserting new records is typically a much less frequent task, although the initial build can also be computationally intensive. With every implementation, there are important design decisions that need to be made depending on the nature of the data, the anticipated queries, and the available infrastructure. Graph databases are, in many ways, a natural fit for knowledge graphs, but other approaches including relational databases can also be used. The graph database software market is still maturing and, unfortunately, some performance issues may only become apparent at scale.

A more fundamental issue than those of technical implementation is the knowledge representation that is chosen for a graph. Describing all of the objects and relationships relevant to drug discovery in as granular detail as possible is intellectually appealing but impractical, for a number of reasons. The first of these is the limitations of the available data. For example, as illustrated in Fig. 2, multiple transcripts are typically generated from a single gene, each of which can be translated into a different protein isoform. In a cell, those proteins then undergo a variety of posttranslational modifications and engage in complex formation with other cellular components. If we wish to accurately represent the interaction between a small molecule drug candidate and a protein in a knowledge graph, we would therefore need to use a representation that captures all of the different levels of detail. However, due to cost and technological constraints, the experimental measurements that would be used to populate the graph typically do not include the necessary level of detail to fill such a representation. Beyond this general limitation on our current state of knowledge, particular data sources can introduce further constraints by virtue of the information structure that is already in place. For example, in GNBR [7] interactions between small molecules and proteins are annotated at the gene level, presumably due to the difficultly in distinguishing genes, transcripts, and proteins using NLP. Therefore, if it is desired to use GNBR as a source of information when constructing a knowledge graph, this underlying limitation will be reflected in the knowledge representation as a whole.

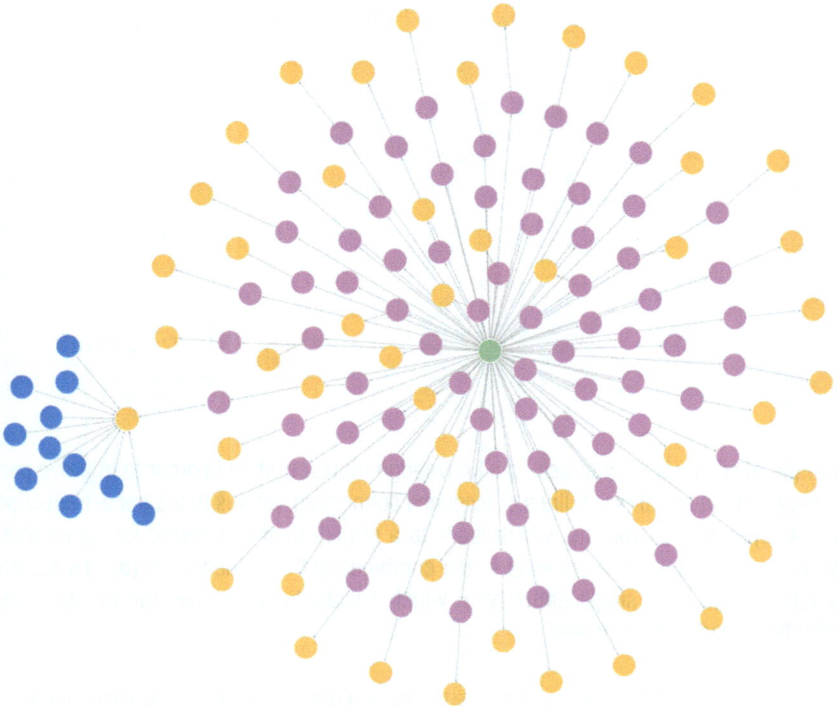

Fig. 2 A knowledge graph in which the different transcripts (purple circles) and proteins (yellow circles) listed in the Ensembl database [22] for the GNAS gene (green circle) are represented explicitly, together with the corresponding gene-TRANSCRIBED_TO- > transcript and transcript-TRANSLATED_TO- > protein edges. Such a representation is necessary if the intended use of the graph is to assist in understanding splice variation and gene transcriptional regulation, but likely highly inefficient if the main purpose is to explore links between small molecules (blue circles) that interact with that protein and the downstream effects of such modulation

A second factor constraining the knowledge representation in a knowledge graph is performance, in terms of the speed with which interpretable information (data or inferences) can be extracted from that graph. For a given query, efficiency can be improved by using the most abstract or coarse-grained representation compatible with addressing that query. As illustrated in Fig. 3, if the main purpose of a knowledge graph is to explore the relationship between genes and diseases, it may be appropriate to collapse genes, transcripts, and proteins into a single node type. This would streamline inference across multiple lines of evidence such as genome-wide association studies (GWAS), CRISPR, siRNA, RNAseq, and drug side effect data, which otherwise would have relationships to different biological entities in the gene-transcript-protein axis.

One approach to resolving the conflict between maximizing both the level of detail included in a knowledge graph representation and the efficiency of processing of the stored data is to use a hierarchy of connected representations. At the base of the hierarchy is a general or "parent" knowledge graph employing the maximum

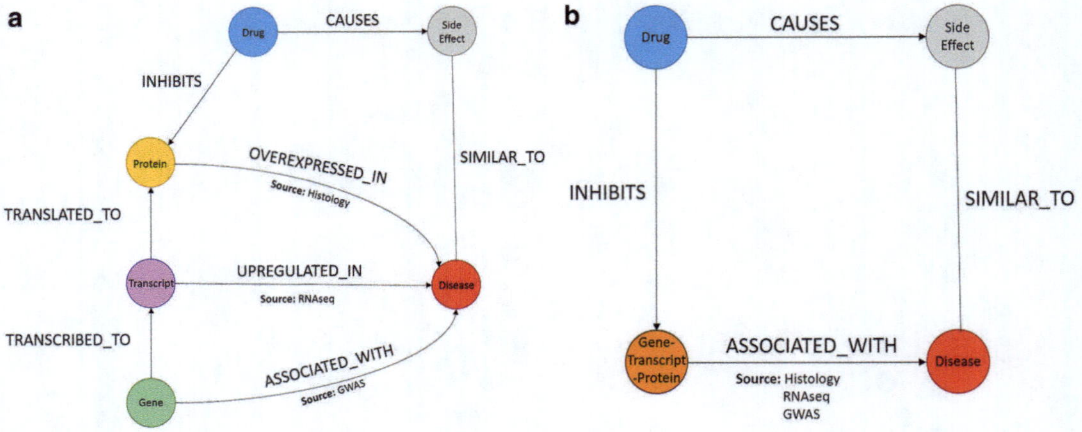

Fig. 3 An example of collapsing/compressing a knowledge representation in order to improve performance on a particular query type. In (**a**), different lines of evidence connecting entities in the gene-transcript-protein axis to a disease are represented explicitly, which leads to a requirement to traverse multiple paths through the graph to summarize and make inferences over the combined set of evidence. In (**b**), genes, transcripts and proteins are represented as a single entity type, which substantially reduces the number and diversity of evidence paths that need to be followed

level of granularity in terms of representation that is desirable and/or achievable given the underlying data sources. Specialized or "child" knowledge graphs can then be built on top of this "parent" using a simplified/condensed representation and potentially only exposing a subset or aggregated form of the data that is tailored toward a specific query type or inference task. Provided that the logic for deriving a specialized graph from a general one can be automated, the maintenance burden of such an implementation is substantially reduced because only the general knowledge graph needs to be actively updated.

2 Applications of Knowledge Graphs in Drug Discovery

In our experience, there are three main areas in which knowledge graphs can make an impact on drug discovery. The first is in democratizing access to the broad range of data types that are relevant to biomedical research. The second area is in visualizing that heterogeneous data. The third is in generating insights from the graph by mining the data in an automated fashion.

3 Democratizing Access to Biomedical Data

Pharmaceutical research and development spans a broad range of scientific disciplines, physical scales, and time frames. At one extreme, experiments are performed on recombinant single proteins in simple buffer solutions over the course of minutes. At the other, phase 4 post-marketing surveillance studies involve observations of human subjects, potentially over many years. Connecting and navigating through the resulting spectrum of data types has historically required specialized expertise. Even then, this expertise has typically been limited to certain domains such as chem(o)informatics or bioinformatics, with truly cross-domain analyses remaining the exception rather than the norm. By encapsulating the logic required to connect different data types into the construction process, knowledge graphs represent an opportunity to democratize access to the whole corpus of information relevant to drug discovery scientists. This can lead to substantial improvements in the efficiency and consistency with which this corpus can be mined. In one recent example in one of our projects, a genetic screen was performed to identify targets that modulate the activity of a specific transcriptional regulator. Due to high experimental variance, more than 200 potential targets were initially flagged. Even after removing clear false positives, manual assessment of 30 targets was still required. This represented a substantial investment of time in terms of assessing the literature data on each candidate target and the plausibility of a link to the transcriptional activity under investigation. Subsequent to this exercise, querying our general biomedical knowledge graph retrieved largely comparable results in a matter of minutes. Facilitating access to information in this way does not necessarily present the opportunity to produce high-profile publications in the same way as, for example, graph inference using deep learning approaches. However, in terms of practical impact on the day-to-day work of scientists, in our experience, there is much value to this application of knowledge graphs.

4 Visualizing and Contextualizing Biomedical Data

Once information has been assembled into a knowledge graph, the challenge often shifts from simply providing harmonized access to that data to controlling the volume and complexity of what is returned to an end user. Graph visualizations represent a potentially valuable way to contextualize information, but striking the appropriate balance between interpretability and scalability is difficult. Examples of biomedical graphs at different points on this interpretability/scalability scale are shown in Fig. 4. Figure 4a is a depiction of a signaling network. This type of representation is commonly used to summarize a particular aspect of cellular biology and is designed to be highly interpretable. However, the creation of such images is largely a manual process and they are, therefore, time consuming to maintain, difficult to scale up, and challenging to extend or combine with other data. Figure 4b is a visualization of

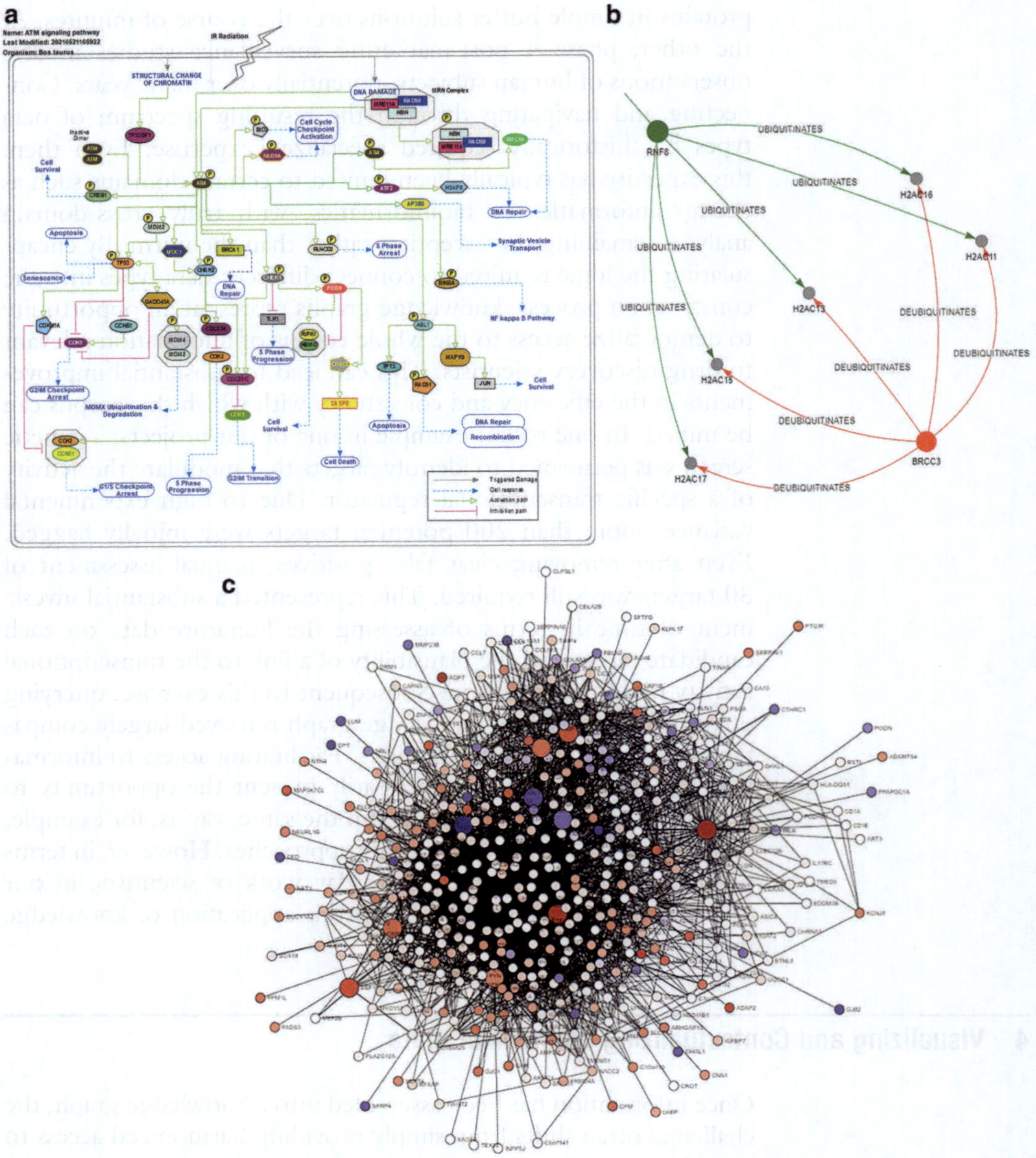

Fig. 4 Examples of graph visualizations of biomedical data with different interpretability/scalability levels. (**a**) A manually curated representation of a cellular signaling cascade, taken from WikiPathways [23]. (**b**) A network showing a sample of the human ubiquitination regulation network. Each node represents a protein, colored according to whether it participates in the ubiquitination (green) or deubiquitination (red) of a substrate (gray). (**c**) Differential gene expression data from human cancer cells [24] mapped onto a protein-interaction network taken from BioGRID [25]. Each node represents a gene and is colored according to the direction of differential expression—red for upregulation and blue for downregulation. Only the top ten most differentially expressed genes (highlighted with larger nodes) and other genes up to two connections away are displayed

a small section of the human ubiquitination regulatory network and is intermediate in terms of interpretability vs. scalability. The use of different display elements, including labels, colors, and arrows, illustrates the potential to exploit the heterogeneous nature of knowledge graph data to provide interpretable context. Additional, connected data, such as the availability of crystal structure evidence for the specified protein-protein interactions, can be layered onto such a representation in a dynamic manner. This graphic was generated starting from an automatic layout, to which minor manual adjustments were made to improve the interpretability. However, as illustrated by Fig. 4c, it is currently difficult to scale such an approach. Large, homogeneous networks such as this can be automatically visualized and certain types of compatible information can be added by, for example, adjusting the size or color of the nodes. However, despite the development of a number of algorithms that attempt to generate interpretable layouts in an automated fashion [26], it is typically very difficult to draw detailed insights from them. Indeed, at least in the bioinformatics domain, they are commonly referred to as "hairballs."

5 Generating Insights from Knowledge Graphs through Automated Data Mining

Once a knowledge graph is constructed, curated, and visualized, the question arises: How to mine the graph to gain insights for drug discovery? Typical inference tasks involving knowledge graphs are link prediction (i.e., predicting new links for a knowledge graph given the existing links among the entities) and node classification. In contrast to other fields such as computer vision, where dedicated benchmark data sets (e.g., ImageNet) have become a standard, widely adopted "standardized" benchmarks in knowledge graphs are still a challenge because of the inherent diversity of data types and graph sizes. Different benchmarks have emerged for measuring algorithmic progress in making inferences from knowledge graphs. Examples include a benchmarking framework for large-scale biomedical link prediction called OpenBioLink, which has been developed to transparently and reproducibly evaluate novel algorithms [27]. For link prediction and triple classification, CODEX, a knowledge graph completion benchmark consisting of datasets extracted from Wikidata and Wikipedia, has been built [28]. Also, PharmKG, a dedicated knowledge graph benchmark for biomedical data mining, has recently been proposed [10].

There are a variety of recent applications of knowledge graphs in drug discovery. Examples include drug repurposing [8], such as for Covid-19 [29, 30], drug-drug interactions [31], and polypharmacy side effects [32]. A knowledge graph-based recommendation framework has been proposed to identify drivers of resistance in EGFR mutant non-small-cell lung cancer [33]. Also DDR, a computational method to predict drug–target interactions using graph mining and machine learning approaches, has been developed [34].

A number of approaches to extracting insights from knowledge graphs have been developed, including rule mining, distributed representation-based reasoning, and neural network-based reasoning [35]. Rule mining involves the identification of meaningful patterns (sometimes referred to as subgraphs or metapaths) in the form of rules from large collections of background knowledge. Rule mining is a form of symbolic learning, which can learn symbolic models—i.e., logical formulae in the form of rules or axioms—from a graph in a self-supervised manner. These rules can be constructed using both bottom-up/agglomerative approaches [36], in which an initial set of specific rules are progressively generalized, or top-down/divisive approaches [37], in which the opposite process occurs. An advantage of rule mining is the interpretability of the results, as the rules extracted by an algorithm are often directly understandable by humans. For an introduction to symbolic learning and rule mining see, for example, Hogan et al. [38].

The majority of recent knowledge graph mining publications have focused on the development of representation-based approaches. Such approaches endeavor to generate latent-space representations—also known as embedding—of the entities and relationships in the graph. This latent space can then be investigated to, for example, predict new triples that do not appear in the original graph. Early examples of distributed representation-based reasoning included matrix and tensor factorization techniques [39, 40]. Matrix factorization techniques attempt to recover missing or corrupted entries by assuming that the matrix can be written as the product of two low-rank matrices. More recently, research has turned to the use of deep learning approaches for generating knowledge graph embeddings. This will be the topic of the following section.

6 Machine Learning on Knowledge Graphs for Drug Discovery

A variety of embedding models have been developed for mining the information contained in knowledge graphs through machine learning algorithms with high accuracy and scalability [41–43]. Knowledge graph embedding models operate by learning a low-dimensional representation of graph nodes and edges while preserving the semantic meaning and the graph's inherent structure [43]. For a list of representative knowledge graph embedding models, see Zeng et al. [41]. For knowledge representation learning, graph neural networks (GNN) are currently a favored architecture due to their ability to capture long-distance

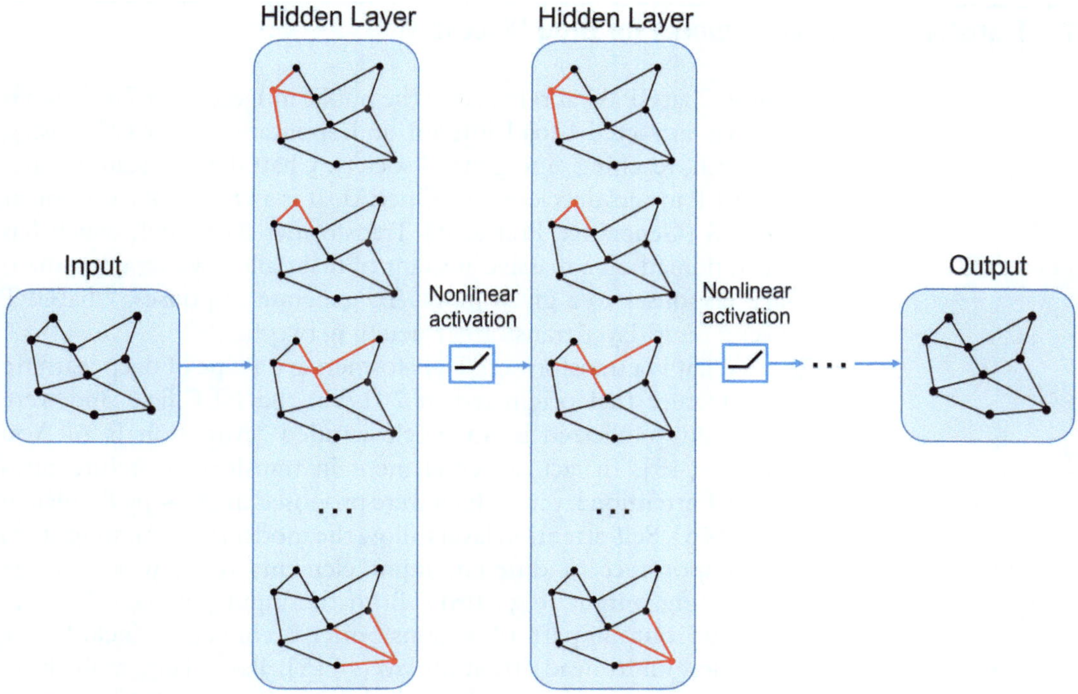

Fig. 5 Illustration of graph neural networks (GNN) for knowledge representation learning. GNN can capture long-distance correlations between nodes. At each iteration, for each node the embedding from all its neighbors are summed up, which will produce a new embedding. This new embedding contains the information of the node plus the information of all the neighbors. In the next iteration, it will also contain the information of its second-order neighbors. The process can, in principle, continue until each of embedding has information from all the other nodes. An optional final step is to collect all embeddings, which yields a single embeddings for the whole graph

correlations between nodes [41]. For instance, polypharmacy side effects have been modeled with GNN [32]. A schematic illustration of generating an embedding with a GNN is shown in Fig. 5. For libraries and GitHub repositories for deep learning on graphs with GNN, see, for example, the following:

- Graph Nets, which is DeepMind's library for building graph networks in TensorFlow https://github.com/deepmind/graph_nets
- The Deep Graph Library (for different frameworks such as PyTorch, TensorFlow, etc.) https://www.dgl.ai/

Based on GNN, an end-to-end framework called Knowledge Graph Neural Networks (KGNN) has been developed for drug-drug interaction prediction. This method aggregates information from neighborhood entities to learn the representation of the entity of interest. KGNN automatically captures both high-order relations and semantic relations (e.g., between drug pairs) in knowledge graphs [44].

7 Transformer Neural Networks for Drug Discovery

Since ChatGPT was released to the public in December 2022, it has quickly attracted broad interest and applications across industry, research, teaching, and general society. ChatGPT is a state-of-the-art NLP model developed by OpenAI. It is a variant of the popular GPT-3 (Generative Pretrained Transformer 3) model, which has been trained on a massive amount of text data to generate human-like responses to a given input. To generate responses, ChatGPT uses a multi-layer transformer neural network.

What is a transformer? Transformers are a type of deep learning architecture that originated in 2017 in the NLP field and were originally publicized in an article entitled "Attention Is All You Need" [45]. In fact, a key element in transformer architectures are self-attention layers, which were proposed in 2014 by Bahdanau et al. [46]. Self-attention layers allow the model to assign weights to the importance of different input elements (e.g., words) when generating output (e.g., text). Both the input part (see Fig. 6a) and the output part of a transformer architecture include the so-called multi-head attention layers [45]. Essentially, multi-head attention means running through the attention mechanism several times. To provide an intuitive understanding for what self-attention can capture, an example of a self-attention output is shown in

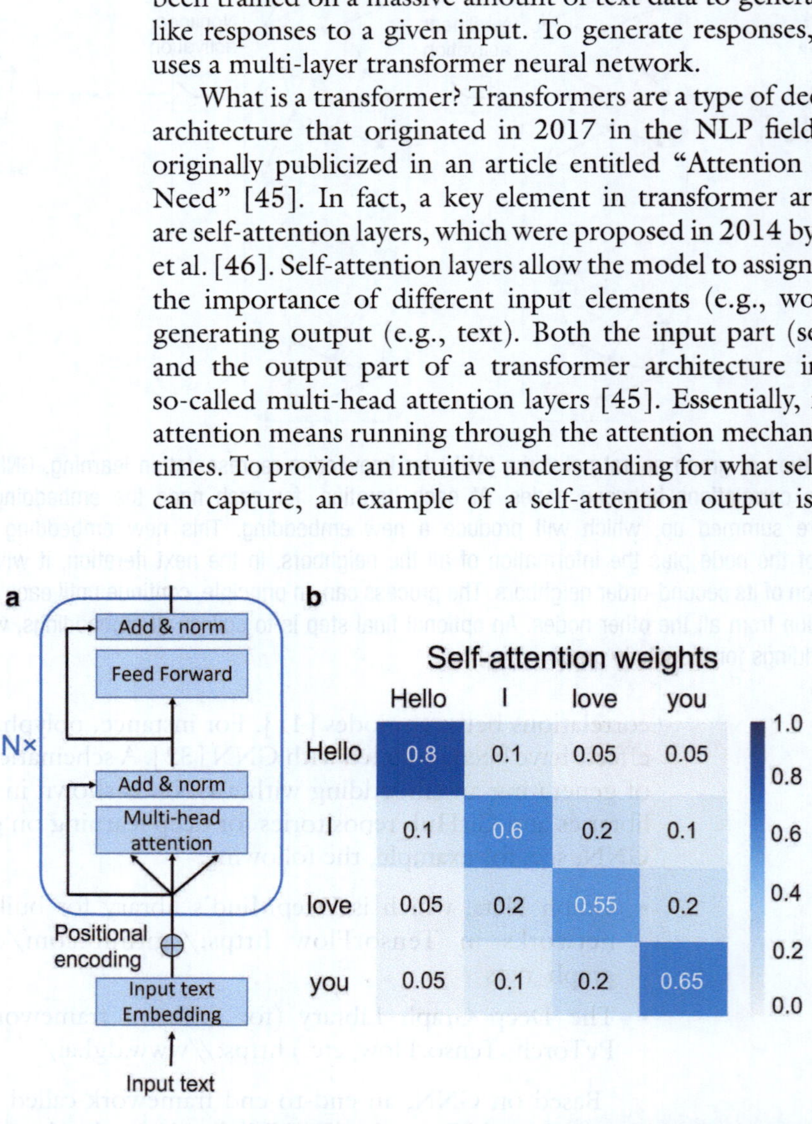

Fig. 6 Overview of the architecture of transformer neural networks, such as ChatGPT. (**a**) Transformer input architecture, figure adapted from Vaswani et al. [45]. A key component in the architecture is the multi-head attention layer. (**b**) An example of a self-attention layer output for the phrase "Hello I love you." A trained self-attention layer will associate the word "love" with the words "I" and "you" with a higher weight than the word "Hello." This is an intuitive way to understand what self-attention will capture, as these words share a subject-verb-object relationship

Fig. 6b for the phrase "Hello I love you." For a detailed introduction on transformer architectures, see, e.g., https://theaisummer.com/transformer/.

How can transformer models be applied to biomedical knowledge graphs? Recently a promising framework called STonKGs ("Sophisticated Transformer trained on biomedical text and Knowledge Graphs") has been developed [42]. This multimodal transformer uses combined input sequences of structured information from knowledge graphs and unstructured text data from biomedical literature to learn joint representations in a shared embedding space.

Transformer models not only excel at NLP tasks but also are making an impact on computer vision. For a recent survey on the current spectrum of applications of vision transformers, see ref. [47]. The core components of vision transformers are self-attention layers, just like in transformer architectures for NLP. While convolutional neural networks revolutionized computer vision applications in research and industry in the last decade, transformers are promising candidates to significantly advance computer vision even further. The wide applicability of transformer neural networks may help bridge the gap between the NLP and computer vision communities. With the development of frameworks for applying transformer neural networks to biomedical knowledge graphs [42], transformers have the potential to become a valuable tool to gain insights in knowledge graphs in drug discovery.

8 Explainable AI for Drug Discovery

Explainability of a machine learning model is usually inversely related to its prediction accuracy—the higher the prediction accuracy, the lower the model explainability [48]. Deep neural networks are comparably weak in explaining their inference processes and final results and they are often considered a black box by developers and users. Interpretability of an AI method, however, is crucial for its adoption, which is even more important in sensitive domains, including the biomedical field and drug discovery. An application for interpreting a deep learning classification has been shown by Richmond et al. in identifying phototoxicity from label-free microscopy images of cells [49]. The authors used the method of Gradient Class Activation Mapping (Grad-CAM) to highlight regions of images that contributed significantly to their classification, see Fig. 7. Grad-CAM is a form of attention visualization. Explainability will likely play an essential role in applying transformer neural networks, such as ChatGPT and vision transformers. Recently, transformer interpretability beyond attention visualization has been proposed by Facebook AI Research [50].

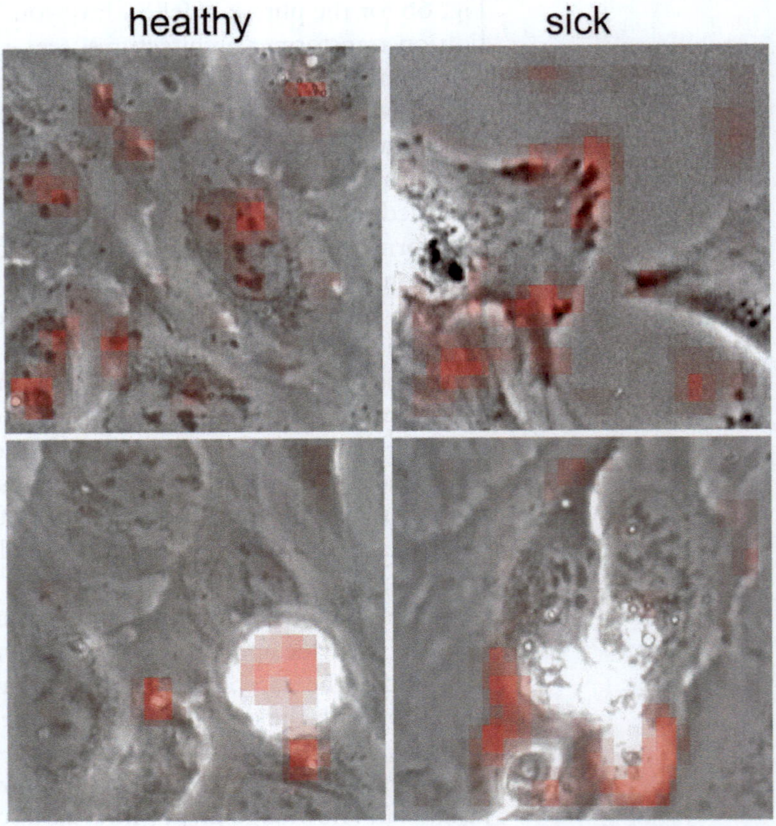

Fig. 7 Example for explainable AI to identify phototoxicity from label-free microscopy images of cells [49]. Regions in the images that contributed significantly to their classification are marked in red by the Grad-CAM algorithm. In healthy cells, often nucleoli (upper left panel) and large isolated vacuoles (lower left panel) are often highlighted. In sick cells, Grad-CAM highlighted retracting cell edges (top right) and the edges of rounded up cells (lower right). Image courtesy of Hunter Elliott [49]

Knowledge graphs, which naturally provide domain background knowledge in a machine-readable format, could be integrated in explainable machine learning approaches to help them provide more meaningful, insightful, and trustworthy explanations [51]. For example, an object classification task can be made explainable by connecting it to an underlying knowledge graph [52]. An overview on the role of knowledge graphs in explainable AI can be found in refs. [51–53].

Why is explainable AI relevant? The more complex and sensitive tasks and decisions are shifted over to AI, the more important it will be to understand as humans how and why the decision was made. Explainability is expected to become part of policies and regulations in a variety of scenarios across the globe. In the General Data Protection Regulation, the European Union grants their citizens a right to explanation if they are affected by algorithmic decision-making [54]. Explainable AI will become increasingly important to all groups of stakeholders, including the users, the affected people, and the developers of AI systems [48].

Prominent examples for the need of explainability are counter-intuitive predictions. The work of [55] reports a model counterintuitively predicting that asthmatic patients have a lower risk of dying from pneumonia. In order to explain such decisions, a doctor's medical expertise is required to reveal that these patients were admitted directly to the intensive care unit, receiving aggressive care that indeed not only lowered their risk of death but also caused incorrect machine-driven conclusions. Such decisions could have been more understandable if the model also provided evidence in the form of explanations found in external, machine-readable knowledge sources e.g., hospital databases providing patients' history or drug-disease interaction datasets such as DisGeNET.

In the field of treatment diagnosis, a user-centered AI system with an explanation ontology can be employed for diagnosis recommendations [56]. A user-centered AI system for diagnosis recommendation requires clinicians to complement the intelligent agent with their own explanations about the patient's case. A user study is carried out to identify the different types of explanations required at the different steps of the automated reasoning, i.e., "everyday explanations" for diagnosis, "trace-based explanations" for planning the treatment, "scientific explanations" to provide scientific evidence from existing studies, and "counterfactual explanations" to allow clinicians to add/edit information to view a change in the recommendation [56]. Ontologies are used to model the components necessary for the AI system to automatically compose explanations exposing different forms of knowledge and address different tasks performed by the agent.

9 Outlook

Knowledge graphs are currently a popular area of research within the biomedical community. As with any new (or newly re-popularized) technology, it seems likely that a subset of genuinely impactful use cases will eventually be distilled from this wave of enthusiasm. Many recent publications focus on the development of deep learning approaches to generate inferences from graphs. With the rise of transformer architectures in NLP—a prominent example is ChatGPT—and in computer vision [47], it seems probable that the application of transformer-based architectures will lead to a statistical improvement in model performance for automated inferencing over biomedical data. Whether such performance improvements will be practically meaningful for knowledge graph mining is, perhaps, a different question.

In our opinion, inferences based on knowledge graphs are likely to offer the greatest value over alternative approaches when the solution requires the synthesis of multiple lines of reasoning, particularly those involving "fuzzy" or qualitative evidence. Drug repurposing is an example of such an application, since this can conceivably take into account diverse factors including compound pharmacology, protein similarity, biological pathways, genetic evidence, clinical and non-clinical observations, and side effect information. Physics-based questions, such as the prediction of compound binding affinities, are less likely to benefit from a knowledge graph–based approach in our experience. Insights based on single, key observations—especially those that contradict trends in the bulk of the data—can also be difficult to capture with automated inference approaches. The development of standardized benchmarks is another challenge in the knowledge graph area because of the interdependence between the data, the data representation, and the inference algorithm [47]. Isolating the exact contribution of these factors from manual post-processing steps can be very difficult for publications describing bespoke applications.

Although algorithmic development maintains a high profile, experience with other machine learning approaches suggests that optimizing the inputs to these algorithms is likely to be more impactful in terms of meaningfully improving their prospective performance. As discussed above, with knowledge graphs both the data and the representation of that data are key inputs. Additional data types that are currently rarely used in knowledge graphs, such as high content imaging, are likely to be incorporated. In the biomedical domain, however, the number of nodes that can reasonably be identified with physical entities is large but finite. By contrast, the number of edges that one can create is essentially infinite. Just to take one example, protein similarity can be quantified in a number of ways and the most appropriate metric will be highly dependent on the question to be addressed. If one is seeking to investigate phylogenetic relationships, then sequence-based similarity edges are likely to provide the most relevant information in the graph. On the other hand, if the focus is on identifying sources of compound polypharmacology, edges based on the 3D similarity of protein binding pockets may be more informative. Simply calculating every conceivable edge type and relying on the inference algorithms to determine the most relevant ones is neither practical nor desirable. Improving the explainability of predictions remains a key topic across machine learning and, by providing a very large number of edges, the probability of identifying chance correlations and/or proxy relationships will increase dramatically. In general, with a deep neural network, the reasons for a certain prediction are often obscure or simply not presented to the typical end user.

Several methodologies for explainable AI are currently being developed. Knowledge graphs appear to offer a valuable basis to obtain trustworthy, understandable, and explainable deep learning results, which are key for drug discovery.

References

1. Boolell M, Allen MJ, Ballard SA et al (1996) Sildenafil: an orally active type 5 cyclic GMP-specific phosphodiesterase inhibitor for the treatment of penile erectile dysfunction. Int J Impot Res 8:47–52
2. Bodenreider O (2004) The Unified Medical Language System (UMLS): integrating biomedical terminology. Nucleic Acids Res 32: D267–D270. https://doi.org/10.1093/nar/gkh061
3. Walsh B, Mohamed SK, Nováček V (2020) BioKG: a knowledge graph for relational learning on biological data. In: Proceedings of the 29th ACM international conference on information & knowledge management. Association for Computing Machinery, New York, NY, USA, pp 3173–3180
4. Fernández-Torras A, Duran-Frigola M, Bertoni M et al (2022) Integrating and formatting biomedical data as pre-calculated knowledge graph embeddings in the Bioteque. Nat Commun 13:5304. https://doi.org/10.1038/s41467-022-33026-0
5. Santos A, Colaço AR, Nielsen AB, et al (2020) Clinical knowledge graph integrates proteomics data into clinical decision-making. 2020.05.09.084897
6. (2023) Drug Repurposing Knowledge Graph (DRKG)
7. Percha B, Altman RB (2018) A global network of biomedical relationships derived from text. Bioinformatics 34:2614–2624. https://doi.org/10.1093/bioinformatics/bty114
8. Himmelstein DS, Lizee A, Hessler C et al (2017) Systematic integration of biomedical knowledge prioritizes drugs for repurposing. elife 6:e26726. https://doi.org/10.7554/eLife.26726
9. Bachman JA, Gyori BM, Sorger PK (2023) Automated assembly of molecular mechanisms at scale from text mining and curated databases. Mol Syst Biol 19:e11325. https://doi.org/10.15252/msb.202211325
10. Zheng S, Rao J, Song Y et al (2021) PharmKG: a dedicated knowledge graph benchmark for bomedical data mining. Brief Bioinform 22: bbaa344. https://doi.org/10.1093/bib/bbaa344
11. Chandak P, Huang K, Zitnik M (2022) Building a knowledge graph to enable precision medicine. 2022.05.01.489928
12. Nelson CA, Butte AJ, Baranzini SE (2019) Integrating biomedical research and electronic health records to create knowledge-based biologically meaningful machine-readable embeddings. Nat Commun 10:3045. https://doi.org/10.1038/s41467-019-11069-0
13. Chen Y-A, Tripathi LP, Fujiwara T et al (2019) The TargetMine Data Warehouse: enhancement and updates. Front Genet 10:934
14. Bonner S, Barrett IP, Ye C, et al (2021) A review of biomedical datasets relating to drug discovery: a knowledge graph perspective. ArXiv210210062 Cs
15. The UniProt Consortium (2019) UniProt: a worldwide hub of protein knowledge. Nucleic Acids Res 47:D506–D515. https://doi.org/10.1093/nar/gky1049
16. Gaulton A, Bellis LJ, Bento AP et al (2012) ChEMBL: a large-scale bioactivity database for drug discovery. Nucleic Acids Res 40:D1100–D1107. https://doi.org/10.1093/nar/gkr777
17. Lo Surdo P, Iannuccelli M, Contino S et al (2023) SIGNOR 3.0, the SIGnaling network open resource 3.0: 2022 update. Nucleic Acids Res 51:D631–D637. https://doi.org/10.1093/nar/gkac883
18. Kuhn M, Letunic I, Jensen LJ, Bork P (2016) The SIDER database of drugs and side effects. Nucleic Acids Res 44:D1075–D1079. https://doi.org/10.1093/nar/gkv1075
19. Kilicoglu H, Shin D, Fiszman M et al (2012) SemMedDB: a PubMed-scale repository of biomedical semantic predications. Bioinformatics 28:3158–3160. https://doi.org/10.1093/bioinformatics/bts591
20. Martin B, Jacob HJ, Hajduk P, et al (2022) Leveraging a billion-edge knowledge graph for drug re-purposing and target prioritization using genomically-informed subgraphs. Bioinformatics
21. Neo4j Neo4j Breaks Scale Barrier with Trillion + Relationship Graph. https://www.prnewswire.com/news-releases/neo4j-breaks-

scale-barrier-with-trillion-relationship-graph-301314720.html. Accessed 19 Jan 2023
22. Cunningham F, Allen JE, Allen J et al (2022) Ensembl 2022. Nucleic Acids Res 50:D988–D995. https://doi.org/10.1093/nar/gkab1049
23. Martens M, Ammar A, Riutta A et al (2021) WikiPathways: connecting communities. Nucleic Acids Res 49:D613–D621. https://doi.org/10.1093/nar/gkaa1024
24. Slyper M, Porter CBM, Ashenberg O et al (2020) A single-cell and single-nucleus RNA--Seq toolbox for fresh and frozen human tumors. Nat Med 26:792–802. https://doi.org/10.1038/s41591-020-0844-1
25. Oughtred R, Rust J, Chang C et al (2021) The BioGRID database: a comprehensive biomedical resource of curated protein, genetic, and chemical interactions. Protein Sci Publ Protein Soc 30:187–200. https://doi.org/10.1002/pro.3978
26. Kobourov SG (2012) Spring embedders and force directed graph drawing algorithms
27. Breit A, Ott S, Agibetov A, Samwald M (2020) OpenBioLink: a benchmarking framework for large-scale biomedical link prediction. Bioinformatics 36:4097–4098. https://doi.org/10.1093/bioinformatics/btaa274
28. Safavi T, Koutra D (2020) CoDEx: a comprehensive knowledge graph completion benchmark
29. Zeng X, Song X, Ma T et al (2020) Repurpose open data to discover therapeutics for COVID-19 using deep learning. J Proteome Res 19: 4624–4636. https://doi.org/10.1021/acs.jproteome.0c00316
30. Al-Saleem J, Granet R, Ramakrishnan S et al (2021) Knowledge graph-based approaches to drug repurposing for COVID-19. J Chem Inf Model 61:4058–4067. https://doi.org/10.1021/acs.jcim.1c00642
31. Celebi R, Uyar H, Yasar E et al (2019) Evaluation of knowledge graph embedding approaches for drug-drug interaction prediction in realistic settings. BMC Bioinform 20: 726. https://doi.org/10.1186/s12859-019-3284-5
32. Zitnik M, Agrawal M, Leskovec J (2018) Modeling polypharmacy side effects with graph convolutional networks. Bioinformatics 34:i457–i466. https://doi.org/10.1093/bioinformatics/bty294
33. Gogleva A, Polychronopoulos D, Pfeifer M, et al (2021) Knowledge graph-based recommendation framework identifies novel drivers of resistance in EGFR mutant non-small cell lung cancer. Cancer Biol
34. Olayan RS, Ashoor H, Bajic VB (2018) DDR: efficient computational method to predict drug–target interactions using graph mining and machine learning approaches. Bioinformatics 34:1164–1173. https://doi.org/10.1093/bioinformatics/btx731
35. Chen X, Jia S, Xiang Y (2020) A review: knowledge reasoning over knowledge graph. Expert Syst Appl 141:112948. https://doi.org/10.1016/j.eswa.2019.112948
36. Galárraga L, Teflioudi C, Hose K, Suchanek FM (2015) Fast rule mining in ontological knowledge bases with AMIE+. VLDB J 24: 707–730. https://doi.org/10.1007/s00778-015-0394-1
37. Meilicke C, Chekol MW, Fink M, Stuckenschmidt H (2020) Reinforced anytime bottom up rule learning for knowledge graph completion
38. Hogan A, Blomqvist E, Cochez M et al (2021) Knowledge Graphs. ACM Comput Surv 54:1–37. https://doi.org/10.1145/3447772
39. Nickel M, Tresp V, Kriegel H-P (2011) A three-way model for collective learning on multi-relational data. Icml 11(10.5555): 3104482–3104584
40. Paliwal S, de Giorgio A, Neil D et al (2020) Preclinical validation of therapeutic targets predicted by tensor factorization on heterogeneous graphs. Sci Rep 10:1–19. https://doi.org/10.1038/s41598-020-74922-z
41. Zeng X, Tu X, Liu Y et al (2022) Toward better drug discovery with knowledge graph. Curr Opin Struct Biol 72:114–126. https://doi.org/10.1016/j.sbi.2021.09.003
42. Balabin H, Hoyt CT, Birkenbihl C et al (2022) STonKGs: a sophisticated transformer trained on biomedical text and knowledge graphs. Bioinformatics 38:1648–1656. https://doi.org/10.1093/bioinformatics/btac001
43. Mohamed SK, Nounu A, Nováček V (2021) Biological applications of knowledge graph embedding models. Brief Bioinform 22: 1679–1693. https://doi.org/10.1093/bib/bbaa012
44. Lin X, Quan Z, Wang Z-J et al (2020) KGNN: knowledge graph neural network for drug-drug interaction prediction. In: Proceedings of the twenty-ninth international joint conference on artificial intelligence. International Joint Conferences on Artificial Intelligence Organization, Yokohama, Japan, pp 2739–2745
45. Vaswani A, Shazeer N, Parmar N et al (2017) Attention is all you need. In: Advances in neural information processing systems. Curran Associates, Inc

46. Bahdanau D, Cho K, Bengio Y (2016) Neural machine translation by jointly learning to align and translate
47. Han K, Wang Y, Chen H et al (2023) A survey on vision transformer. IEEE Trans Pattern Anal Mach Intell 45:87–110. https://doi.org/10.1109/TPAMI.2022.3152247
48. Xu F, Uszkoreit H, Du Y et al (2019) Explainable AI: a brief survey on history, research areas, approaches and challenges. In: Tang J, Kan M-Y, Zhao D et al (eds) Natural language processing and chinese computing. Springer International Publishing, Cham, pp 563–574
49. Richmond D, Jost AP-T, Lambert T, et al (2017) Deadnet: identifying phototoxicity from label-free microscopy images of cells using deep convnets. arXiv preprint arXiv:170106109
50. Chefer H, Gur S, Wolf L (2021) Transformer interpretability beyond attention visualization. pp 782–791
51. Tiddi I, Schlobach S (2022) Knowledge graphs as tools for explainable machine learning: a survey. Artif Intell 302:103627. https://doi.org/10.1016/j.artint.2021.103627
52. Lecue F (2020) On the role of knowledge graphs in explainable AI. Semantic Web 11: 41–51. https://doi.org/10.3233/SW-190374
53. Rajabi E, Etminani K (2022) Knowledge-graph-based explainable AI: a systematic review. J Inf Sci 016555152211128. https://doi.org/10.1177/01655515221112844
54. Goodman B, Flaxman S (2017) European Union regulations on algorithmic decision-making and a "right to explanation". AI Mag 38:50–57. https://doi.org/10.1609/aimag.v38i3.2741
55. Caruana R, Lou Y, Gehrke J et al (2015) Intelligible models for HealthCare: predicting pneumonia risk and hospital 30-day readmission. In: Proceedings of the 21th ACM SIGKDD international conference on knowledge discovery and data mining. ACM, Sydney NSW Australia, pp 1721–1730
56. Chari S, Seneviratne O, Gruen DM et al (2020) Explanation ontology: a model of explanations for user-centered AI. In: Pan JZ, Tamma V, d'Amato C et al (eds) The semantic web – ISWC 2020. Springer International Publishing, Cham, pp 228–243

Chapter 10

Natural Language Processing for Drug Discovery Knowledge Graphs: Promises and Pitfalls

J. Charles G. Jeynes, Tim James, and Matthew Corney

Abstract

Building and analyzing knowledge graphs (KGs) to aid drug discovery is a topical area of research. A salient feature of KGs is their ability to combine many heterogeneous data sources in a format that facilitates discovering connections. The utility of KGs has been exemplified in areas such as drug repurposing, with insights made through manual exploration and modeling of the data. In this chapter, we discuss promises and pitfalls of using natural language processing (NLP) to mine "unstructured text"— typically from scientific literature— as a data source for KGs. This draws on our experience of initially parsing "structured" data sources—such as ChEMBL—as the basis for data within a KG, and then enriching or expanding upon them using NLP. The fundamental promise of NLP for KGs is the automated extraction of data from millions of documents—a task practically impossible to do via human curation alone. However, there are many potential pitfalls in NLP-KG pipelines, such as incorrect named entity recognition and ontology linking, all of which could ultimately lead to erroneous inferences and conclusions.

Key words NLP, Named entity recognition, Named entity linking, Normalization, Ontologies, DrugBank, SemMedDB, PubTator, Database, Heterogeneous data, Unstructured text

1 Introduction

The explosion in biomedical data over the last few decades has led to a pressing need for a coherent framework to manage the information. This is particularly true for drug discovery where many factors influence the success or failure of a program as it passes through the various stages, from target prioritization, hit identification, lead optimization to clinical trials. Knowledge graphs [1] are a promising framework for this task as heterogeneous data sources can be combined and analyzed, with entities such as drugs or genes represented as "nodes" and relationships or links between nodes represented as "edges."

Within the biomedical domain, there are several examples of publicly available KGs. Generally, these have been constructed by parsing publicly available "structured" (mostly manually curated)

Alexander Heifetz (ed.), *High Performance Computing for Drug Discovery and Biomedicine*, Methods in Molecular Biology, vol. 2716, https://doi.org/10.1007/978-1-0716-3449-3_10,

databases into a KG compatible format known as a semantic triple (subject-verb-object or subject-predicate-object) such as drug-TREATS-disease. These databases include STRING, Uniprot, ChEMBL, DrugBank, and Reactome, all of which focus on a particular biomedical domain (*see* [2] and [3] for reviews of biomedical databases relevant to constructing KGs). Examples of publicly available KGs include CROssBAR [4], Hetionet [5], and Cornell Universities' KG [6]. These KGs have comparable but different "schemas" (how nodes and edges are named and arranged) and incorporate a different number of underlying datasets. Also, biological entities (nodes) can be normalized to equally valid but different ontologies (e.g., ENTREZ [7] vs ENSEMBL [8] gene IDs) in comparable KGs, making direct comparisons non-trivial. Strikingly, none of these KGs include arguably the largest data source available—unstructured text in scientific articles like journals, patents, and clinical notes. That said, there are examples of KGs that combine NLP-derived triples with those from structured databases. For instance, the majority of triples (~80%) in AstraZeneca's "Biomedical Insights Knowledge Graph" are derived from their NLP pipelines, with the rest of the data coming from 39 structured datasets [9]. Other companies that have built KGs combining data from their NLP pipelines and public databases include, Euretos, Tellic [10], and Benevolent AI [11].

To our knowledge, there are very few articles that directly investigate the expansion of biomedical knowledge graphs using NLP. In one example, Nicholson et al. compared their NLP approach of extracting associations of biomedical entities from sentences in the scientific literature [12] to Hetionet v1 [5], which is a biomedical KG. They compared four data subsets, including "disease-ASSOCIATED-gene" and "compound-TREATS-disease", and found that their NLP method recalled around 20–30% of the edges compared to Hetionet but added thousands of novel ones. For example, with "compound-TREATS-disease" triples, NLP recalled 30% of existing edges from Hetionet but added 6282 new ones. Their NLP methodology showed an AUROC of around 0.6–0.85 depending on the type of triple, meaning that a relatively large proportion of the novel finds are wrong in some way. Nevertheless, they concluded that NLP could incorporate novel triples into their source KG (Hetionet).

In this chapter, we highlight some examples of how we are using NLP to extract information from unstructured scientific text to expand Evotec's KG. We currently use commercial NLP software to create rules-based queries that extract specific information from text, which can be then parsed into the KG. This article is not meant to be a description of Evotec's knowledge graph, however, nor an explanation of our NLP pipelines. How Evotec's KG is constructed and queried will be the subject of another article.

To highlight some pitfalls, we have used examples from the Semantic Medline database (SemMedDB) [13] and PubTator [14] databases. However, it is worth noting that the examples picked are purely illustrative of the variety of issues that must be considered when supplementing structured data sources with NLP-derived insights. In general, both databases are highly valuable with excellent accompanying tools and research.

As this chapter is in an edition focusing on "high-performance computing" (HPC), it is worth noting that all the results and examples explained herein would simply not be practically possible without HPC technology. NLP pipelines are generally "embarrassingly parallel" and optimized to return results from millions of documents within seconds. The technicality of how NLP algorithms and pipelines are constructed and optimized is an active area of research. However, it is not the focus of this article, where instead we discuss the results that certain current NLP implementations can offer and possible considerations to ensure the greatest post pipeline data quality.

2 Promises

The essential promise of NLP is the automated coverage of millions of documents from many sources such as scientific literature, clinician notes, and patents. In the next few sections, we highlight some of these promises with specific examples drawn from Evotec's own biomedical knowledge graph.

2.1 Relationship Verbs and/or Causality Between Entities

The following example of a protein-protein interaction highlights how NLP-derived data can provide additional information and extra insights compared to available structured data sources into this domain.

2.1.1 Example: Protein-Protein Interactions

Figure 1 illustrates the relationship between two proteins: MAP2K1 and MAPK3. These two proteins play a part in the MAPK signaling pathway, which is important in controlling cell division and is often aberrated in cancers [15]. Data from STRING (a structured data source [16]) shows that these two proteins physically bind to each other; data in STRING is derived from biochemical experiments and co-occurrences in the literature (among other measures), but generally there is little information describing *how* proteins affect each other. We have used NLP to expand the relationship (and many other protein relationships like it) to extract verbs linking the two proteins. In this case, we have evidence from NLP that MAP2K1 phosphorylates and activates MAPK3 from sentences such as "Mek1/2 are direct substrates of Raf kinases and phosphorylated Mek1/2 activate downstream Erk1/2" (MEK2 = MAP2K2, ERK2 = MAPK1) [17]. Understanding the mechanistic relationship between proteins is crucial

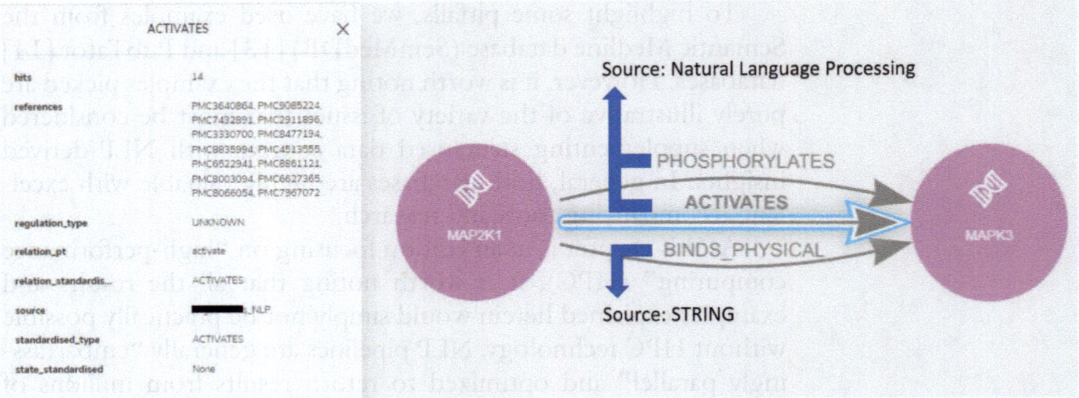

Fig. 1 An example of protein-protein interaction data in Evotec's knowledge graph. This example illustrates how triples (subject-predicate-object) extracted by NLP add valuable information (like the verbs "activates" and "phosphorylates") to structured data (here from the STRING protein-protein interaction database). From the left-hand panel titled "ACTIVATES," it can be seen that there are 14 articles ("hits" and "references") that contain phrases that equate to "MAP2K1 activates MAPK3." Future work involves displaying sentences and links to the user

in many aspects of drug discovery. For instance, if we wanted to indirectly inhibit the activity of MAPK3 (perhaps because it had proven difficult to drug directly), we could infer from this NLP information that inhibiting MAP2K1 could be an avenue to pursue. Such an inference would be much harder to make without the causative directionality added by NLP.

2.1.2 Case Study: Prioritizing Protein Targets Based on Their Association with a Specific Protein, Cancer, and/or Arthritis

In this section, we illustrate the promise that knowledge graphs hold, exemplified by a project at Evotec. In this project, an experiment had identified a number of proteins that were associated with the regulation of a specific protein (for confidentiality we shall call this proteinX), with the caveat that there were likely to be many false positives in this list due to experimental uncertainty. From this list of proteins we wanted to understand the plausibility of their involvement in regulation of proteinX, but also which had associations with cancer and/or arthritis. Ranking the proteins for relevance was highly desirable as further experimentation was costly and only one or two candidates could feasibly be studied in detail.

Figure 2 shows a network that was created to analyze the associations of the list of proteins. As part of our analysis, we used a "betweenness centrality" algorithm [18] on the network to indicate which proteins were most connected in terms of shortest paths between pairs of nodes. The results of this algorithm are incorporated into Fig. 2, where protein nodes with a higher "betweenness centrality" score are larger. Other algorithms such as VoteRank [19] were used for different perspectives on the rank-

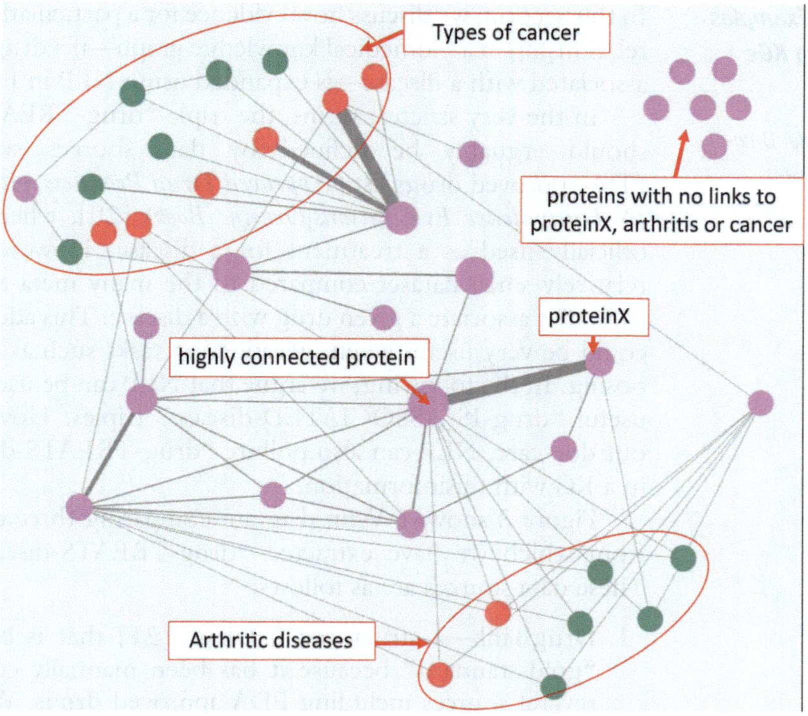

Fig. 2 An example of a mini-network or sub-graph made using data from Evotec's KG, focusing on protein-protein interactions between a specific list of proteins and their connections with cancer or arthritis. The motivation was to explore which proteins were likely regulators of "proteinX," but were also connected to cancer and/or arthritis. The purple nodes represent proteins, with the node size scaled to reflect their "betweenesss centrality" scores. The thickness of the network edges reflects the amount of evidence supporting the corresponding relationship. One protein (here labeled "highly connected protein") might be a suitable candidate for further experimental testing as it has the highest score and substantial NLP-derived evidence that it activates "proteinX"

ing of the proteins. We made several observations from this network. Firstly, we can see that there is a group of six proteins that have no links to any other entity in the network; these are probably false positives from the experiment. Secondly, proteinX is only directly linked to three other proteins. Thirdly, one of the directly connected proteins is a "hub" protein that has the highest "betweeness centrality" score in the network. Fourthly, the thick line between the "highly connected protein" and "proteinX" indicate that that this relationship is highly annotated by way of NLP; closer inspection shows there is strong evidence from the literature that the "highly connected protein" ACTIVATES "proteinX." Overall, out of the initial list of proteins, at least one—the "highly connected protein"—has a strong association with proteinX, cancer, and arthritis and so might be a suitable candidate to take forward for further experiments.

2.2 Other Examples of Expanding KGs Using NLP

2.2.1 Example: Drug-TREATS-Disease

In this section, we discuss how evidence for a particularly important relationship in a biomedical knowledge graph—if a drug treats or is associated with a disease—is expanded using NLP in Evotec's KG.

In the very strictest terms, the triple "drug-TREATS-disease" should arguably be defined by data sources such as the FDA-approved drugs list (*Approved Drug Products with Therapeutic Equivalence Evaluations|Orange Book*) [20], where a drug is officially used as a treatment for a disease. However, this is a relatively small dataset compared to the many meta-analyses and trials that associate a given drug with a disease. This additional data could be very useful when investigating tasks such as drug repurposing. In the following, we argue that NLP can be used to extract useful "drug-IS_ASSOCIATED-disease" triples. However, without due care, NLP can also pollute "drug-TREATS-disease" data in a KG with misinformation.

Figure 3 shows a Venn diagram comparing three data sources from which we have extracted "drug-TREATS-disease" triples. These data sources are as follows:

1. DrugBank—a structured database [21] that is considered a "gold-standard" because it has been manually curated from several sources including FDA-approved drugs. We extracted all drugs that had an indication for a disease.
2. A "strict NLP query" that we applied to the entirety of MEDLINE using commercial NLP software. The query extracted all

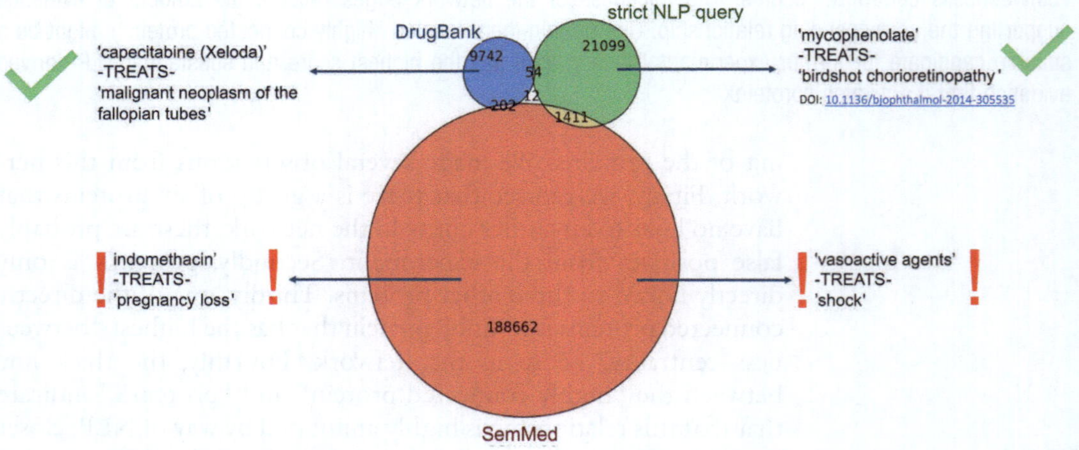

Fig. 3 A Venn diagram comparing "drug-TREATS-disease" triples extracted from three sources: (1). DrugBank, (2). a "strict NLP query" that we constructed and used to extract information from the entirety of MEDLINE (~33 million abstracts), and (3). SemMedDB. The numbers show how many unique triples each of the datasets have and how many intersect with the other datasets. The text shows examples from each of the datasets with ticks indicating that we felt the triple represented a true reflection of the source material or sentence, while exclamation marks indicate triples that were wrong or misleading (*see* Subheading 3.2 for more discussion on these)

proteins and diseases that were within one word of "treat*" (* wildcard term for derivatives like "treatment" and treated) and within twenty words of "conclusion*".

3. The subset "subject-TREATS-object" of the SemMed database [13].

To allow for comparison, each entity in the datasets were mapped to a Unified Medical Language System (UMLS) Concept Unique Identifier (CUI) for normalization [22]. We avoided comparing the raw text of entities as this can be misleading due to the occurrence of multiple synonyms.

In Fig. 3, it can be seen that there are 9742 unique triples from the DrugBank dataset, such as "carpecitabine-TREATS-malignant neoplasm of the fallopian tubes." In comparison, from our "strict NLP query," there are 21,099 unique triples, like "mycophenolate-TREATS-birdshot chorioretinopathy."

We manually inspected several of these triples from the "strict NLP query" and many came from the conclusion section of a retrospective or meta-analysis abstract. One example, "mycophenolate-TREATS-birdshot chorioretinopathy," is drawn from sentences such as "Conclusions: Derivatives of mycophenolic acid are effective and safe drugs for the treatment of BSCR [birdshot chorioretinopathy]" [23].

Interestingly, there are only 54 triples that intersect with the DrugBank dataset. One reason why there are few NLP-derived triples that overlap with DrugBank could be because NLP tends to find sentences in retrospective or meta-analysis abstracts. This contrasts with DrugBank which takes sources such as the FDA drug listings as its primary data. It is expected that we will find many drugs associated with a disease from the literature that do not make it through to clinical application.

However, it is surprising that we find so few DrugBank triples using NLP, as we might have expected there to have been many mentions of a drug with a disease in the literature prior to it becoming a treatment in the clinic. The scarce overlap is probably due to the specificity of the disease that the drug is eventually indicated for and the "strictness" of our NLP query not recalling all the possible combinations of the drug with a specific disease. So, for instance, in DrugBank the drug "Capecitabine" is indicated for several cancers including "malignant neoplasm of the fallopian tube [CUI:C0153579]". Similarly, from our "strict NLP query," the same drug is associated with over twenty types of cancer but not that specific cancer ("malignant neoplasm of the fallopian tube"). Because there are hundreds of different types of specific cancers each with their own CUI, and we are comparing the intersection using these CUIs, it becomes less surprising that there is little overlap. A combination of the "strictness" of our query missing information from the literature and/or the evolution of the drugs' use in terms of disease specificity from preclinical to a clinical setting could be an explanation.

For many KG tasks like link prediction or other hypothesis generation, we argue that triples from the "strict NLP query" add extra useful information. Appropriate tagging of NLP-derived triples is, however, critical to enable easy identification and filtering. For example, we use "drug-IS_ASSOCIATED-disease" rather than the more specific "drug-TREATS-disease" assertion, which may be an important distinction. It is likely that most of these novel triples are sufficiently accurate to provide new insights into tasks like drug repurposing. Such insights would be much harder to attain if the NLP data was omitted from the knowledge graph.

In contrast, the data in SemMedDB exemplifies many of the potential pitfalls associated with adding NLP-derived data to KGs. Briefly—as this is described in more detail in Subheading 3—from Fig. 3, it can be seen that SemMedDB has 188,662 unique "drug-TREATS-disease" triples. However, many of these are unhelpful or misleading in some way. Around 40,000 of the "drug-TREATS-disease" triples have a "drug" which is in fact a drug classification rather than a specific agent, for example, "vasoactive agents-TREATS-shock."

2.2.2 Example: gene-HAS_FEATURE-ProteinFeature

NLP has the promise of adding crucial contextual information so that a KG is ultimately more accurate and useful. We demonstrate this using the "gene-HAS_FEATURE-ProteinFeature" triple that we have included in the Evotec KG.

Figure 4a shows the record for the AHI1 gene, where the majority of ProteinFeatures have been extracted from the literature using our NLP pipelines (other sources of information for ProteinFeatures include UniProt). One ProteinFeature is highlighted as an example; here the variant T304fs*309 is reported to be associated with mutagenesis. This information was extracted and normalized from the sentence shown in the field "text_mentions" from the article shown in "text_references."

This example is closely related to a pitfall described later (*see* Subheading 3.2). There we describe how the NLP-derived triple "TP53 gene-CAUSES-neuroblastoma" is inaccurate because it is the *dysregulation* of this gene that can lead to diseases, not the gene itself. With Evotec's KG schema, we mitigate against this sort of error by including ProteinFeature nodes that can, in turn, be linked to diseases. Figure 4b shows an example of this:

"TP53 -> HAS_FEATURE -> TP53_gene mutation -> MUTATION_LINKED_TO -> Keratosis."

Here, the "TP53 gene mutation" is the ProteinFeature. The "Relationship properties" box shows the evidence for this link in the sentence (*see* "text_mentions"), the article ("text_reference"),

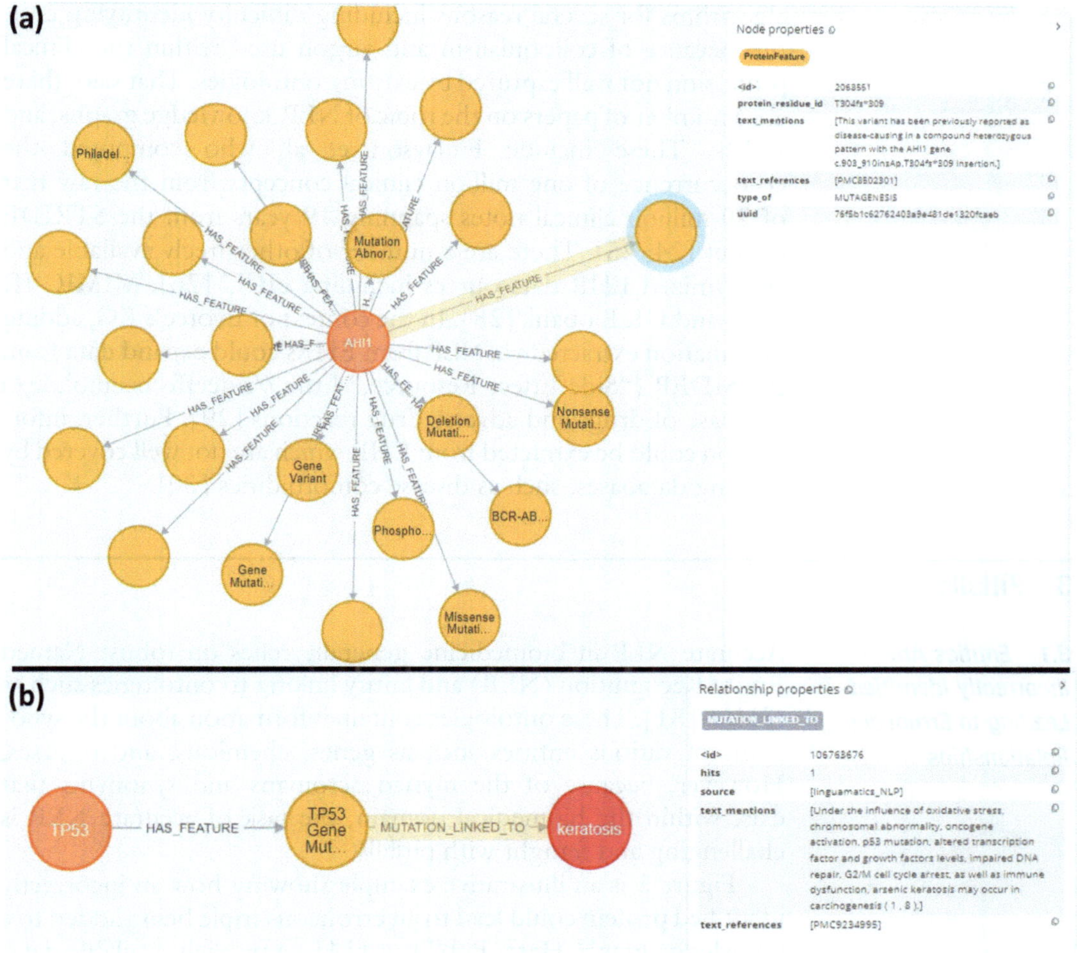

Fig. 4 (**a**) An example of how NLP can add contextual information that is crucial for biological interpretation. In Evotec's KG, proteins can have links to a node called "ProteinFeature" that has attributes such as "Mutagenesis." These attributes have been extracted from sentences in the scientific literature (as shown in the field "text_mentions") or from other sources such as Uniprot. (**b**) Protein features can be linked to diseases. Here we show an example of a mutation in the TP53 gene that is associated with the disease keratosis—a link derived from the sentence shown in the "text_mentions" box. This example is closely related to a pitfall in using NLP to construct KGs, where genes can be linked to diseases but omitting the context, such as a mutation (*see* Subheading 3.2)

and shows that NLP was the source of the evidence ("source"). Using this schema it is clear to the user that it a mutation of TP53 that is associated with keratosis.

2.2.3 Inclusion of Data Sources Such as Electronic Health Records (EHR)

Besides the scientific literature, there are many other sources of unstructured text, including electronic health records (EHR) and patents, which hold much promise for mining using NLP.

EHRs are important as they capture details such as drug administration regimes, comorbidities, and side-effects within day-to-day practice in the clinic. They can be particularly challenging for NLP algorithms for several reasons including difficulty identifying entities because of colloquialism and jargon used within the clinical profession not well captured by existing ontologies. That said, there are a number of papers on the topic of NLP, knowledge graphs, and EHRs. These include Finlayson et al. who computed the co-occurrence of one million clinical concepts from the raw text of 20 million clinical notes spanning 19 years from the STRIDE dataset [24, 25]. There are a number of other freely available and anonymized EHR datasources including eICU [26], MIMIC-III [27], and UK Biobank [28]. In the context of Evotec's KG, adding information extracted via NLP from EHRs could expand data from the SIDER ("Side Effect Resource," http://sideeffects.embl.de/) database of drugs and adverse drug reactions [29]. Further, information could be extracted from EHR which are not well covered by existing databases, such as disease comorbidities [30].

3 Pitfalls

3.1 Entities Are Incorrectly Identified Leading to Erroneous Relationships

Accurate NLP in biomedicine generally relies on robust Named Entity Recognition (NER) and entity linking to ontologies such as UMLS [31]. These ontologies contain information about the synonyms of various entities such as genes, chemicals, and diseases. However, because of the myriad acronyms and synonyms that exist within the biomedical domain, the task of accurate NER is challenging and fraught with pitfalls.

Figure 5 is an illustrative example showing how an incorrectly identified protein could lead to an erroneous triple being added to a knowledge graph. Here, PubTator [14]—a publicly available dataset made possible using various NER engines including Tagger One [32]—has been used to search for "COX1" and "aspirin." Unfortunately, "COX1" is a synonym for two completely different proteins; "cytochrome c oxidase 1" (Uniprot accession number P00395) and "cyclo-oxygenase 1" (aka prostaglandin synthase I, Uniprot accession number P23219). In this instance, PubTator has normalized "COX1" to "cytochrome c oxidase 1,' which in this context is an incorrect assignment.

This is important as PubTator has been used as the underlying data for several knowledge graphs. For example, [33] built a KG from a dataset called the Global Network of Biomedical Relationships [34], which in turn used PubTator as the basis for its entity recognition. It is likely that many downstream tasks in KG analysis such as link prediction are less accurate because of these sorts of errors.

There are many other ambiguous biomedical acronyms. For example, "TTF-1" ("thyroid transcription factor 1") can be confused with "TTF1" ("transcription termination factor 1"), which

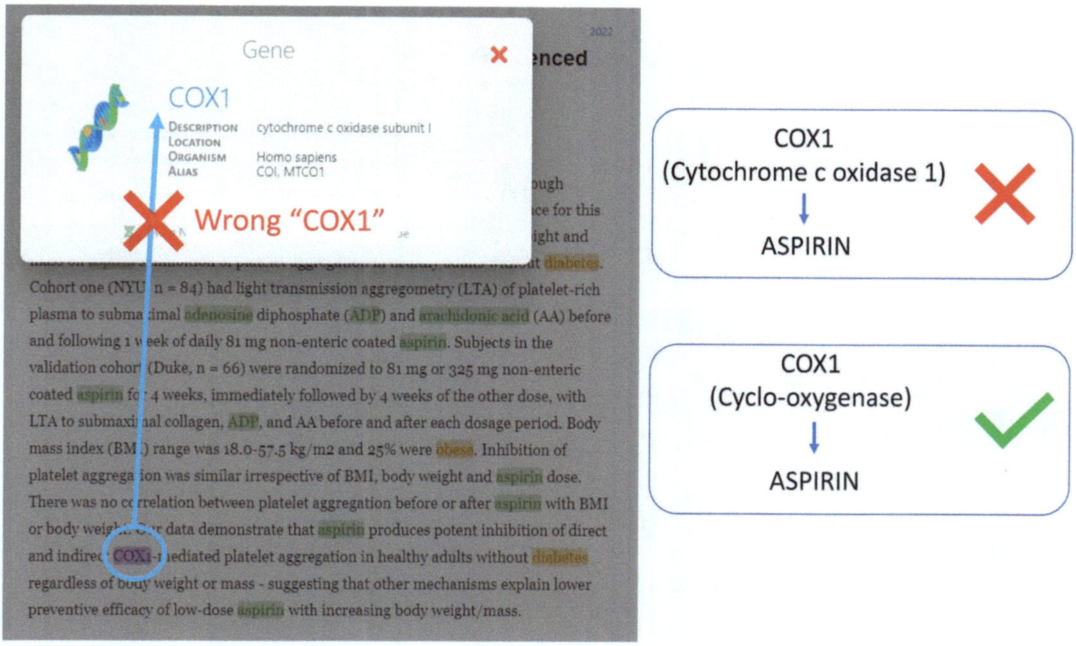

Fig. 5 An example of a named entity recognition (NER) error that could have serious implications if such data was included in a knowledge graph. Here, the mention "COX1" in the context of "aspirin" has been incorrectly identified as the gene "cytochrome c oxidase subunit I" in PubTator. The correct gene association with aspirin in this context is "cyclo-oxygenase 1," otherwise known as prostaglandin G/H synthase 1 (Uniprot accession number P23219). The screenshot is taken from the PubTator user interface (https:/www.ncbi.nlm.nih.gov/research/pubtator/)

are two distinct proteins [35]. In a wider medical setting, "RA" can stand for "Right Atrium," "Rheumatoid Arthritis," or "Room Air" depending on the context.

The prevalence of NER errors, such as that illustrated in Fig. 5, is likely dependent on the NLP tool used. Authors employing state-of-the-art neural network models claim disambiguation accuracy >0.8 [36]. The commercial NLP software we currently use at Evotec has a "disambiguation score," which can be tuned for precision versus recall. Usually, when we run our queries we set the disambiguation threshold high to ensure that the entities returned are accurate. There are occasions where the "hits" returned for a particular entity are so few or ambiguous in nature that they warrant the high recall returned by using a lower disambiguation threshold. Typically, we would then manually check the results for false positives and filter the data appropriately.

3.2 Relationships Are Wrong Because They Lack Context

This example comes from the SemMedDB database [13], which uses a NER engine called "meta-mapper" to identify biomedical entities and a subject-predicate-object extraction tool called "SemRep" [37]. The entirety of MEDLINE is parsed to create the SemMedDB dataset. Figure 6 shows an example of one such triple

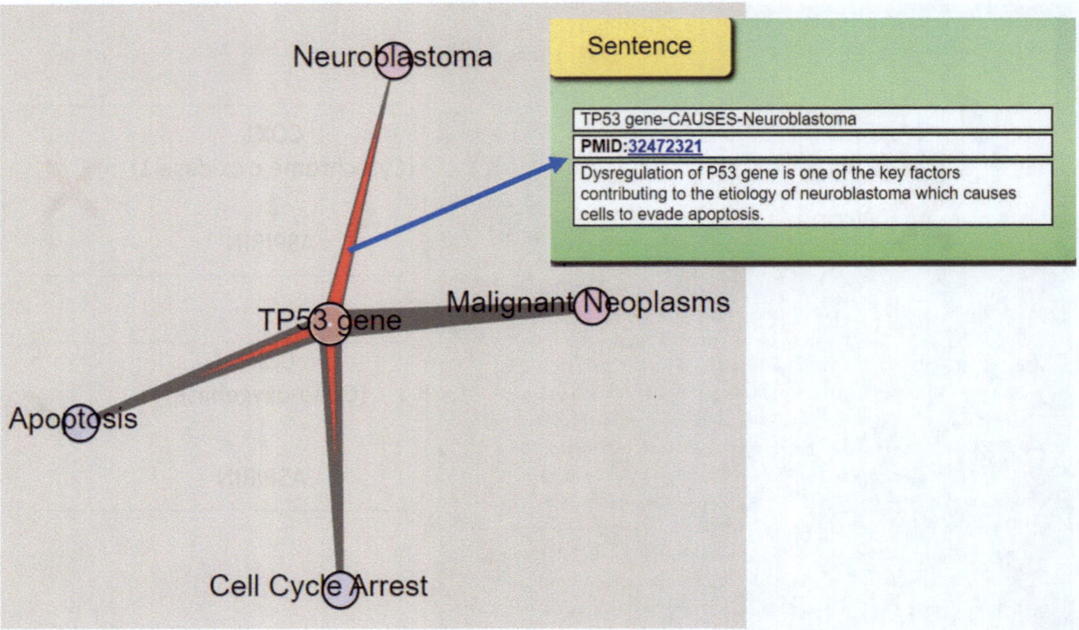

Fig. 6 An illustration of how assertions derived from NLP can be wrong because context is missing. Here, SemMedDB asserts that "TP53 gene-CAUSES-Neuroblastoma," whereas, in reality, it is the dysregulation of TP53 that is a factor causing neuroblastoma. The same error also occurs with "Malignant Neoplasms." The screenshot is from the SemMedDB browser after searching for "TP53" and filtering on "CAUSES" (https:/ii.nlm.nih.gov/SemMed/semmed.html—a UMLS licence is required)

that is, at best, highly misleading because crucial context is omitted. The bare triple "TP53 gene-CAUSES-Neuroblastoma" is incorrect as TP53 regulates the cell cycle and so—when functioning correctly—prevents cancer rather than causes it [38]. In the sentence from which the triple is extracted, the key context word is "dysregulation." Thus, a more accurate representation of the sentence would be "TP53[dysregulated]-CAUSES-Neuroblastoma."

Drawing conclusions from networks with this sort of error could lead to bad decisions. Taken literally, if "TP53 causes neuroblastoma," a reasonable hypothesis might be that inhibiting TP53 could treat the disease. In reality, the reverse is true; if we were to choose TP53 as a target to treat neuroblastoma we would probably want to restore its wild-type function rather than inhibiting it. The molecular biology of TP53 is well studied so this error is easily identified. However, there are many proteins whose functions are unclear and where this type of error could easily go unnoticed.

To mitigate against the pitfall raised above, Evotec's knowledge graph has a "ProteinFeature" node that allows us to assess more accurately which features of a protein are related to diseases and whether that feature might be a mutation or a modification. Here, we refer readers back to Subheading 2.2.2 for a detailed example.

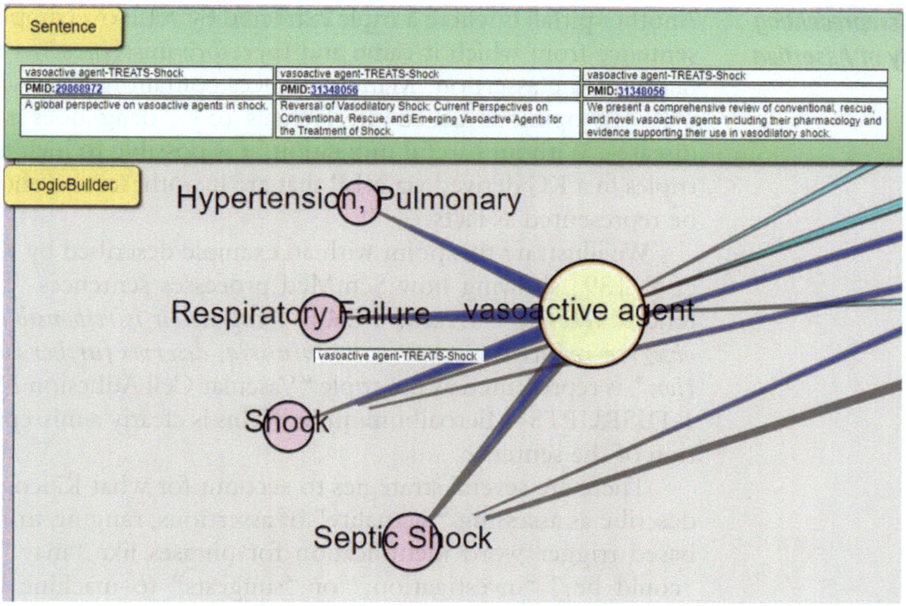

Fig. 7 An illustration of how NLP can derive assertions that can be considered "noise"—that is, not untrue but arguably unhelpful. This example from SemMedDB asserts the generic triple "vasoactive agent-TREATS-shock" (and also hypertension, respiratory failure, and septic shock). While the assertion is undoubtably true, the term "vasoactive agent" is so general as to be unhelpful for most drug discovery purposes. The screenshot is from the SemMedDB browser after searching for "vasoactive agent" and filtering on "TREATS" (https:/ii.nlm.nih.gov/SemMed/semmed.html—a UMLS licence is required)

3.3 Adding Noise (Assertions Are Not Incorrect But Are Generally Unhelpful Due to Insufficient Granularity)

Another pitfall is adding "noise" to the knowledge graph, which we define as data that is not necessarily incorrect but just generally unhelpful. Figure 7 shows an example of this from SemMedDB. Here, we have the triple "Vasoactive agent-TREATS-shock," which is derived from sentences such as "We present a comprehensive review of conventional, rescue and novel vasoactive agents including their pharmacology and evidence supporting their use in vasodilatory shock." The entity "vasoactive agent" is a legitimate biological entity with an associated UMLS CUI [22], which is why it has been identified by SemMed's NER algorithm. However, we argue that for most, if not all, knowledge graph tasks these so-called "parent terms" (other examples of which include "pharmaceutical substances," "vaccines," and "antibiotic agents") in an ontology like UMLS are just adding noise. Almost always, we are interested in which *specific* vasoactive agents treat a disease. Fortunately, it is relatively simple to filter out these terms by manually compiling a list—made easier as many nuisance terms often contain keywords such as "agent"—or by filtering using the hierarchical structure of the ontology. Thus, SemMed can be filtered for more specific triples. Of note, we use commercial NLP software that has a "leaf node only" option, so that "parent terms" can be easily excluded.

3.4 Misrepresenting Certainty of Assertion

Another pitfall is where a triple extracted by NLP oversimplifies the sentence from which it came and therefore misrepresents the certainty of the assertion. Many sentences contain hypotheses, speculations, opinions or counterfactuals (e.g., drug does *not* treat disease). Without careful mitigation, it is possible to include many triples in a KG derived via NLP that are hypotheses and should not be represented as facts.

We illustrate this point with an example described by Kilicoglu et al. [39] showing how SemMed processes sentences. The sentence "*Whether decreased VCAM-1 expression is responsible for the observed reduction in microalbuminuria, deserves further investigation*". is represented as the triple "Vascular Cell Adhesion Molecule 1-DISRUPTS- Microalbuminuria." This is clearly a misrepresentation of the sentence.

There are several strategies to account for what Kilicoglu et al. describe as assessing "factuality" of assertions, ranging from rules-based trigger word identification for phrases like "may," "not," "could be," "investigation," or "suggests" to machine learning approaches. Kilicoglu et al. developed a scoring system to categorize assertions as "certain_not," "doubtful," "possible," "probable", or "certain."

At Evotec we are working toward incorporating such a system into our pipeline for extracting triples. Currently, we exclude sentences that contain trigger words like "not" or "investigated." This serves for the short term as it ensures we have high precision (certainty of an assertion) but it does also mean we potentially decrease recall. One could envisage including all assertions into a knowledge graph with a "factuality" score, which could be filtered depending on the users' need.

4 Discussion

In this chapter, we have provided several examples of where we believe NLP holds the most promise in terms of expansion of KGs. At the same time, we have illustrated a number of pitfalls that can easily be encountered if one is not cautious when using NLP.

We have used NLP to expand Evotec's knowledge graph in an iterative and evolving process and will continue to do so as the need for new data becomes apparent. These needs are identified by using graph to try to answer real-world questions that arise in the projects we are working on, as well as building up robust datasets for AI–/ML-based approaches.

This workflow is illustrated in our first example where we have expanded the data relating to protein-protein interactions. We were not entirely satisfied with the data we were able to retrieve from structured data sources such as STRING because we felt useful relationship types were missing. Using NLP we could add biologically relevant verbs like "phosphorylates" to describe the relationships between proteins. We also found we could add directional or causative verbs like "activates" and "inhibits," which describe the effects one protein has on another. We have found that, when exploring relationships in the KG, having this information readily available made the graph much more useful in the analysis and decision-making process. Prioritizing proteins as potential targets is an example use case.

We have discussed several pitfalls we have encountered when thinking about expanding the KG using NLP or NLP-derived datasets. For example, one might be tempted to expand a KG using the SemMedDB [13] dataset, which extracts "subject-predicate-object" triples from the entirety of MEDLINE. However, as we have illustrated, such a "broad-brush" NLP engine can result in many misrepresentations and errors that could corrupt a KG. That is not to say that SemMedDB is not useful or to be trusted, rather it needs careful filtering before any of its data is incorporated into a KG. Our approach at Evotec to circumnavigate many of the pitfalls we have highlighted is to create highly stringent "domain-specific" NLP queries. These ensure that extracted data is of high quality and can be trusted (high precision), perhaps at the expense of some recall.

Unfortunately, the pitfalls faced when attempting to expand KGs using NLP are compounded by the fact that biomedical KG "schemas" (aka metagraphs or KG model) are not standardized within the field. Published biomedical KGs vary in the data they use and the "schema" they choose to represent the data. This is particularly relevant to NLP as the possibilities of including various aspects and combinations of entities and relationships are numerous, with a corresponding potential for variation between knowledge graphs. For example, we have used the schema "Gene-HAS_FEATURE-ProteinFeature," where attributes of the ProteinFeature node have fields such as "type_of: mutagenesis" that can then be linked to a disease. However, it could be equally valid to use a "Gene-MUTANT_VERSION-Disease" representation, with the details of the mutation being an attribute of the mutant version edge.

It is possible that in the coming years there may be a coalescence around a standardized method of constructing, analyzing, and interpreting a biomedical KG from publically available data sources. This would include data extracted from the literature using NLP pipelines. Standardization efforts are already underway with examples such as BIOLINK [40], where a canonical schema or model for how a KG should be constructed is proposed. That said, the task of "reducing" complex biological data down to the basic framework of the KG—the sematic triple—is a challenging task.

There may be no "correct" model for this, as various representations could be equally valid. However, some unification in the biomedical KG field is probably needed on the basis that, currently, two separate KGs constructed from broadly the same public sources can give wildly different predictions after embedding algorithms are applied to them. For example, in a drug repurposing task for COVID-19, only one drug overlapped from the top 30 predictions made by two separate authors using their own KGs [41, 42], even though similar embedding models were used. In related work, Ratajckzak et al. showed that the relative ranking of drugs predicted to treat SARS-CoV2 depended on the KG (Hetionet vs DRKG) and the subset of data the embedding models were trained on [43].

Related to the above example is the fact that different users often require distinct levels of detail in a particular domain. For instance, those working in proteomics may want details on gene transcript isoforms (gene->transcript->protein), while for others working in synthetic chemistry, this granularity may just obscure rather than reveal. One potential route around this (and one we use with Evotec's KG) is to expose different subsets of the data to different users. As Ratajckzak et al. demonstrated, domain specific models are more accurate, so creating predictive models for use in a particular domain is likely necessary. Overall, we expect that through a trial-and-error process, and as more scientists work with KGs, their usefulness to the drug discovery endeavor will improve year by year.

5 Conclusion

The process of biomedical KG construction, analysis, and interpretation is still a relatively new field. The promise of using NLP to expand or even form the basis of KGs for drug discovery is huge. However, there are also many pitfalls that should be considered and could, if ignored, compromise the integrity of a KG. Here, we have discussed a few examples, including how NLP can add highly valuable descriptions and context to relationships between entities. We have also demonstrated some common pitfalls that include errors relating to named entity recognition and misrepresenting the true meaning of sentences so that opinions are introduced into the KG as facts. At Evotec, we have been updating the data and modifying the schema of our KG in an iterative fashion as we use it to answer various drug discovery–related questions. We have found using domain-specific NLP queries very useful in expanding the KG where data from structured sources was lacking. Overall, we believe that carefully introduced data derived from NLP should be an essential part of a biomedical KG data curation pipeline.

References

1. Ehrlinger L, Wöß W (2016) Towards a definition of knowledge graphs. In: CEUR Workshop Proceedings
2. Bonner S, Barrett IP, Ye C, Swiers R, Engkvist O, Bender A, et al (2021) A review of biomedical datasets relating to drug discovery: a knowledge graph perspective
3. Nicholson DN, Greene CS (2020) Constructing knowledge graphs and their biomedical applications. Comput Struct Biotechnol J 18: 1414–1428
4. Doğan T, Atas H, Joshi V, Atakan A, Rifaioglu AS, Nalbat E et al (2021) CROssBAR: comprehensive resource of biomedical relations with knowledge graph representations. Nucleic Acids Res 49(16):e96–e96
5. Himmelstein DS, Lizee A, Hessler C, Brueggeman L, Chen SL, Hadley D et al (2017) Systematic integration of biomedical knowledge prioritizes drugs for repurposing. elife 6:e26726
6. Su C, Hou Y, Guo W, Chaudhry F, Ghahramani G, Zhang H, et al (2021) CBKH: The Cornell Biomedical Knowledge Hub. medRxiv
7. Maglott D, Ostell J, Pruitt KD, Tatusova T (2007) Entrez Gene: gene-centered information at NCBI. Nucleic Acids Res 35(Database issue):D26
8. Cunningham F, Allen JE, Allen J, Alvarez-Jarreta J, Amode MR, Armean IM et al (2022) Ensembl 2022. Nucleic Acids Res 50 (D1):D988–D995
9. Geleta D, Nikolov A, Edwards G, Gogleva A, Jackson R, Jansson E, et al (2021) Biological insights knowledge graph: an integrated knowledge graph to support drug development. bioRxiv. 2021.10.28.466262
10. Martin B, Jacob HJ, Hajduk P, Wolfe E, Chen L, Crosby H, et al (2022) Leveraging a billion-edge knowledge graph for drug re-purposing and target prioritization using genomically-informed subgraphs. bioRxiv. 2022.12.20.521235
11. Paliwal S, de Giorgio A, Neil D, Michel JB, Lacoste AM (2020) Preclinical validation of therapeutic targets predicted by tensor factorization on heterogeneous graphs. Sci Rep 10(1):18250
12. Nicholson DN, Himmelstein DS, Greene CS (2022) Expanding a database-derived biomedical knowledge graph via multi-relation extraction from biomedical abstracts. BioData Min 15(1):26
13. Kilicoglu H, Shin D, Fiszman M, Rosemblat G, Rindflesch TC (2012) SemMedDB: a PubMed-scale repository of biomedical semantic predications. Bioinformatics 28(23):3158
14. Wei CH, Allot A, Leaman R, Lu Z (2019) PubTator central: automated concept annotation for biomedical full text articles. Nucleic Acids Res 47(W1):W587–W593
15. Braicu C, Buse M, Busuioc C, Drula R, Gulei D, Raduly L et al (2019) A comprehensive review on MAPK: a promising therapeutic target in cancer. Cancers (Basel) 11(10)
16. Szklarczyk D, Gable AL, Nastou KC, Lyon D, Kirsch R, Pyysalo S et al (2021) The STRING database in 2021: customizable protein-protein networks, and functional characterization of user-uploaded gene/measurement sets. Nucleic Acids Res 49(D1):D605–D612
17. Xiong C, Liu X, Meng A (2015) The kinase activity-deficient isoform of the protein araf antagonizes Ras/Mitogen-activated protein kinase (Ras/MAPK) signaling in the zebrafish embryo. J Biol Chem 290(42):25512
18. Brandes U (2010) A faster algorithm for betweenness centrality. J Math Sociol 25(2): 163–177
19. Zhang J-X, Chen D-B, Dong Q, Zhao Z-D (2016) Identifying a set of influential spreaders in complex networks. Sci Rep 6(1):27823
20. Approved Drug Products with Therapeutic Equivalence Evaluations | Orange Book [Internet]. Available from: https://www.fda.gov/drugs/drug-approvals-and-databases/approved-drug-products-therapeutic-equivalence-evaluations-orange-book
21. Wishart DS, Feunang YD, Guo AC, Lo EJ, Marcu A, Grant JR et al (2018) DrugBank 5.0: a major update to the DrugBank database for 2018. Nucleic Acids Res 46(D1):D1074–D1082
22. Bodenreider O (2004) The Unified Medical Language System (UMLS): integrating biomedical terminology. Nucleic Acids Res 32 (suppl_1):D267–D270
23. Doycheva D, Jägle H, Zierhut M, Deuter C, Blumenstock G, Schiefer U et al (2015) Mycophenolic acid in the treatment of birdshot chorioretinopathy: long-term follow-up. Br J Ophthalmol 99(1):87–91
24. Finlayson SG, LePendu P, Shah NH (2014) Building the graph of medicine from millions of clinical narratives. Sci Data 1(1):140032
25. Lowe HJ, Ferris TA, Hernandez PM, Weber SC (2009) STRIDE – An Integrated

Standards-Based Translational Research Informatics Platform. AMIA Annu Symp Proc 2009:391

26. Pollard TJ, Johnson AEW, Raffa JD, Celi LA, Mark RG, Badawi O (2018) The eICU Collaborative Research Database, a freely available multi-center database for critical care research. Sci Data 5(1):180178
27. Johnson AEW, Pollard TJ, Shen L, Lehman LH, Feng M, Ghassemi M et al (2016) MIMIC-III, a freely accessible critical care database. Sci Data 3(1):160035
28. Malki MA, Dawed AY, Hayward C, Doney A, Pearson ER (2021) Utilizing large electronic medical record data sets to identify novel drug–gene interactions for commonly used drugs. Clin Pharmacol Ther 110(3):816–825
29. Kuhn M, Letunic I, Jensen LJ, Bork P (2016) The SIDER database of drugs and side effects. Nucleic Acids Res 44(Database issue):D1075
30. Koskinen M, Salmi JK, Loukola A, Mäkelä MJ, Sinisalo J, Carpén O et al (2022) Data-driven comorbidity analysis of 100 common disorders reveals patient subgroups with differing mortality risks and laboratory correlates. Sci Rep 12(1):1–9
31. Bodenreider O (2004) The Unified Medical Language System (UMLS): integrating biomedical terminology. Nucleic Acids Res 32 (Database issue):D267
32. Leaman R, Lu Z (2016) TaggerOne: joint named entity recognition and normalization with semi-Markov Models. Bioinformatics 32(18):2839–2846
33. Sosa DN, Derry A, Guo M, Wei E, Brinton C, Altman RB (2020) A literature-based knowledge graph embedding method for identifying drug repurposing opportunities in rare diseases. Pac Symp Biocomput 2020(25): 463–474
34. Percha B, Altman RB (2018) A global network of biomedical relationships derived from text. Bioinformatics 34(15):2614–2624
35. Fujiyoshi K, Bruford EA, Mroz P, Sims CL, O'Leary TJ, Lo AWI et al (2021) Standardizing gene product nomenclature-a call to action. Proc Natl Acad Sci U S A 118(3): e2025207118
36. Skreta M, Arbabi A, Wang J, Drysdale E, Kelly J, Singh D et al (2021) Automatically disambiguating medical acronyms with ontology-aware deep learning. Nat Commun 12(1):1–10
37. Kilicoglu H, Rosemblat G, Fiszman M, Shin D (2020) Broad-coverage biomedical relation extraction with SemRep. BMC Bioinform 21(1):1–28
38. Mantovani F, Collavin L, Del Sal G (2018) Mutant p53 as a guardian of the cancer cell. Cell Death Differ 26(2):199–212
39. Kilicoglu H, Rosemblat G, Rindflesch TC (2017) Assigning factuality values to semantic relations extracted from biomedical research literature. PLoS One 12(7):e0179926
40. Unni DR, Moxon SAT, Bada M, Brush M, Bruskiewich R, Caufield JH et al (2022) Biolink Model: a universal schema for knowledge graphs in clinical, biomedical, and translational science. Clin Transl Sci 15(8):1848–1855
41. Zeng X, Song X, Ma T, Pan X, Zhou Y, Hou Y et al (2020) Repurpose open data to discover therapeutics for COVID-19 using deep learning. J Proteome Res 19(11):4624–4636
42. Zhang R, Hristovski D, Schutte D, Kastrin A, Fiszman M, Kilicoglu H (2020) Drug Repurposing for COVID-19 via Knowledge Graph Completion. J Biomed Inform 115:103696
43. Ratajczak F, Joblin M, Ringsquandl M, Hildebrandt M (2022) Task-driven knowledge graph filtering improves prioritizing drugs for repurposing. BMC Bioinform 23(1):84

Chapter 11

Alchemical Free Energy Workflows for the Computation of Protein-Ligand Binding Affinities

Anna M. Herz, Tahsin Kellici, Inaki Morao, and Julien Michel

Abstract

Alchemical free energy methods can be used for the efficient computation of relative binding free energies during preclinical drug discovery stages. In recent years, this has been facilitated further by the implementation of workflows that enable non-experts to quickly and consistently set up the required simulations. Given the correct input structures, workflows handle the difficult aspects of setting up perturbations, including consistently defining the perturbable molecule, its atom mapping and topology generation, perturbation network generation, running of the simulations via different sampling methods, and analysis of the results. Different academic and commercial workflows are discussed, including FEW, FESetup, FEPrepare, CHARMM-GUI, Transformato, PMX, QLigFEP, TIES, ProFESSA, PyAutoFEP, BioSimSpace, FEP+, Flare, and Orion. These workflows differ in various aspects, such as mapping algorithms or enhanced sampling methods. Some workflows can accommodate more than one molecular dynamics (MD) engine and use external libraries for tasks. Differences between workflows can present advantages for different use cases, however a lack of interoperability of the workflows' components hinders systematic comparisons.

Key words AFE, Alchemical free energy methods, FEP, Free energy perturbation, RBFE, Relative binding free energies, Workflows

1 Introduction to AFE

1.1 Recent History

For the past 30 years, alchemical free energy (AFE) methods, also often referred to as free energy perturbation (FEP) methods, have been constantly developed and improved for the purpose of computer-aided drug design. The widespread availability of graphical processing units (GPUs) in recent years has made the use of FEP for drug design easier by enabling the efficient parallelization of molecular dynamics (MD) simulations [1–5].

These continuous advancements in AFE methods include novel proprietary OPLS force fields developed by the company Schrödinger and publicly available force fields, such as the general Amber force field 2 (GAFF2) and force fields from the Open Forcefield

Alexander Heifetz (ed.), *High Performance Computing for Drug Discovery and Biomedicine*, Methods in Molecular Biology, vol. 2716, https://doi.org/10.1007/978-1-0716-3449-3_11,

Initiative [6–10]. The implementation of enhanced sampling techniques, such as replica exchange with solute tempering (REST/REST2) and alchemical enhanced sampling (ACES), has further advanced the field [11–13].

With the combination of increased computing power and improved methods, AFE methods have become viable for practical drug discovery applications, particularly during the hit-to-lead optimization stages of a drug discovery program. Examples of this include studies on imidazotriazines, KRAS, norovirus, phosphodiesterase 2A inhibitors, and polycomb repressive complex 2 [14–18].

1.2 Using Alchemical Methods to Calculate Relative Binding Free Energies

AFE methods are a family of statistical mechanics-based techniques used for computing the free energy differences between two distinct thermodynamic states. The states are represented by a collection of microscopic states, which are typically sampled through MD or metropolis Monte Carlo (MC) algorithms. AFE methods have been used to model various physiochemical properties relevant to drug discovery, such as hydration free energies, buried water displacement, and covalent inhibitors [19–21]. The focus of this chapter is on AFE workflows used to compute relative binding free energies (RBFE), one of the most mature applications of AFE methods. Calculating the binding energy is crucial for drug discovery as drugs typically act by forming a stoichiometric non-covalent complex with a biomolecule. Accurate predictions of binding affinities can reduce costs in the preclinical drug discovery stages by only advancing to the expensive synthesis and biological characterization stage for ligands with a high likelihood of achieving the desired affinity profile [3, 22].

In RBFE calculations, the two thermodynamic states of interest only differ in the nature of the ligand. For numerical reasons, it is essential that a ligand is transformed into another gradually. This is accomplished by gradually modifying the potential energy function that describes the interactions of the ligand atoms with their surrounding environment. During this process, non-physical intermediate states are introduced, and thus the method is referred to as "alchemical." The initial and final states of this transformation are referred to as endstates, and they correspond to chemically plausible molecules. This transformation can be mathematically expressed through the introduction of a coupling parameter, λ, with $\lambda = 0$ representing the initial state and $\lambda = 1$ representing the final state. In practice, the transformation is divided into discrete λ values, and MD simulations are carried out at each λ value. The analysis of the sampled microstates for ligand or protein-ligand structures then enables the computation of protein-ligand binding affinities. The RBFE is computed by subtracting the free energy change of the solvated ligand's transformation from the free energy change of the bound ligand's transformation. This approach is substantially more computationally efficient than trying to directly simulate the reversible binding/unbinding of the ligand to the target protein. There are several excellent papers that explain the theory behind AFE calculations in detail [23–28].

1.3 Other Binding AFE Methods

Most commonly, RBFE is used to rank a congeneric series of ligands in terms of binding affinity and, therefore, it is the ligands' R groups that differ. A variation of RBFE is scaffold hopping. During scaffold hopping, the R groups of a ligand important for the binding to a protein receptor are conserved whilst the core is changed. This strategy is popular to improve the solubility of a compound and its pharmacokinetic properties, or to bypass patent claims around existing structures [29]. A scaffold hopping FEP approach with a soft bond stretch potential has been implemented with initial success using FEP+, and the alternative dual-topology approach used in the QLigFEP workflow has been shown to overcome issues associated with changing bond topologies [30–32].

Other AFE methods include calculating the binding affinities for different protein mutations, where a protein residue is perturbed instead of the ligand, such as in the PMX workflow [33–35]. For absolute binding free energies (ABFE), it is the complete disappearance of the ligand instead of the perturbation into another. Although in theory similar to RBFE, there are many practical challenges to consider, such as the necessity of restraining the ligand to the binding site and the introduction of correction terms to account for the influence of the restraints [1, 3, 36]. Some ABFE workflows have started to appear, but these will not be discussed further here [37–39].

2 Introduction to RBFE Workflows

2.1 Running RBFE Simulations

A significant amount of technical knowledge is required to run an RBFE simulation from scratch, as in-depth knowledge of the MD engine, the setup, the parameters, and sensible default settings is needed [40–42]. In workflows, this is carried out via sophisticated scripts that handle many of these features, particularly the generation of all the required input files. Often python scripts are used, and some workflows, especially proprietary ones, have intuitive GUIs. The default values provided by workflows aim to be robust and suitable for a diverse range of inputs. Although there is still always some required user input, depending on the workflow this can be reduced significantly, making it a lot easier for non-experts to run RBFE simulations.

Typical drug discovery programs can be usefully driven by binding free energy predictions when the number of compounds that are computationally assessed exceeds experimental capability to prepare and test compounds by at least one order of magnitude.

This means that an FEP campaign can contain tens or even hundreds of ligands. For many drug discovery organizations, this translates into a time limit of a few hours per compound at most [43]. With modern GPUs, the actual execution time of a typical RBFE calculation is under this target, and other considerations such as the degree of automation of the setup and analysis steps of the workflow become important. For example, in the FEW workflow released in 2013, the estimated time for setting up an initial TI simulation by a familiar user is 2 days, while the time taken with the support of the workflow and some user intervention is 30 minutes [40]. This is a huge increase in efficiency, which has only improved further with more recent workflows. For example, previously setting up NAMD/FEP input files would take an experienced user approximately a full day of work, whereas the FEPrepare workflow achieves this in minutes or even seconds [44]. For the PyAutoFEP workflow, it takes about half a minute for system building and input preparation per perturbation on a desktop computer [45].

Additionally, the process of setting up the simulations is also tedious and can easily result in errors as the user has to edit many files and reorder inputs; using a workflow alleviates much of this [42, 44]. The chance of finding a molecule better than a lead when using the PyAutoFEP pipeline was investigated, and this probability was found to increase up to sevenfold when 100 molecules were screened [45].

2.2 Components of an RBFE Workflow

The general features of an RBFE workflow include parameterization of the protein and the ligands, including the choice of forcefields and solvation model; generation of perturbation networks; mapping the atoms and topology generation; input file generation; equilibration and running the MD simulation based on a certain sampling scheme; and finally extracting and analyzing the data, as well as analyzing the perturbation network. Here we include as a RBFE workflow anything that covers more than one of these steps. For example, there could be a setup workflow that generates input files such as CHARMM-GUI, while other workflows can run all the steps, such as PyAutoFEP or the Flare software [45–47].

Below will be a more detailed breakdown of the workflow and the steps involved in setting up an FEP simulation.

2.3 Preparing for an RBFE Workflow—Parameterizing Inputs and Ligand Poses

Before starting the RBFE workflow, suitable ligand and protein files must be selected. What is meant by this is that the protein must be well resolved, with no missing residues and appropriate protonation states. Ligand structures must adopt the expected bioactive conformation, and chirality and tautomerization states must be clearly defined. For prospective applications of RBFE, if the bioactive conformation of the ligand is not known it must be inferred, for instance, via docking calculations [48, 49]. The docking is usually followed by in-pocket optimization or short MD equilibration simulations. It is important to note that, with the simulation timescales typically used, conformational sampling is usually insufficient to enable relaxation of an incorrect pose into the correct bioactive conformation, so the starting structure is important [23, 50]. Generating the starting poses is outside of the scope of what is considered to be part of an RBFE workflow here, as this presents a set of challenges entirely separate from any alchemical considerations; however, some of the later introduced workflows do have methods for preparing the ligand poses [45, 47, 51–53].

2.4 Defining the Perturbable Molecule

For the RBFE simulation, the perturbable molecule needs to be defined; it needs to be specified how atoms in the initial state map into atoms in the final state, and how their transformation will proceed. As the number of atoms in the canonical ensemble must stay constant, this frequently involves the creation of dummy atoms [54]. Different approaches to defining the perturbable molecule will be described below.

2.4.1 Topology

For RBFE simulations, the topology used plays a large part during the setup of a simulation, particularly how the atoms are defined and represented in the input files. There are three different coupling methodologies that will be discussed: single, dual, and a hybrid approach.

A single topology approach makes use of the fact that similar molecules share a common substructure. This common substructure includes any transforming atoms and is augmented with the additional atoms of each ligand represented by dummy atoms, which retain their bonded terms but have no non-bonded interactions with the system at their respective endstates. During the transformation, the parameters of these dummy atoms, as well as those of transforming atoms, are gradually adjusted between their start/end state to either turn on/off their non-bonded interactions or to change their type [23]. A limitation of this approach is that the explicit modeling of dummy atoms requires careful handling of their energy terms, so that their contribution to the free energy will overall cancel, while their position and orientation remain well defined [54, 55]. This is particularly limiting for the case of breaking/forming ring systems, where the contributions of the dummy atoms to the free energy of the system often do not cancel out [56].

A dual topology approach may be preferred for scaffold hopping approaches, such as in the QLigFEP workflow, or if no common substructure is shared between the ligands. In a strict dual topology approach, both molecules are present at the same time but do not interact with each other, and are either fully coupled/decoupled from their surrounding environment at their respective start/end state with the rest of the system. The surrounding system sees a mixture of both states, as the total potential energy is a linear combination of the potential energy of the two states across the whole perturbation [32, 57]. Complications can arise as the end-state molecules, when fully decoupled from their environment, can drift apart which necessitates the use of restraints to keep both molecules spatially coupled [58]. The shared atom pairs in the ligands can be coupled to each other to avoid the non-interacting ligand leaving the binding site, and as the energetic term of the restraints cancel, no additional correction needs to be applied unlike what needs to be done for ABFE [32].

The final discussed topology is a hybrid dual topology approach. Although "dual topology" may also sometimes be used to describe this hybrid approach, here we strictly differentiate them. In the hybrid approach, the non-common atoms between the end states are identified and implicitly treated as dummy atoms. All changing atoms are dummy atoms, including those that in the single topology approach would only be changing their type. The common substructure, which in this case must have all the same atom types, strictly maintains identical coordinates throughout the transformation, or may even be shared depending on the chosen scheme and MD engine input format [36, 55].

For example, the pmemd MD engine from the AMBER software suite implements a hybrid approach, where the shared "single" topology region serves to keep the ligands in place as all the coordinates are shared, although each ligand has its own set of atoms for the shared region. The atoms in the "single" region are continuously present and strictly maintain the same coordinates while being non-interacting, and all the perturbing atoms are represented as dummy atoms in their respective end states [22, 41]. A single topology approach is implemented for example in SOMD, and a strict dual topology is adopted in the QLigFEP workflow for use with the Q MD engine [32, 55]. Traditionally, CHARMM and NAMD have supported dual topology implementations, although modern versions also allow a hybrid approach [44, 57, 58].

2.4.2 Atom Mappings

Before creating the topology for a system, it is necessary to map the ligands to each other. This process can be extremely time-consuming during an FEP campaign. However, it can be made more efficient through automation, which also ensures consistent treatment of the mapping decisions, regardless of the user.

The most widely used method to perform the mapping computationally is by utilizing a maximum common substructure (MCS) approach or its variations. This step can often be incorporated into a workflow by utilizing external dependencies, such as RDKit [41, 43]. The MCS search is essentially a technique to maximize the shared atoms between two perturbing ligands, thus minimizing the number of atoms that are changed. In general, the more the atoms are perturbed, the more challenging the perturbation is considered to be, and the longer it may take to converge. It is crucial to include hydrogen atoms for the mapping to be successful.

Different workflows utilize different algorithms to find the MCS. For example, FESetup uses the "fmcs" implementation from RDKit to achieve maximum overlap [41, 59]. PyAutoFEP uses either a 3D-guided or graph-based MCS, with the 3D-guided method designed specifically for perturbations that involve the inversion of asymmetric centers, while the graph-based method is used for all other cases [45]. ProFESSA (Production Free Energy Simulation Setup and Analysis) can utilize the MCS algorithm from RDKit, as well as an extended-MCS method (MCS-E) that excludes atoms from the overlap region, such as those with different hybridization states, in order to prevent issues with cycle closure or unstable simulations. It also implements MCS-E_{nw}, which is a variation of MCS-E that ensures that all common and soft-core regions of a ligand molecule are identical for all its perturbations within a given network [60]. Overall, these different MCS algorithms aim to ensure mappings that lead to good convergence of simulations without losing important structural features of the ligands. This can be seen by the improvement to a 3D-guided MCS search, as a solely graph-based search can lead to loss of chiral information of the molecule due to the inherent 2D nature of the molecular graphs [41].

In many of these algorithms, such as FESetup, there is an additional criterion that prevents rings from being broken, due to the previously mentioned issues that arise when this happens [22, 61]. In BioSimSpace, it is possible to specify whether a ring can be broken during the mapping. Recently, a recursive joint-traversal superimposition algorithm has been implemented in TIES (Thermodynamic Integration with Enhanced Sampling) to allow for partial ring perturbation, only matching them if they have the same atom type. This improved control over the alchemical region enables a smaller number of perturbed atoms without negatively impacting the results [62].

An alternative approach to the conventional topology and atom mapping is the common core approach. Instead of transforming one ligand into another, all ligands are transformed to a common core [63]. An example of its usage can be found in the Transformato workflow [64].

2.5 Network Generation

The decision of which ligand to perturb into which other ligand is usually determined by a perturbation network. In practice, the chosen perturbations are not arbitrary, as the network should minimize the number of calculations required while providing stable, well-converged perturbation pathways. Many network generation tools are currently based on the Lead Optimization Mapper

(LOMAP) software, which focuses on minimizing atom insertions/deletions while avoiding ring-breaking/forming perturbations [65]. The workflows described here often use an adjusted LOMAP for network generation, as LOMAP can also take arbitrary similarity scores as input and its open-source code can be modified [45, 47, 51, 61, 66]. In the case of a common core approach, such as in Transformato, a network is not required as the relative binding affinities for all possible perturbations can be computed.

2.6 Running the Simulations

A variety of implementations are available for conducting the MD/Monte Carlo simulations necessary for AFE simulations, including but not limited to AMBER, SOMD, OpenMM, GROMACS, CHARMM, NAMD, FEP+ (Desmond), and Q. Each implementation has its own unique features and differences, such as the handling of constraints, treatment of long-range non--bonded interactions, topologies, and available enhanced sampling schemes, all of which are critical considerations when preparing suitable input files. The creation of these input files is managed by workflows, along with the generation of scripts for running the simulations, either locally or on high-performance computing (HPC) systems.

2.7 Sampling Methods

Most workflows use MD simulations for sampling, and some use enhanced sampling methods such as REST, REST2, and/or ACES. These are all permutations of Hamiltonian Replica Exchange, of which there is in-depth discussion in various papers [11–13, 67]. Enhanced sampling techniques are used to be able to access conformational states that are not accessible during regular MD, for example due to high-energy barriers, and thereby improve the sampling efficiency of regions of interest and the predicted free energy result [68]. Although not an "enhanced sampling" method per se, TIES makes use of an ensemble averaging of MD simulations to improve sampling and reproducibility [43, 62].

2.8 Analysis

Most modern AFE implementations use thermodynamic integration (TI), Bennett acceptance ratio (BAR), or multistate Bennett acceptance ratio (MBAR) for estimating free energies from the trajectories generated during by the simulations [69–72]. MBAR has been demonstrated to be a more reliable estimator of free energy changes, but TI remains popular. One advantage of TI is that additional simulations at different values of the coupling parameter can be run at a later stage to improve the precision of a free energy estimate. Several analysis libraries are available to implement such analyses, such as the PyMBAR and alchemlyb libraries, and often an MD package will have some form of analysis integrated [72, 73].

In addition to the analysis of the individual perturbations, there is the whole network analysis to consider as well. This frequently involves using a reference compound to comparatively rank the ligands in the network based on their relative free energy changes. Given that free energy is a thermodynamic state function, cycle closures in a network should theoretically be equal to zero. However, hysteresis is observed in practice due to random and/or systematic errors and different paths between two ligands may not give the same result. For choosing a path to use, this can be done for example by either using the shortest path, the average of the shortest paths, the average of all possible paths, or a weighted average of all paths. It can also be desirable to incorporate experimental reference values if available.

More advanced methods, such as cycle closure correction, have been used with FEP+ to achieve optimal estimates and evaluate their reliability [68]. Another module for the analysis of the network, BARnet/MBARnet, carries out a non-linear optimization of an objective function to enforce cycle closure conditions and can include experimental reference values, showing a reduction of error in the six systems considered [74]. Additionally, an MBARnet network analysis using Lagrange multipliers in the ProFESSA workflow for six perturbations between 4 ligands of CDK2 provided new information that allowed for the identification of "uncertain" ligands [60].

3 A Survey of Current RBFE Workflows

Currently available RBFE workflows can be broadly divided into academic or commercially available categories. This section does not aim to provide a comprehensive list, instead highlighting the key differences in how workflows operate, the approaches they can take, and their underlying principles. The workflows will be introduced roughly chronologically to demonstrate the advancements in the field and its potential for future growth.

An outline of all the discussed workflows is shown in Fig. 1.

3.1 FEW

FEW (free energy workflow) is an older academic workflow written in Perl and is designed to be run through the command line [40]. It is included with AmberTools from version 14 onwards. It covers the setup for a linear interaction energy (LIE), an implicit solvent molecular mechanics (MM-PBSA/MM-GBSA), or a thermodynamic integration (TI) approach. This includes automated ligand parameterization, checking input parameters, selecting the procedure, and setting up the MD simulation. As this is an older workflow, users need to specify which atoms are part of the softcore themselves.

The FEW workflow has a "gray box" approach, which is common in many academic workflows. It provides default settings, for example in template files, but the parameter settings can still be

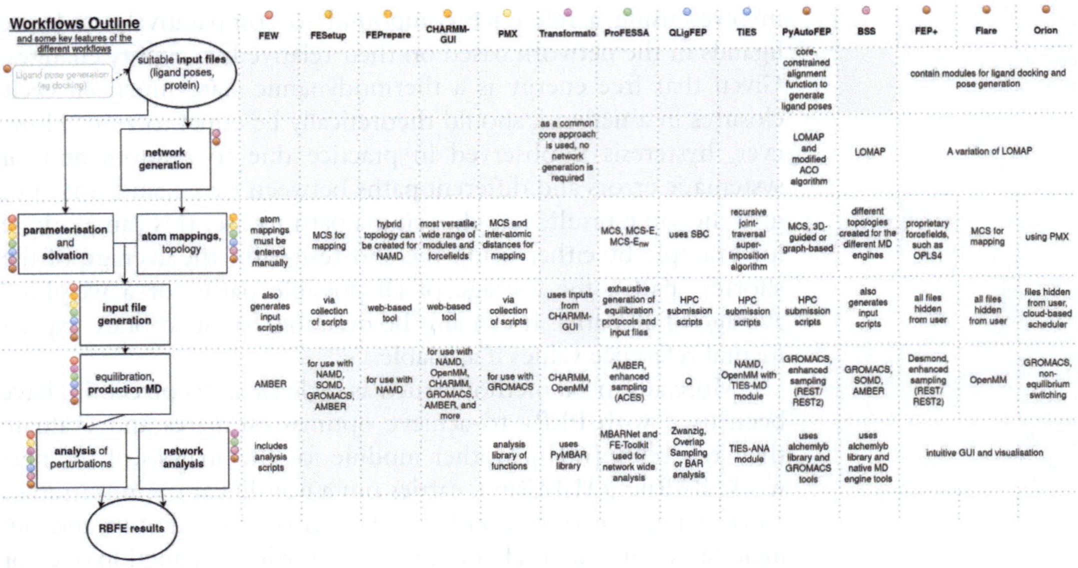

Fig. 1 An outline of the discussed AFE workflows for protein-ligand binding affinities (FEW, FESetup, FEPrepare, CHARMM-GUI, PMX, Transformato, ProFESSA, QLigFEP, TIES, PyAutoFEP, BSS, FEP+, Flare, Orion), showing some of their key features during each stage of the workflow

changed if desired. To run the simulations, the scripts must be modified to match the computing environment, which is another common feature of many academic workflows.

Although this workflow is no longer being developed, it highlights important design features of RBFE workflows, such as time improvement compared to manual setup, modularity for future expansion and adaptation, checking of provided inputs, and the "gray box" approach that allows experts to modify the workflow as needed.

3.2 FESetup

The next few introduced workflows focus on the setup stage only, which is the most labor-intensive part of running an RBFE simulation manually, and as such was the main focus of workflows initially. FESetup is an example of such a workflow, and it was designed as a component to be included in larger workflows [41]. Although it is deprecated and is no longer being maintained, it highlights some improvements and unique features in the construction of workflows. Firstly, a notable improvement compared to FEW is that the atom mappings are done automatically using an MCS approach, with the option for the user to add specific mappings for purposes such as preserving binding modes. FESetup has support for more than one MD engine, including AMBER, GROMACS, SOMD, and NAMD, making it very flexible. FESetup also introduced the concept of a molecular dynamics engine-agnostic API for processing inputs, which was achieved through the use of the Sire library to represent molecules in an engine-independent manner [75].

This highlights many desirable features in FEP workflows, such as the ability to support multiple MD engines, support multiple force fields, and provide suitable error and failure reporting, with the possibility of expanding the workflow to include more features. FESetup also has another common feature that is found among academic workflows, namely relying on external software tools—in this case, OpenBabel, RDKit, and AmberTools—to complete certain tasks and combining them into an efficient workflow via Python. FESetup was used successfully to set up alchemical perturbations in several studies, including Farsenoid X receptor inhibitors, 180 ligands of HSP90-α protein, bromodomain-containing protein 4, and cyclophilins, demonstrating the success of this approach [76–79].

3.3 FEPrepare

FEPrepare is a newer web-based workflow, and this offers the advantage of generally not having local hardware requirements or system-specific demands, making it easier to use immediately. For web-based workflows no installation is necessary, and developers do not have to worry about making the workflow available on different platforms. However, experienced users cannot edit or expand upon its functionality as they do not have access to the code locally. Due to the computational demands of running production simulations, the online aspect of workflows is usually limited to setup and input file generation, and the files must be downloaded and run locally. FEPrepare sets up files specifically for NAMD simulations using the OPLS-AA force field topology and parameter files [44].

A drawback of this workflow is that ligands must be aligned before running, unlike FESetup, which performs the alignment during mapping. The LigParGen server is used to parameterize the ligands based on a user-defined charge model and molecular charge, in case the charge needs to be redistributed for the hybrid protocol used [80]. This workflow uses a novel approach to create a hybrid file from the dual topology file, which reduces the number of perturbations needed for FEP/MD calculations and speeds up simulation convergence.

A local copy of VMD is required for solvation and is interfaced with the server via Tcl scripts. Currently, the setup only works for single ligand pairs and has been tested and found to work successfully for a CDK8 perturbation and an actin-related protein 2/3 perturbation [44].

3.4 CHARMM-GUI

CHARMM-GUI is a web-based setup workflow that was originally developed in 2006 and first published in a 2008 paper; since then, it has undergone significant development both in-house and through collaboration [42, 46]. The workflow is designed to be accessible to both inexperienced and experienced users, as they can optimize or adapt the protocols as needed. Currently, it can generate input files for various simulation engines, including CHARMM, NAMD,

GROMACS, AMBER, GENESIS, Tinker, LAMMPS, Desmond, and OpenMM.

One of the key benefits of the web-based interface of CHARMM-GUI is its interactivity. If a user encounters any issues or sees any problems during visualization, they can go back to a previous step and regenerate the system. Each stage allows for user input, and the setup is split into modules, including but not limited to a PDB Reader & Manipulator, Ligand Reader & Modeler, Quick MD Simulator, Membrane Builder, Nanodisc Builder, HMMM Builder, Monolayer Builder, Micelle Builder, Hex Phase Builder, and Ligand Binder for ligand solvation and binding energy simulations [46, 81–95]. It can also accommodate a range of forcefields and repartition masses for hydrogen mass repartitioning (HMR) [96, 97].

The modular approach of CHARMM-GUI allows it to be one of the most versatile workflow discussed in this review, as this approach has inherently allowed for new features to be incorporated. The different modules can be easily combined. For example, the use of the Membrane Builder module with the Relative Ligand Binder for FEP simulations on the adenosine A2A membrane receptor showed consistency with both experimental results and previously published free energy results [89]. In conclusion, CHARMM-GUI is a versatile and user-friendly workflow that provides a wide range of input generators for various computational modeling processes, with many additional modules available as well.

3.5 Transformato

Transformato is a Python package that covers the running and analysis of FEP simulations and is designed specifically to be used with outputs from CHARMM-GUI [64]. It supports the running of simulations with OpenMM and CHARMM by generating input scripts. It is the only described workflow here to use the common core approach for perturbing atoms. The use of the serial atom insertion (SAI) approach in combination with the common core approach in Transformato eliminates the need for customized softcore potentials and has the potential to be expanded to different MD engines [63, 98]. Four out of five model applications the workflow was tested on performed similarly to previous studies, with the system that did not perform as well being attributed to the forcefield rather than the common core approach. For analysis, Transformato generates scripts for all post-processing of results and uses the PyMBAR library to calculate the free energy differences [72].

3.6 PMX

PMX was initially released as a collection of python scripts for generating hybrid structures for amino acid mutations to run in GROMACS. It has since expanded to include a webserver for generating input files and setting up ligand perturbations as well with scripts available on Github [33, 99, 100]. It is generally used as a workflow tool within larger custom workflows. PMX has been used to set up a complete workflow for a non-equilibrium TI approach for 13 different protein-ligand datasets and 482 perturbations. PMX used an MCS approach for mapping between ligands and then compared it to an approach where the proposed mapping is based on the inter-atomic distances between the two ligands. The chosen mapping, with the most atoms identified for direct morphing, is then used to generate the single topology input files required for GROMACS. The mean absolute error (MAE) of this workflow for thirteen different systems was found to be 0.87 ± 0.03 kcal/mol which was comparable with that of FEP+ [100].

PMX has also been utilized to develop an automated workflow for use with a cloud-based computer cluster provided via Amazon Web Services (AWS). This workflow consisted of 19,872 independent simulations that were capable of running on up to 3000 GPUs simultaneously and completed in just 2 days [101]. This was made possible by the abstraction offered by the AceCloud service, which allows users to run MD simulations with ACEMD, NAMD, AMBER, and GROMACS on AWS without directly interacting with cloud services [102]. This ease of use eliminates the difficulties sometimes encountered when utilizing HPC, as each HPC environment typically has its own specific requirements that may differ from a user's local setup.

3.7 QLigFEP

Unlike previous workflows that focused on atomistic simulations in periodic boundary conditions (PBC), the QLigFEP workflow, introduced in 2019, makes use of spherical boundary conditions (SBC) and truncated biomolecular structures [32]. The use of SBC offers an advantage of reduced computational cost as they are smaller and simulations can be run faster. This automated workflow, written in Python, covers the setup, running, and analysis of the FEP workflow using a modular approach and utilizes the Q MD package for its simulations. However, it should be noted that the workflow is not easily expandable to other MD engines due to its highly specific features.

The workflow begins by taking in PDB coordinate files for the first two steps, which are the ligand parameterization and complex preparation. The ligand must be in the correct protonation and tautomerization state, as is common in most workflows. The workflow only takes the coordinates and converts them into a format readable by Q. Protein coordinates are used to prepare SBC around the binding site. The center of the sphere is typically defined based on a reference ligand, but can also be defined based on cartesian coordinates or a protein/ligand atom. The module also automatically solvates using a pre-generated water grid and removes any waters overlapping with the ligands. The ligands are parameterized and a dual topology representation is created, with half-harmonic distance restraints applied to maintain pairs of equivalent non-dummy atoms during the MD simulations. The FEP simulation input parameters are also generated by this module. The workflow includes options to write input files for use on a HPC cluster, and the final module analyzes results using one of three methods, Zwanzig, Overlap Sampling, or BAR, and includes statistical tests.

QLigFEP has been used for 16 cyclin-dependent kinase 2 (CDK2) ligands with different forcefields and showed good agreement with experimental values. It has also been used for a series of ligands for the adenosine A2A receptor, where the advantages of using SBC for simulation speed become notable, as only approximately 7400 atoms, instead of around 42,000 atoms, needed to be modeled. This showed a similar MAE as a previous approach using PBC in FEP+. A unique feature of the workflow is its ability to perform scaffold hopping calculations, thanks to its dual-topology implementation. This was exemplified by executing calculations for 5 inhibitors of checkpoint kinase 1.

3.8 TIES

The TIES workflow was first introduced in 2017 and was later updated as TIES20 [43, 62]. Both versions have been referred to as TIES throughout this review. TIES is a python-based workflow that has a modular, object-oriented design and includes unit testing and continuous integration. It is designed to be used with NAMD/OpenMM using OpenMMTools.

TIES uses ensemble averaging for enhanced sampling. During this process, replicas are performed at each lambda window with different initial velocities obtained from a Maxwell-Boltzmann distribution. The results from the replicas are then averaged, and the average differential is used for carrying out TI. In TIES20, the process of identifying the MCS was improved by changing from RDKit to an inbuilt recursive joint-traversal superimposition algorithm.

TIES is a highly automated workflow, with the ligand and protein being uploaded to a web server, parameterized, merged using a hybrid topology, and the generated input files then being downloaded for running on a cluster using TIES-MD. After the production run, the results can be analyzed using the TIES-ANA module and python packages such as MDAnalysis [103, 104]. Like for many academic workflows, the protocol can be adjusted, but a default protocol is provided.

Five protein systems were investigated using TIES and showed reproducible results between TIES17 and TIES20, or moderate improvements, with overall good agreement with experimental data. The runs were carried out on CPUs as NAMD was used, although there is now a GPU version of NAMD available [5]. All other workflows use GPU accelerated MD due to the vast speed improvements it provides.

3.9 ProFESSA

The ProFESSA workflow is a recent, end-to-end pipeline for use with the AMBER Drug Discovery Boost package, which has been included in AMBER22 and later versions [60]. This pipeline provides top-level control over the calculation through the use of a simplified input file, which defines the parameters for the simulations. The input file is used to carry out an automated setup of the file infrastructure, including an exhaustive setup of equilibration protocols with up to ten steps of preparation for the systems before setting up the lambda windows. Mapping can be done via the MCS, MCS-E, or MCS-E_{nw} algorithms, and enhanced sampling techniques such as ACES can also be set up and run [13]. Network-wide analysis is possible using MBARnet methods and the FE-Toolkit to improve the robustness of predictions [74].

The workflow has been tested on TYK2 with 16 ligands and 24 edges. After analyzing the binding energy of the ligands, an MAE of 0.58 kcal/mol was found, which dropped to 0.55 kcal/mol with the inclusion of cycle closure constraints.

3.10 PyAutoFEP

PyAutoFEP is an open-source Python workflow available on GitHub that automates the generation of perturbation maps, as well as the setup of multiple forcefields and input file generation, MD runs, and analysis [45]. It uses a cost function based on LOMAP and a modified ant colony optimization (ACO) algorithm to minimize this cost function for the generation of perturbation maps, which are parallelized to reduce computation time. The MCS can be guided by either the 3D-guided or graph-based methods, and the data from this is retained for later use in the setup.

PyAutoFEP also has the capability to generate ligand poses using a core-constrained alignment function in addition to reading in user-provided structures. This function superimposes the ligands onto a reference structure using a common core and samples conformations of the flexible group to optimize the overlap volume. When applied to performing RBFE on FXR ligands, 500 trial poses were generated and ranked according to the shape-Tanimoto algorithm to find the structure most similar to the reference ligand from the crystal structure.

The MD simulations are run with GROMACS and the workflow incorporates REST/REST2 as enhanced sampling methods. Preparatory scripts for Slurm clusters are included to facilitate migration between computer infrastructures for running the simulations. The analysis incorporated in PyAutoFEP uses MBAR via alchemlyb, as well as interfacing with GROMACS analysis tools for RMSD, SASA, RMSF, and distance analysis [72]. When validated using 14 Farnesoid X receptor ligands, the workflow was found to be comparable to top performers of the Top Grand Challenge.

3.11 BioSimSpace

BioSimSpace is a flexible Python library under active development designed to promote interoperability between biomolecular simulation software [61]. The library is built on the engine-agnostic Sire simulation framework and supports a range of input file formats [75]. BioSimSpace can parameterize molecular structures using commonly used force fields, solvate assemblies of molecules, perform mappings between ligands for RBFE calculations, and set up input files to run FEP calculations (currently supporting SOMD, GROMACS, and AMBER). The library interfaces with external tools such as RDKit, LOMAP, AmberTools, and GROMACS functionality [65]. BioSimSpace also supports use of HMR to achieve a 4 fs integration timestep. Equilibration protocols and best practices for processing are provided, and analysis functionality is available through native methods or through alchemlyb [73]. BioSimSpace is available as a Conda package and can also be accessed through Jupyter-Hub servers. A suite of tutorials, available on GitHub, covers the use of BioSimSpace for various simulations, including steered MD simulations, metadynamics simulations, RBFE, and ABFE simulations.

3.12 FEP+

FEP+ is a well-established commercial workflow by Schrödinger, known for its accuracy and often used as a benchmark for comparison when evaluating new workflows [51]. However, as it is proprietary and closed-source software, its implementation details are not publicly available. FEP+ offers a complete end-to-end solution, with features such as input parameterization, network generation, file creation, MD running through Desmond, and comprehensive network-wide analysis, all within a user-friendly interface. It is compatible with the proprietary OPLS forcefield series, with OPLS4 being the most recent iteration [8]. The default protocol for FEP+ has long involved the use of enhanced sampling techniques such as REST/REST2 and has been applied to various systems, including membrane proteins such as GPCRs, and scaffold hopping [30, 105, 106].

3.13 Flare

Flare is a structure-based drug design software commercialized by Cresset. A wide range of functionality is included, such as visualization, ligand alignment, QSAR models, docking, and QM calculations, to name a few [47, 107–109]. Flare has an FEP module that covers all steps from setup to analysis, making use of the open-source OpenMM to run its simulations. The pipeline is built upon BioSimSpace and Sire, which means that the code generated for running these simulations is available. Users interact with these open-source components indirectly through an easy-to-use GUI that abstracts implementation details, making it accessible for non-experts to run FEP simulations and obtain usable results. Flare can also be used via a Python application programming interface. The implemented protocols for reliability include hysteresis of forward and backward perturbations, checking cycle closure errors, and lambda overlap windows. Benchmarking results for 14 protein systems have shown Flare-FEP to have a performance comparable to other leading workflows, and it has also been shown to capture the trends in changes in binding affinity for scaffold hopping perturbations [109].

3.14 Orion

Orion, a cloud-based workflow developed by OpenEye Scientific, has been designed to make use of a cloud-native design and scheduler [110]. As already highlighted in the PMX workflow section, cloud-based computing has the potential for astronomical computation speeds, and the platform also utilizes AWS to efficiently perform simulations. Similar to other commercial software, it offers various tools to support additional workflow features such as docking. The platform also supports the integration of third-party code, with an API that enables customization. For instance, a non-equilibrium switching (NES) feature was implemented in Orion using the PMX package [111].

4 The Future for RBFE Workflows

It is important to note that much of the variability between workflows is not necessarily attributed to the workflow itself, but rather to the chosen forcefield, sampling scheme, and other factors, as the workflow is simply a means of utilizing these features. Even when using the same workflow, significant variability in prediction accuracy was reported, depending on the forcefield, sampling regime, or number of lambda windows used [45]. The TIES workflow found that a different charge model, AM1-BCC vs REST, improved agreement with experimental data by about 10% [62]. When selecting a workflow, it is therefore important to consider which of these features are available and best suited to the system and perturbations being carried out, as well as the available computational infrastructure and resources.

In conclusion, using a workflow makes it easier, faster, and more consistent to run RBFE simulations. The development of RBFE workflows is a rapidly evolving field in drug discovery, with commercial workflows benefiting from well-developed GUIs, user support, and additional tools. However, it is challenging to maintain academic software as there is usually only a small team of developers behind it. As the complexity of the workflows increases, so does their flexibility for modeling more applications and systems. Initial work on RBFE workflows focused mainly on setting up inputs and scripts, which was the most time-consuming part for the user. This has expanded to include more robust mapping, workflows suited to different MD engines, and the analysis of the output, incorporating external libraries as needed. More recent work has focused on enhanced sampling and incorporating network setup and analysis considerations, as these can significantly improve the quality of the final free energy predictions.

In the future, systematic comparison of the different methods for each stage of a RBFE workflow will be essential to determine where performance differences occur. Many workflow components currently lack compatibility with other workflows, leading to significant duplication of efforts and slower uptake of new improved algorithms. Toolkits that promote greater interoperability such as BioSimSpace have the potential to facilitate the development of modular RBFE workflows that capture state-of-the art and best practices.

References

1. Song LF, Merz KM Jr (2020) Evolution of alchemical free energy methods in drug discovery. J Chem Inf Model 60(11): 5308–5318. https://doi.org/10.1021/acs.jcim.0c00547
2. Friedrichs MS, Eastman P, Vaidyanathan V, Houston M, Legrand S, Beberg AL, Ensign DL, Bruns CM, Pande VS (2009) Accelerating molecular dynamic simulation on graphics processing units. J Comput Chem 30(6): 864–872. https://doi.org/10.1002/jcc.21209
3. Cournia Z, Allen B, Sherman W (2017) Relative binding free energy calculations in drug discovery: recent advances and practical considerations. J Chem Inf Model 57(12): 2911–2937. https://doi.org/10.1021/acs.jcim.7b00564
4. Salomon-Ferrer R, Case DA, Walker RC (2013) An overview of the Amber biomolecular simulation package. WIREs Comput Mol Sci 3(2):198–210. https://doi.org/10.1002/wcms.1121
5. Chen H, Maia JDC, Radak BK, Hardy DJ, Cai W, Chipot C, Tajkhorshid E (2020) Boosting free-energy perturbation calculations with GPU-accelerated NAMD. J Chem Inf Model 60(11):5301–5307. https://doi.org/10.1021/acs.jcim.0c00745
6. Shivakumar D, Harder E, Damm W, Friesner RA, Sherman W (2012) Improving the prediction of absolute solvation free energies using the next generation OPLS force field. J Chem Theory Comput 8(8):2553–2558. https://doi.org/10.1021/ct300203w
7. Harder E, Damm W, Maple J, Wu C, Reboul M, Xiang JY, Wang L, Lupyan D, Dahlgren MK, Knight JL, Kaus JW, Cerutti DS, Krilov G, Jorgensen WL, Abel R, Friesner RA (2016) OPLS3: a force field providing broad coverage of drug-like small molecules and proteins. J Chem Theory Comput 12(1): 281–296. https://doi.org/10.1021/acs.jctc.5b00864
8. Lu C, Wu C, Ghoreishi D, Chen W, Wang L, Damm W, Ross GA, Dahlgren MK, Russell E, Von Bargen CD, Abel R, Friesner RA, Harder ED (2021) OPLS4: improving force field accuracy on challenging regimes of chemical space. J Chem Theory Comput 17(7): 4291–4300. https://doi.org/10.1021/acs.jctc.1c00302
9. Qiu Y, Smith DGA, Boothroyd S, Jang H, Hahn DF, Wagner J, Bannan CC, Gokey T, Lim VT, Stern CD, Rizzi A, Tjanaka B, Tresadern G, Lucas X, Shirts MR, Gilson MK, Chodera JD, Bayly CI, Mobley DL, Wang LP (2021) Development and benchmarking of open force field v1.0.0-the Parsley small-molecule force field. J Chem Theory Comput 17(10):6262–6280. https://doi.org/10.1021/acs.jctc.1c00571
10. Mobley DL, Bannan CC, Rizzi A, Bayly CI, Chodera JD, Lim VT, Lim NM, Beauchamp KA, Slochower DR, Shirts MR, Gilson MK, Eastman PK (2018) Escaping atom types in force fields using direct chemical perception. J Chem Theory Comput 14(11):6076–6092. https://doi.org/10.1021/acs.jctc.8b00640
11. Liu P, Kim B, Friesner RA, Berne BJ (2005) Replica exchange with solute tempering: a method for sampling biological systems in explicit water. Proc Natl Acad Sci U S A 102(39):13749–13754. https://doi.org/10.1073/pnas.0506346102
12. Wang L, Friesner RA, Berne BJ (2011) Replica exchange with solute scaling: a more efficient version of replica exchange with solute tempering (REST2). J Phys Chem B 115(30):

9431–9438. https://doi.org/10.1021/jp204407d

13. Lee TS, Tsai HC, Ganguly A, York DM (2023) ACES: optimized alchemically enhanced sampling. J Chem Theory Comput. https://doi.org/10.1021/acs.jctc.2c00697
14. Lovering F, Aevazelis C, Chang J, Dehnhardt C, Fitz L, Han S, Janz K, Lee J, Kaila N, McDonald J, Moore W, Moretto A, Papaioannou N, Richard D, Ryan MS, Wan ZK, Thorarensen A (2016) Imidazotriazines: spleen tyrosine kinase (Syk) inhibitors identified by Free-Energy Perturbation (FEP). ChemMedChem 11(2):217–233. https://doi.org/10.1002/cmdc.201500333
15. Mortier J, Friberg A, Badock V, Moosmayer D, Schroeder J, Steigemann P, Siegel F, Gradl S, Bauser M, Hillig RC, Briem H, Eis K, Bader B, Nguyen D, Christ CD (2020) Computationally empowered workflow identifies novel covalent allosteric binders for KRAS(G12C). ChemMedChem 15(10):827–832. https://doi.org/10.1002/cmdc.201900727
16. Freedman H, Kundu J, Tchesnokov EP, Law JLM, Nieman JA, Schinazi RF, Tyrrell DL, Gotte M, Houghton M (2020) Application of molecular dynamics simulations to the design of nucleotide inhibitors binding to norovirus polymerase. J Chem Inf Model 60(12):6566–6578. https://doi.org/10.1021/acs.jcim.0c00742
17. Tresadern G, Velter I, Trabanco AA, Van den Keybus F, Macdonald GJ, Somers MVF, Vanhoof G, Leonard PM, Lamers M, Van Roosbroeck YEM, Buijnsters P (2020) [1,2,4]Triazolo[1,5-a]pyrimidine phosphodiesterase 2A inhibitors: structure and free-energy perturbation-guided exploration. J Med Chem 63(21):12887–12910. https://doi.org/10.1021/acs.jmedchem.0c01272
18. O'Donovan DH, Gregson C, Packer MJ, Greenwood R, Pike KG, Kawatkar S, Bloecher A, Robinson J, Read J, Code E, Hsu JH-R, Shen M, Woods H, Barton P, Fillery S, Williamson B, Rawlins PB, Bagal SK (2021) Free energy perturbation in the design of EED ligands as inhibitors of polycomb repressive complex 2 (PRC2) methyltransferase. Bioorg Med Chem Lett 39: 127904. https://doi.org/10.1016/j.bmcl.2021.127904
19. Jorgensen WL, Ravimohan C (1985) Monte Carlo simulation of differences in free energies of hydration. J Chem Phys 83(6): 3050–3054. https://doi.org/10.1063/1.449208
20. Ross GA, Russell E, Deng Y, Lu C, Harder ED, Abel R, Wang L (2020) Enhancing water sampling in free energy calculations with grand canonical Monte Carlo. J Chem Theory Comput 16(10):6061–6076. https://doi.org/10.1021/acs.jctc.0c00660
21. Yu HS, Gao C, Lupyan D, Wu Y, Kimura T, Wu C, Jacobson L, Harder E, Abel R, Wang L (2019) Toward atomistic modeling of irreversible covalent inhibitor binding kinetics. J Chem Inf Model 59(9):3955–3967. https://doi.org/10.1021/acs.jcim.9b00268
22. Wang L, Wu Y, Deng Y, Kim B, Pierce L, Krilov G, Lupyan D, Robinson S, Dahlgren MK, Greenwood J, Romero DL, Masse C, Knight JL, Steinbrecher T, Beuming T, Damm W, Harder E, Sherman W, Brewer M, Wester R, Murcko M, Frye L, Farid R, Lin T, Mobley DL, Jorgensen WL, Berne BJ, Friesner RA, Abel R (2015) Accurate and reliable prediction of relative ligand binding potency in prospective drug discovery by way of a modern free-energy calculation protocol and force field. J Am Chem Soc 137(7): 2695–2703. https://doi.org/10.1021/ja512751q
23. Mey ASJS, Allen BK, Macdonald HEB, Chodera JD, Hahn DF, Kuhn M, Michel J, Mobley DL, Naden LN, Prasad S, Rizzi A, Scheen J, Shirts MR, Tresadern G, Xu H (2020) Best practices for alchemical free energy calculations [Article v1.0]. Living J Comput Mol Sci 2(1). https://doi.org/10.33011/livecoms.2.1.18378
24. Bash PA, Singh UC, Langridge R, Kollman PA (1987) Free energy calculations by computer simulation. Science 236(4801): 564–568. https://doi.org/10.1126/science.3576184
25. Bash PA, Singh UC, Brown FK, Langridge R, Kollman PA (1987) Calculation of the relative change in binding free energy of a protein-inhibitor complex. Science 235(4788): 574–576. https://doi.org/10.1126/science.3810157
26. Kollman PA (1993) Free energy calculations: applications to chemical and biochemical phenomena. Chem Rev 93(7):2395–2417. https://doi.org/10.1021/cr00023a004
27. Gilson MK, Given JA, Bush BL, McCammon JA (1997) The statistical-thermodynamic basis for computation of binding affinities: a critical review. Biophys J 72(3):1047–1069. https://doi.org/10.1016/S0006-3495(97)78756-3
28. Michel J, Essex JW (2010) Prediction of protein-ligand binding affinity by free energy simulations: assumptions, pitfalls and

expectations. J Comput Aided Mol Des 24(8):639–658. https://doi.org/10.1007/s10822-010-9363-3

29. Böhm H-J, Flohr A, Stahl M (2004) Scaffold hopping. Drug Discov Today Technol 1(3):217–224. https://doi.org/10.1016/j.ddtec.2004.10.009
30. Wang L, Deng Y, Wu Y, Kim B, LeBard DN, Wandschneider D, Beachy M, Friesner RA, Abel R (2017) Accurate modeling of scaffold hopping transformations in drug discovery. J Chem Theory Comput 13(1):42–54. https://doi.org/10.1021/acs.jctc.6b00991
31. Wu D, Zheng X, Liu R, Li Z, Jiang Z, Zhou Q, Huang Y, Wu XN, Zhang C, Huang YY, Luo HB (2022) Free energy perturbation (FEP)-guided scaffold hopping. Acta Pharm Sin B 12(3):1351–1362. https://doi.org/10.1016/j.apsb.2021.09.027
32. Jespers W, Esguerra M, Aqvist J, Gutierrez-de-Teran H (2019) QligFEP: an automated workflow for small molecule free energy calculations in Q. J Cheminform 11(1):26. https://doi.org/10.1186/s13321-019-0348-5
33. Gapsys V, Michielssens S, Seeliger D, de Groot BL (2015) pmx: automated protein structure and topology generation for alchemical perturbations. J Comput Chem 36(5):348–354. https://doi.org/10.1002/jcc.23804
34. Boukharta L, Gutierrez-de-Teran H, Aqvist J (2014) Computational prediction of alanine scanning and ligand binding energetics in G-protein coupled receptors. PLoS Comput Biol 10(4):e1003585. https://doi.org/10.1371/journal.pcbi.1003585
35. Keranen H, Aqvist J, Gutierrez-de-Teran H (2015) Free energy calculations of A (2A) adenosine receptor mutation effects on agonist binding. Chem Commun (Camb) 51(17):3522–3525. https://doi.org/10.1039/c4cc09517k
36. Lee TS, Allen BK, Giese TJ, Guo Z, Li P, Lin C, McGee TD Jr, Pearlman DA, Radak BK, Tao Y, Tsai HC, Xu H, Sherman W, York DM (2020) Alchemical binding free energy calculations in AMBER20: advances and best practices for drug discovery. J Chem Inf Model 60(11):5595–5623. https://doi.org/10.1021/acs.jcim.0c00613
37. Heinzelmann G, Gilson MK (2021) Automation of absolute protein-ligand binding free energy calculations for docking refinement and compound evaluation. Sci Rep 11(1):1116. https://doi.org/10.1038/s41598-020-80769-1
38. Santiago-McRae E, Ebrahimi M, Sandberg JW, Brannigan G, Hénin J (2022) Computing absolute binding affinities by Streamlined Alchemical Free Energy Perturbation. bioRxiv:2022.2012.2009.519809. https://doi.org/10.1101/2022.12.09.519809
39. Fu H, Chen H, Cai W, Shao X, Chipot C (2021) BFEE2: automated, streamlined, and accurate absolute binding free-energy calculations. J Chem Inf Model 61(5):2116–2123. https://doi.org/10.1021/acs.jcim.1c00269
40. Homeyer N, Gohlke H (2013) FEW: a workflow tool for free energy calculations of ligand binding. J Comput Chem 34(11):965–973. https://doi.org/10.1002/jcc.23218
41. Loeffler HH, Michel J, Woods C (2015) FESetup: automating setup for alchemical free energy simulations. J Chem Inf Model 55(12):2485–2490. https://doi.org/10.1021/acs.jcim.5b00368
42. Jo S, Cheng X, Lee J, Kim S, Park SJ, Patel DS, Beaven AH, Lee KI, Rui H, Park S, Lee HS, Roux B, MacKerell AD Jr, Klauda JB, Qi Y, Im W (2017) CHARMM-GUI 10 years for biomolecular modeling and simulation. J Comput Chem 38(15):1114–1124. https://doi.org/10.1002/jcc.24660
43. Bhati AP, Wan S, Wright DW, Coveney PV (2017) Rapid, accurate, precise, and reliable relative free energy prediction using ensemble based thermodynamic integration. J Chem Theory Comput 13(1):210–222. https://doi.org/10.1021/acs.jctc.6b00979
44. Zavitsanou S, Tsengenes A, Papadourakis M, Amendola G, Chatzigoulas A, Dellis D, Cosconati S, Cournia Z (2021) FEPrepare: a web-based tool for automating the setup of relative binding free energy calculations. J Chem Inf Model 61(9):4131–4138. https://doi.org/10.1021/acs.jcim.1c00215
45. Carvalho Martins L, Cino EA, Ferreira RS (2021) PyAutoFEP: an automated free energy perturbation workflow for GROMACS integrating enhanced sampling methods. J Chem Theory Comput 17(7):4262–4273. https://doi.org/10.1021/acs.jctc.1c00194
46. Jo S, Kim T, Iyer VG, Im W (2008) CHARMM-GUI: a web-based graphical user interface for CHARMM. J Comput Chem 29(11):1859–1865. https://doi.org/10.1002/jcc.20945
47. Cresset® (2022) Flare™. V6 edn., Litlington
48. Morris GM, Huey R, Lindstrom W, Sanner MF, Belew RK, Goodsell DS, Olson AJ (2009) AutoDock4 and AutoDockTools4: automated docking with selective receptor flexibility. J Comput Chem 30(16):

2785–2791. https://doi.org/10.1002/jcc.21256

49. Bieniek MK, Cree B, Pirie R, Horton JT, Tatum NJ, Cole DJ (2022) An open-source molecular builder and free energy preparation workflow. Commun Chem 5(1):136. https://doi.org/10.1038/s42004-022-00754-9
50. Suruzhon M, Bodnarchuk MS, Ciancetta A, Viner R, Wall ID, Essex JW (2021) Sensitivity of binding free energy calculations to initial protein crystal structure. J Chem Theory Comput 17(3):1806–1821. https://doi.org/10.1021/acs.jctc.0c00972
51. Schrödinger (2021) FEP+. Release 2023-1 edn., New York
52. Sayle RA (2010) So you think you understand tautomerism? J Comput Aided Mol Des 24(6–7):485–496. https://doi.org/10.1007/s10822-010-9329-5
53. Hu Y, Sherborne B, Lee TS, Case DA, York DM, Guo Z (2016) The importance of protonation and tautomerization in relative binding affinity prediction: a comparison of AMBER TI and Schrodinger FEP. J Comput Aided Mol Des 30(7):533–539. https://doi.org/10.1007/s10822-016-9920-5
54. Fleck M, Wieder M, Boresch S (2021) Dummy atoms in alchemical free energy calculations. J Chem Theory Comput 17(7):4403–4419. https://doi.org/10.1021/acs.jctc.0c01328
55. Loeffler HH, Bosisio S, Duarte Ramos Matos G, Suh D, Roux B, Mobley DL, Michel J (2018) Reproducibility of free energy calculations across different molecular simulation software packages. J Chem Theory Comput 14(11):5567–5582. https://doi.org/10.1021/acs.jctc.8b00544
56. Liu S, Wang L, Mobley DL (2015) Is ring breaking feasible in relative binding free energy calculations? J Chem Inf Model 55(4):727–735. https://doi.org/10.1021/acs.jcim.5b00057
57. Pearlman DA (1994) A comparison of alternative approaches to free energy calculations. J Phys Chem 98(5):1487–1493. https://doi.org/10.1021/j100056a020
58. Jiang W, Chipot C, Roux B (2019) Computing relative binding affinity of ligands to receptor: an effective hybrid single-dual-topology free-energy perturbation approach in NAMD. J Chem Inf Model 59(9):3794–3802. https://doi.org/10.1021/acs.jcim.9b00362
59. Dalke A, Hastings J (2013) FMCS: a novel algorithm for the multiple MCS problem. J Chem 5(1):O6. https://doi.org/10.1186/1758-2946-5-S1-O6
60. Ganguly A, Tsai HC, Fernandez-Pendas M, Lee TS, Giese TJ, York DM (2022) AMBER drug discovery boost tools: automated workflow for Production Free-Energy Simulation Setup and Analysis (ProFESSA). J Chem Inf Model 62(23):6069–6083. https://doi.org/10.1021/acs.jcim.2c00879
61. Hedges LO, Mey AS, Laughton CA, Gervasio FL, Mulholland AJ, Woods CJ, Michel J (2019) BioSimSpace: an interoperable Python framework for biomolecular simulation. J Open Source Softw 4(43):1831. https://doi.org/10.21105/joss.01831
62. Bieniek MK, Bhati AP, Wan S, Coveney PV (2021) TIES 20: relative binding free energy with a flexible superimposition algorithm and partial ring morphing. J Chem Theory Comput 17(2):1250–1265. https://doi.org/10.1021/acs.jctc.0c01179
63. Wieder M, Fleck M, Braunsfeld B, Boresch S (2022) Alchemical free energy simulations without speed limits. A generic framework to calculate free energy differences independent of the underlying molecular dynamics program. J Comput Chem 43(17):1151–1160. https://doi.org/10.1002/jcc.26877
64. Karwounopoulos J, Wieder M, Boresch S (2022) Relative binding free energy calculations with transformato: a molecular dynamics engine-independent tool. Front Mol Biosci 9:954638. https://doi.org/10.3389/fmolb.2022.954638
65. Liu S, Wu Y, Lin T, Abel R, Redmann JP, Summa CM, Jaber VR, Lim NM, Mobley DL (2013) Lead optimization mapper: automating free energy calculations for lead optimization. J Comput Aided Mol Des 27(9):755–770. https://doi.org/10.1007/s10822-013-9678-y
66. Scheen J, Mackey M, Michel J (2022) Data-driven generation of perturbation networks for relative binding free energy calculations. Digit Discov 1(6):870–885. https://doi.org/10.1039/D2DD00083K
67. Hénin J, Lelièvre T, Shirts MR, Valsson O, Delemotte L (2022) Enhanced sampling methods for molecular dynamics simulations [Article v1.0]. Living J Comput Mol Sci 4(1):1583. https://doi.org/10.33011/livecoms.4.1.1583
68. Wang L, Deng Y, Knight JL, Wu Y, Kim B, Sherman W, Shelley JC, Lin T, Abel R (2013) Modeling local structural rearrangements using FEP/REST: Application to relative binding affinity predictions of CDK2 inhibitors. J Chem Theory Comput 9(2):

1282–1293. https://doi.org/10.1021/ct300911a

69. Kirkwood JG (1935) Statistical mechanics of fluid mixtures. J Chem Phys 3(5):300–313. https://doi.org/10.1063/1.1749657
70. Jorge M, Garrido NM, Queimada AJ, Economou IG, Macedo EA (2010) Effect of the integration method on the accuracy and computational efficiency of free energy calculations using thermodynamic integration. J Chem Theory Comput 6(4):1018–1027. https://doi.org/10.1021/ct900661c
71. Bennett CH (1976) Efficient estimation of free energy differences from Monte Carlo data. J Comput Phys 22(2):245–268. https://doi.org/10.1016/0021-9991(76)90078-4
72. Shirts MR, Chodera JD (2008) Statistically optimal analysis of samples from multiple equilibrium states. J Chem Phys 129(12): 124105. https://doi.org/10.1063/1.2978177
73. Beckstein O, Dotson D, Wu Z, Wille D, Marson D, Kenney I, Shuail, Lee H, trje, Lim V, Schlaich A, Alibay I, Hénin J, Barhaghi MS, Merz P, Joseph T, Hsu W-T (2022) alchemistry/alchemlyb: 2.0.0. 2.0.0 edn. Zenodo. https://doi.org/10.5281/zenodo.7433270
74. Giese TJ, York DM (2021) Variational method for networkwide analysis of relative ligand binding free energies with loop closure and experimental constraints. J Chem Theory Comput 17(3):1326–1336. https://doi.org/10.1021/acs.jctc.0c01219
75. OpenBioSim (2023) Sire. 2023.1.0 edn
76. Mey ASJS, Jimenez JJ, Michel J (2018) Impact of domain knowledge on blinded predictions of binding energies by alchemical free energy calculations. J Comput Aided Mol Des 32(1):199–210. https://doi.org/10.1007/s10822-017-0083-9
77. Mey ASJS, Juarez-Jimenez J, Hennessy A, Michel J (2016) Blinded predictions of binding modes and energies of HSP90-alpha ligands for the 2015 D3R grand challenge. Bioorg Med Chem 24(20):4890–4899. https://doi.org/10.1016/j.bmc.2016.07.044
78. Aldeghi M, Heifetz A, Bodkin MJ, Knapp S, Biggin PC (2016) Accurate calculation of the absolute free energy of binding for drug molecules. Chem Sci 7(1):207–218. https://doi.org/10.1039/c5sc02678d
79. De Simone A, Georgiou C, Ioannidis H, Gupta AA, Juarez-Jimenez J, Doughty-Shenton D, Blackburn EA, Wear MA, Richards JP, Barlow PN, Carragher N, Walkinshaw MD, Hulme AN, Michel J (2019) A computationally designed binding mode flip leads to a novel class of potent tri-vector cyclophilin inhibitors. Chem Sci 10(2): 542–547. https://doi.org/10.1039/c8sc03831g
80. Dodda LS, Cabeza de Vaca I, Tirado-Rives J, Jorgensen WL (2017) LigParGen web server: an automatic OPLS-AA parameter generator for organic ligands. Nucleic Acids Res 45 (W1):W331–W336. https://doi.org/10.1093/nar/gkx312
81. Park S-J, Kern N, Brown T, Lee J, Im W (2023) CHARMM-GUI PDB manipulator: various PDB structural modifications for biomolecular modeling and simulation. J Mol Biol 167995. https://doi.org/10.1016/j.jmb.2023.167995
82. Kim S, Lee J, Jo S, Brooks CL 3rd, Lee HS, Im W (2017) CHARMM-GUI ligand reader and modeler for CHARMM force field generation of small molecules. J Comput Chem 38(21):1879–1886. https://doi.org/10.1002/jcc.24829
83. Jo S, Lim JB, Klauda JB, Im W (2009) CHARMM-GUI membrane builder for mixed bilayers and its application to yeast membranes. Biophys J 97(1):50–58. https://doi.org/10.1016/j.bpj.2009.04.013
84. Wu EL, Cheng X, Jo S, Rui H, Song KC, Davila-Contreras EM, Qi Y, Lee J, Monje-Galvan V, Venable RM, Klauda JB, Im W (2014) CHARMM-GUI membrane builder toward realistic biological membrane simulations. J Comput Chem 35(27):1997–2004. https://doi.org/10.1002/jcc.23702
85. Lee J, Patel DS, Stahle J, Park SJ, Kern NR, Kim S, Lee J, Cheng X, Valvano MA, Holst O, Knirel YA, Qi Y, Jo S, Klauda JB, Widmalm G, Im W (2019) CHARMM-GUI membrane builder for complex biological membrane simulations with glycolipids and lipoglycans. J Chem Theory Comput 15(1):775–786. https://doi.org/10.1021/acs.jctc.8b01066
86. Qi Y, Lee J, Klauda JB, Im W (2019) CHARMM-GUI nanodisc builder for modeling and simulation of various nanodisc systems. J Comput Chem 40(7):893–899. https://doi.org/10.1002/jcc.25773
87. Qi Y, Cheng X, Lee J, Vermaas JV, Pogorelov TV, Tajkhorshid E, Park S, Klauda JB, Im W (2015) CHARMM-GUI HMMM builder for membrane simulations with the highly mobile membrane-mimetic model. Biophys J 109(10):2012–2022. https://doi.org/10.1016/j.bpj.2015.10.008

88. Cheng X, Jo S, Lee HS, Klauda JB, Im W (2013) CHARMM-GUI micelle builder for pure/mixed micelle and protein/micelle complex systems. J Chem Inf Model 53(8): 2171–2180. https://doi.org/10.1021/ci4002684
89. Kim S, Oshima H, Zhang H, Kern NR, Re S, Lee J, Roux B, Sugita Y, Jiang W, Im W (2020) CHARMM-GUI free energy calculator for absolute and relative ligand solvation and binding free energy simulations. J Chem Theory Comput 16(11):7207–7218. https://doi.org/10.1021/acs.jctc.0c00884
90. Jo S, Cheng X, Islam SM, Huang L, Rui H, Zhu A, Lee HS, Qi Y, Han W, Vanommeslaeghe K, MacKerell AD, Roux B, Im W (2014) Chapter Eight – CHARMM-GUI PDB manipulator for advanced modeling and simulations of proteins containing nonstandard residues. In: Karabencheva-Christova T (ed) Advances in protein chemistry and structural biology, vol 96. Academic, Oxford, pp 235–265. https://doi.org/10.1016/bs.apcsb.2014.06.002
91. Jo S, Jiang W, Lee HS, Roux B, Im W (2013) CHARMM-GUI ligand binder for absolute binding free energy calculations and its application. J Chem Inf Model 53(1):267–277. https://doi.org/10.1021/ci300505n
92. Guterres H, Park S-J, Zhang H, Perone T, Kim J, Im W (2022) CHARMM-GUI high-throughput simulator for efficient evaluation of protein–ligand interactions with different force fields. Protein Sci 31(9):e4413. https://doi.org/10.1002/pro.4413
93. Suh D, Feng S, Lee H, Zhang H, Park S-J, Kim S, Lee J, Choi S, Im W (2022) CHARMM-GUI enhanced sampler for various collective variables and enhanced sampling methods. Protein Sci 31(11):e4446. https://doi.org/10.1002/pro.4446
94. Guterres H, Park SJ, Cao Y, Im W (2021) CHARMM-GUI ligand designer for template-based virtual ligand design in a binding site. J Chem Inf Model 61(11): 5336–5342. https://doi.org/10.1021/acs.jcim.1c01156
95. Lee J, Cheng X, Swails JM, Yeom MS, Eastman PK, Lemkul JA, Wei S, Buckner J, Jeong JC, Qi Y, Jo S, Pande VS, Case DA, Brooks CL 3rd, MacKerell AD Jr, Klauda JB, Im W (2016) CHARMM-GUI input generator for NAMD, GROMACS, AMBER, OpenMM, and CHARMM/OpenMM simulations using the CHARMM36 additive force field. J Chem Theory Comput 12(1):405–413. https://doi.org/10.1021/acs.jctc.5b00935
96. Gao Y, Lee J, Smith IPS, Lee H, Kim S, Qi Y, Klauda JB, Widmalm G, Khalid S, Im W (2021) CHARMM-GUI supports hydrogen mass repartitioning and different protonation states of phosphates in lipopolysaccharides. J Chem Inf Model 61(2):831–839. https://doi.org/10.1021/acs.jcim.0c01360
97. Lee J, Hitzenberger M, Rieger M, Kern NR, Zacharias M, Im W (2020) CHARMM-GUI supports the Amber force fields. J Chem Phys 153(3):035103. https://doi.org/10.1063/5.0012280
98. Boresch S, Bruckner S (2011) Avoiding the van der Waals endpoint problem using serial atomic insertion. J Comput Chem 32(11): 2449–2458. https://doi.org/10.1002/jcc.21829
99. Gapsys V, de Groot BL (2017) pmx webserver: a user friendly interface for alchemistry. J Chem Inf Model 57(2):109–114. https://doi.org/10.1021/acs.jcim.6b00498
100. Gapsys V, Perez-Benito L, Aldeghi M, Seeliger D, van Vlijmen H, Tresadern G, de Groot BL (2019) Large scale relative protein ligand binding affinities using non-equilibrium alchemy. Chem Sci 11(4): 1140–1152. https://doi.org/10.1039/c9sc03754c
101. Kutzner C, Kniep C, Cherian A, Nordstrom L, Grubmuller H, de Groot BL, Gapsys V (2022) GROMACS in the cloud: a global supercomputer to speed up alchemical drug design. J Chem Inf Model 62(7): 1691–1711. https://doi.org/10.1021/acs.jcim.2c00044
102. Harvey MJ, De Fabritiis G (2015) AceCloud: molecular dynamics simulations in the cloud. J Chem Inf Model 55(5):909–914. https://doi.org/10.1021/acs.jcim.5b00086
103. Michaud-Agrawal N, Denning EJ, Woolf TB, Beckstein O (2011) MDAnalysis: a toolkit for the analysis of molecular dynamics simulations. J Comput Chem 32(10):2319–2327. https://doi.org/10.1002/jcc.21787
104. Gowers RJ, Linke M, Barnoud J, Reddy TJE, Melo MN, Seyler SL, Dotson DL, Domanski J, Buchoux S, Kenney IM, Beckstein O (2016) MDAnalysis: a Python package for the rapid analysis of molecular dynamics simulations. Paper presented at the proceedings of the 15th Python in science conference, 2016
105. Lenselink EB, Louvel J, Forti AF, van Veldhoven JPD, de Vries H, Mulder-Krieger T, McRobb FM, Negri A, Goose J, Abel R, van Vlijmen HWT, Wang L, Harder E, Sherman W, Ijzerman APIJ, Beuming T (2016) Predicting binding affinities for

GPCR ligands using free-energy perturbation. ACS Omega 1(2):293–304. https://doi.org/10.1021/acsomega.6b00086

106. Wang L, Chambers J, Abel R (2019) Protein-ligand binding free energy calculations with FEP. Methods Mol Biol 2022:201–232. https://doi.org/10.1007/978-1-4939-9608-7_9

107. Cheeseright T, Mackey M, Rose S, Vinter A (2006) Molecular field extrema as descriptors of biological activity: definition and validation. J Chem Inf Model 46(2):665–676. https://doi.org/10.1021/ci050357s

108. Bauer MR, Mackey MD (2019) Electrostatic complementarity as a fast and effective tool to optimize binding and selectivity of protein–ligand complexes. J Med Chem 62(6):3036–3050. https://doi.org/10.1021/acs.jmedchem.8b01925

109. Kuhn M, Firth-Clark S, Tosco P, Mey A, Mackey M, Michel J (2020) Assessment of binding affinity via alchemical free-energy calculations. J Chem Inf Model 60(6):3120–3130. https://doi.org/10.1021/acs.jcim.0c00165

110. OpenEye (2022) Orion®. 2022.3 edn. Cadence Molecular Sciences, Santa Fe

111. Scientific O (2022) Relative binding free energy with non-equilibrium switching in orion. OpenEye Scientific

Chapter 12

Molecular Dynamics and Other HPC Simulations for Drug Discovery

Martin Kotev and Constantino Diaz Gonzalez

Abstract

High performance computing (HPC) is taking an increasingly important place in drug discovery. It makes possible the simulation of complex biochemical systems with high precision in a short time, thanks to the use of sophisticated algorithms. It promotes the advancement of knowledge in fields that are inaccessible or difficult to access through experimentation and it contributes to accelerating the discovery of drugs for unmet medical needs while reducing costs. Herein, we report how computational performance has evolved over the past years, and then we detail three domains where HPC is essential. Molecular dynamics (MD) is commonly used to explore the flexibility of proteins, thus generating a better understanding of different possible approaches to modulate their activity. Modeling and simulation of biopolymer complexes enables the study of protein-protein interactions (PPI) in healthy and disease states, thus helping the identification of targets of pharmacological interest. Virtual screening (VS) also benefits from HPC to predict in a short time, among millions or billions of virtual chemical compounds, the best potential ligands that will be tested in relevant assays to start a rational drug design process.

Key words HPC, High Performance Computing, Supercomputer, GPU, Drug discovery, Drug design, MD, Molecular dynamics, MC, Monte Carlo, Simulation, Protein complexes, VS, Virtual screening

Abbreviations

A1AR	A1 Adenosine Receptor
ACE2	Angiotensin-Converting Enzyme 2
AF	AlphaFold
AI	Artificial Intelligence
AMBER	Assisted Model Building with Energy Refinement
CADD	Computer-Aided Drug Design
CAPRI	Critical Assessment of Prediction of Interactions
CASP14	Critical Assessment of Structure Prediction, round 14
CGenFF	Charmm General Force Field
CHARMM	Chemistry at HARvard Molecular Mechanics
COVID-19	COrona VIrus Disease 2019
CPU	Central Processing Unit

Alexander Heifetz (ed.), *High Performance Computing for Drug Discovery and Biomedicine*, Methods in Molecular Biology, vol. 2716, https://doi.org/10.1007/978-1-0716-3449-3_12,

Cryo-EM	Cryo-Electron Microscopy
DMTA	Design, Make, Test, Analyze
DNA	Deoxyribonucleic Acid
DNN	Deep Neural Network
DUD-E	Database of Useful Decoys Enhanced
ECL2	Extra Cellular Loop 2
FEP	Free Energy Perturbation
FF	Force Field
FGFR2	Fibroblast Growth Factor Receptor 2
GAFF	Generalized Amber Force Field
GAMD	Gaussian Accelerated Molecular Dynamics
GPCR	G-Protein Coupled Receptor
GPU	Graphical Processing Unit
GROMACS	GROningen Machine for Chemical Simulations
H2L	Hit to Lead
HPC	High-Performance Computing
IC50	Half Maximal Inhibitory Concentration
IL2	InterLeukin 2
JAK1	Janus Kinase 1
JAK2	Janus Kinase 2
LO	Lead Optimization
MC	Monte Carlo
MD	Molecular Dynamics
MixMD	Mixed-Solvent Molecular Dynamics
MMFF	Merck Molecular Force Field
MM-GBSA	Molecular Mechanics Generalized Born Surface Area
MM-PBSA	Molecular Mechanics Poisson-Boltzmann Surface Area
MSA	Multiple Sequence Alignment
MSM	Markov State Model
NAMD	Nanoscale Molecular Dynamics
NCATS	National Center for Advancing Translational Sciences
ns	nano second
OPLS4	Optimized Potentials for Liquid Simulation, Version 4
PAM	Positive Allosteric Modulator
PDB	Protein Data Bank
PELE	Protein Energy Landscape Exploration
PFLOP	Peta FLoating-point Operations Per second
PMF	Potential of Mean Force
PPI	Protein-Protein Interactions
PRC	Pose Ranking Consensus
QCP	Quaternion-Based Characteristic Polynomial
QM	Quantum Mechanics
QSAR	Quantitative Structure Activity Relationship
RBD	Receptor Binding Domain
RdRp	RNA-Dependent RNA polymerase
RMSD	Root Mean Square Deviation
RNA	Ribonucleic Acid
SARS-CoV-2	Severe Acute Respiratory Syndrome CoronaVirus 2
SAXS	Small-Angle X-Ray Scattering

SBVS	Structure-Based Virtual Screening
TPU	Tensor Processing Unit
TREMD	Temperature Replica Exchange Molecular Dynamics
VS	Virtual Screening
μOR	μ-Opioid Receptor
μs	micro second

1 Introduction

Drug discovery is an expensive and slow process, and the need to shorten its duration has recently been highlighted by the Covid-19 pandemic [1]. Designing highly selective modulators for a disease-related target is a multi-objective problem where various criteria must be satisfied simultaneously. To this end, computational modeling and experimental activities are used in a synergistic way. High-performance computing (HPC) helps explore, prepare, and prioritize experimental work, with the latter generating evidence for further modeling studies in new DMTA (Design, Make, Test, Analyze) cycles.

HPC is increasingly becoming a fundamental tool for drug discovery as its computational capacity has increased approximately 1000-fold every decade [2], Fig. 1. Nowadays massive parallel architectures containing hundreds or thousands of graphical processing units (GPUs), specialized supercomputers like Anton, and cloud computing are commonly used by drug discovery pipelines [3–8].

In this chapter, we give some examples in which HPC is essential: molecular dynamics of proteins for the study of their flexibility, construction of multi-protein complexes for the identification of targets with pharmacological interest, and virtual screening of huge libraries of chemical compounds for the identification of modulators for disease-related targets.

2 HPC and MD

In the recent years, HPC was converted as an important tool for chemists just as their daily lab work. Accurate simulations are possible to predict the outcome of many conventional experiments. Simulations of how a drug binds to its target protein, applying different techniques, are part of the machinery of many academic groups, biotech, and pharma companies. Regarding the use of molecular mechanics force fields (FF) and their implementation for MD simulations, the Nobel Prize in Chemistry in 2013 recognized the work of Martin Karplus, Michael Levitt, and Arieh Warshel "for the development of multiscale models for complex

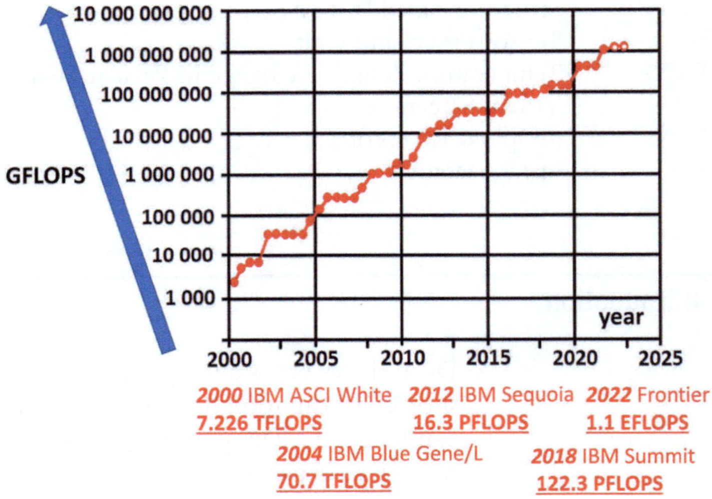

Fig. 1 The fastest supercomputer/year for the last 20 years. The y-axis in GFLOPS (10^9 floating point operations per second) is logarithmic (data from the top500.org)

chemical systems." Their studies date back to the 1970s, when computers were available for fairly limited applications and the majority of chemists were used to create models of molecules using balls and sticks made of plastic, paper or wood.

For the purpose of drug discovery studies, MD simulations have emerged as HPC-based approaches applied for analyzing the physical motions of atoms and molecules over a period of time, thus gaining insight in the dynamic evolution of the system of interest (i.e., the complex of a protein and a small drug molecule). MD trajectories are determined by numerically solving Newton's equations of motion, and forces between the atoms and their potential energies are calculated using molecular mechanics FF.

FF are developed and trained to work with all types of known molecules including drug-like small molecules, proteins, RNA, DNA, lipids, and other systems of interest. The parameters for the FF energy functions used in the drug discovery are derived from experiments and quantum mechanics (QM) calculations [9], usually performed at high level of theory. FF for small molecules implemented in different software for computer-aided drug design (CADD) simulations are the OPLS4 [10], GAFF [11], CGenFF [12], and MMFF [13] and for proteins include AMBER ff19sb [14], and CHARMM [15]. Some of the most relevant and used MD software accelerated by GPU implementation is shown in Table 1.

The reliability of MD simulations is linked to several key points. First and most important is the proper and realistic representation of the system of interest, second is the accuracy of simulated molecular interactions, and last but not least is the sufficient sampling of

Table 1
Ten of the most used GPU accelerated software for MD simulations in drug discovery

Software for MD	License[a]	Internet Link
AMBER	P, FOSS	https://ambermd.org/
CHARMM	P, C	https://www.charmm.org/
Desmond	P, C, A	https://www.deshawresearch.com/ https://www.schrodinger.com/
Discovery Studio	P	https://www.computabio.com/
GROMACS	FOSS	https://www.gromacs.org/
GROMOS	P, C	http://www.gromos.net/
NAMD/VMD	P, A	https://www.ks.uiuc.edu/
TeraChem	P	http://petachem.com/
TINKER	P	https://dasher.wustl.edu/tinker/
YASSARA	P	http://yasara.org/

[a]License types include academic (A), commercial (C), proprietary (P), and free and open-source software (FOSS)

the conformational space. All three are deeply related, need approximations, and may be greatly improved using the HPC environment. General MD workflow applied for drug discovery using an HPC facility is shown in Fig. 2.

3 Domains of Application

3.1 HPC MD in Drug Discovery

3.1.1 HPC MD Support for the Refinement of Cryo-Electron Microscopy Structures

The Nobel Prize in Chemistry in 2017 was awarded for the development of cryo-electron microscopy (cryo-EM), unveiling structures of proteins at high enough resolution to use computational drug design solutions, including MD and Monte Carlo (MC) simulations [16]. MD simulations appear to be an important support for the cryo-EM structures refinement, helping to better define the drug–target interactions. Zhuang and co-workers [17] obtained and studied cryo-EM structures of several opioid drugs bound to their GPCR target (μ-opioid receptor, μOR) applying MD simulations in lipid bilayer. Simulations reveal stable binding poses of fentanyl and morphine in the orthosteric binding pocket of μOR with an RMSD of up to 1.5 Å for the ligand in the cryo-EM structure.

Due to the size and complexity of bilayer target systems, they often need HPC simulations. Important targets of interest belong to the machinery of *Mycobacterium tuberculosis* (*Mtb*), the causative agent of tuberculosis [18, 19]. Using the potential of mean force

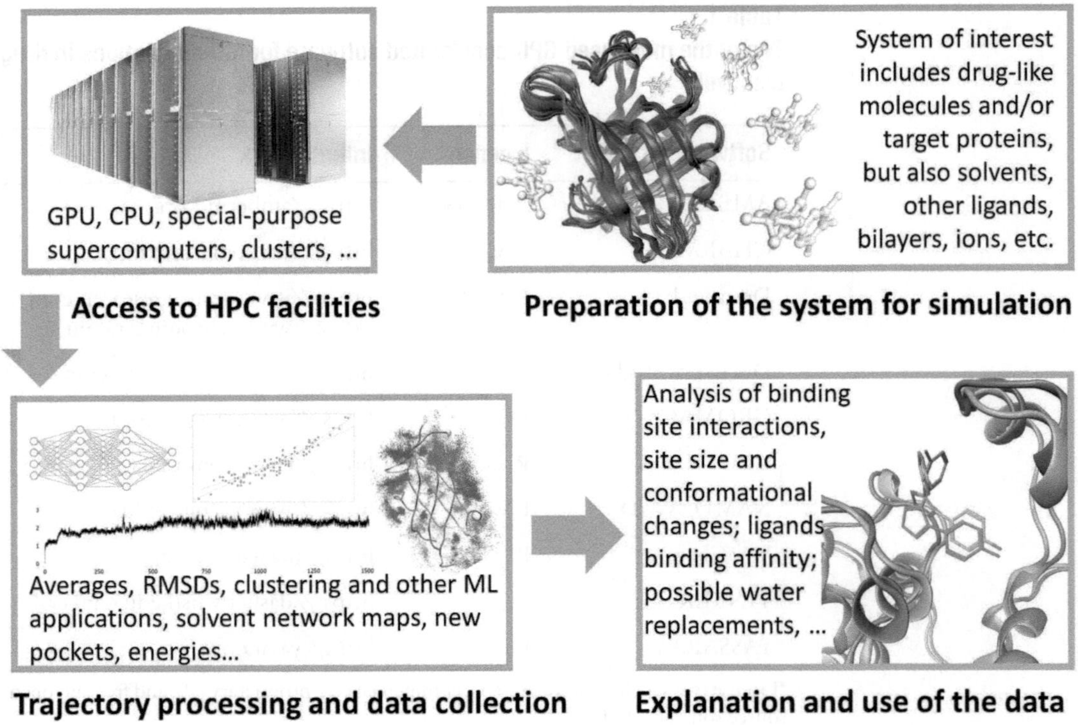

Fig. 2 Part of the computational drug discovery machinery today often includes MD simulations of the target, target/ligand, target/organic solvent, and other interactions

(PMF) calculations, Matthieu Chavent et al. [19] studied the interactions of two antibiotics (isoniazid and bedaquiline) with the *Mtb* membrane and proteins. Applying multiscale MD simulations, the team made an effort to understand the structure-function relationship of the lipid self-organization that drive biophysical properties of the *Mtb* plasma membrane [19].

Even more tangled tasks are simulations of ErmBL-stalled ribosomes with complex structures such as macrolide antibiotics [20]. Recent articles reveal the stalling mechanism with cryo-EM structures, including the nascent peptide conformation, which is supported by MD simulations. Mutual interactions between the drug, peptide, and ribosome unit give an insight into the basic mechanism of polypeptide synthesis [21, 22].

3.1.2 Special-Purpose HPC MD Simulations—The Anton Machines

A pioneer in the HPC simulations involving MD is the D.E. Shaw group with the construction and use of Anton and later Anton 2 special-purpose machines [23, 24]. Anton is a massively parallel supercomputer that dates back to 2008, running as a special-purpose system for MD simulations of biological macromolecules. Recently Huafeng Hu and co-authors, using Anton 2, explored human pandemic influenza strains and binding preference of the avian viral glycoprotein hemagglutinin (HA) to human sialic acid

(SA) receptors. Their MD-generated HA–SA conformational ensembles could support the predictions of human-adaptive mutations, in the context of emerging pandemic threats [25].

Anton supports different challenging strategies in drug discovery, such as the exploration of drugs binding to cryptic pockets for protein-protein interaction (PPI) inhibition. Yibing Shan et al. [26] performed more than hundred μs, unbiased, all-atom MD simulations of the binding process of several small-molecule PPI inhibitors to interleukin 2 (IL2). In multiple binding events, one of the small molecule inhibitors was found in a stable binding pose in the cryptic site of IL2, very accurately reproducing the existing X-ray crystal [26].

Christos Adamopoulos and co-workers identified a class of RAF inhibitors [27] that selectively inhibits dimeric BRAF over monomeric BRAF. Using extensive MD simulations in the μs time scale, they studied the movement of the BRAF αC-helix revealing it as the most important for the inhibitor selectivity. The authors showed that rationally designed MEK inhibitors according to their conformational selectivity increased efficacy when used to target BRAF V600E tumors [27].

3.1.3 HPC MD for Cryptic Pockets

Acquiring data and knowledge from HPC MD simulations of protein flexibility can be used for the discovery and understanding of hidden pockets in protein structures. Such pockets, visible or detectable only in the bound state (also known as holo-state) but not in the unbound state (apo, without ligand), are called cryptic pockets. The discovery and validation of cryptic pockets is an important drug design field especially for allosteric modulation, de novo design, and potential modulation of PPI. Several different MD methods for cryptic pockets discovery include classical and enhanced sampling approaches and/or the use of a non-natural environment for the protein [28].

In an MD study, Zuzic et al. [29] used solvents with benzene probes to investigate the membrane-embedded SARS-CoV-2 spike glycoprotein. Simulations revealed several ligand entry routes and a novel potentially druggable cryptic pocket which is located under the 617–628 loop. The flexible nature of the 617–628 loop observed in the MD simulations is validated by hydrogen-deuterium exchange mass spectrometry experiments. Curiously, the pocket was found in the site of mutations associated with increased transmissibility of Omicron [29].

Smith and Carlson [30] performed an interesting test of mixed-solvent MD (MixMD) for mapping the surface of 12 protein targets with different complexity of conformational changes of the cryptic site. Three of them require reorganization of only side chains, five would be uncovered by loop movements, and four need more significant structure rearrangements. They simulated the unbound protein solvated in a box containing 5% of probe molecules and explicit water. Authors found that in five cases, the standard MixMD simulations could map the cryptic binding site. In other cases, the accelerated MixMD with enhanced sampling of torsional angles was a necessary addition [30].

Meller and co-authors [31] studied the myosin II inhibitor blebbistatin (an allosteric modulator), and its binding site, using Markov state models (MSMs) created from an impressive 2 milliseconds of total MD simulations in explicit solvent. They found that the blebbistatin's cryptic binding pocket opens mainly in simulations of blebbistatin-sensitive myosin isoforms and correctly identified which isoforms are most sensitive to blebbistatin inhibition. Furthermore, in a docking study, the authors quantitatively predicted blebbistatin binding affinities [31].

3.1.4 An Alternative to MD: HPC Monte Carlo Simulations

One alternative to HPC MD simulations for the drug discovery is the Monte Carlo (MC) approach of the Protein Energy Landscape Exploration (PELE) [32–34]. Briefly, the MC procedure in PELE involves ligand and receptor perturbations, followed by a relaxation step comprising the side-chain sampling and minimization steps. Typical simulations for drug discovery involve tens to thousands of processors and tens to thousands of MC steps.

Liang et al. [35] investigated SARS-CoV-2 papain-like protease and the potential antiviral effects of hypericin compared to the well-known noncovalent inhibitor GRL-0617. Molecular dynamics and PELE Monte Carlo simulations highlight favorable binding of hypericin and GRL-0617 to the naphthalene binding pocket of PL^{pro}.

Perez and co-authors [36] developed an interesting in silico drug design tool—FragPELE for potential use in the hit-to-lead phase. FragPELE can grow a fragment from a bound core, while exploring the protein–ligand conformations. The authors proposed that such an approach also can be useful for finding new cavities and novel binding modes, as well as to rank ligand activities in a reasonable amount of time and with an acceptable precision. They tested it on predictions of crystallographic data sets, with cases of cryptic sub-pockets, and evaluated the potential of the software on growing and scoring compared to other known software and approaches like FEP+, Glide SP and Induced-Fit Glide, and MM-GBSA simulations [36].

3.1.5 SARS-CoV-2 Studies with HPC MD

A vast number of studies related to the COVID-19 pandemic, including many HPC MD studies, were published in the past 3 years. Simulations are exploring main parts of the machinery of the virus such as the SARS-CoV-2 main protease (Mpro), the Coronavirus spike protein, and the RNA-dependent RNA polymerase (RdRp) involving inhibitors or PPI. All MD studies became possible due to the huge amount of structural biology work that rapidly produced hundreds of good-resolution X-ray crystal structures and many Cryo-EM ones.

Menéndez and co-authors performed several microseconds of simulations to understand the preferred binding sites of the inhibitor ebselen to the Mpro [37]. Mehdipour and Hummer [38] studied the interaction of the human angiotensin-converting enzyme 2 (ACE2) receptor with the receptor binding domain (RBD) of the SARS-CoV-2 spike protein with all-atom explicit solvent MD simulations in lipid bilayers ($\sim$1,000,000 atoms per system). Their extensive MD simulation work shed light on the role of the ACE2 receptor glycosylation in the binding of the SARS-CoV-2 and could be used for the rational development of either antibodies and/or small molecules targeting the N322-glycan binding site in the RBD of the SARS-CoV-2 spike protein [38].

Byléhn and co-workers [39] studied the binding of Remdesivir, Ribavirin, and Favilavir with the RdRp complex of SARS-CoV-2 in microseconds simulations. Using free energy calculations, the authors showed that all the three nucleotide analogue inhibitors showed a strong binding capacity to the active site (the strongest being Remdesivir). Interestingly, Remdesivir demonstrated strong binding capacity to its complementary base pair, while Ribavirin and Favilavir showed more dynamic hydrogen bond network [39].

3.1.6 The Ultimate Future: Combination of HPC MD and AI/ML

Half a century ago, Levinthal [40] estimated that an unfolded protein molecule has an extremely large number of possible conformations, due to the available degrees of freedom in its polymer chain. An accurate solution to this problem has come recently from DeepMind and their AI program AlphaFold (AF). Designed as a deep learning system, AF performs the most accurate predictions of the 3D protein structures to date. AF with its latest version [41] has achieved a level of accuracy much higher than any other approach in the CASP14 competition (Nov. 2020). However, despite this advancement in the prediction of protein structures, the routine implementation of the method in the context of small-molecule drug discovery needs to be extensively explored in the future. Yin et al. [42] studied whether the AI-based models can reliably reproduce the 3D structures of protein–ligand complexes. They picked a challenging target in terms of obtaining its three-dimensional model and compared in detail the conformation of several binding pockets generated by the AI models with experimental structures. The authors explored further by molecular docking and results indicated that AI-predicted protein structures combined with molecular dynamics simulations offer a promising approach in small-molecule drug discovery [42].

3.2 Protein-Protein Interactions

Proteins interact with one or more other proteins to organize and regulate multiple biological functions [43]. Understanding their interactions at the molecular level is essential for studying healthy and disease-related conditions and finding therapeutic modalities like protein-protein inhibitors. The human interactome is estimated to contain 650,000 protein interactions [44]. However, the number of experimental complexes reported in Interactome3D in 2022 is around 8300 [45]. Computational docking methods are therefore needed to help build the molecular human interactome, Fig. 3.

Predicting the molecular structure of protein-protein complexes is a big challenge in computational biology. It requires two steps: searching and docking the partners in a 3D space and scoring the poses to identify the best complex [46]. The problem is made even more difficult by the flexibility of proteins both on the side chains and on the backbone [47, 48]. Analysis of protein-protein complexes revealed that they interact through complementary patches on their surfaces which are rather flat. The contact surfaces are relatively large, ranging from 1000 to 4000 $Å^2$ [49].

3.2.1 Conventional Approaches

Three main approaches are available for the prediction of protein-protein complexes: free docking, template-based modeling, and integrative modeling [47, 50]. Free docking methods generate molecular complexes without prior knowledge of similar entities or experimental data. Template-based modeling relies on the availability of a similar complex to build a model by homology without exploring beyond the binding mode of the template. The similarity can be described in terms of protein sequences or surface patches on the proteins [51]. Template-based modeling generates the most accurate models when good templates are found, as shown in CAPRI (Critical Assessment of Prediction of Interactions) blind prediction challenges [52]. However for most known interactions only templates with low homology are available [53]. Integrative approaches use experimental information to narrow the range of possible solutions [54]. Information can be generated by a variety of sources, including small-angle X-ray scattering (SAXS) [55], electron microscopy (EM) [56], chemical cross-linking [57], and co-evolutionary bioinformatics [58].

Free docking is difficult, and recent tests show that the best methods find acceptable models among the top 10 predictions for about 40% of the targets [48]. Most of the algorithms consider the backbone of the unbound protein structures as rigid, and they use in particular Fast Fourier Transform sampling [59], Monte-Carlo algorithm [60], or shape and physicochemical complementarity [61]. Ensemble docking can be used if a set of conformations is available for each protein.

Often, flexibility is considered for top-scored complexes, especially by sampling side-chain angles from rotamer libraries [62] or

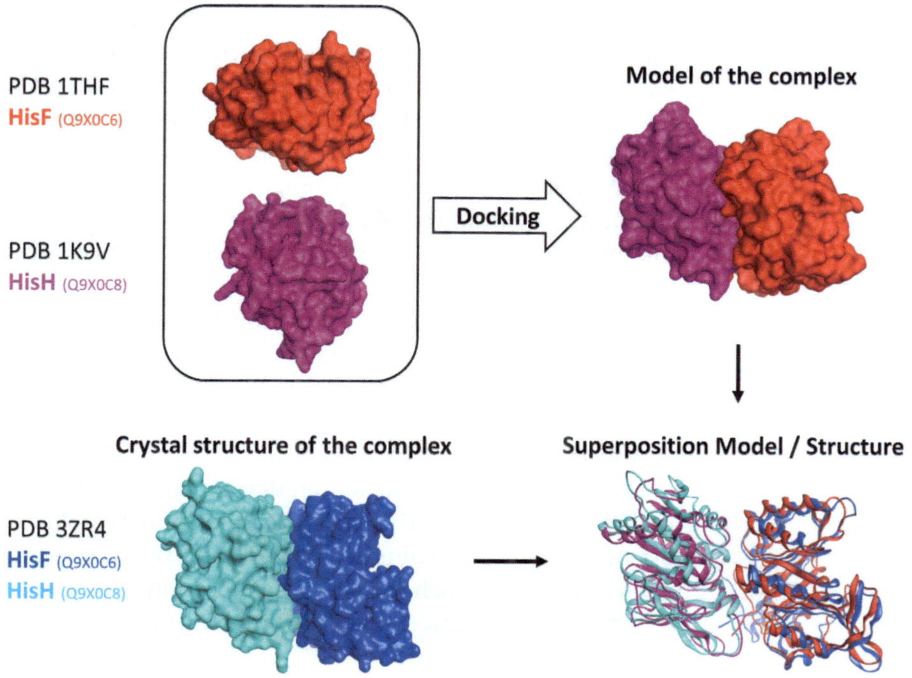

Fig. 3 General scheme of the protein-protein docking method for an example heterodimeric target. PDB structures were used for each monomer: they are colored in red and magenta. The model of the complex was built with ClusPro. An experimental structure of the complex is shown in light blue and dark blue. For this example, the comparison of the predicted model of the complex with the experimental structure shows a good superposition (ribbon representation)

by using MD simulations [63]. As an example, Jandova and colleagues [64] used MD of 100 ns to discriminate between native and non-native complexes. They found that native models exhibited greater stability compared to non-native models. Coarse-grained models are frequently used to reduce the complexity of flexible protein docking [65]. However, taking the flexibility of proteins into account increases the computation time very sharply as the size of the proteins increases, and requires high-performance computational resources in terms of processor and storage.

Some of the most popular methods are those of Haddock [66], ClusPro [67], SwarmDock [68], RosettaDock [60], PatchDock [69], ZDock [70], and Gramm-X [71].

3.2.2 Artificial Intelligence (AI) Methods

Very recently, a breakthrough in protein structure prediction has been achieved by Deep Learning approaches including AlphaFold [41, 72] and RoseTTAFold [73]. These predictors use sophisticated neural networks trained by considering both multiple sequence alignments (MSA) that contain information about the co-evolution of amino acid pairs, suggesting their spatial proximity, and protein structures in the Protein Data Bank (PDB). In CASP14 (Critical Assessment of protein Structure Prediction, round 14), a blind test for protein structure prediction, AF achieved an unprecedented accuracy in modeling the structure of monomer proteins [74]. AF training used several TPU (Tensor Processing Unit) during weeks, and the time to infer a model is from minutes to hours on V100 GPU, depending on the protein length. The memory must be carefully managed as its usage is approximately quadratic in the number of residues [41]. A database (https://alphafold.ebi.ac.uk) was built, with 360,000 predicted structures, in the first release [75].

Immediately, the idea of predicting the structure of heterodimers with AF was explored. A first approach consisted of building a single sequence with the two protein sequences linked by a long poly-glycine segment and letting the software build the 3D model as if it were a single protein [76, 77].

AlphaFold-Multimer was then developed by Evans and colleagues [78] for modeling the structure of protein complexes. It used most of AF's deep learning framework and was trained on multimeric data. On a benchmark of 17 heterodimer proteins without templates, the prediction accuracy was improved compared to AF with the flexible linker trick.

3.2.3 Toward the Simulation of the Cytoplasm

Protein-protein docking does not inform about the kinetics of the association. In a cell, several copies of proteins are immersed in a crowded environment containing water, ions, metabolites, macromolecules, and lipids at different concentrations. MD simulations of cytoplasmic environments were used to understand how they affect protein folding, influence enzyme activity, modulate diffusion, and act on complexes specificity and duration. Realistic simulations of environments containing millions of atoms, for a few microseconds or more, pose technical challenges. Most reported studies use GPU platforms, large-scale supercomputers, or the MD special supercomputer Anton 2. Another approach is to reduce the number of atoms in the simulated system using coarse-grained models.

A small volume of *E. coli* cytoplasm containing multiple all-atom copies of more than a dozen biopolymers including enzymes and tRNA was constructed by Rickard et al. [79]. The biopolymers were selected from the most abundant cytoplasmic components, and some were monomers of proteins known to oligomerize. Metabolites were included with concentrations consistent with experimental measures, as well as ions and water molecules. The system, containing about 200,000 atoms, was simulated using CHARMM-derived force fields, for a timescale of 20 μs on the Anton 2 supercomputer, and the final 5 μs was analyzed. They observed protein-rich and water-rich clusters. Contacts were highly dynamic for all protein-protein pairs. Oligomeric proteins interacted with their partners, but monomeric proteins also interacted with other proteins in the system. The lifetime of protein-protein contacts was between 0.03 and 3 μs, with most contacts lasting less than 1 μs.

The cytosol-membrane interface has been studied by Nawrocki et al. [80]. They performed MD simulation of a system containing a neutral phospholipid bilayer and a mixture of proteins not known to specifically interact with phospholipid membranes, for 10 μs, on Anton 2, with NAMD. The main finding was the creation of a depletion area near the membrane surface, where protein-membrane contacts were infrequent and of short duration.

3.3 Virtual Screening

Experimental high throughput screening (HTS) is commonly used for the discovery of biologically active hits, but the method is time consuming, expensive, and not suitable for screening libraries containing billions of molecules [81]. Another approach used in recent years to accelerate the drug discovery process is to perform virtual screening (VS) of large virtual libraries of compounds, taking advantage of the computing power of high-performance computing (HPC) systems. This approach selects small sets of chemical compounds predicted to bind to a biological target for experimental testing. VS is particularly attractive when biological assays are expensive or their miniaturization for HTS is difficult.

Structure-Based Virtual Screening (SBVS) involves docking each chemical compound from a virtual library into the relevant site of a three-dimensional model of the target of interest and calculating its affinity, Fig. 4. The docking and the affinity evaluation for each compound do not depend on the other compounds and, therefore, the process is parallelizable. Preparation of VS is a key step, and the use of automatic workflows, developed for example in Knime, makes it possible to choose the best parameters from a wide range of possibilities in a short time [82, 83].

An important property that must be considered in SBVS is the flexibility of proteins and their docking sites, which occur especially in loops and for some side chains. Molecular dynamics (MD) has been recognized as a powerful method to identify conformations explored by proteins or conformations stabilized in protein-ligand complexes and is particularly appealing for targets with limited number of crystallographic structures. It has been used to prepare proteins before docking or to improve protein-ligand complexes after docking. Examples of the inclusion of MD in SBVS processes have demonstrated improved virtual screening performance for highly flexible targets [84, 85].

3.3.1 Protein Preparation for Ensemble Docking

Ensemble docking considers different conformations exhibiting three-dimensional diversity of the protein site and performs docking of the compounds from a database in each conformation of the site [86, 87]. This approach simulates the conformation selection

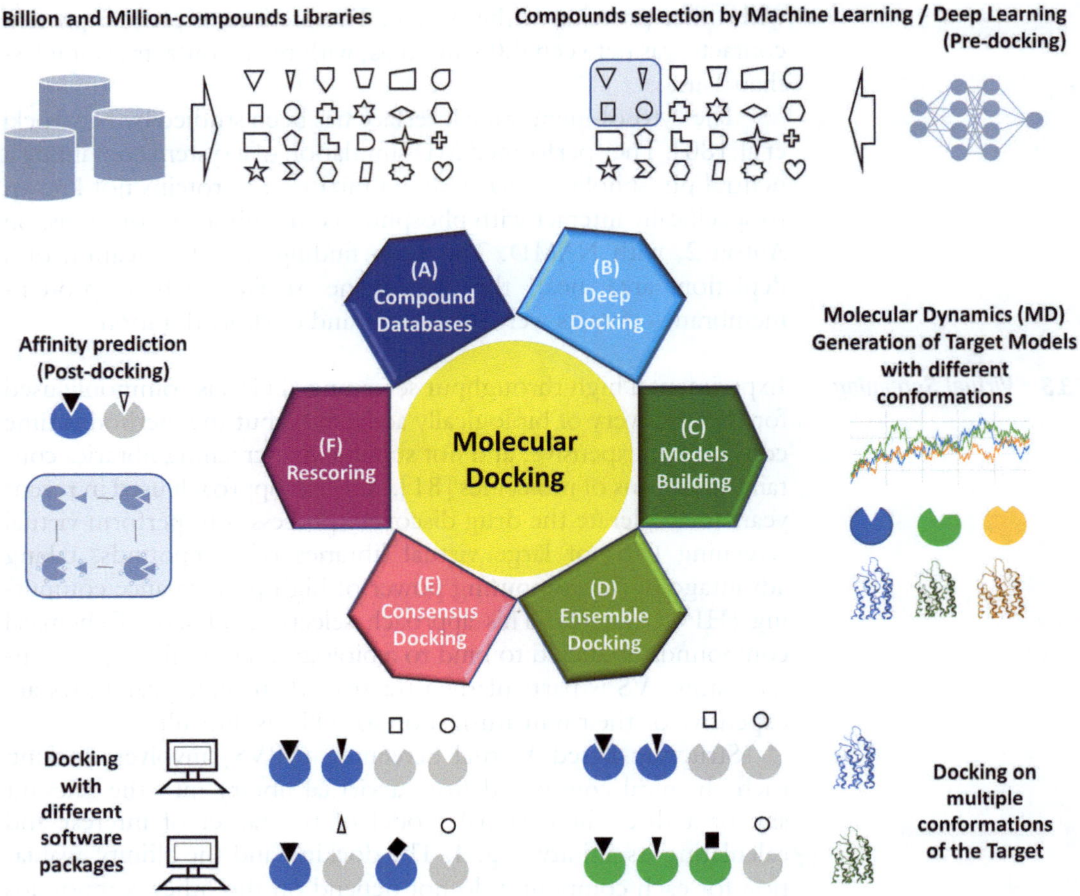

Fig. 4 Main modules used in molecular docking. (**a**) Compound databases contain millions of virtual compounds to consider in structure-based virtual screening campaigns. (**b**) When the number of compounds to be docked is very large, deep docking approaches including machine learning or deep learning methods are used to select a small number of compounds for molecular docking. (**c**) The construction of an ensemble of models for a considered target includes experimental structures, models built by homology modeling and by deep learning approaches, and models generated by molecular dynamics that sample different conformations of target. (**d**) Ensemble docking performs molecular docking on a diversity of target conformations. (**e**) Consensus docking uses molecular docking with different docking software to select compounds that show good docking with the majority of software. (**f**) Rescoring of a top-list of compounds selected by molecular docking, with computationally demanding methods, allows better prediction of compound affinity

binding process, different from induced fit binding which is more limited around a given conformation.

The process involves building a set of protein conformations representative of protein flexibility. This requires the use of methods that go beyond classical molecular dynamics in which the protein is frequently trapped in a conformational subspace [88]. Many authors have used temperature replica exchange molecular dynamics (TREMD) to improve the sampling of the conformational space of proteins and RNA [89–91]. TREMD involves performing parallel simulations of multiple copies of a target (replicas), each at a different temperature, and periodically exchanging conformations between replicas at different temperatures. The exchange of conformations allows the exploration of intermediates between relatively stable conformations and leads to an increase in conformational diversity.

Acharya and coworkers [1] employed TREMD within the GROMACS software suite [92] to prepare protein model ensembles for ensemble docking. They used the Summit supercomputer at the Oak Ridge National Laboratory to perform massive parallel supercomputing and provide rapid conformational space sampling for 8 SARS-CoV-2 proteins in various states of protonation and oligomerization, leading to a total of 24 protein systems. The number of replicates ranged from 20 to 60 for each system. The simulations' duration was 750 ns per replicate, and in total, about 0.6 ms of TREMD was done. The reported theoretical performance of the Summit supercomputer is ~1 ms/day if fully dedicated to this task, so less than a day would suffice for conformational sampling of all 24 protein systems. During production time, frames were recorded every 10 ps. At the end, pairwise RMSD matrices were calculated for each protein system using the QCP algorithm [93], and hierarchical clustering was used to select cluster configurations showing structural diversity of the binding pocket for ensemble docking.

3.3.2 Ensemble Docking

Acharya and colleagues performed ensemble docking of repositioning databases of approximately 14 k compounds into 10 configurations of each of the 24 SARS-CoV-2 systems, using Autodock Vina on HPC clusters. In a separate screening campaign, 2900 compounds were tested in two protein assays by the National Center for Advancing Translational Sciences (NCATS). For the two targets, they analyzed the ensemble docking hit lists, taking into account the experimentally proven active compounds, and they found remarkably high hit rates, mostly in the top-rated tranches of compounds.

In a retrospective study, Bhattarai and coworkers used Gaussian accelerated molecular dynamics (GaMD) simulations to generate an ensemble of conformations for the ECL2 allosteric site of the A_1AR receptor [94]. Next, they considered 25 known positive allosteric modulators (PAMs) of the A_1AR receptor and 2475 decoys obtained from the DUD-E website [95]. They found that ensemble docking, using Autodock, outperformed single docking into a cryo-EM structure of A_1AR in the agonist-Gi-bound conformation.

For the Jak2 receptor, Bajusz and colleagues performed ensemble docking of ~100 k kinase-like compounds into five Jak2 structures, producing 429 docking hits [96]. For each compound, the highest-scoring pose across the five protein structures was taken into account. After diversity selection and visual inspection, 54 compounds were purchased and tested in vitro. Six compounds were identified as hits. In a similar study for Jak1, they used two X-ray structures and three frames from MD simulation for ensemble docking of a ~100 k library [97]. Ten top-scoring compounds were tested in an enzyme-based in vitro Jak1 inhibition assay. Five hits were confirmed to inhibit Jak1 with IC50 values in the single-digit micromolar and sub-micromolar range.

Diaz and coworkers used the sequence-based Switch-P algorithm to predict a secondary structure switch in the β5 strand of D3 extracellular domain of FGFR2. MD carried out on the protein D3 domain, in the presence of an allosteric modulator, revealed the change in secondary structure and a modification of the 3D structure of the site between β5 and the rest of the domain. They docked a library with 1.4 million compounds in an X-ray structure and in an MD model and performed consensus ranking of the compounds. Thirty-two compounds were selected, and in an FGFR2-dependent Erk phosphorylation assay, 13 compounds showed activating or inhibiting action [98].

3.3.3 Consensus Scoring, Consensus Docking, and Mixed Consensus Scoring

Many virtual screening programs are available, each having strengths and limitations, and their performance depends on the characteristics of the target docking sites [99, 100]. It is not possible to know a priori if a VS software, composed of a docking method and a scoring algorithm, is more suitable than another for a given target. Thus, the idea emerged to combine the results of different VS programs for the same target structure.

- *Consensus Scoring*

  A first approach is consensus scoring. Park and colleagues [101] performed VS of a 260 k compound library in an X-ray crystal structure of AMPK2, with AutoDock and with FlexX. They listed 1000 top-scored compounds with each software and identified 118 compounds included in both sets. All compounds selected from the consensus scoring approach were tested in an enzyme inhibition assay, which resulted in seven structurally diverse AMPK2 inhibitors with micromolar activity.
- *Consensus Docking*

  Houston and Walkinshaw proposed a consensus docking procedure [102]. As an example, they used 228 protein-ligand complexes from the PDBbind-CN database, and they docked 228 ligands into their corresponding protein structures with Autodock and Vina separately. For each VS software and each compound, only the best-scored pose was retained. A pose was considered accurate if the RMSD between the docked and the crystallographic pose was less than 2 Å. With this success condition, the success rate for each program alone was 55% for Autodock and 64% for Vina. Next, they explored consensus docking. Similar docking poses (RMSD ≤2.0 Å) were generated by Autodock and by Vina for 118 ligands, and 97 of these ligands show correct poses. Therefore, the success rate of consensus docking with Autodock and Vina was 82%. Finally, they introduced a third VS program Dock, which standalone produced a success rate of 58%. Extending the consensus docking method to three VS software increased the rate of ligands with correct predicted poses in the retained set to 92%.

- *Mixed Consensus Scoring*

  Scardino and colleagues [103] have explored the combination of pose and ranking consensus. The method, named Pose/Ranking Consensus (PRC), was tested with four VS programs (ICM, rDock, Autodock, and PLANTS) on 34 targets containing binding sites with different properties. For each target, chemical libraries were prepared by merging known actives and decoys with different structures but similar physicochemical properties. For a compound docked with a given VS program, only the best scored pose was selected, resulting in four docking poses for every molecule in the database. The method calculated all RMSD among the four poses and retained poses with RMSD less than 2.0 Å. Finally, consensus scoring was performed for the remaining poses. The method revealed improved performance over consensus scoring alone or consensus docking alone.

3.3.4 Rescoring and Affinity Calculations

Accurate prediction of protein-ligand binding affinities clearly affects SBVS performance. However, to achieve docking and scoring of libraries of several million compounds in a reasonable time, the scoring functions usually neglect important terms that contribute to binding affinity, like solvation. Therefore, a popular approach is to calculate the binding energies of top-ranked docking poses (protein-ligand complexes) using more accurate and more computationally demanding methods. They are mainly of two types: end-point and pathway methods.

In the first group, MM-PBSA (molecular mechanics Poisson-Boltzmann surface area) and MM-GBSA (molecular mechanics generalized Born surface area) are commonly used [104] and successfully applied in the post-processing of VS campaigns [105–109]. Usually, short MD simulations of protein-ligand systems are performed before binding energy calculations.

The second group includes free energy perturbation (FEP) and thermodynamic integration (TI). These methods are more precise, but they are much more computationally demanding. They have been used with success to select compounds for experimental testing from short lists of compounds top-ranked by VS [110–113]. As these methods require high-performance computing resources, they are mainly used in drug design in the downstream phases of hit to lead (H2L) and lead optimization (LO) for congeneric chemical series [114, 115].

3.3.5 Billion-Compound Databases

Popular databases containing thousands to billions of compounds are listed in Table 2, revealing a wide variety of data sources and library sizes for VS. The GDB-17 database contains 166.4 billion molecules of up to 17 atoms of C, N, O, S, and halogens [116]. These molecules are synthetically plausible, but most are not commercially available and therefore must be synthetized after VS. The Enamine REAL database contains 4.5 billion compounds that are readily available, albeit not in stock, with shipment within 3 weeks of purchase [117]. The ZINC-20 database includes 1.3 billion compounds that can be purchased from the catalogs of different companies [118]. The database is regularly updated: more than 90% of the catalogs are updated every 90 days, and more than 90% of the compounds were purchasable in the last three months.

The estimated size of the chemical space for organic compounds with molecular mass <500 daltons is 10^{60}, suggesting future growth in database size.

3.3.6 Docking Ultra-Large Databases

Docking large databases requires parallel computing architectures. Additionally, the recent use of GPUs instead of CPUs for docking has further reduced docking time in a very significant way. As an example, the program AutoDock-GPU achieves a 50× speedup compared to the previous version AutoDock4 which runs on CPUs.

Acharya and colleagues have modified AutoDock-GPU to run optimally on the Summit supercomputer, and they demonstrated to be able to dock over 1 billion compounds in less than 1 day when scaled to the totality of the computer [1]. Gadioli and coworkers deployed the Exascale docking platform on two top European supercomputers (CINECA-Marconi100 and ENI-HPC5) with a combined computational performance of 81 PFLOPS, and they docked with LiGen 70 billion ligands into 15 binding sites of 12 SARS-Cov2 proteins, thus representing 1 trillion dockings, in 2.5 days [119]. For a review on the most efficient docking algorithms and their performance on various computer architectures, the reader is referred to [3].

3.3.7 Deep Docking

The docking of compound databases in a reasonable time is threatened by several factors: ensemble docking which docks compounds in multiple conformations of the same target site is increasingly employed; the sizes of compound databases grow exponentially; and supercomputers are not accessible to everyone.

Table 2
Popular compound databases used in virtual screening (data as of July 25, 2022)

Database	# Compounds	Website
GDB-17	166 billion	https://gdb.unibe.ch/downloads/
Enamine REAL	4.5 billion	https://enamine.net
ZINC-20	1.3 billion	https://zinc20.docking.org/
ChemSpider	114 million	http://www.chemspider.com/
PubChem	111 million	https://pubchem.ncbi.nlm.nih.gov/
LifeChemicals	2.5 million	https://lifechemicals.com/
ChEMBL	2.2 million	https://www.ebi.ac.uk/chembl/
ChemDiv	1.6 million	https://www.chemdiv.com/
ChemBridge	1.3 million	https://www.chembridge.com/
Cambridge Structural Database (CSD)	1.2 million	https://www.ccdc.cam.ac.uk/
Binding DB	1.1 million	https://www.bindingdb.org/bind/index.jsp
Asinex	575,302	https://www.asinex.com/
Protein Data Bank	37,294	https://www.rcsb.org/
DUD-E	22,886	http://dude.docking.org/
Drug Bank	14,677	https://www.drugbank.com/

To circumvent these difficulties, the deep docking approach has been developed. It is based on the observation that only a very small percentage of the docked compounds is interesting for a considered target and that docking the whole database is too computationally expensive and slow. The deep docking method performs docking of a small randomly selected portion of the database, uses the compounds in the subset and their docking scores to build a quantitative structure-activity relationship (QSAR) model for prediction of docking scores based on the chemical descriptors of the compounds, discards compounds in the database that show unfavorable predicted scores (without docking those compounds), and finally docks the remaining compounds. As a result, the method filters out most of the compounds in the database without investing time and resources to dock them.

Ton and colleagues considered the SARS-CoV2 Mpro protease, and 1.3 billion compounds from the ZINC15 library. They used the docking software Glide, 2D compound descriptors, and a deep neural network (DNN) for the QSAR modeling [120]. The DNN model was iteratively trained, starting with a random selection of 3 million docked compounds, divided into training, validation, and test sets. At the end, 3 million compounds with the best predicted scores were docked to the protease site, and 1000 compounds were selected. Their docking scores were better than protease inhibitors from BindingDB docked at the same site. The process took a week using 13 Intel(R) Xeon(R) Gold 6130 CPUs for docking and 40 Nvidia Tesla V100 GPUs for deep learning.

Gentile and colleagues used a database with 40 billion compounds, and five docking software (Autodock-GPU, Glide SP, FRED, ICM and QuickVina2), for deep docking of the SARS-Cov2 Mpro protein [121]. They started by using Autodock-GPU, then sequentially switched to a different program (Glide SP, then FRED, then ICM, and finally QuickVina2) when the iterative training of the corresponding QSAR model was completed.

The number of compounds was reduced to 24 million compounds, or 0.06% of the original database. After various additional post-deep docking steps, 1700 compounds were ordered, among which 1283 were available and tested in SARS-CoV2 Mpro enzymatic assay. Experimental testing revealed 117 compounds with >10% inhibition at 5 mM (9% of compounds tested). The deep docking protocol is detailed in [122].

4 Conclusion and Outlook

Today, HPC has become an essential tool in drug discovery. It performs immense amounts of calculations required for the numerical simulation of complex biochemical processes, such as interactions between proteins, protein flexibility, and the binding of chemical compounds to targets of therapeutic interest. Computing performance has increased from millions of operations per second to billions, and we are entering the era of trillions of operations per second. HPC makes it possible to use sophisticated algorithms to generate accurate simulations in a reduced time [9]. It contributes to rational drug design and accelerates drug discovery by limiting or replacing experiments and shortening testing phases. HPC has also fostered the emergence of deep neural networks and machine learning techniques that can extract information from very large datasets to build accurate predictive models [123].

At the end of 2022, the fastest known supercomputer is Frontier with a speed of 1.1 exaflops. The main challenges of HPC are the control of electricity consumption, which can be close to that of a small town for the largest supercomputers, and the optimization of algorithms for distributed architectures, which requires the joint work of applied mathematicians and computer science experts in parallel systems to eliminate useless calculations.

Cloud computing has started to democratize access to supercomputing, which was previously restricted to certain universities, institutes, and large companies.

The development of quantum computers and the first reported uses suggest an imminent paradigm shift especially for simulation and optimization. Although it is still too early to predict the real impact, the use of quantum computers should lead to future improvements in the drug discovery process.

Acknowledgments

The authors thank Brice Sautier and Gaurao Dhoke (Evotec (France) SAS, Toulouse, France) for valuable suggestions to improve the manuscript.

References

1. Acharya A, Agarwal R, Baker MB, Baudry J et al (2020) Supercomputer-based ensemble docking drug discovery pipeline with application to Covid-19. J Chem Inf Model 60: 5832–5852
2. Mann A (2020) Core concept: nascent exascale supercomputers offer promise, present challenges. Proc Natl Acad Sci U S A 117(37):22623–22625
3. Murugan NA, Podobas A, Vitali E, Gadioli D, Palermo G, Markidis S (2022) A review on parallel virtual screening softwares for high-performance computers. Pharmaceuticals 15(1):63
4. Jung J, Kobayashi C, Kasahara K, Tan C, Kuroda A, Minami K, Ishiduki S, Nishiki T, Inoue H, Ishikawa Y, Feig M, Sugita Y (2020) New parallel computing algorithm of molecular dynamics for extremely huge scale biological systems. J Comput Chem 42(4): 231–241
5. Jones D, Allen JE, Yang Y, Drew Bennett WF, Gokhale M, Moshiri N, Rosing TS (2022) Accelerators for classical molecular dynamics simulations of biomolecules. J Chem Theory Comput 18(7):4047–4069
6. Vermaas JV, Sedova A, Baker MB, Boehm S, Rogers DM, Larkin J, Glaser J, Smith MD, Hernandez O, Smith JC (2020) Supercomputing pipelines search for therapeutics against COVID-19. Comput Sci Eng 23(1): 7–16
7. Kutzner C, Kniep C, Cherian A, Nordstrom L, Grubmüller H, de Groot BL, Gapsys V (2022) GROMACS in the cloud: a global supercomputer to speed up alchemical drug design. J Chem Inf Model 62(7): 1691–1711
8. Puertas-Martín S, Banegas-Luna AJ, Paredes-Ramos M, Redondo JL, Ortigosa PM, Brovarets OO, Pérez-Sánchez H (2020) Is high performance computing a requirement for novel drug discovery and how will this impact academic efforts? Expert Opin Drug Discov 15(9):981–986
9. Kotev M, Sarrat L, Diaz Gonzalez C (2020) User-friendly quantum mechanics: applications for drug discovery. Methods Mol Biol 2114:231–255
10. Lu C, Wu C, Ghoreishi D, Chen W, Wang L, Damm W, Ross GA, Dahlgren MK, Russell E, Von Bargen CD, Abel R, Friesner RA, Harder ED (2021) OPLS4: improving force field accuracy on challenging regimes of chemical space. J Chem Theory Comput 17:4291–4300
11. Wang J, Wolf RM, Caldwell JW, Kollman PA, Case DA (2004) Development and testing of a general Amber Force Field. J Comput Chem 25(9):1157–1174
12. Vanommeslaeghe K, Hatcher E, Acharya C, Kundu S, Zhong S, Shim J, Darian E, Guvench O, Lopes P, Vorobyov I, Mackerell AD Jr (2010) CHARMM general force field: a force field for drug-like molecules compatible with the CHARMM all-atom additive biological force fields. J Comput Chem 31(4):671–690
13. Halgren TA (1999) MMFF VI. MMFF94s option for energy minimization studies. J Comput Chem 20(7):720–729

14. Tian C, Kasavajhala K, Belfon KAA, Raguette L, Huang H, Migues AN, Bickel J, Wang Y, Pincay J, Wu Q, Simmerling C (2020) ff19SB: amino-acid-specific protein backbone parameters trained against quantum mechanics energy surfaces in solution. J Chem Theory Comput 16(1):528–552
15. Brooks BR, Brooks CL III, MacKerell AD Jr, Nilsson L, Petrella RJ, Roux B, Won Y, Archontis G, Bartels C, Boresch S, Caflisch A, Caves L, Cui Q, Dinner AR, Feig M, Fischer S, Gao J, Hodoscek M, Im W, Kuczera K, Lazaridis T, Ma J, Ovchinnikov V, Paci E, Pastor RW, Post CB, Pu JZ, Schaefer M, Tidor B, Venable RM, Woodcock HL, Wu X, Yang W, York DM, Karplus M (2009) CHARMM: the biomolecular simulation program. J Comput Chem 30(10):1545–1614
16. Kotev M, Pascual R, Almansa C, Guallar V, Soliva R (2018) Pushing the limits of computational structure-based drug design with a cryo-EM structure: the Ca2+ channel α2δ-1 subunit as a test case. J Chem Inf Model 58(8):1707–1715
17. Zhuang Y, Wang Y, He B, He X, Zhou XE, Guo S, Rao Q, Yang J, Liu J, Zhou Q, Wang X, Liu M, Liu W, Jiang X, Yang D, Jiang H, Shen J, Melcher K, Chen H, Jiang Y, Cheng X, Wang MW, Xie X, Xu HE (2022) Molecular recognition of morphine and fentanyl by the human μ-opioid receptor. Cell 185(23):4361–4375
18. Lopez Quezada L, Silve S, Kelinske M, Liba A, Diaz Gonzalez C, Kotev M, Goullieux L, Sans S, Roubert C, Lagrange S, Bacqué E, Couturier C, Pellet A, Blanc I, Ferron M, Debu F, Li K, Aubé J, Roberts J, Little D, Ling Y, Zhang J, Gold B, Nathan C (2019) Bactericidal disruption of magnesium metallostasis in Mycobacterium tuberculosis is counteracted by mutations in the metal ion transporter CorA. MBio 10(4):e01405–e01419. https://doi.org/10.1128/mBio.01405-19
19. Brown CM, Corey RA, Gao Y, Choi YK, Gilleron M, Destainville N, Fullam E, Im W, Stansfeld PJ, Chavent M (2022) From molecular dynamics to supramolecular organization: the role of PIM lipids in the originality of the mycobacterial plasma membrane, bioRxiv. https://doi.org/10.1101/2022.06.29.498153
20. Kotev MI, Ivanov PM (2008) Molecular Mechanics (MM3(pi)) conformational analysis of molecules containing conjugated pi-electron fragments: leucomycin-V. Chirality 20:400–410
21. Beckert B, Leroy EC, Sothiselvam S, Bock LV, Svetlov MS, Graf M, Arenz S, Abdelshahid M, Seip B, Grubmüller H, Mankin AS, Innis CA, Vázquez-Laslop N, Wilson DN (2021) Structural and mechanistic basis for translation inhibition by macrolide and ketolide antibiotics. Nat Commun 12(1):4466
22. Arenz S, Bock LV, Graf M, Innis CA, Beckmann R, Grubmüller H, Vaiana AC, Wilson DN (2016) A combined cryo-EM and molecular dynamics approach reveals the mechanism of ErmBL-mediated translation arrest. Nat Commun 7:12026
23. Shaw DE, Deneroff MM, Dror RO, Kuskin JS, Larson RH, Salmon JK et al (2008) Anton, a special-purpose machine for molecular dynamics simulation. Commun ACM 51(7):91–97. https://doi.org/10.1145/1364782.1364802
24. Shaw DE, Grossman JP, Bank JA, Batson B, Butts JA, Chao JC et al (2014) Anton 2: raising the Bar for performance and programmability in a special-purpose molecular dynamics supercomputer. In: International conference for high performance computing, networking, storage and analysis, SC. IEEE, New York City, pp 41–53. https://doi.org/10.1109/SC.2014.9
25. Xu H, Palpant T, Weinberger C, Shaw DE (2022) Characterizing receptor flexibility to predict mutations that lead to human adaptation of influenza hemagglutinin. J Chem Theory Comput 18(8):4995–5005
26. Shan Y, Mysore VP, Leffler AE, Kim ET, Sagawa S, Shaw DE (2022) How does a small molecule bind at a cryptic binding site? PLoS Comput Biol 18(3):e1009817
27. Adamopoulos C, Ahmed TA, Tucker MR, Ung PMU, Xiao M, Karoulia Z, Amabile A, Wu X, Aaronson SA, Ang C, Rebecca VW, Brown BD, Schlessinger A, Herlyn M, Wang Q, Shaw DE, Poulikakos PI (2021) Exploiting allosteric properties of RAF and MEK inhibitors to target therapy-resistant tumors driven by oncogenic BRAF signaling. Cancer Discov 11(7):1716–1735
28. Kuzmanic A, Bowman GR, Juarez-Jimenez J, Michel J, Gervasio FL (2020) Investigating cryptic binding sites by molecular dynamics simulations. Acc Chem Res 53(3):654–661
29. Zuzic L, Samsudin F, Shivgan AT, Raghuvamsi PV, Marzinek JK, Boags A, Pedebos C, Tulsian NK, Warwicker J, MacAry P, Crispin M, Khalid S, Anand GS, Bond PJ (2022) Uncovering cryptic pockets in the SARS-CoV-2 spike glycoprotein. Structure 30(8):1062–1074

30. Smith RD, Carlson HA (2021) Identification of cryptic binding sites using MixMD with standard and accelerated molecular dynamics. J Chem Inf Model 61(3):1287–1299
31. Meller A, Lotthammer JM, Smith LG, Novak B, Lee LA, Kuhn CC, Greenberg L, Leinwand LA, Greenberg MJ, Bowman GR (2023) Drug specificity and affinity are encoded in the probability of cryptic pocket opening in myosin motor domains. elife 12: e83602
32. Kotev M, Lecina D, Tarragó T, Giralt E, Guallar V (2015) Unveiling prolyl oligopeptidase ligand migration by comprehensive computational techniques. Biophys J 108(1): 116–125
33. Kotev M, Soliva R, Orozco M (2016) Challenges of docking in large, flexible and promiscuous binding sites. Bioorg Med Chem 24(20):4961–4969
34. Kotev M, Manuel-Manresa P, Hernando E, Soto-Cerrato V, Orozco M, Quesada R, Pérez-Tomás R, Guallar V (2018) Inhibition of human enhancer of zeste homolog 2 with tambjamine analogs. J Chem Inf Model 57(8):2089–2098
35. Liang JJ, Pitsillou E, Ververis K, Guallar V, Hung A, Karagiannis TC (2022) Investigation of small molecule inhibitors of the SARS-CoV-2 papain-like protease by all-atom microsecond modelling, PELE Monte Carlo simulations, and in vitro activity inhibition. Chem Phys Lett 788:139294
36. Perez C, Soler D, Soliva R, Guallar V (2020) FragPELE: dynamic ligand growing within a binding site. A novel tool for hit-to-Lead drug design. J Chem Inf Model 60(3):1728–1736
37. Menéndez CA, Byléhn F, Perez-Lemus GR, Alvarado W, de Pablo JJ (2020) Molecular characterization of ebselen binding activity to SARS- CoV-2 main protease. Sci Adv 6(37):eabd0345. https://doi.org/10.1126/sciadv.abd0345
38. Mehdipour AR, Hummer G (2021) Dual nature of human ACE2 glycosylation in binding to SARS-CoV-2 spike. Proc Natl Acad Sci U S A 118(19):e2100425118. https://doi.org/10.1073/pnas.2100425118
39. Byléhn F, Menéndez CA, Perez-Lemus GR, Alvarado W, De Pablo JJ (2021) Modeling the binding mechanism of remdesivir, favilavir, and ribavirin to SARS-CoV-2 - RNA-dependent RNA polymerase. ACS Cent Sci 7(1):164–174. https://doi.org/10.1021/acscentsci.0c01242
40. Levinthal C (1969) How to fold graciously. Mossbauer Spectroscopy in Biological Systems Proceedings 67(41):22–26. http://www-miller.ch.cam.ac.uk/levinthal/levinthal.html
41. Jumper J, Evans R, Pritzel A, Green T, Figurnov M, Ronneberger O, Tunyasuvunakool K, Bates R, Žídek A, Potapenko A, Bridgland A, Meyer C, Kohl SAA, Ballard AJ, Cowie A, Romera-Paredes B, Nikolov S, Jain R, Adler J, Back T, Petersen S, Reiman D, Clancy E, Zielinski M, Steinegger M, Pacholska M, Berghammer T, Bodenstein S, Silver D, Vinyals O, Senior AW, Kavukcuoglu K, Kohli P, Hassabis D (2021) Highly accurate protein structure prediction with AlphaFold. Nature 596(7873):583–589
42. Yin J, Lei J, Yu J, Cui W, Satz AL, Zhou Y, Feng H, Deng J, Su W, Kuai L (2022) Assessment of AI-based protein structure prediction for the NLRP3 target. Molecules 27(18): 5797
43. Jones S, Thornton JM (1996) Principles of protein-protein interactions. Proc Natl Acad Sci U S A 93(1):13–20
44. Stumpf MP, Thorne T, de Silva E, Stewart R, An HJ, Lappe M, Wiuf C (2008) Estimating the size of the human interactome. Proc Natl Acad Sci U S A 105(19):6959–6964
45. Mosca R, Céol A, Aloy P (2013) Interactome3D: adding structural details to protein networks. Nat Methods 10(1):47–53
46. Ruiz Echartea ME, Chauvot de Beauchêne I, Ritchie DW (2019) EROS-DOCK: protein-protein docking using exhaustive branch-and-bound rotational search. Bioinformatics 35(23):5003–5010
47. Soni N, Madhusudhan MS (2017) Computational modeling of protein assemblies. Curr Opin Struct Biol 44:179–189
48. Porter KA, Desta I, Kozakov D, Vajda S (2019) What method to use for protein-protein docking? Curr Opin Struct Biol 55: 1–7
49. Sable R, Jois S (2015) Surfing the protein-protein interaction surface using docking methods: application to the design of PPI inhibitors. Molecules 20(6):11569–11603
50. Rosell M, Fernández-Recio J (2020) Docking approaches for modeling multi-molecular assemblies. Curr Opin Struct Biol 64:59–65
51. Baspinar A, Cukuroglu E, Nussinov R, Keskin O, Gursoy A (2014) PRISM: a web server and repository for prediction of protein-protein interactions and modeling their 3D complexes. Nucleic Acids Res 42: W285–W289
52. Lensink MF, Velankar S, Kryshtafovych A, Huang SY, Schneidman-Duhovny D, Sali A,

Segura J, Fernandez-Fuentes N et al (2016) Prediction of homoprotein and heteroprotein complexes by protein docking and template-based modeling: a CASP-CAPRI experiment. Proteins 84(Suppl 1):323–348

53. Negroni J, Mosca R, Aloy P (2014) Assessing the applicability of template-based protein docking in the twilight zone. Structure 22(9):1356–1362
54. Koukos PI, Bonvin AMJJ (2020) Integrative modelling of biomolecular complexes. J Mol Biol 432(9):2861–2881
55. Mertens HD, Svergun DI (2010) Structural characterization of proteins and complexes using small-angle X-ray solution scattering. J Struct Biol 172(1):128–141
56. Thalassinos K, Pandurangan AP, Xu M, Alber F, Topf M (2013) Conformational States of macromolecular assemblies explored by integrative structure calculation. Structure 21(9):1500–1508
57. Zeng-Elmore X, Gao XZ, Pellarin R, Schneidman-Duhovny D, Zhang XJ, Kozacka KA, Tang Y, Sali A, Chalkley RJ, Cote RH, Chu F (2014) Molecular architecture of photoreceptor phosphodiesterase elucidated by chemical cross-linking and integrative modeling. J Mol Biol 426(22):3713–3728
58. Uguzzoni G, John Lovis S, Oteri F, Schug A, Szurmant H, Weigt M (2017) Large-scale identification of coevolution signals across homo-oligomeric protein interfaces by direct coupling analysis. Proc Natl Acad Sci U S A 114(13):E2662–E2671
59. Katchalski-Katzir E, Shariv I, Eisenstein M, Friesem AA, Aflalo C, Vakser IA (1992) Molecular surface recognition: determination of geometric fit between proteins and their ligands by correlation techniques. Proc Natl Acad Sci U S A 89(6):2195–2199
60. Lyskov S, Gray JJ (2008) The RosettaDock server for local protein-protein docking. Nucleic Acids Res 36:W233–W238
61. Axenopoulos A, Daras P, Papadopoulos GE, Houstis EN (2013) SP-dock: protein-protein docking using shape and physicochemical complementarity. IEEE/ACM Trans Comput Biol Bioinform 10(1):135–150. https://doi.org/10.1109/TCBB.2012.149
62. Shapovalov MV, Dunbrack RL Jr (2011) A smoothed backbone-dependent rotamer library for proteins derived from adaptive kernel density estimates and regressions. Structure 19(6):844–858
63. Smith GR, Sternberg MJE, Bates PA (2005) The relationship between the flexibility of proteins and their conformational states on forming protein-protein complexes with an application to protein–protein docking. J Mol Biol 347:1077–1101
64. Jandova Z, Vargiu AV, Bonvin AMJJ (2021) Native or non-native protein-protein docking models? Molecular dynamics to the rescue. J Chem Theory Comput 17(9):5944–5954
65. Harmalkar A, Gray JJ (2021) Advances to tackle backbone flexibility in protein docking. Curr Opin Struct Biol 67:178–186
66. Van Zundert GCP, Rodrigues JPGLM, Trellet M, Schmitz C, Kastritis PL, Karaca E, Melquiond ASJ, van Dijk M, de Vries SJ, Bonvin AMJJ (2016) The HADDOCK2.2 web server: user-friendly integrative modeling of biomolecular complexes. J Mol Biol 428(4):720–725
67. Kozakov D, Hall DR, Xia B, Porter KA, Padhorny D, Yueh C, Beglov D, Vajda S (2017) The ClusPro web server for protein-protein docking. Nat Protoc 12(2):255–278
68. Torchala M, Moal IH, Chaleil RA, Fernandez-Recio J, Bates PA (2013) SwarmDock: a server for flexible protein-protein docking. Bioinformatics 29(6):807–809
69. Schneidman-Duhovny D, Inbar Y, Nussinov R, Wolfson HJ (2005) PatchDock and SymmDock: servers for rigid and symmetric docking. Nucleic Acids Res 33: W363–W367. https://doi.org/10.1093/nar/gki481
70. Pierce BG, Wiehe K, Hwang H, Kim BH, Vreven T, Weng Z (2014) ZDOCK server: interactive docking prediction of protein-protein complexes and symmetric multimers. Bioinformatics 30(12):1771–1773
71. Tovchigrechko A, Vakser IA (2006) GRAMM-X public web server for protein-protein docking. Nucleic Acids Res 34: W310–W314. https://doi.org/10.1093/nar/gkl206
72. Tunyasuvunakool K, Adler J, Wu Z, Green T, Zielinski M, Žídek A, Bridgland A, Cowie A, Meyer C, Laydon A, Velankar S, Kleywegt GJ, Bateman A, Evans R, Pritzel A, Figurnov M, Ronneberger O, Bates R, Kohl SAA, Potapenko A, Ballard AJ, Romera-Paredes B, Nikolov S, Jain R, Clancy E, Reiman D, Petersen S, Senior AW, Kavukcuoglu K, Birney E, Kohli P, Jumper J, Hassabis D (2021) Highly accurate protein structure prediction for the human proteome. Nature 596(7873):590–596
73. Baek M, DiMaio F, Anishchenko I, Dauparas J, Ovchinnikov S, Lee GR, Wang J, Cong Q, Kinch LN, Schaeffer RD, Millán C, Park H, Adams C, Glassman CR,

DeGiovanni A, Pereira JH, Rodrigues AV, van Dijk AA, Ebrecht AC, Opperman DJ, Sagmeister T, Buhlheller C, Pavkov-Keller T, Rathinaswamy MK, Dalwadi U, Yip CK, Burke JE, Garcia KC, Grishin NV, Adams PD, Read RJ, Baker D (2021) Accurate prediction of protein structures and interactions using a three-track neural network. Science 373(6557):871–876

74. Kryshtafovych A, Schwede T, Topf M, Fidelis K, Moult J (2021) Critical assessment of methods of protein structure prediction (CASP)-Round XIV. Proteins 89(12): 1607–1617

75. Varadi M, Anyango S, Deshpande M, Nair S, Natassia C, Yordanova G, Yuan D, Stroe O, Wood G, Laydon A, Žídek A, Green T, Tunyasuvunakool K, Petersen S, Jumper J, Clancy E, Green R, Vora A, Lutfi M, Figurnov M, Cowie A, Hobbs N, Kohli P, Kleywegt G, Birney E, Hassabis D, Velankar S (2022) AlphaFold Protein Structure Database: massively expanding the structural coverage of protein-sequence space with high-accuracy models. Nucleic Acids Res 50(D1): D439–D444. https://doi.org/10.1093/nar/gkab1061

76. Ko J, Lee J (2021) Can AlphaFold2 predict protein-peptide complex structures accurately? bioRxiv. https://doi.org/10.1101/2021.07.27.453972

77. Zhao Y, Rai J, Xu C, He H, Li H (2022) Artificial intelligence-assisted cryoEM structure of Bfr2-Lcp5 complex observed in the yeast small subunit processome. Commun Biol 5(1):523

78. Evans R, O'Neill M, Pritzel A, Antropova N, Senior A, Green T, Žídek A, Bates R, Blackwell S, Yim J, Ronneberger O, Bodenstein S, Zielinski M, Bridgland A, Potapenko A, Cowie A, Tunyasuvunakool K, Jain R, Clancy E, Kohli P, Jumper J, Hassabis D (2022) Protein complex prediction with AlphaFold-Multimer. bioRxiv. https://doi.org/10.1101/2021.10.04.463034

79. Rickard MM, Zhang Y, Gruebele M, Pogorelov TV (2019) In-cell protein-protein contacts: transient interactions in the crowd. J Phys Chem Lett 10(18):5667–5673

80. Nawrocki G, Im W, Sugita Y, Feig M (2019) Clustering and dynamics of crowded proteins near membranes and their influence on membrane bending. Proc Natl Acad Sci U S A 116(49):24562–24567

81. LeGrand S, Scheinberg A, Tillack AF, Thavappiragasam M, Vermaas JV, Agarwal R, Larkin J, Poole D, Santos-Martins D, Solis-Vasquez L, Koch A, Forli S, Hernandez O, Smith JC, Sedova A (2020) GPU-accelerated drug discovery with docking on the summit supercomputer: porting, optimization, and application to COVID-19 research. arXiv:2007.03678

82. Pihan E, Kotev M, Rabal O, Beato C, Diaz Gonzalez C (2021) Fine tuning for success in structure-based virtual screening. J Comput Aided Mol Des 35(12):1195–1206

83. David L, Mdahoma A, Singh N, Buchoux S, Pihan E, Diaz C, Rabal O (2022) A toolkit for covalent docking with GOLD: from automated ligand preparation with KNIME to bound protein-ligand complexes. Bioinform Adv 2(1):vbac090

84. Spyrakis F, Benedetti P, Decherchi S, Rocchia W, Cavalli A, Alcaro S, Ortuso F, Baroni M, Cruciani G (2015) A pipeline to enhance ligand virtual screening: integrating molecular dynamics and fingerprints for ligand and proteins. J Chem Inf Model 55: 2256–2274

85. Wang YY, Li L, Chen T, Chen W, Xu Y (2013) Microsecond molecular dynamics simulation of Ab42 and identification of a novel dual inhibitor of Ab42 aggregation and BACE1 activity. Acta Pharmacol Sin 34:1243–1250

86. Amaro RE, Baudry J, Chodera J, Demir O, McCammon JA, Miao Y, Smith JC (2018) Ensemble docking in drug discovery. Biophys J 114(10):2271–2278

87. Korb O, Olsson TS, Bowden SJ, Hall RJ, Verdonk ML, Liebeschuetz JW, Cole JC (2012) Potential and limitations of ensemble docking. J Chem Inf Model 52:1262–1274

88. Mitsutake A, Mori Y, Okamoto Y (2013) Enhanced sampling algorithms. Methods Mol Biol 924:153–195

89. Ravindranathan KP, Gallicchio E, Friesner RA, McDermott AE, Levy RM (2006) Conformational equilibrium of cytochrome P450 BM-3 complexed with N-Palmitoylglycine: a replica exchange molecular dynamics study. J Am Chem Soc 128(17):5786–5791

90. Turner M, Mutter ST, Kennedy-Britten OD, Platts JA (2019) Replica exchange molecular dynamics simulation of the coordination of Pt(ii)-Phenanthroline to amyloid-β. RSC Adv 9(60):35089–35097. https://doi.org/10.1039/c9ra04637b

91. Ke Y, Jin H, Sun L (2019) Revealing conformational dynamics of 2'-O-methyl-RNA guanine modified G-quadruplex by replica exchange molecular dynamics. Biochem Biophys Res Commun 520(1):14–19

92. Abraham MJ, Murtola T, Schulz R, Páll S, Smith JC, Hess B, Lindahl E (2015)

GROMACS: high performance molecular simulations through multi-level parallelism from laptops to supercomputers. SoftwareX 1:19–25. https://doi.org/10.1016/j.softx.2015.06.001

93. Theobald DL (2005) Rapid calculation of RMSDs using a quaternion-based characteristic polynomial. Acta Crystallogr, Sect A 61: 478–480

94. Bhattarai A, Wang J, Miao Y (2020) Retrospective ensemble docking of allosteric modulators in an adenosine Gprotein-coupled receptor. Biochim Biophys Acta Gen Subj 1864(8):129615. https://doi.org/10.1016/j.bbagen.2020.129615

95. Mysinger MM, Carchia M, Irwin JJ, Shoichet BK (2012) Directory of Useful Decoys, Enhanced (DUD-E): better ligands and decoys for better benchmarking. J Med Chem 55(14):6582–6594. https://doi.org/10.1021/jm300687e

96. Bajusz D, Ferenczy GG, Keserű GM (2016) Discovery of subtype selective Janus Kinase (JAK) inhibitors by structure-based virtual screening. J Chem Inf Model 56(1):234–247

97. Bajusz D, Ferenczy GG, Keserű GM (2016) Ensemble docking-based virtual screening yields novel spirocyclic JAK1 inhibitors. J Mol Graph Model 70:275–283

98. Diaz C, Herbert C, Vermat T, Alcouffe C, Bozec T, Sibrac D, Herbert JM, Ferrara P, Bono F, Ferran E (2014) Virtual screening on an α-helix to β-strand switchable region of the FGFR2 extracellular domain revealed positive and negative modulators. Proteins 82(11):2982–2997

99. Li Y, Liu ZH, Han L, Li J, Liu J, Zhao ZX, Wang RX (2014) Comparative assessment of scoring functions on an updated benchmark: 1. Compilation of the test set. J Chem Inf Model 54(6):1700–1716

100. Li Y, Han L, Liu Z, Wang R (2014) Comparative assessment of scoring functions on an updated benchmark: 2. Evaluation methods and general results. J Chem Inf Model 54(6): 1717–1736

101. Park H, Eom JW, Kim YH (2014) Consensus scoring approach to identify the inhibitors of AMP-activated protein kinase a2 with virtual screening. J Chem Inf Model 54:2139–2146

102. Houston DR, Walkinshaw MD (2013) Consensus docking: improving the reliability of docking in a virtual screening context. J Chem Inf Model 53(2):384–390

103. Scardino V, Bollini M, Cavasotto CN (2021) Combination of pose and rank consensus in docking-based virtual screening: the best of both worlds. RSC Adv 11(56): 35383–35391. https://doi.org/10.1039/d1ra05785e

104. Wang E, Sun H, Wang J, Wang Z, Liu H, Zhang JZH, Hou T (2019) End-point binding free energy calculation with MM/PBSA and MM/GBSA: strategies and applications in drug design. Chem Rev 119:9478–9508

105. Zhang X, Wong SE, Lightstone FC (2014) Toward fully automated high performance computing drug discovery: a massively parallel virtual screening pipeline for docking and molecular mechanics/generalized Born surface area rescoring to improve enrichment. J Chem Inf Model 54(1):324–337

106. Poli G, Granchi C, Rizzolio F, Tuccinardi T (2020) Application of MM-PBSA methods in virtual screening. Molecules 25(8):1971. https://doi.org/10.3390/molecules25081971

107. Yau MQ, Emtage AL, Loo JSE (2020) Benchmarking the performance of MM/PBSA in virtual screening enrichment using the GPCR-bench dataset. J Comput Aided Mol Des 34(11):1133–1145

108. Zhou Y, Lu X, Du C, Liu Y, Wang Y, Hong KH, Chen Y, Sun H (2021) Novel BuChE-IDO1 inhibitors from sertaconazole: virtual screening, chemical optimization and molecular modeling studies. Bioorg Med Chem Lett 34:127756. https://doi.org/10.1016/j.bmcl.2020.127756

109. Mittal L, Kumari A, Srivastava M, Singh M, Asthana S (2021) Identification of potential molecules against COVID-19 main protease through structure-guided virtual screening approach. J Biomol Struct Dyn 39(10): 3662–3680

110. Lee HS, Jo S, Lim HS, Im W (2012) Application of binding free energy calculations to prediction of binding modes and affinities of MDM2 and MDMX inhibitors. J Chem Inf Model 52(7):1821–1832

111. Park H, Jung HY, Mah S, Hong S (2018) Systematic computational design and identification of low Picomolar inhibitors of Aurora Kinase. J Chem Inf Model 58(3):700–709

112. Li Z, Li X, Huang YY, Wu Y, Liu R, Zhou L, Lin Y, Wu D, Zhang L, Liu H, Xu X, Yu K, Zhang Y, Cui J, Zhan CG, Wang X, Luo HB (2020) Identify potent SARS-CoV-2 main protease inhibitors via accelerated free energy perturbation-based virtual screening of existing drugs. Proc Natl Acad Sci U S A 117(44): 27381–27387

113. Leit S, Greenwood JR, Mondal S, Carriero S, Dahlgren M, Harriman GC, Kennedy-Smith

JJ, Kapeller R, Lawson JP, Romero DL, Toms AV, Shelley M, Wester RT, Westlin W, McElwee JJ, Miao W, Edmondson SD, Masse CE (2022) Potent and selective TYK2-JH1 inhibitors highly efficacious in rodent model of psoriasis. Bioorg Med Chem Lett 73:128891. https://doi.org/10.1016/j.bmcl.2022.128891

114. Deflorian F, Perez-Benito L, Lenselink EB, Congreve M, van Vlijmen HWT, Mason JS, Graaf C, Tresadern G (2020) Accurate prediction of GPCR ligand binding affinity with free energy perturbation. J Chem Inf Model 60(11):5563–5579

115. Cappel D, Hall ML, Lenselink EB, Beuming T, Qi J, Bradner J, Sherman W (2016) Relative binding free energy calculations applied to protein homology models. J Chem Inf Model 56(12):2388–2400

116. Ruddigkeit L, van Deursen R, Blum LC, Reymond JL (2012) Enumeration of 166 billion organic small molecules in the chemical universe database GDB-17. J Chem Inf Model 52(11):2864–2875

117. Grygorenko O (2021) Enamine LTD.: the science and business of organic chemistry and beyond. Eur J Org Chem 2021(47): 6474–6477. https://doi.org/10.1002/ejoc.202101210

118. Irwin JJ, Tang KG, Young J, Dandarchuluun C, Wong BR, Khurelbaatar M, Moroz YS, Mayfield J, Sayle RA (2020) ZINC20-a free ultralarge-scale chemical database for ligand discovery. J Chem Inf Model 60(12):6065–6073

119. Gadioli D, Vitali E, Ficarelli F, Latini C, Manelfi C, Talarico C, Silvano C, Cavazzoni C, Palermo G, Beccari AR (2021) EXSCALATE: An extreme-scale in-silico virtual screening platform to evaluate 1 trillion compounds in 60 h on 81 PFLOPS supercomputers. arXiv:2110.11644. https://doi.org/10.48550/arXiv.2110.11644

120. Ton AT, Gentile F, Hsing M, Ban F, Cherkasov A (2020) Rapid identification of potential inhibitors of SARS-CoV-2 main protease by Deep Docking of 1.3 billion compounds. Mol Inform 39(8):e2000028. https://doi.org/10.1002/minf.202000028

121. Gentile F, Fernandez M, Ban F, Ton AT, Mslati H, Perez CF, Leblanc E, Yaacoub JC, Gleave J, Stern A, Wong B, Jean F, Strynadka N, Cherkasov A (2021) Automated discovery of noncovalent inhibitors of SARS-CoV-2 main protease by consensus Deep Docking of 40 billion small molecules. Chem Sci 12(48):15960–15974

122. Gentile F, Yaacoub JC, Gleave J, Fernandez M, Ton AT, Ban F, Stern A, Cherkasov A (2022) Artificial intelligence-enabled virtual screening of ultra-large chemical libraries with deep docking. Nat Protoc 17(3):672–697

123. Muller C, Rabal O, Diaz Gonzalez C (2022) Artificial intelligence, machine learning, and deep learning in real-life drug design cases. Methods Mol Biol 2390:383–407. https://doi.org/10.1007/978-1-0716-1787-8_16

Chapter 13

High-Throughput Structure-Based Drug Design (HT-SBDD) Using Drug Docking, Fragment Molecular Orbital Calculations, and Molecular Dynamic Techniques

Reuben L. Martin, Alexander Heifetz, Mike J. Bodkin, and Andrea Townsend-Nicholson

Abstract

Structure-based drug design (SBDD) is rapidly evolving to be a fundamental tool for faster and more cost-effective methods of lead drug discovery. SBDD aims to offer a computational replacement to traditional high-throughput screening (HTS) methods of drug discovery. This "virtual screening" technique utilizes the structural data of a target protein in conjunction with large databases of potential drug candidates and then applies a range of different computational techniques to determine which potential candidates are likely to bind with high affinity and efficacy. It is proposed that high-throughput SBDD (HT-SBDD) will significantly enrich the success rate of HTS methods, which currently fluctuates around ~1%. In this chapter, we focus on the theory and utility of high-throughput drug docking, fragment molecular orbital calculations, and molecular dynamics techniques. We also offer a comparative review of the benefits and limitations of traditional methods against more recent SBDD advances. As HT-SBDD is computationally intensive, we will also cover the important role high-performance computing (HPC) clusters play in the future of computational drug discovery.

Key words Structure-based drug design, Drug development, Ligand docking, Fragment molecular orbitals, FMO, Molecular dynamics, High-performance computing, Virtual screening

1 Introduction

Structure-based drug design (SBDD) is a powerful approach used in the discovery and development of novel drug candidates for chosen protein targets [1]. In SBDD, an interaction between candidate and target is a measure of electrostatic, hydrophilic, and steric compatibility [2]. For a small molecule to bind to a target protein, it must be composed of moieties complementary to that of the target's binding region, such that the drug candidate sterically aligns with the binding residues of the protein [3]. Fortunately, this chemical relationship can be broken down into a set of physical

Alexander Heifetz (ed.), *High Performance Computing for Drug Discovery and Biomedicine*, Methods in Molecular Biology, vol. 2716, https://doi.org/10.1007/978-1-0716-3449-3_13,

equations, which can then be used by computational software to predict if, and how well, a drug candidate will bind to a target. This is the principle upon which SBDD is based [3].

SBDD involves the application of computational methods to understand the structure and function of target proteins and to design drugs that bind with high affinity and specificity to modulate the activity of these proteins [1]. There has been a rapid increase in the potential of SBDD in recent years, where large high-performance clusters (HPCs) have been utilized for high-throughput virtual screening of millions of potential candidates [4]. Several computational techniques have been developed to aid in SBDD, including drug docking, molecular dynamics (MD), and fragment molecular orbital (FMO) calculations (Fig. 1) [5, 6].

This chapter will cover an overview of the respective strengths and limitations of drug docking, MD, FMO, and the application of HPC for HT-SBDD. Docking describes computational software that attempts to predict the optimal binding between a receptor and a given ligand. The "Molecular Docking Problem" is defined as predicting the biologically correct bound association of a ligand to a protein, given only their respective atomic coordinates [7]. Fundamentally, no extra information is given. In practice, however, additional biologically relevant information is often incorporated into the software, such as known binding sites [7]. Drug docking utilizes the information of the atomic coordinates of the protein and the ligand with known electrostatic information about each component to calculate and predict the electrostatic interactions the drug candidate is likely to make with the target protein [7]. The output is a set of different potential poses the drug candidate may make with the target protein, which can provide information on the potential affinity and mode of binding of different drug candidates, along with providing insight into their potential efficacy. This can be used to virtually screen large libraries of small molecules to a target protein to identify potential drug candidates. However, it is important to note that the drug-protein interaction is a dynamic relationship, which is a significant limitation for most drug docking software [8].

MD simulations allow for the exploration of protein-ligand interactions in this dynamic environment. FMO then utilizes the application of quantum mechanical calculations that provide accurate energy estimates for protein-ligand interactions [5, 6]. For the application of HT-SBDD, these techniques require considerable computational power, which is why HPC is essential for the future development of SBDD (Fig. 2) [4].

A case study targeting the adenosine A_1 receptor, a member of the GPCR superfamily, will be used to demonstrate a real-world application of these techniques. We will use this exemple to show how the integration of these techniques is essential to minimize the innate uncertainty present with computational biological systems.

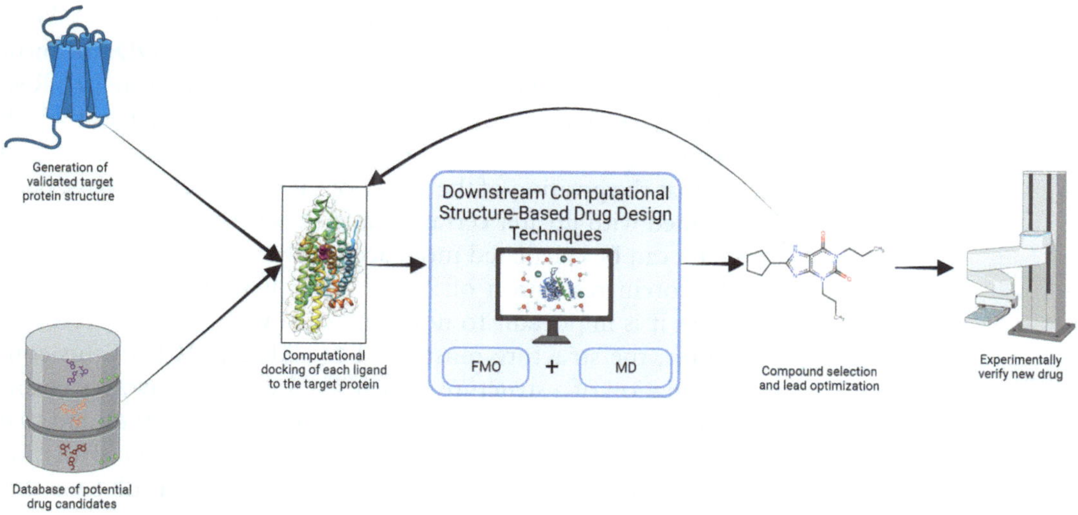

Fig. 1 Schematic representation of the general workflow of structure-based drug design, created with BioRender.com.

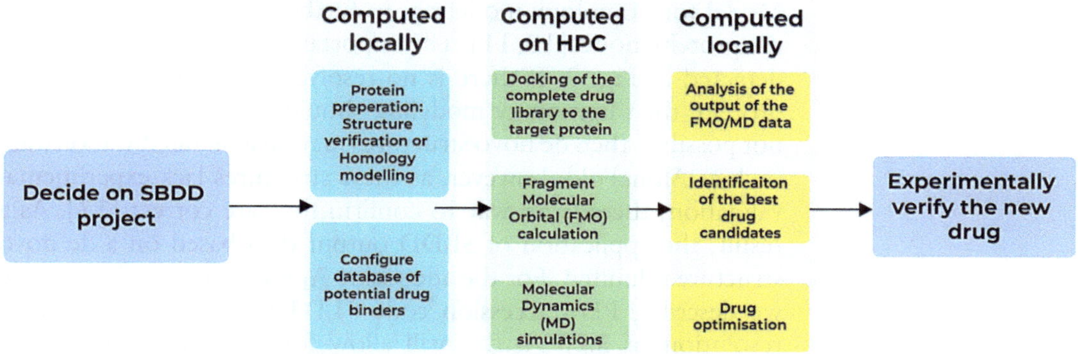

Fig. 2 The SBDD workflow overview, stressing the importance of HPC application

A novel protocol for SBDD will be explored, including the combination of drug docking, molecular dynamics (MD) simulations, and fragment molecular orbital (FMO) calculations. This worked example will provide an understanding of how these different techniques can be used for structure-based drug design (SBDD) and will describe how to reduce uncertainty in the target system.

2 Developing the Input Files

The first step for the computational study of a target protein is to source a structure file of the target protein. There are numerous means by which to accomplish this, though the most viable option is to identify a structure from the protein data bank (PDB). With any crystal structure that is intended to be used for SBDD, the most important aspect to check is the resolution of the structure [9]. Resolution of a crystal structure refers to the smallest distance at which the crystallographic data can be determined. A higher resolution means that the position of the atoms of the target protein have been determined with greater certainty, and thus protein-ligand interaction data can be calculated more accurately.

A protein resolution of 2Å or lower is preferred for SBDD; however, it is important to note that even with very well-resolved structures, the structure must still be validated [10]. In this context, validation of a structure to be used for SBDD involves ensuring that all amino acid residues of the protein are present and that there are no steric hindrances or inconsistencies with the structure. The wwPDB website (wwpdb.org) has many helpful tools for validation. Validation is essential for structures ranging from resolutions of 2–3.5 Å [10]. Lower resolution structures, i.e. those with resolution values greater than 3.5 Å, are generally considered to be of too low a resolution for immediate SBDD. In this case protein modeling should be considered instead, typically using Swiss-Model for homology modeling, to further increase the validity of the protein model [10, 11]. The importance of protein resolution is depicted in Fig. 3. If there is no resolved structure of the target protein, then homology modeling should be attempted. If this is not possible, then de novo structures may be generated via software such as AlphaFold; however, as these structures lack experimental validation, there is no way to confirm they are correct [12]. As a result, the application of SBDD output data based on a de novo structure is limited. For the adenosine A_1 receptor protein exemple we describe, PDB accession code 7LD3 has been used. With a resolution of 3.20 Å, this will allow for exploration of protein validation [13].

7LD3 is a structure of the adenosine A1 receptor (A_1R) in complex with a Positive Allosteric Modulator (PAM) and the native agonist adenosine [13]. The A_1R was isolated by removing the PAM and adenosine computationally, as the GPCR is the target protein. The missing residues of intracellular loop 3 were modeled using Modeller [14]. For SBDD it is important that the structure is then minimized to account for any conformational impact of PAM [15]. The validated protein (shown in Fig. 4) is now ready for SBDD.

Once a protein structure has been found and validated, drug candidates must then be identified. In the instance of high-throughput SBDD, a repository of candidate drug molecules must first be compiled. This repository can be populated by possible compounds listed in databases such as ZINC, ChEMBL, PubChem, DrugBank, and MolPort [5]. The next stage is to dock these drug candidates to the target protein.

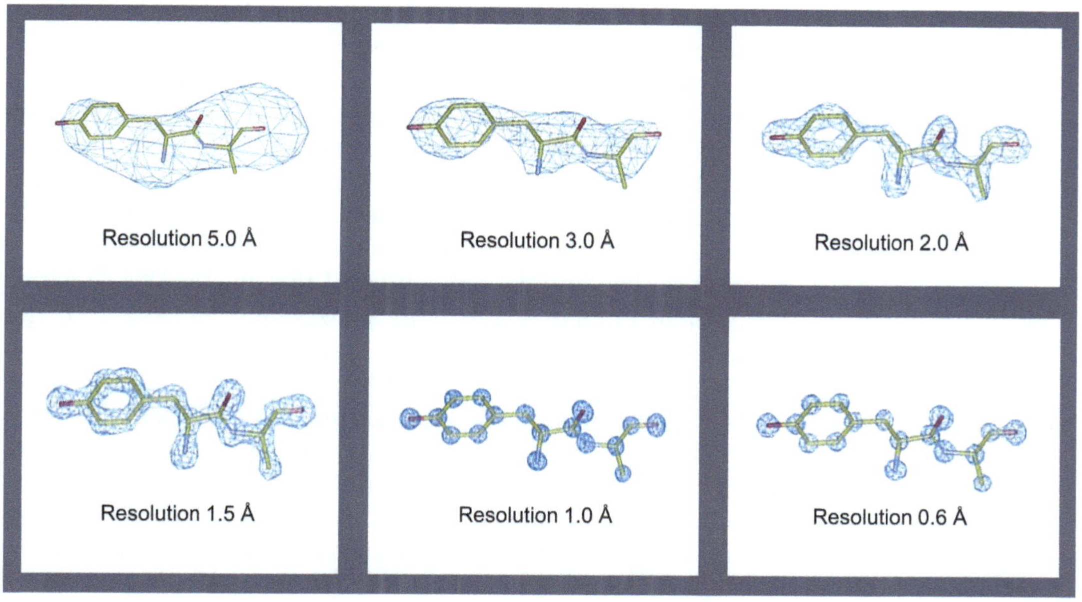

Fig. 3 Visualization of the impact resolution has on the certainty of the structure obtained from the experimental design. (Image adapted from Kuster et al. [9])

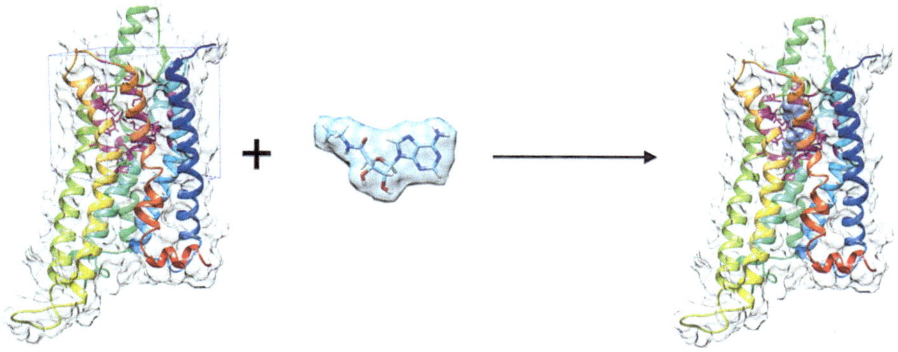

Fig. 4 Ligand docking using AutoDock Vina integrated with Chimera. The A_1 adenosine receptor is shown here as an example, with a drug candidate (NECA) shown for simplicity (light blue). The GPCR has regions that the potential drug candidate would not be able to physically access, such as the intermembrane regions, and the surfaces that are embedded in the membrane. The binding pocket is known and colored (magenta) for clarity. AutoDock Vina allows for the binding sight to be specifically chosen for the docking process, shown by the blue box on the left-hand image. This not only ensures the docking to be in the correct location of the protein but also reduces the computational load, as calculations of potential interactions for the non-selected regions of the proteins are no longer computed. (These images were made using Chimera [20])

3 Ligand Docking

Ligand docking is a computational technique that is used to predict the optimal binding pose of small molecules, such as drug candidates, to the target protein. This technique typically takes place in two steps: first the drug candidate is minimized, and conformational sampling is performed to generate a range of ligand conformers; these conformers are then docked to the most energetically favorable pose on the targets binding site [7]. The docking process involves calculating the interaction energies between the small molecule and the protein, typically using scoring functions that evaluate factors such as Van der Waals interactions, hydrogen bonding, and electrostatic interactions [16]. The docking software searches for the optimal orientation of the small molecule within the protein binding site that maximizes the interaction energy, aligning complementary groups to maximize attraction, and minimize repulsive forces [16]. Typically, the most energetically favorable complex is used for downstream analysis; however, a selection of top poses may be selected if there is significant variation, such as a complete inversion of the drug candidate.

The output obtained from docking a large compound library to a target protein is information on each complex with details of the binding mode and a measure of an estimated binding energy of each given drug; this is denoted as the scoring function [7, 8, 16] and can be used to identify potential binders from the library, as well as key interacting residues of the target protein. This data can either be used to optimize the structure of the drug candidate or as input for further downstream SBDD techniques. It is important to note that there are significant limitations in ligand docking as ligand-protein interactions are both dynamic and delicate [8]. Ligand docking assumes a relatively static interaction when performing docking; this is a limitation that can be overcome by the application of molecular dynamics [8]. Furthermore, ligand docking does not account for interactions that cannot be calculated via the Newtonian-based calculations that most drug docking software uses to determine interaction energies; this limitation can be overcome by applying the quantum-mechanical calculations used in Fragment Molecular Orbital (FMO) processing [17, 18].

There is a large range of docking software available, such as AutoDock-Vina, GOLD, MegaDock, and SwissDock [16]. Each software has its own benefits. For example, AutoDock-Vina is integrated into molecular editing software such as Chimera, which is particularly helpful when targeting membrane-bound proteins (such as GPCRs) as the binding pocket can be easily defined by the user; this prevents docking to incorrect regions of the protein (Fig. 4) [19]. AutoDock-Vina, much like MegaDock, is also open source, so it can be used as a command-line program without the need for a license. SwissDock also provides a helpful user experience as it utilizes a web-based interface, so no installation is necessary.

In summary, molecular docking is a powerful tool in SBDD that can provide insight into the binding mode and relative interaction of small molecules to target proteins. The technique involves generating multiple conformations of a small molecule and docking them onto the protein binding site. Although incredibly powerful, drug docking still has its limitations. Some of these limitations will be explored next.

4 Fragment Molecular Orbitals and FMO-HT

It is important to recognize that biological interactions at an atomic level are not solely governed by Newtonian physics and that quantum mechanics is responsible for a significant proportion of interaction energies [21]. Fragment molecular orbital (FMO) calculations apply quantum mechanical calculations to biologically relevant targets to determine the nature and relative strength of these non-intuitive interactions. We will briefly introduce the theory behind FMO and describe the limitations it overcomes, including some surprising recent advances that have led to the development of high-throughput FMO (FMO-HT).

During the Fragment Molecular Orbital (FMO) process, a large system is fragmented into smaller systems, typically with the drug candidate as one fragment and individual residues of the target protein as other fragments (Fig. 5) [22]. This fragmentation is the only way that allows quantum mechanical calculations to be performed on a system that is large enough to be biologically interesting, as the method is based on the theory that the interactions between different fragments of a large molecule can be approximated as the sum of the interactions between the individual fragments. FMO calculates interaction energies that can correlate with binding affinities, a current limitation for computational drug design [23]. FMO may, therefore, negate the necessity for high-throughput experimental affinity assays, which require significantly more time, funding, and human input.

Several software packages are available for FMO calculations, including GAMESS and Q-Chem [17, 24]. Selection of the appropriate software depends on the specific research question and the properties of the target protein and ligand, though modern FMO protocols may combine multiple packages to ensure wide sampling of the available data. FMO has traditionally been arduous, time consuming, and incredibly computationally expensive [25]. Recent advancements, driven by researchers at Evotec, have led to the development of an automated pipeline for the automation of FMO, which has recently been optimized to enable high-throughput FMO (FMO-HT). A novel FMO acceleration framework has recently been developed that functions by offloading four-index two-electron repulsion integrals to graphical processing units (GPUs) using OpenMP [26]. This evolution is especially exciting as it has been shown to present linear scalability of up to 4608 NVIDIA V100 GPUs. This level of scalability shows

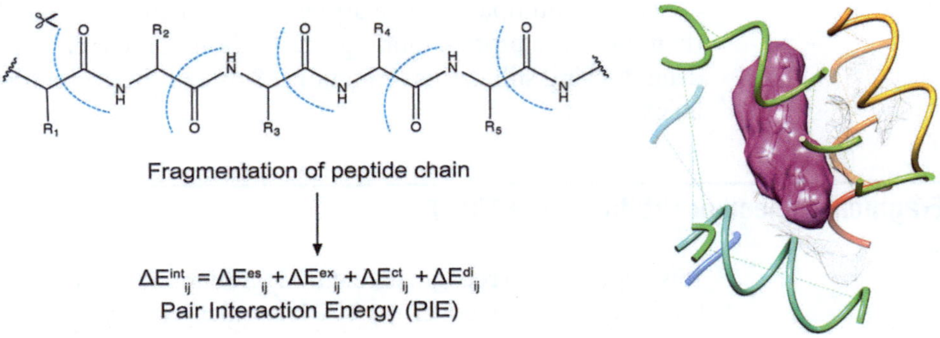

Fig. 5 FMO Fragmentation. Shown in schematic representation (left) is a figure from Heifetz A. [18]. An example of the FMO input file is shown for the adenosine A_1 receptor: drug target complex (right). In the case of drug-ligand complexes, the ligand will consist of one fragment and all protein amino acids that are within a pre-set distance, typically 4.5 Å, as the other fragments

impressive promise for the integration of GPU acceleration with Evotec's FMO-HT, a technique previously reserved for the fine-tuning of drug discovery, and to allow the commercial application of FMO-HT for drug design. The potential bottlenecking effect FMO-HT/GPU integration may have on identifying drug candidates will be discussed following the description of a key method to overcome limitations in the dynamic interactions of protein-ligand interactions.

5 Molecular Dynamics

Arguably the most computationally intensive technique covered in this chapter is all-atom molecular dynamics (MD). MD simulations employ a set of Newtonian-based step-by-step calculations to represent the movement of proteins and ligands at an atomic scale [27]. MD simulations can be thought of as akin to stop motion, where each frame is an expression of mathematically derived interactions between all atoms of any given system.

In its simplified form, molecular dynamics is a result of applying Newton's second law of motion, $F = ma$, to biologically relevant systems. In essence, the force on each atom in a system is calculated on a static system, every atom of the system then moves as a result of its calculated acceleration by a given timestep (typically 2 femtoseconds), and the force on each atom is recalculated on the new static positions of each atom in the system (Fig. 6). By repeating this process millions of times, a trajectory of how a system will interact dynamically can be generated. These simulations can provide detailed information on the stability of protein-ligand complexes and on the chemical nature of their dynamic interaction.

Forcefields are used to define the types of interactions atoms will make, where each substituent of a system to be simulated are

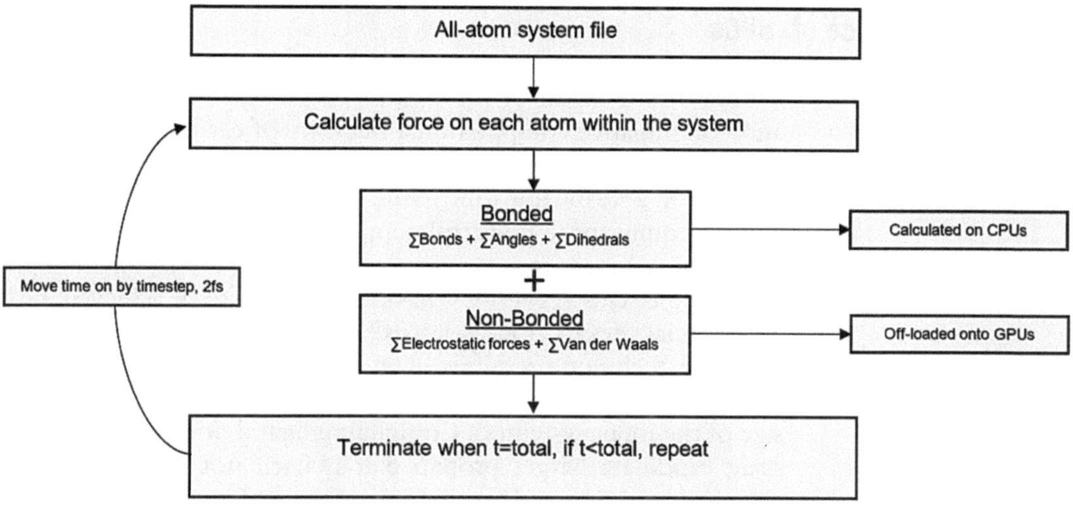

Fig. 6 Schematic representation of an MD simulation using GROMACS. This flow chart describes the process of an all-atom MD simulation, where these fundamental processes are followed during energy minimization, equilibration, and during the production run. A timestep of 2fs is applied to ensure sufficient interaction resolution [28]. When available, non-bonded interactions are offloaded to GPUs to increase the simulation speed. (Adapted from Figs. 2 and 3 in [29])

parametrized so that the MD software engine knows how each atom is likely to interact with other atoms [30]. Common substituents are well-parameterized, such as proteins, water, typical ions, and the components of membrane bilayers [30]. Small molecule drugs, however, must be parameterized on a case-by-case basis [31]. This is an essential step in MD simulations, as the MD software will only be able to process information that is given; if the input file is a set of atoms that it does not recognize or know how to interpret, then the simulation will fail [32]. Furthermore, since MD simulations initially apply a random velocity to each atom in the system, each simulation will be distinct from each other. This means that one simulation must be repeated in a set of ensembles to ensure full coverage of the conformational space, further increasing the computational load required to run MD simulations. Due to the necessity for individual ligand parametrization and for repeated simulations, MD simulations are traditionally used for fine-tuning the understanding of ligand-binding modalities and for the final stages of SBDD and/or lead compound optimization [33].

To summarize, MD is a powerful tool used in SBDD and can provide important insight into the dynamic interaction of biological systems at an atomic level. The technique involves the calculation of the variation of force on each atom of a given system over a given time, calculated in a repetitive process to generate a trajectory of the simulation. This trajectory is then used for analysis that can highlight key residues within the target protein and can identify poses that were potentially sterically inaccessible prior to MD simulations.

6 The Importance of HPCs

An important caveat of these SBDD techniques is the combined and accumulative computational intensity of each step when considered at a commercial level [34]. To run SBDD at a high throughput level in a reasonable time frame (hours or days), each technique would require more powerful computers than personal local configurations. Advances in the development and accessibility of high-performance clusters (HPCs) have allowed this to be achieved, be it commercial sites or cloud services [34–36].

Each technique requires many calculations to be performed, with the computational load varying depending on the number and size of the input system(s). Completing ligand docking for a single drug candidate/target protein pair is itself not computationally intensive, and these calculations are often performed locally. However, performing ligand docking for a larger drug candidate library of 100,000+ potential candidates is very intensive and requires significant computational acceleration. High-performance computers enable calculations like these to be completed in a reasonable timeframe [34]. The output data of all docked structures can then be used as input for other downstream techniques, such as MD and FMO-HT.

FMO is innately computationally intensive, much like MD simulations, both of which are similarly performed on HPC facilities. FMO calculations are computationally intensive due to the large number of electronic interactions that must be evaluated between the individual fragments. HPC enables the efficient use of parallel computing to accelerate FMO calculations and reduce the computational time required, which could be further accelerated by up to 52× by offloading subsets of calculations to GPUs, showing the aforementioned linear acceleration for up to 4608 V100 Nvidia GPUs [26, 35]. Molecular dynamics simulations require the integration of the equations of motion for each atom or molecule of a system, which becomes computationally demanding as the size of the system and the length of the simulation increase. HPC allows for the parallelization of MD simulations across multiple processors or nodes, reducing the time required for calculations. This can be further accelerated by offloading non-bonded calculations onto GPUs, which has been shown to reduce simulation time to a third of the non-accelerated time (Fig. 6) [29, 37].

In summary, HPC are critical for ligand docking, molecular dynamics simulations, and fragment molecular orbital calculations in SBDD. These methods are computationally intensive, and HPC enables the use of sophisticated algorithms to generate and evaluate large numbers of conformations and interactions in a reasonable amount of time. HPCs also allows for the parallelization of calculations across multiple processors or nodes, reducing the computational time required for these methods. Since the evolution and continued growth of power of HPCs, novel SBDD protocols have been developed, which hold promise for the high-throughput commercial application of SBDD. We will finalize this chapter by comparatively exploring traditional drug design techniques with a novel SBDD protocol.

7 The Integration of SBDD Techniques to Develop an Automated Pipeline

Before the evolution of powerful computational resources, the timeline for identifying and developing new drug candidates had been dependent on running high throughput screening (HTS) of several hundred thousand to a few million random small molecule drugs, with the hopes of identifying well-binding candidates experimentally [38]. Current rates of HTS can test up to 100,000 drug candidates per day, but are incredibly expensive to invest in as they require a factory of robots, detectors, and personell to manage and run the facility [39]. Furthermore, this technique leads to excessive redundancy in wasted materials as there is typically only a ~1% hit rate [40]. We propose the application of a novel high-throughput SBDD (HT-SBDD) protocol that negates the requirement for experimental HTS until a significantly reduced pool of drug candidates has been selected to result in an enriched hit rate and negate the need for largy HTS facilities (Fig. 7).

In this novel HT-SBDD, initial ligand docking is applied to a given drug candidate library. The docked structures are then be piped into FMO-HT, which will provide the total interaction energies of each ligand-target protein complex. The best binders will then be identified by ranking the total interaction energies of the complexes in the output of the FMO-HT. From this ranking, the top binders will be selected, such as the top 5–10%, dependent on how broad the binding energies are. These complexes can then be used for further analysis to map the binding pocket of the target protein. For this, molecular dynamics will be applied to allow the drug candidates to rotate and dynamically move to a more energetically favorable pose. Analysis of the trajectory data of the different MD simulations will yield an accurate map of the binding pocket of the target protein, which will be useful for downstream drug optimization. FMO analysis of the final frame of each of the MD simulations could also be applied to identify any changes in the optimal binding candidates for the target protein, as this will account for dynamic interactions that are missed in docking and the initial FMO-HT processing. The top 5–10% of this subset of drug candidates would then have to have the binding affinities experimentally validated. Further drug optimization may also be performed if a novel, highly specified and efficacious drug is needed

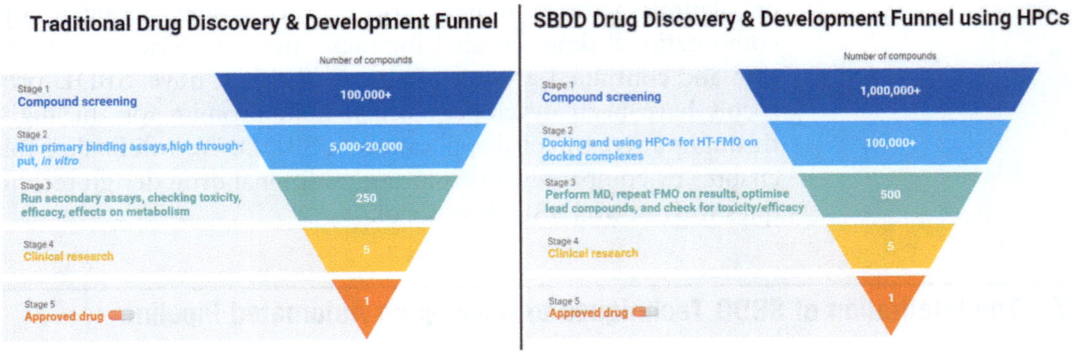

Fig. 7 Comparative representation of the different drug discovery pathways, showing the steps of traditional (left) and the novel SBDD (right) drug discovery funnels. Computational methods of SBDD allow for significantly more drug compounds to be tested in the same time frame as traditional methods. Not only will this increase the chance of identifying the optimal drug candidate, but it will also reduce the risk of competitor compounds being identified in the future. Furthermore, the SBDD processes can be highly and efficiently automated, further reducing time and money wasted in pre-clinical stages of drug development. Adapted from 'Drug Discovery and evelopment Funnel', by BioRender.com (2020). Retrieved from https:/app.biorender.com/biorender-templates

(Fig. 1). This is often a necessity if the target protein is of a family of other very similar proteins [41], as off target interactions may result in undesired toxicological side effects of the drug [42].

In conclusion, the access to computational power is ever increasing and drug/protein databases are ever expanding. Although SBDD is currently in the relatively early stages of development, it is likely that heavily automated HT-SBDD techniques will be paramount in targeting many of the as-yet-undrugged proteins for which high resolutions structures are currently available. However, although the use of multiple computational techniques may reduce this uncertainty in the identification of candidate drugs, the results must always be validated experimentally. Traditional experimental validation methods must, therefore, still be used in conjunction with the novel SBDD protocols to confirm any SBDD finding. The use of HPC and HT-SBDD, however, dramatically reduces the amount of experimental work needed and affords the promise of significantly increasing the current success rate of experimental HTS methods.

References

1. Batool M, Ahmad B, Choi S (2019) A structure-based drug discovery paradigm. Int J Mol Sci 20(11). https://doi.org/10.3390/ijms20112783
2. Ferreira de Freitas R, Schapira M (2017) A systematic analysis of atomic protein-ligand interactions in the PDB. Medchemcomm 8(10):1970–1981. https://doi.org/10.1039/c7md00381a
3. Costanzo LD, Ghosh S, Zardecki C, Burley SK (2016) Using the tools and resources of the RCSB protein data bank. Curr Protoc Bioinformatics 55(1):1.9.1–1.9.35. https://doi.org/10.1002/cpbi.13

4. Liu T, Lu D, Zhang H, Zheng M, Yang H, Xu Y, Luo C, Zhu W, Yu K, Jiang H (2016) Applying high-performance computing in drug discovery and molecular simulation. Natl Sci Rev 3(1):49–63. https://doi.org/10.1093/nsr/nww003
5. Yu W, MacKerell AD Jr (2017) Computer-aided drug design methods. Methods Mol Biol 1520:85–106. https://doi.org/10.1007/978-1-4939-6634-9_5
6. Heifetz A, Aldeghi M, Chudyk EI, Fedorov DG, Bodkin MJ, Biggin PC (2016) Using the fragment molecular orbital method to investigate agonist-orexin-2 receptor interactions. Biochem Soc Trans 44(2):574–581. https://doi.org/10.1042/bst20150250
7. Halperin I, Ma B, Wolfson H, Nussinov R (2002) Principles of docking: an overview of search algorithms and a guide to scoring functions. Proteins Struct Funct Genet 47(4): 409–443. https://doi.org/10.1002/prot.10115
8. Gioia D, Bertazzo M, Recanatini M, Masetti M, Cavalli A (2017) Dynamic docking: a paradigm shift in computational drug discovery. Molecules 22(11). https://doi.org/10.3390/molecules22112029
9. Kuster DJ, Liu C, Fang Z, Ponder JW, Marshall GR (2015) High-resolution crystal structures of protein helices reconciled with three-centered hydrogen bonds and multipole electrostatics. PLoS One 10(4):e0123146. https://doi.org/10.1371/journal.pone.0123146
10. Lee S, Seok C, Park H (2023) Benchmarking applicability of medium-resolution cryo-EM protein structures for structure-based drug design. J Comput Chem 44(14):1360–1368. https://doi.org/10.1002/jcc.27091
11. Waterhouse A, Bertoni M, Bienert S, Studer G, Tauriello G, Gumienny R, Heer FT, de Beer TAP, Rempfer C, Bordoli L, Lepore R, Schwede T (2018) SWISS-MODEL: homology modelling of protein structures and complexes. Nucleic Acids Res 46(W1):W296–W303. https://doi.org/10.1093/nar/gky427
12. Borkakoti N, Thornton JM (2023) AlphaFold2 protein structure prediction: implications for drug discovery. Curr Opin Struct Biol 78:102526. https://doi.org/10.1016/j.sbi.2022.102526
13. Draper-Joyce CJ, Bhola R, Wang J, Bhattarai A, Nguyen ATN, Cowie-Kent I, O'Sullivan K, Chia LY, Venugopal H, Valant C, Thal DM, Wootten D, Panel N, Carlsson J, Christie MJ, White PJ, Scammells P, May LT, Sexton PM, Danev R, Miao Y, Glukhova A, Imlach WL, Christopoulos A (2021) Positive allosteric mechanisms of adenosine A1 receptor-mediated analgesia. Nature 597(7877): 571–576. https://doi.org/10.1038/s41586-021-03897-2
14. Fiser A, Sali A (2003) Modeller: generation and refinement of homology-based protein structure models. Methods Enzymol 374: 461–491. https://doi.org/10.1016/s0076-6879(03)74020-8
15. Jabeen A, Mohamedali A, Ranganathan S (2019) Protocol for protein structure modelling. In: Ranganathan S, Gribskov M, Nakai K, Schönbach C (eds) Encyclopedia of bioinformatics and computational biology. Academic, Oxford, pp 252–272. https://doi.org/10.1016/B978-0-12-809633-8.20477-9
16. Kishor D, Deweshri N, Vijayshri R, Ruchi S, Ujwala M (2022) Molecular docking: metamorphosis in drug discovery. In: Erman Salih I (ed) Molecular docking. IntechOpen, Rijeka, p Ch. 3. https://doi.org/10.5772/intechopen.105972
17. Fedorov DG (2017) The fragment molecular orbital method: theoretical development, implementation in GAMESS, and applications. WIREs Comput Mol Sci 7(6):e1322. https://doi.org/10.1002/wcms.1322
18. Heifetz A, Chudyk EI, Gleave L, Aldeghi M, Cherezov V, Fedorov DG, Biggin PC, Bodkin MJ (2016) The fragment molecular orbital method reveals new insight into the chemical nature of GPCR–ligand interactions. J Chem Inf Model 56(1):159–172. https://doi.org/10.1021/acs.jcim.5b00644
19. Eberhardt J, Santos-Martins D, Tillack AF, Forli S (2021) AutoDock Vina 1.2.0: new docking methods, expanded force field, and python bindings. J Chem Inf Model 61(8): 3891–3898. https://doi.org/10.1021/acs.jcim.1c00203
20. Pettersen EF, Goddard TD, Huang CC, Couch GS, Greenblatt DM, Meng EC, Ferrin TE (2004) UCSF Chimera – a visualization system for exploratory research and analysis. J Comput Chem 25(13):1605–1612. https://doi.org/10.1002/jcc.20084
21. Arodola OA, Soliman ME (2017) Quantum mechanics implementation in drug-design workflows: does it really help? Drug Des Devel Ther 11:2551–2564. https://doi.org/10.2147/dddt.S126344
22. Fedorov DG, Nagata T, Kitaura K (2012) Exploring chemistry with the fragment molecular orbital method. Phys Chem Chem Phys 14(21):7562–7577. https://doi.org/10.1039/C2CP23784A

23. Handa C, Yamazaki Y, Yonekubo S, Furuya N, Momose T, Ozawa T, Furuishi T, Fukuzawa K, Yonemochi E (2022) Evaluating the correlation of binding affinities between isothermal titration calorimetry and fragment molecular orbital method of estrogen receptor beta with diarylpropionitrile (DPN) or DPN derivatives. J Steroid Biochem Mol Biol 222:106152. https://doi.org/10.1016/j.jsbmb.2022.106152
24. Kong J, White C, Krylov A, Sherrill C, Adamson R, Furlani T, Lee M, Lee A, Gwaltney S, Adams T, Ochsenfeld C, Gilbert A, Kedziora G, Rassolov V, Maurice D, Nair N, Shao Y, Besley N, Maslen P, Pople J (2000) Q-Chem 2.0: a high-performance Ab initio electronic structure program package. J Comput Chem 21:1532–1548. https://doi.org/10.1002/1096-987X(200012)21:163.0.CO;2-W
25. Wannipurage D, Deb I, Abeysinghe E, Pamidighantam S, Marru S, Pierce M, Frank AT (2022) Experiences with managing data parallel computational workflows for high-throughput fragment molecular orbital (FMO) Calculations. arXiv preprint arXiv:220112237
26. Pham BQ, Alkan M, Gordon MS (2023) Porting fragmentation methods to graphical processing units using an OpenMP application programming interface: offloading the Fock build for low angular momentum functions. J Chem Theory Comput 19(8):2213–2221. https://doi.org/10.1021/acs.jctc.2c01137
27. Lindahl E, Hess B, Van Der Spoel D (2001) GROMACS 3.0: a package for molecular simulation and trajectory analysis. Mol Model Ann 7(8):306–317
28. Huang J, Rauscher S, Nawrocki G, Ran T, Feig M, de Groot BL, Grubmüller H, MacKerell AD (2017) CHARMM36m: an improved force field for folded and intrinsically disordered proteins. Nat Methods 14(1):71–73. https://doi.org/10.1038/nmeth.4067
29. Schmidt C, Robinson CV (2014) Dynamic protein ligand interactions – insights from MS. FEBS J 281(8):1950–1964. https://doi.org/10.1111/febs.12707
30. Huang J, MacKerell AD Jr (2013) CHARMM36 all-atom additive protein force field: validation based on comparison to NMR data. J Comput Chem 34(25):2135–2145. https://doi.org/10.1002/jcc.23354
31. Lin FY, MacKerell AD Jr (2019) Force fields for small molecules. Methods Mol Biol 2022:21–54. https://doi.org/10.1007/978-1-4939-9608-7_2
32. Bauer P, Hess B, Lindahl E (2022) GROMACS 2022.1, Manual. Zenodo. https://doi.org/10.5281/zenodo.6451567
33. De Vivo M, Masetti M, Bottegoni G, Cavalli A (2016) Role of molecular dynamics and related methods in drug discovery. J Med Chem 59(9):4035–4061. https://doi.org/10.1021/acs.jmedchem.5b01684
34. Ocaña K, Benza S, De Oliveira D, Dias J, Mattoso M (2014) Exploring large scale receptor-ligand pairs in molecular docking workflows in HPC clouds. In: 2014 IEEE international parallel & distributed processing symposium workshops. IEEE, pp 536–545
35. Pham BQ, Gordon MS (2019) Hybrid distributed/shared memory model for the RI-MP2 method in the fragment molecular orbital framework. J Chem Theory Comput 15(10):5252–5258. https://doi.org/10.1021/acs.jctc.9b00409
36. Kutzner C, Páll S, Fechner M, Esztermann A, de Groot BL, Grubmüller H (2019) More bang for your buck: improved use of GPU nodes for GROMACS 2018. J Comput Chem 40(27):2418–2431
37. Kohnke B, Kutzner C, Grubmüller H (2020) A GPU-accelerated fast multipole method for GROMACS: performance and accuracy. J Chem Theory Comput 16(11):6938–6949. https://doi.org/10.1021/acs.jctc.0c00744
38. Attene-Ramos MS, Austin CP, Xia M (2014) High throughput screening. In: Wexler P (ed) Encyclopedia of toxicology, 3rd edn. Academic, Oxford, pp 916–917. https://doi.org/10.1016/B978-0-12-386454-3.00209-8
39. Szymański P, Markowicz M, Mikiciuk-Olasik E (2012) Adaptation of high-throughput screening in drug discovery-toxicological screening tests. Int J Mol Sci 13(1):427–452. https://doi.org/10.3390/ijms13010427
40. Fox S, Farr-Jones S, Sopchak L, Boggs A, Nicely HW, Khoury R, Biros M (2006) High-throughput screening: update on practices and success. J Biomol Screen 11(7):864–869. https://doi.org/10.1177/1087057106292473
41. Wang J, Miao Y (2019) Mechanistic insights into specific G protein interactions with adenosine receptors. J Phys Chem B 123(30):6462–6473. https://doi.org/10.1021/acs.jpcb.9b04867
42. Rudmann DG (2013) On-target and off-target-based toxicologic effects. Toxicol Pathol 41(2):310–314. https://doi.org/10.1177/0192623312464311

Chapter 14

HPC Framework for Performing in Silico Trials Using a 3D Virtual Human Cardiac Population as Means to Assess Drug-Induced Arrhythmic Risk

Jazmin Aguado-Sierra, Renee Brigham, Apollo K. Baron, Paula Dominguez Gomez, Guillaume Houzeaux, Jose M. Guerra, Francesc Carreras, David Filgueiras-Rama, Mariano Vazquez, Paul A. Iaizzo, Tinen L. Iles, and Constantine Butakoff

Abstract

Following the 3 R's principles of animal research—replacement, reduction, and refinement—a high-performance computational framework was produced to generate a platform to perform human cardiac in-silico clinical trials as means to assess the pro-arrhythmic risk after the administrations of one or combination of two potentially arrhythmic drugs. The drugs assessed in this study were hydroxychloroquine and azithromycin. The framework employs electrophysiology simulations on high-resolution three-dimensional, biventricular human heart anatomies including phenotypic variabilities, so as to determine if differential QT-prolongation responds to drugs as observed clinically. These simulations also reproduce sex-specific ionic channel characteristics. The derived changes in the pseudo-electrocardiograms, calcium concentrations, as well as activation patterns within 3D geometries were evaluated for signs of induced arrhythmia. The virtual subjects could be evaluated at two different cycle lengths: at a normal heart rate and at a heart rate associated with stress as means to analyze the proarrhythmic risks after the administrations of hydroxychloroquine and azithromycin. Additionally, a series of experiments performed on reanimated swine hearts utilizing Visible Heart® methodologies in a four-chamber working heart model were performed to verify the arrhythmic behaviors observed in the in silico trials.

The obtained results indicated similar pro-arrhythmic risk assessments within the virtual population as compared to published clinical trials (21% clinical risk vs 21.8% in silico trial risk). Evidence of transmurally heterogeneous action potential prolongations after providing a large dose of hydroxychloroquine was found as the observed mechanisms for elicited arrhythmias, both in the in vitro and the in silico models. The proposed workflow for in silico clinical drug cardiotoxicity trials allows for reproducing the complex behavior of cardiac electrophysiology in a varied population, in a matter of a few days as compared to the months or years it requires for most in vivo human clinical trials. Importantly, our results provided evidence of the common phenotype variants that produce distinct drug-induced arrhythmogenic outcomes.

Key words Drug-induced arrhythmia, Computational electrophysiology, Cardiac population, Cardiac safety

Alexander Heifetz (ed.), *High Performance Computing for Drug Discovery and Biomedicine*, Methods in Molecular Biology, vol. 2716, https://doi.org/10.1007/978-1-0716-3449-3_14,

1 Introduction

The fast and safe development of pharmacologic therapies is key for providing patients with better treatments at hopefully lower costs. The same applies for repurposed drugs or off-labeled uses, particularly when one or more potentially pro-arrhythmic drugs might be administered simultaneously. These therapeutic issues were more frequent during the COVID-19 pandemic, with the combined uses of potentially pro-arrhythmic drugs employed as urgent attempts to repurpose existing treatments against the disease progression [1]. For example, this was the case of the combined uses of hydroxychloroquine (HCQ) and azithromycin (AZM) [2, 3]. In this context and other clinical situations requiring high-throughput testing, the creation of novel methodologies capable of providing urgent information concerning the cardiotoxic risks of using potentially pro-arrhythmic drugs will provide both scientists and physicians with a powerful tool to support clinical decision-making. Importantly, there remains today no highly effective clinical predictors, other than assessments of QT-interval prolongation and less commonly employed J-T peak intervals, that can provide critical information regarding the potential risks for certain patents with normal QTc intervals to develop QT-prolongation. The function of various cardiac ion channels can be significantly modified by environmental conditions (i.e. hormones, electrolyte concentrations or pH), which can in turn have substantial effects on their overall electrical profiles. Males and females present different risks for drug-induced arrhythmias and QT-interval prolongations due to sex-specific hormones [4, 5]. In the case of COVID-19, hypokalemia was identified as a prevalent condition in many patients, which may in turn increase further the risk of QT prolongation [6]. Furthermore, the combined administrations of several drugs further increase the complexities of understanding the associated clinical implications, especially given the limited information relative to potential drug interactions at the start of the pandemic.

Computational methods have become important tools to study drug-induced arrhythmias [7–11]. However, most of the reported current studies have focused on single-cell population analyses [12, 13]. Only a few full 3D biventricular anatomies have been described in the literature [14–16]. After the start of the COVID-19 pandemic, a variety of single cell, tissue, and biventricular modeling methodologies were employed to assess the potential pro-arrhythmic risk after the administrations of HCQ and AZM [1, 16–19]. In one study, Okada et al. (2021) [17] employed a computational biventricular model to assess the therapeutic concentrations of HCQ, chloroquine (CQ), and AZM that could produce Torsades des Pointes (TdP). Yet importantly, studies on a normal human population described using a highly detailed, sex-specific human ventricular anatomies is yet to be reported. Further, the uses of full biventricular anatomies are crucial to assess drug effects on the human heart and understand the mechanisms of the arrhythmic risks beyond a single-cell behavior. Furthermore, the uses of the full heart anatomies allow one to include other risk factors like associated myocardial infarction, ischemia, or tissue heterogeneity.

In this work, we employed a novel, computational, high-performance computing framework with the following characteristics:

- A methodology to define a "normal" virtual population of 3D hearts for in silico cardiotoxicity trials using diverse ion channel phenotypes and including sex-specific ion channel properties in order to reproduce gender-specific risks to drug QTc prolongation [11, 20–22].
- The use of published IC_{50} and drug effective plasma concentrations from a variety of available sources at the time of the COVID-19 pandemic employed following published data [10], while assuming the combination of the two drugs to have additive effects [9].
- Develop a reliable methodology to identify the risks of cardiotoxicity in silico using virtual stress tests and classify the responses within the virtual population to the administered drugs. This arrhythmic risk evaluation is based on clinical case reports of exercise-induced arrhythmias related to the administration of antimalarial drugs [23].

This computational framework can be used to quantify the pro-arrhythmic risks produced by the administrations of one or more potentially arrhythmic drugs. The virtual human population was used to reproduce the arrhythmic risks observed clinically after the administrations of HCQ and AZM. Published effective plasma concentrations and IC_{50} values of HCQ and AZM employed as means to administer a variety of doses to the human virtual population provided meaningful information regarding clinically observed pro-arrhythmic risks. Finally, we employed reanimated swine hearts to observe acute arrhythmic features induced by the administrations of high doses of drugs and compare the observed arrhythmic mechanisms to the ones observed within our virtual population.

We consider here that our present work represents a critical leap from existing experimental single cardiomyocyte data to full 3D cardiac physiologic responses that reproduce the behaviors of humans, which opens the possibilities of full organ-scale, translational, in silico human clinical trials to assess drug-induced arrhythmic risks.

2 Materials and Methods

The detailed biventricular anatomy of a female (24yo, BMI 31.2) ex-vivo, perfusion-fixed human heart, was segmented from high-resolution magnetic resonance imaging (MRI) data. This heart was obtained from an organ donor whose heart was deemed not viable for transplant, via the procurement organization LifeSource (Minneapolis, MN, USA): all such donors had consented to donating their organs for transplant prior to death. This specimen remains a part of the Visible Heart® Laboratories' Human Heart Library at the University of Minnesota. The use of these heart specimens for research was appropriately consented and signed by the donor families and witnessed by LifeSource, and all research protocols for use were reviewed by the University of Minnesota's Institutional Review Board (IRB) and LifeSource's Research Committee. All data were deindentified from their source and are medically exempt from IRB. The heart obtained was cannulated and perfusion-fixed under a pressure of 40–50 mmHg, with 10% phosphate buffered formalin to preserve it in an end-diastolic state. High-resolution images of this specimen were acquired via a 3T Siemens scanner: with 0.44 times 0.44 mm in-plane resolution and slice thickness of 1–1.7 mm, which provides detailed endocardial trabeculae and false tendons with $\approx 1\ \text{mm}^2$ cross-sectional, which can be observed in Fig. 1.

A volumetric finite element mesh (of 65.5 million tetrahedral elements) was created [24], with a regular element side length of 328 microns using ANSA (BETA CAE Systems USA Inc.), with a rule-based fiber model [25] and transmural cell heterogeneity (endocardial, midmyocardial, and epicardial). The finite element model for electrophysiology was solved using Alya, a high-performance computing software developed at the Barcelona Supercomputing Center, employing the monodomain approximation to the anisotropic electrical propagation [26, 27]. This software was designed from scratch to run efficiently in high-performance computers with a tested scalability of up to 100,000 cores [28–32]. Briefly, the equations for electrical propagation and the system to be solved are as follows:

$$\begin{cases} \nabla \tilde{n} G \nabla \phi = S_{\mathrm{v}} \left(C_{\mathrm{m}} \dfrac{\partial \phi}{\partial t} + I_{\mathrm{ion}}(\phi, p) \right) \\ \dfrac{\partial p}{\partial t} = f(p, \phi, t) \\ \Gamma : (n \tilde{n} D) = 0 \end{cases} \tag{1}$$

where C_{m} refers to the membrane capacitance, ϕ is the transmembrane potential, G is the conductivity tensor which defines the anisotropic conductivity in the cardiac muscle; S_{v} is the surface to volume ratio of the cardiomyocytes, and I_{ion} is the solution of the ODEs that define the cardiomyocyte ion channel model based on the one published by O'Hara-Rudy [33]; f is this system of ODEs,

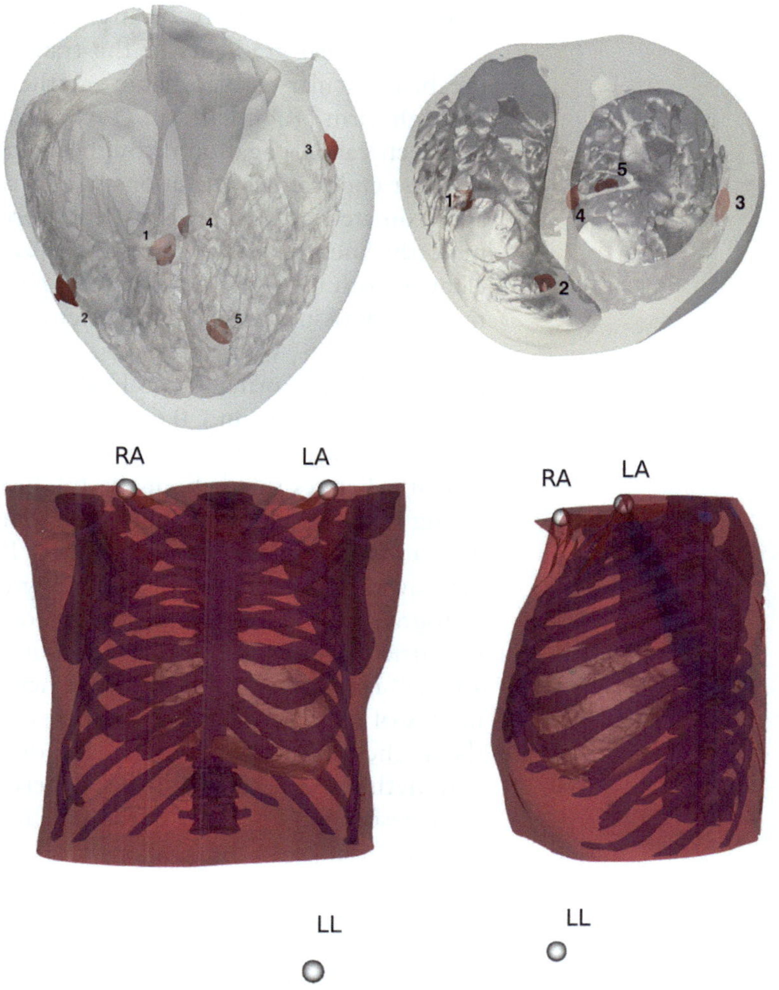

Fig. 1 First row shows the high-resolution human biventricular heart anatomy employed. The initial stimuli locations are shown in red. The activation magnitude was −80 mV applied for 5 ms, reproducing a spherical stimulus of 0.2 cm radius. Below, the torso is employed as reference to spatially locate the pseudo-ECG leads. The heart is shown registered within the torso

which is dependent on the transmembrane potential ϕ and a set of parameters p. Zero flux is defined at the boundary Γ of the domain. A Yanenko operator splitting technique was applied to discretize the system. The ODEs system is solved using a forward Euler explicit scheme with a time step of $1e^{-5}$. The PDE is solved using a Crank-Nicolson scheme with a time step of $1e^{-5}$. The problem domain is partitioned using METIS [34] into several sub-domains and parallelized using MPI.

Pseudo-electrocardiograms (pseudo-ECGs) were calculated by positioning each generated biventricular model within a generic torso (constructed from the computed tomography scan of the given patient) and by recording the cardiac potentials (integrals over the spatial gradients of the transmembrane voltages within the biventricular cardiac tissue) at the approximate locations of the right arm (RA), left arm (LA), and left leg (LL) [35]. Note, the electrical propagation through the torso was not computed.

Fast endocardial activations due to the Purkinje network were approximated by defining a fast endocardial layer with an approximate thickness of 348 microns. The anisotropic diffusion coefficient of both the Purkinje layer (10x myocardial diffusion) and the myocardial tissues was obtained after thorough analyses of the total activation times, QRS durations, and epicardial conduction velocities to obtain normal clinical values (diffusion of 1.85 cm^2/s in the fiber direction and 0.6 cm^2/s in the transverse directions) [24]. Convergence was assessed by quantifying the conduction velocity changes throughout the full heart simulation after subdividing it. This yielded a simulation mesh of 524 million elements, with an average element volume of 5.71^{e-7} cm^3 and an approximate average side length of 169 μm. Figure 2 shows the normalized histogram of the surface apparent conduction velocities of the biventricular anatomies at the resolution of this study (65 M), and at after subdividing each element into 8 tetrahedra. The modes of conduction velocities increased from 85 to 95 cm/s in the highest resolution mesh (10% difference). All comparisons within this work are being done between meshes of the exact same resolutions, so results are not affected by errors in spatial convergence.

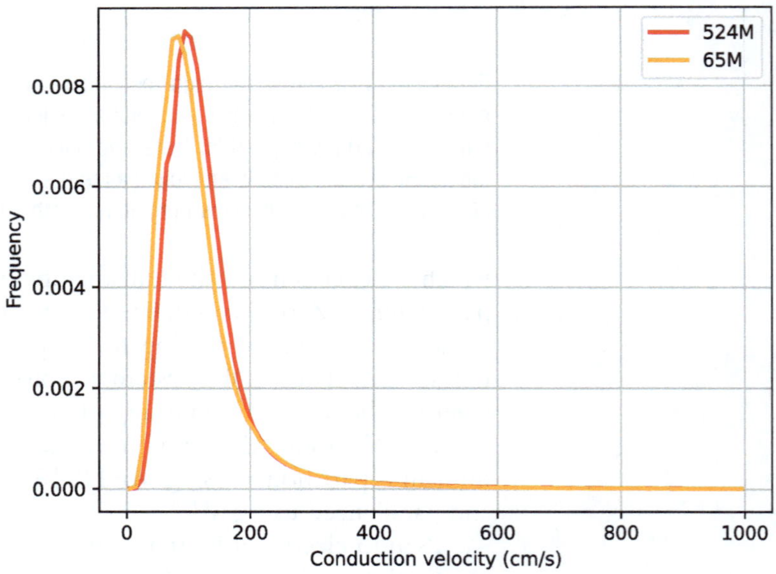

Fig. 2 Electrophysiological model verification. Histogram of the apparent conduction velocity of two simulation meshes at two different resolutions, one resulting in 65 million elements and the other in 534 million elements

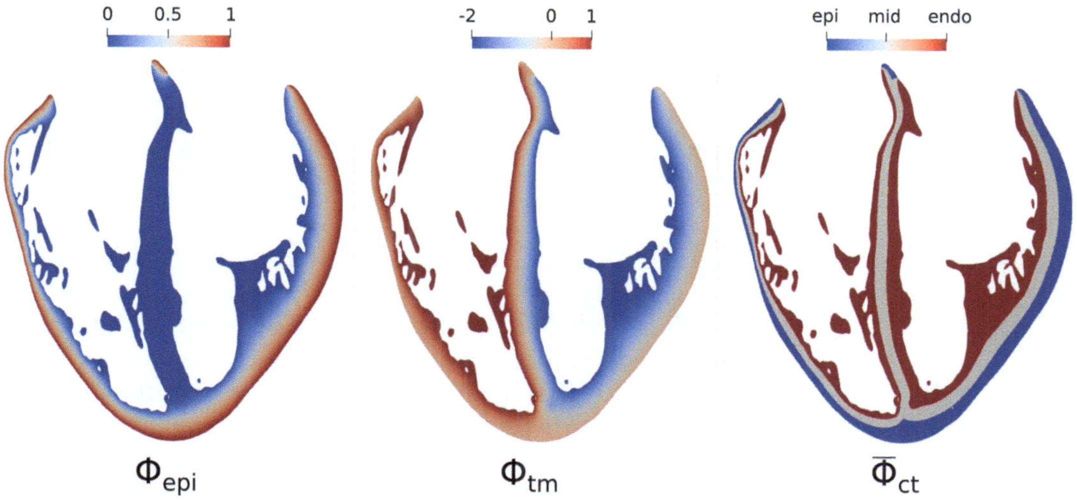

Fig. 3 Cell-type definition Φ_{ct} calculated from the heat equation solutions Φ_{tm} and Φ_{epi}. Shown is an approximately long-axis four-chamber view, RV on the left and LV on the right

Transmural myocyte heterogeneities were added by processing the solutions of the heat equation obtained during fiber generation [25]. Given are two solutions Φtm and Φepi (Φtm obtained by applying temperatures 0 on epicardium, 1 on RV endocardium, –2 on LV endocardium, Φepi by applying 1 on epicardium and 0 on LV and RV endocardium). The celltype was assigned using the following rule (c.f. Fig. 3):

$$\bar{\Phi}_{ct} = \begin{cases} \text{endo}, & \Phi_{ct} \leq 0.33 \\ \text{mid}, & \text{otherwise} \\ \text{epi}, & \Phi_{ct} \geq 0.67 \end{cases} \tag{2}$$

where

$$\Phi_{ct} = \frac{\bar{\Phi}_{tm} - \Phi_{epi}}{2} + 0.5; \quad \bar{\Phi}_{tm} = \begin{cases} 0.5 \cdot \Phi_{tm}, & \Phi_{tm} < 0 \\ \Phi_{tm}, & \text{otherwise} \end{cases} \tag{3}$$

Transmural myocyte heterogeneity was applied following the published ventricular cell model of O'Hara & Rudy [33]. The locations of the activation regions within the biventricular cavities were set following the published work of Durrer [36]. The initial activation regions (IARs) were as follows: (1, 2) two activation areas on the RV wall, near the insertion of the anterior papillary muscle (A-PM); (3) the high anterior para-septal LV area below the mitral valve; (4) the central area on the left surface of the septum; and (5) the posterior LV para-septal area at 1/3 of the apex-base distance, as shown in Fig. 1. The employed stimuli had a magnitude of 5 mA/cm^3 applied for 5 ms at each of the specified locations assuming a cell membrane capacitance of 1 μF and using a half-spherical stimulus of 0.2 cm radius at the required heart rate.

Table 1
Scaling factors by which each ion channel is modified in a combinatorial manner

Quartile	GKr	GKs	GNaL	GNa	GCaL
Q1	0.8	0.75	0.75	0.8	0.75
Q3	1.25	1.25	1.25	1.2	1.25

The O'Hara-Rudy [33] human cardiac ionic model with the conductance modifications to the sodium channel suggested by Dutta et al. (2017) [37], as used by the FDA as the basis for the CiPA initiative for proarrhythmic risk assessment, was employed to compute the cardiomyocyte action potentials. The effects of steroid hormones on the action potential defined the sex-specific ionic channel behaviors within the human heart population. These sex-specific variations of ion channel sub-unit expressions of both endocardial and epicardial cell types were assigned following the work of Yang and Clancy (2012) [22]. No sex-specific definition was applied to the M-cells, since no sex-specific information in such cell types has been reported. The spectrum of phenotypes within the normal population was set up by employing the experimentally calibrated normal population data published by Muszkiewicz et al. (2016) [21]. Five ion channels that have the highest influences on action potential durations were selected: INa, IKr, IKs, ICaL, and INaL. The first and third quartiles of the above channel conductances (see Table 1) were employed as edges to a five-dimensional hypercube that defines the ranges of variations of each of the selected ion channel conductances. To observe a significant amount of variation between subjects, this hypercube was sampled at the edges. Therefore, the problem was reduced to 2^5 permutations for the 5 channels, for a total of 64 subjects: 32 unique male and 32 female phenotypes. In addition to the normal baseline population, a hypokalemic population was obtained by reducing the extracellular potassium concentrations of each of the normal subjects to 3.2 mmol/L. HCQ and AZM were used to illustrate the potential of this framework. The effects of the drugs affecting the ion channel conductances were incorporated following the methodology described by Mirams et al. [38] using a multi-channel conductance-block formulation for each drug assessed. Peak plasma concentrations for clinical oral doses of AZM, HCQ and their IC50 values [39–41] were employed as follows: 800 and 400 mg oral dose of HCQ alone and 200 and 400 mg HCQ and 500 mg AZM in combination. All drug-related concentrations and IC50 values employed in this study are given in Tables 2 and 3. The assumptions regarding the interactions of both drugs were based on the additive effects of each drug on the affected ion channels described by the following equations (with Di, IC_{50i}, and hi being the

Table 2
Plasma concentrations employed for each oral dose tested

Oral doses	HCQ 200 mg	HCQ 400 mg	HCQ 800 mg	AZM 500 (3-day)
Peak plasma Concentration (μ mol/lt)	1.19	2.97	4.7	0.59

Obtained from Refs. [21, 41]

Table 3
IC50 values for each channel affected by HCQ and AZM

Drug	IKr IC50 (μmol)	ICaL IC50 (μmol)	IKs IC50 (μmol)	INa IC50 (μmol)	IK1 IC50 (μmol)	INaL IC50 (μmol)
HCQ	5.57	22	NA	NA	NA	NA
AZM (chronic effect)	219	66.5	184	53.3	43.8	62.2

Data obtained from Refs. [40, 41]

concentration, IC_{50}, and h values for the i-th drug, respectively, and g, g' being the channels conductances before and after the application of drugs):

$$g_i = \left(1 + \left(\frac{D_i}{IC50_i}\right)^{h_i}\right)^{-1} \tag{4}$$

$$g' = g \cdot \left(1 + \sum_i (g_i - 1)\right) \tag{5}$$

The ion channel conductance variations and the drug-blocking effects on the ion channel models were solved in zero-dimensional (0D) models (endocardial, midmyocardial, and epicardial) until steady state of the calcium transient was achieved (root mean square error (RMSE) between consecutive beats smaller than 1e-7(μmol) for 3 consecutive beats). The results of the 0D models provide initial conditions to parameterize the 3D finite element simulations. Simulations on the 3D mesh of 3–5 beats were solved to achieve steady state at two different basic cycle lengths (BCL): 600 (Baseline) and 400 ms (Stress). A total of 896 simulations were run, each one using 640 cores on the Joliot-Curie Rome supercomputer (GENCI, CEA, France). An approximate total of 2.3 million core hours were employed.

The QTc and QRS values were measured using the three pseudo-ECG leads. An example pseudo-ECG for both a male and female subject is shown in Fig. 4, including an example of the calculation of the QRS and QT markers. It is important to clarify that the main aim of our work was not to validate the electrocardiogram. Instead, the pseudo-electrocardiogram was employed to

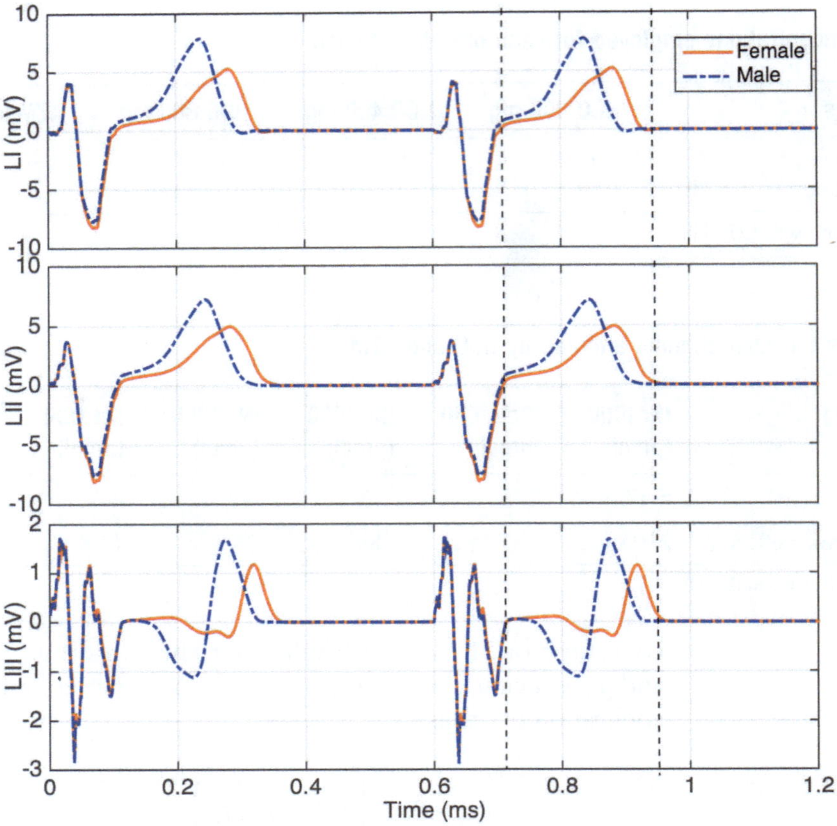

Fig. 4 Pseudo-electrocardiogram calculated at three standard lead locations: LI, LII, and LIII of a male and female subject of our population. QRS and QT intervals as calculated for our analysis are marked within each lead. Both markers were determined using the three leads

obtain markers at baseline and after the administrations of drugs to quantify their effects on the simulations. Both Framingham and Bazett formulae were used to calculate the QTc from the cohort with a BCL of 600 ms. A surrogate marker for contractility was obtained by calculating the integral of the calcium transient throughout the anatomy. The magnitude of the peak calcium, the time to peak calcium, calcium 90, and the magnitude of the T-wave were also quantified in each simulation. Data were analyzed using RStudio and Orange [42]. The relative importance of each phenotype on each marker was assessed using the LMG (Lindeman, Merenda, and Gold's (LMG) method, see [43]), implemented in the relaimpo R package [44]. A risk characterization was used to identify the currents that had an individual contribution to a multiple regression model using the simulations which were characterized as having risk due to any of the two drugs. Two response variables were employed: QTc interval values and QTc-interval prolongations. Multiple linear regressions were performed with the ion channel conductance factors as regressors to either of the response variables.

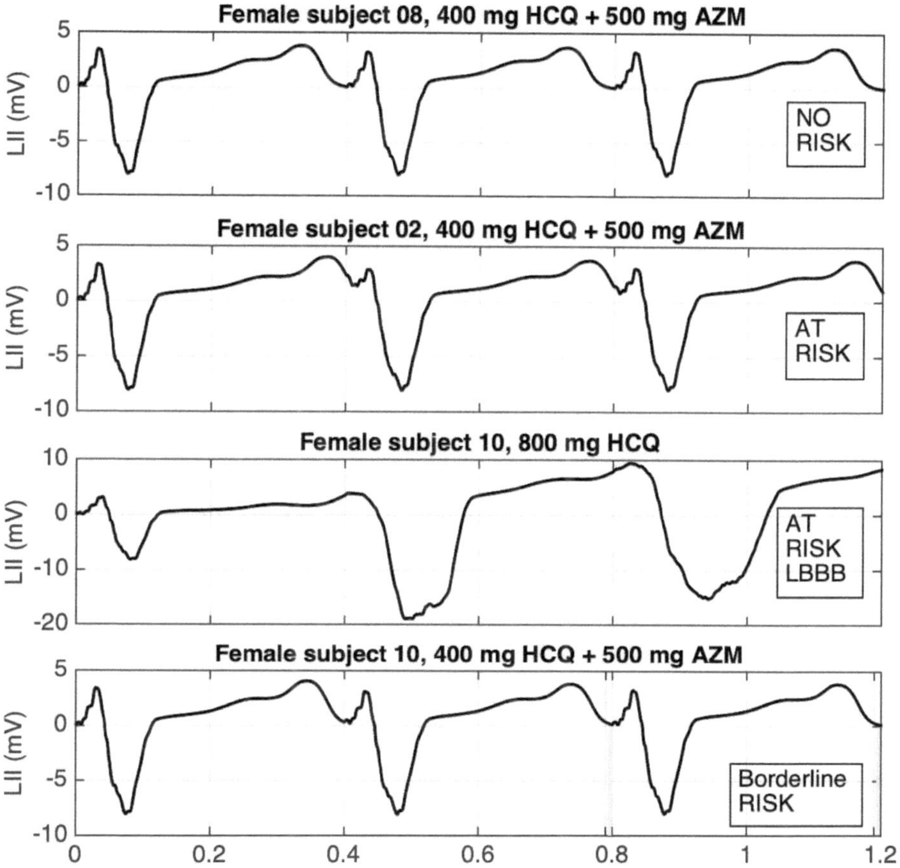

Fig. 5 Analysis of the pseudo-ECG to characterize risk after under a stress test. Four different cases were selected. Risk is characterized as the incomplete repolarization of the heart before the next heart beat occurs. If end of repolarization and activation occurs at the same time (marked by yellow vertical bars), the subsequent beat presents a lengthening of the QRS due to dyssynchronous activation produced in regions where the AP is still in a refractory phase. The 10 ms window between the 390 and 400 ms of the last two cardiac cycles is highlighted in yellow

The stress test to assess the arrhythmic risk of each subject was performed by pacing the population at a BCL of 400 ms (150 bpm). A subject at risk was defined as having spontaneous arrhythmic or dangerous rhythms, which included left bundle branch block, ventricular tachycardia, or QT-interval greater than 390 ms at a 400 ms BCL, and this leads to observable asynchronous activation because the endocardial tissue remains refractory. Figure 5 shows examples of the classifications performed on the pseudo-ECG data after the applied stress test. A semi-automatic algorithm was employed to estimate the QT interval durations of the pseudo-ECGs. Detailed analyses were also performed by observing the results of the electrophysiological propagation in the human heart of interest.

2.1 In Vitro Experimentation on Reanimated Swine Hearts

To validate the potential arrhythmic behaviors of the computational models with regard to electrophysiological and functional effects in response to HCQ administrations, an experimental in vitro setup employing reanimated swine hearts was employed. All animals received humane care in compliance with the "Principles of Laboratory Animal Care," formulated by the National Society for Medical Research, and the Guide for the Care of Laboratory Animals, published by the National Institutes of Health. This research protocol was ethically approved by the University of Minnesota Institutional Animal Care and Use Committee (IACUC). The IACUC protocol number is 2006A38201 "Swine Isolated Heart Model." The precise experimental protocol was published previously in [45]. Electrical potentials and mechanical function were monitored throughout these experiments. These supplemental experiments were performed on a reanimated swine heart utilizing Visible Heart® methodologies. Briefly, swine hearts were continuously perfused with a clear, modified Krebs-Henseleit buffer according to previously described methodologies [46, 47]. After providing a single 30 J shock, each heart elicited and sustained an intrinsic rhythm and associated hemodynamic function. The experimental setup is shown in Fig. 6.

2.1.1 Mechanical and Electrical Data Acquisition

Pressure-volume loops were measured using clinically available conductance catheters and a data acquisition system (CD Leycom, the Netherlands). Conductance catheters were placed in both left and right ventricles and calibrated to the cardiac function, and recordings were sampled at a rate of 250 Hz. Local Monophasic Action Potentials (MAPS) were recorded from both ventricles with MAP4 catheters (Medtronic, Minneapolis, MN, USA) positioned in the RV and LV endocardium and LV anterior epicardium, and the rise time and time to decay were measured for 10 consecutive MAP waveforms for each treatment group. To quantify the dispersion of repolarization, two decapolar catheters were sutured onto the epicardial surfaces of both ventricles to record relative changes in the electrical activations. Three-lead surface electrocardiograms were recorded throughout the duration of these studies as means to measure global electrical activity as shown in Fig. 6. The datasets were collected at four different time points: baseline, during drug infusion of 25 mg of HCQ, after a 10-minute washout period (in which buffer was replaced twice), during a second drug infusion of 25 mg of HCQ, and then a final 50 mg infusion of HCQ after another 10 minutes washout period. The drug was prepared for administration by dissolving 25 mg of HCQ in 10 L of recirculated buffer (equivalent to 7.44 μmol/lt) and allowed to equilibrate for 2 minutes before data collection. Doses were experimentally derived to observe the greatest functional differential from controls to treatment with HCQ.

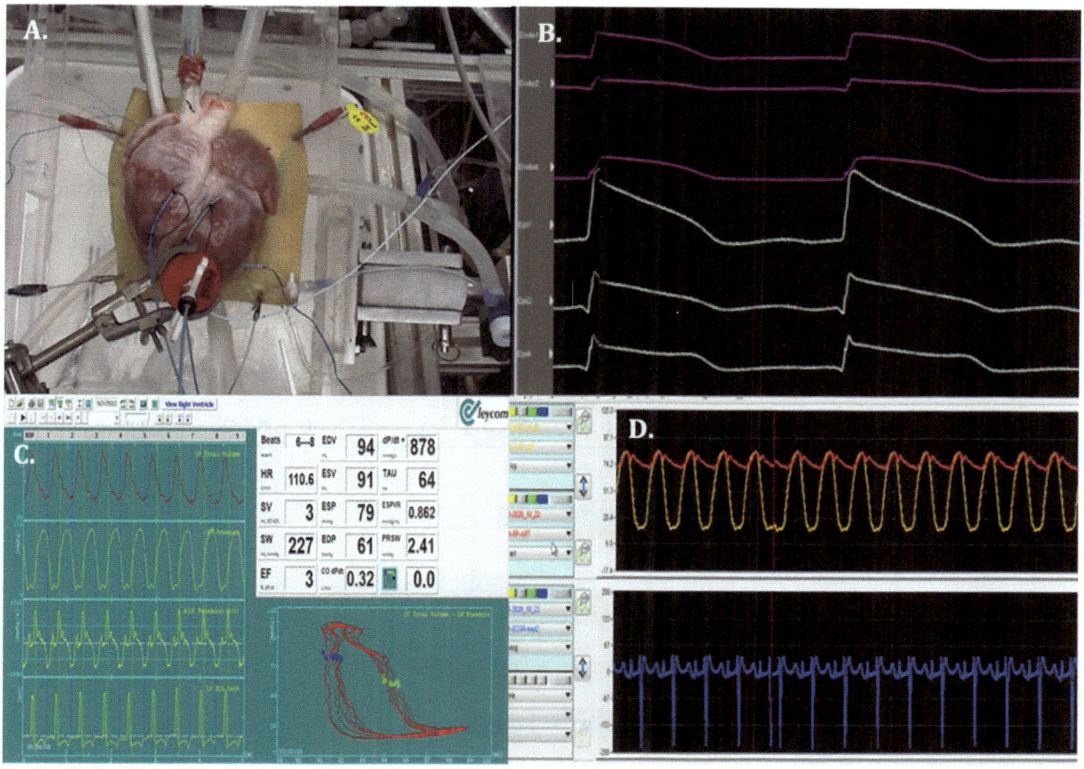

Fig. 6 (**a**) An external view of the reanimated swine heart. Two decapolar catheters were sutured to the epicardial surfaces of both the left and right ventricles and a custom device holds the MAPS4 catheter in constant contact with the LV epicardium. (**b**) Electrical activations in LV and RV as measured from decapolar catheters. (**c**) PV loops are measured from conductance catheters placed within both the LV and RV. (**d**) Arterial (red trace) and LV pressures (yellow trace) are recorded along with the three-lead surface ECG (blue trace)

3 Results

3.1 In Silico Experiments

Medians, interquartile ranges, mean, and standard deviations for all markers measured are shown in Table 4. Means and standard deviations were included in order to compare the results to clinically published data. The QTcFra histograms of the simulated population classified by gender at baseline, hypokalemia, and after the administration of a variety of oral doses and combined administrations of HCQ and AZM are shown in Fig. 7. The QTc-interval prolongation values are color coded by sex of the entire cohort is shown in Fig. 8. A preliminary classification of the virtual clinical trials based on QTc values (at BCL = 600 ms) is shown in Table 5. There were no arrhythmic events in any of the subjects at this BCL. The number of males and females with QTc Framingham values higher than 500 ms are included. It was observed that the virtual female subjects presented with higher likelihoods of increased QTc-interval values and arrhythmic risks. The solutions for the full cohort after a stress test showed numerous spontaneous

Table 4
In vitro functional response to the administration of 7.44 mol/lt of HCQ and after full drug washout (mean ± std)

	BCL (s)	LVP systolic (mmHg)	LVP diastolic (mmHg)	dP/dt (mmHg/s)	Stroke work (mL mmHg)
Baseline	0.604 ± 0.002	76.9 ± 5.56	18.39 ± 0.39	948.17 ± 289.2	179.48 ± 6.81
25 mg dose 1	0.586 ± 0.135	50.3 ± 9.12	19.99 ± 1.3	396.2 ± 107.63	58.2 ± 27.4
Washout	0.578 ± 0.133	60 ± 10.97	16.43 ± 3.83	597.49 ± 107.14	81.45 ± 30.8
25 mg dose 2	0.598 ± 0.06	87 ± 5.73	30.28 ± 5.07	542.55 ± 125.99	60.8 ± 12
50 mg	0.711 ± 0.121	49.14 ± 6.5	18.26 ± 7.5	375.66 ± 88.08	61.5 ± 23.2

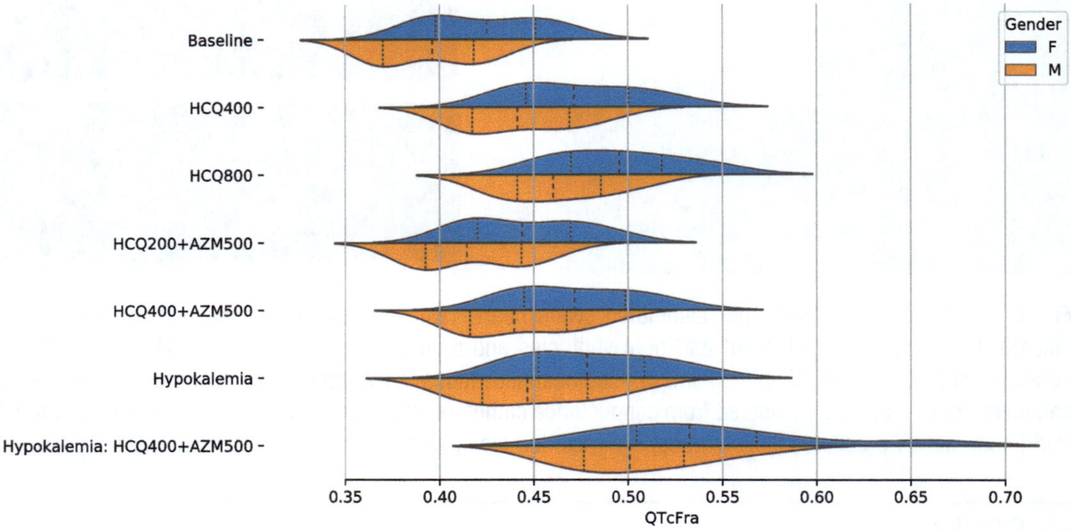

Fig. 7 Histogram of the QTcFra values of the entire population classified by gender during baseline and after the administrations of 400 and 800 mg HCQ and 200 and 400 mg HCQ in combination with 500 mg of AZM under normal conditions and under hypokalemia (3.2 mmol/L). QTcFra corresponds to the baseline BCL of 600 ms

arrhythmic responses that were employed to characterize the risk of drug-induced arrhythmia. For example, left bundle branch block (LBBB) was observed in two cases after the administration of 800 mg of HCQ; one case after the administration of 400 mg of HCQ and 500 mg of AZM; and four in a hypokalemic state after the administrations of HCQ 400 mg and AZM 500 mg. Four cases of ventricular tachycardia were observed at the hypokalemic state after the administration of HCQ 400 mg and AZM 500 mg. The QTcFra interval values of the virtual populations after the administrations of drugs is shown in Fig. 7. The risk classifications according to the QTcFra obtained at 600 BCL are shown in Fig. 9. The quantification of risk, including sex differences, is shown in

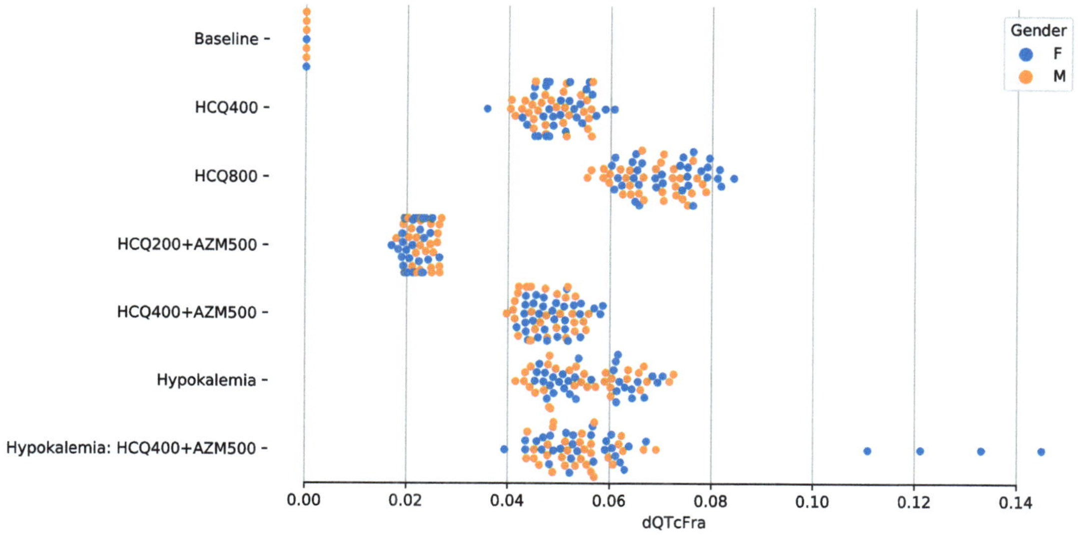

Fig. 8 QTcFra (s) prolongation classified by gender (dQTcFra). Both genders provide approximately similar QTcFra prolongations. Four female subjects with hypokalemia and HCQ400 + AZM500 presented extreme prolongation that led to ventricular tachycardia. The QTcFra corresponds to the baseline BCL of 600 ms

Table 5
Classification based on QTc Framingham values from the virtual clinical trial, corresponding to BCL of 600 ms

Human in silico trial	Subjects with QTcFra >500 (%)	Female $N = 32$	Male $N = 32$
Baseline	0	0	0
HCQ 400 mg	9, (14%)	8	1
HCQ 800 mg	17, (26%)	22	4
HCQ 200 mg + AZM 500 mg	1, (1%)	1	0
HCQ 400 mg + AZM 500 mg	7, (10%)	7	0
Hypokalemia (Ko = 3.2 mmol)	14, (21%)	11	3
Hypokalemia (Ko = 3.2 mmol) + HCQ 400 mg + AZM 500 mg	41, (64%)	24	17

Table 6. It is important to note that the virtual stress test was able to reproduce the complexities of drug-induced arrhythmias, where QTc was not an absolute marker of arrhythmic risk, as it is shown in detail in Fig. 10, but it was useful towards providing clinical guidance. The QTc values for each virtual patient at baseline, with hypokalaemia, and as the progression of the QTc prolongations occurred after the administrations of one or more drugs are plotted in Fig. 10. Those presenting on the right were the QTc prolongations in subjects that present no arrhythmic response, and to the left

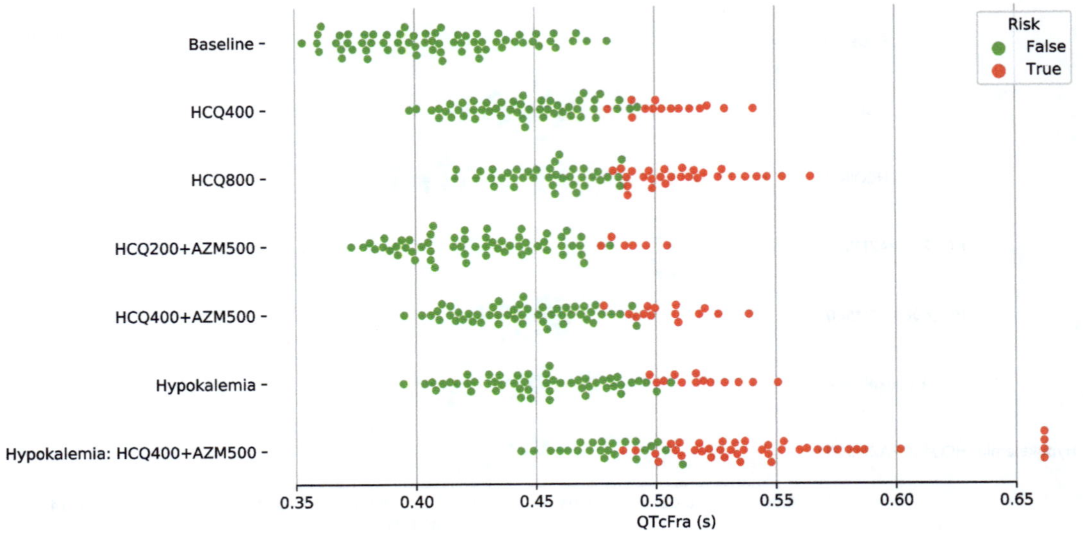

Fig. 9 QTcFra results of all cohorts classified by risk. Subjects at risk of drug-induced arrhythmias are shown in red. QTcFra shown is the one calculated at the baseline BCL of 600 ms

Table 6
Classification based on the stress test (BCL = 400 ms) on the virtual clinical trial

Human in silico trial	Subjects at Risk (%)	Female $N = 32$	Male $N = 32$
Baseline	0	0	0
HCQ 400 mg	14 (21.9%)	13	1
HCQ 800 mg	27 (42.2%)	19	8
HCQ 200 mg + AZM 500 mg	6 (9.4%)	6	0
HCQ 400 mg + AZM 500 mg	14 (21.9%)	13	1
Hypokalemia (Ko = 3.2 mmol)	13 (20.3%)	11	2
Hypokalemia (Ko = 3.2 mmol) + HCQ 400 mg + AZM 500 mg	41 (64%)	26	15

were the subjects that present an arrhythmic risk after being classified by the stress test for each drug intervention. The effect of each drug regime is also clearly indicated by the slope of the drug-induced QTc prolongation. The fact that the QTc interval is not a 100% precise predictor of drug-induced arrhythmias makes drug-induced pro-arrhythmic risk a major unsolved problem clinically. Simulations show that patients with the same QTcFra at baseline could present two distinct responses to drug administration during the stress test. The characteristics of the arrhythmic phenotypes were traced back within the baseline population, in order to identify

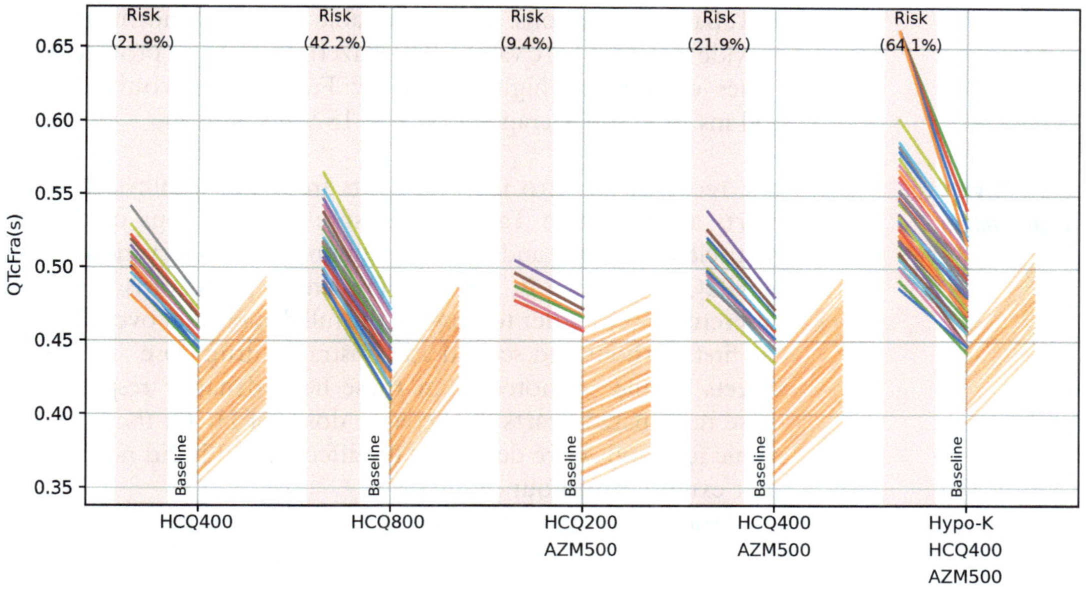

Fig. 10 QtcFra prolongation trajectory of each subject of the normal and hypokalemic virtual population after the administration of various doses of HCQ (oral doses in mg) alone and in combination with AZM (oral doses in mg) classified by risk. Each line connects the values of QTcFra at baseline and treated; on the left are the lines corresponding to the subjects at risk, while on the right are the subjects without arrhythmic events. Note the overlap of the lines on the baseline axis, demonstrating the population that reproduced the complexity of the arrhythmic risk using QTc as marker. The QTcFra corresponds to the baseline BCL of 600 ms

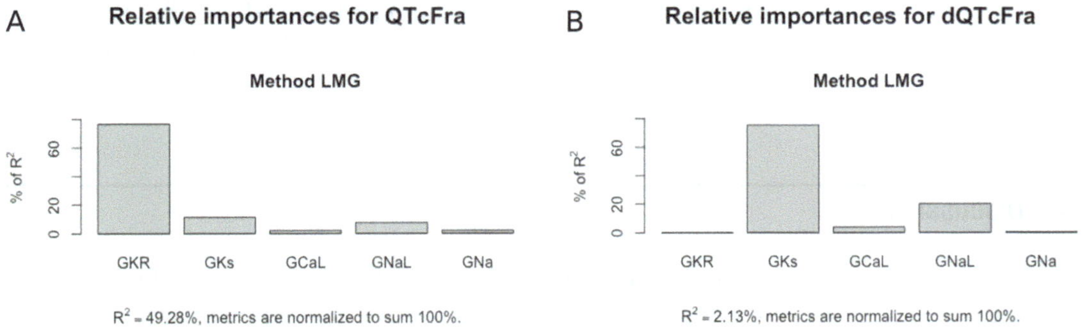

Fig. 11 LMG relative importance of the ion channel current conductances on two of the measured markers: (**a**) QTcFra and (**b**) the prolongation of QTcFra

the phenotypes that might exhibit a propensity for a drug-induced arrhythmic behavior. The most common phenotype in the virtual population at high risk was a low GKr; however, not all of the phenotypes with low GKr elicited these arrhythmic scenarios (see Fig. 11). The relative importance of these currents shows that the main current that determines the length of the QTc-interval was GKr; however, Gas and GNaL had a high relative importance on the QTc interval prolongation. Within this study, QTc was normalized employing both the Framingham (QTcFra) and Bazett

(QTcBaz) conventions. It is noticeable that both normalizations provide different QTc values. The Bazett convention provides QTc values consistently higher than the Framingham convention at 600 ms BCL (on average by 52.5 ± 15.9 ms higher).

3.2 In Vitro Experiments

The recorded in vitro measurements from the reanimated swine hearts are shown in Table 7. Pressure-volume loops of the left ventricle at baseline and after the administrations of various doses of HCQ are shown in Fig. 12. The entire protocol is shown in different plots in order to observe the full dataset that overlaps after the first washout and second administrated drug dose. Two main aspects should be noted from these hemodynamic responses of these reanimated hearts to the large doses of HCQ: the hemodynamic functions were detrimentally affected and it did not recover after extensive washout of the drug.

Monophasic action potentials (MAPs) measured in these ex vivo reanimated swine hearts are shown in Fig. 13. They portray the Action Potential Duration (APD) alterations after the administrations of a large dose of HCQ. Epicardial MAP measurements elicited APD increases after a 25 mg administration of HCQ from 352 to 363 ms (APD prolongation of 11 ms) in the epicardium and an increase from 350 to 481 ms in the endocardium (APD prolongation of 131 ms). For the extremely high dose of 50 mg (14.88 μmol/l), the APDs recorded from the epicardium increased to 589 ms (APD prolongations of 237 ms). Similarly, the APD differences between endocardium and epicardium after the administration of 800 mg of HCQ on a virtual subject with LBBB showed an increase of the AP in the endocardium of an average 0.063 ± 0.0315 s with respect to the epicardium.

4 Discussion

This is the first reported study of drug-induced arrhythmic risks after the administration of one or more potentially cardiotoxic drugs using a 3D biventricular cardiac "virtual patient population." Unlike the work of Okada et al. (2021) [17], the aim of this work was to reproduce a varied human cardiac population so as to identify the percentage of subjects that might develop drug-induced arrhythmias. The developed computational framework employs sex-specific cardiac phenotypes under both normal and hypokalemic conditions. We specifically sought to identify arrhythmic alterations induced by the administrations of HCQ and AZM. The alterations due to HCQ were also identified in a series of experiments on reanimated swine hearts where the cardiac function was also assessed. The heterogeneous action potential prolongation observed within the in vitro reanimated swine hearts after the

Table 7

Median (interquartile range) and mean ± standard deviations of all the markers assessed in the entire population at baseline and after the administration of 400 mg and 800 mg of HCQ and 400 and 200 mg of HCQ combined with 500 mg of AZM

Marker median (iqr), mean ± std	Baseline	Hypokalemia (3.2 mmol/L)	HCQ 400 mg	HCQ 800 mg	HCQ 200 mg + AZM 500 mg	HCQ 400 mg + AZM 500 mg	Hypokalemia + HCQ 400 mg + AZM 500 mg
QRS (ms)	102 (98–108) 103.7 ± 4.9	120.4 (114–126) 119.9 ± 6.1	106 (99–109) 104.5 ± 5.2	103 (99–109) 104.5 ± 5.2	104.9 (99–109) 104.7 ± 5.3	102 (99–108) 104.5 ± 5.1	119.8 (113–125) 122.9 ± 27.5
QTc Fra (ms)	408 (382–434) 409.7 ± 31.9	462.3 (435–494) 465.4 ± 37.4	456 (430–485) 458.8 ± 34.2	476 (455–504) 479.3 ± 34.9	429.7 (405–457) 431.8 ± 32.7	455 (431–483) 457.8 ± 34.2	516.8 (486–547) 523.8 ± 50.8
QTc Baz (ms)	447.5 (413–481) 449.4 ± 41.3	517.2 (482–558) 521.3 ± 48.2	509.3 (476–546) 512.8 ± 44.6	536 (508–571) 539.2 ± 45	475.3 (444–511) 477.9 ± 42.2	508.6 (477–543) 511.5 ± 44.2	587.7 (548–627) 598.4 ± 70
QTc Fra prolongation (ms)	0	54.1 (48–62) 55.7 ± 8.3	48.4 (45–53) 49.1 ± 5.1	70 (64–75) 69.5 ± 7.0	22 (20–23) 22.1 ± 2.4	47.6 (44–51) 48.1 ± 4.7	54.3 (48–60) 58.4 ± 19.2
QTc Baz prolongation (ms)	0	69.9 (62–80) 71.9 ± 10.7	62.4 (58–69) 63.4 ± 6.6	90.3 (82–97) 89.8 ± 9	28.4 (25–30) 28.6 ± 3.1	61.4 (57–66) 62.2 ± 6.1	70.1 (63–78) 75.4 ± 24.8
Peak Calcium Integral (mM)	0.283 (0.24–0.35) 0.293 ± 0.087	0.297 (0.234–0.335) 0.291 ± 0.066	0.262 (0.21–0.3) 0.263 ± 0.065	0.250 (0.22–0.28) 0.249 ± 0.056	0.276 (0.212–0.304) 0.280 ± 0.074	0.260 (0.21–0.3) 0.262 ± 0.065	0.250 (0.203–0.295) 0.253 ± 0.049
Time to Peak Calcium Integral (ms)	74 (72–77) 74.7 ± 3.7	82 (78–84) 81.4 ± 3.8	75 (72–77) 74.8 ± 3.9	75 (72–77) 75 ± 3.8	74.5 (72–77) 74.9 ± 3.9	75 (72–77) 74.9 ± 4	82 (78–84) 82.2 ± 6.1

(continued)

Table 7 (continued)

Marker median (iqr), mean ± std	Baseline	Hypokalemia (3.2 mmol/L)	HCQ 400 mg	HCQ 800 mg	HCQ 200 mg + AZM 500 mg	HCQ 400 mg + AZM 500 mg	Hypokalemia + HCQ 400 mg + AZM 500 mg
Calcium 90 (ms)	349 (341–372) 352.7 ± 18.2	373 (359–380) 373.4 ± 19.2	369 (354–381) 369 ± 17	376.5 (361–384) 373.9 ± 15.4	358 (346–378) 360.8 ± 17.6	369 (354–381) 369 ± 16.8	386.5 (374–396) 385.5 ± 22.2
T-wave Magnitude (mV)	6.71 (5.6–7.2) 6.45 ± 0.99	8.45 (7.78–8.95) 8.38 ± 0.88	6.28 (5.2–6.6) 6.02 ± 0.86	6.08 (5.1–6.5) 5.82 ± 0.86	6.41 (5.35–6.82) 6.13 ± 0.94	6.24 (5.2–6.6) 5.97 ± 0.87	7.91 (7.11–8.38) 7.81 ± 1.4

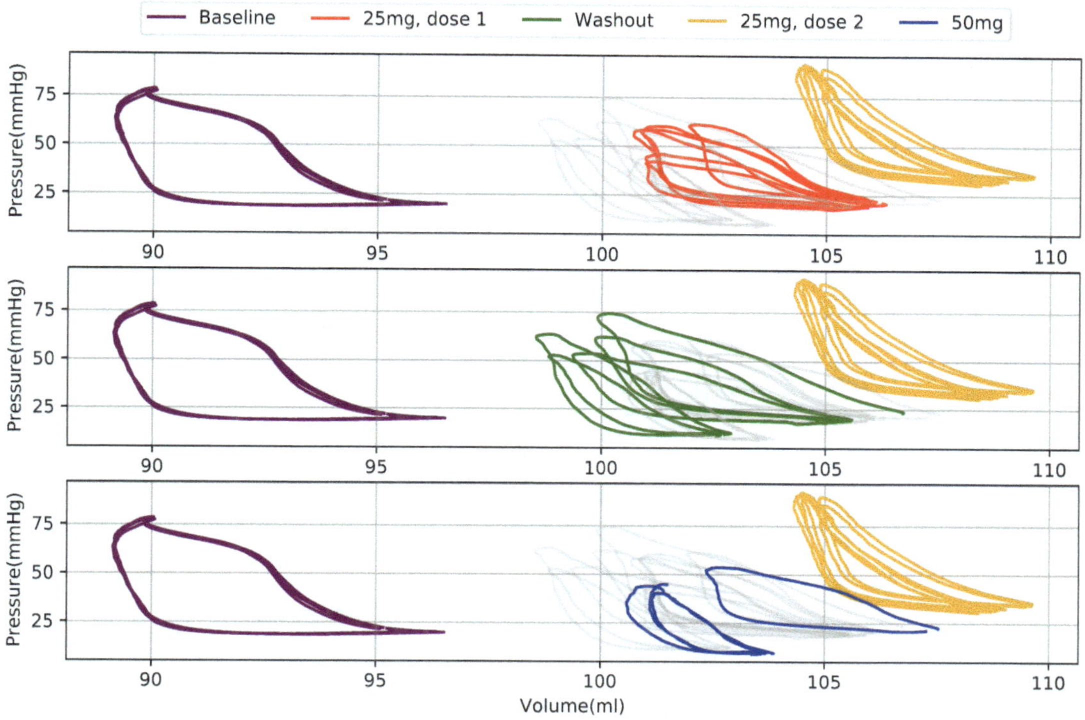

Fig. 12 Characteristic left ventricular pressure-volume loops from an ex vivo perfused, reanimated swine heart at baseline and following administration, washout, and readministration of HCQ. The entire protocol is shown in three panels due to overlapping data (in gray color as reference). Baseline and second dose of 25 mg are included in all panels for ease of comparison. First 25 mg dose is shown in red in the top panel, washout in green in the middle panel, and the second 25 mg dose in blue in the third panel. Notice the variability observed in the hemodynamics after the administration of a large dose of HCQ. The hemodynamic function was not recovered after washout

administration of HCQ was a pro-arrhythmic mechanism also observed computationally.

The QTc values and QTc prolongations obtained from the computational framework fell within the ranges observed on clinical patients. The results were compared to those of clinical studies recently published in [48, 49]. In both clinical studies, the Bazett normalization was employed as shown in Table 8.

The stress test results indicated that 21.9% of the computational cohort exhibited higher arrhythmic risks after the administrations of a single dose of HCQ (400 mg) and combination of HCQ (400 mg) plus AZM (500 mg). For the administrations of 200 mg of HCQ plus 500 mg of AZM, the percentage risk fell to 9.4%. The risk increased significantly in subjects with hypokalemia (3.2 mol K+ concentration) to 64.1%, within this virtual population. At a peak plasma concentration after the administration of 800 mg of HCQ, the risk for arrhythmic events was calculated to be 42.2%. This virtual clinical trial was able to confirm that the use of

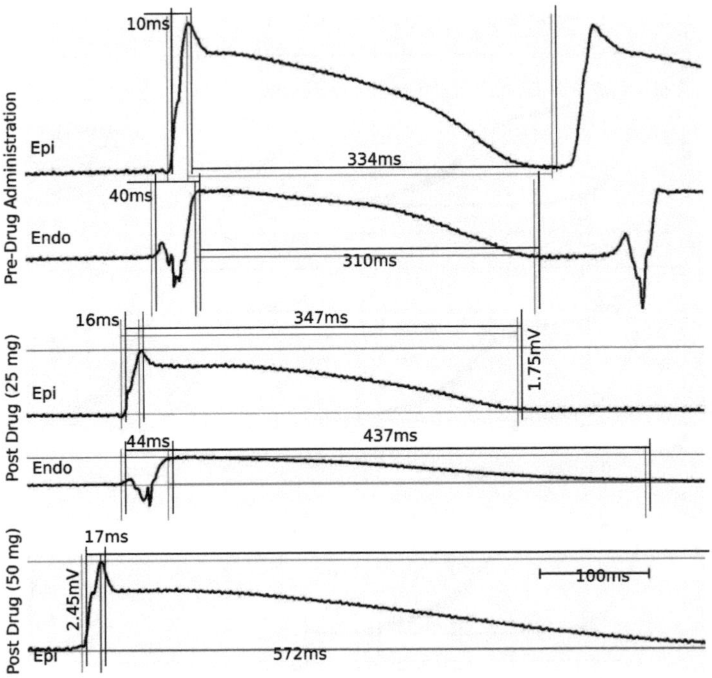

Fig. 13 MAPs recorded from the endocardial and epicardial walls of a reanimated swine heart. Note that only epicardial MAPs were able to be recorded for the 50 mg HCQ dose

HCQ was associated with a proarrhythmic risk, but it did not worsen when it was used in combination with AZM as has been observed clinically, i.e., experimentally and in silico [9, 48, 49]. It is important to stress that the concentrations tested within these virtual healthy patients were peak plasma concentrations. These predicted values may change according to the relative metabolisms of these drugs in a clinical setting and according to the interactions with other anti-arrhythmic medications. Nevertheless, this proposed framework allows for the testing of any plasma concentration of a given drug on cardiac electrophysiology. The cohort statistics reported by Saleh et al. [48] include data of patients with a variety of potassium concentrations, so direct comparison to the computational data is difficult. However, it was observed that ventricular tachycardia occurred in 7 out of 8 of the reported patients with a potassium concentration lower than normal, in agreement with the higher risks shown in the virtual hypokalemic population (Table 9). The match between the percentages of subjects at arrhythmic risk after the administration of 400 mg of HCQ and the combination of 400 mg of HCQ and 500 mg of AZM was remarkable as compared to previously published data [49]. The in vitro studies on the reanimated swine hearts provided evidence for transmural heterogeneous effects of HCQ on action potential durations as shown in Fig. 13. This was evidenced by heightened drug-induced

Table 8
Comparison of QTc Bazett values between the virtual clinical trial and published clinical trials

QTc Bazett (ms)	Virtual clinical trial $N = 64$ Median[iqr] Mean ± std	Mercuro et al. (2020) [49] $N = 90$ Median[iqr]	Saleh et al. (2020) [48] $N = 200$ Mean ± std
Baseline	447.5[413–481] 449.4 ± 41.3	455[430–474]	440.6 ± 24.9
HCQ 400 mg	509[476–546] 512.8 ± 44.6	473[454–487]	NA
HCQ 200 mg + AZM 500 mg	475[444–511] 477.9 ± 42.2	NA	470.4 ± 45.0
HCQ 400 mg + AZM 500 mg	508[477–543] 511.5 ± 44.2	442[427–461]	NA
Hypokalemia (Ko = 3.2 mmol) + HCQ 400 mg + AZM 500 mg	587 [548–627] 598.4 ± 70	NA	NA

Table 9
Comparison of the percentage of subjects at risk between the virtual clinical trial and the clinical trials published in the literature

Subjects at risk (%)	Virtual clinical trial $N = 64$	Mercuro et al. (2020) [49] $N = 90$	Saleh et al. (2020) [48] $N = 200$
HCQ 400 mg	21.8%	19%	NA
HCQ 200 mg + AZM 500 mg	9.3%	NA	3.5%
HCQ 400 mg + AZM 500 mg	21.8%	21%	NA
Hypokalemia (Ko = 3.2 mmol) + HCQ 400 mg + AZM 500 mg	64%	NA	NA

transmural heterogeneities, as has been previously associated with multi-channel blocking drugs [50]. Enhanced transmural heterogeneities of local APs provide an important arrhythmic risk, as observed in one of the virtual subjects that presented an LBBB after the administration of 800 mg of HCQ. This further confirms the capabilities of the computational models to exhibit arrhythmic behaviors observed both experimentally and clinically. Lastly, the long-lasting effect on cardiac function after a complete washout of a large dose of HCQ was an important observation toward its clinical use. One limitation of the electrophysiology simulations in this study is the lack of an electro-mechanical simulation that could provide mechanistic information regarding the detrimental hemodynamic effects of the large dose of HCQ as observed in the in vitro swine study. An extension to the solution of an electromechanical model to assess drugs in a large population like the one proposed here would require an approximate 6.5 times more computation time. While running electro-mechanics was not an aim of this work, it constitutes a part of the future work.

A potential limitation of this study is the absence of an incorporated conduction system. The addition of a realistic conduction system would likely provide a more physiological behavior of the global cardiac activation sequence, which in turn may provide a more realistic pseudo-ECG. Future work is ongoing to integrate a one-dimensional conduction system description within detailed human anatomies. Another potential limitation of our study is related to the IC50 values obtained from the literature [39, 41]. Unlike in the work of Okada et al. (2021) [17], we did not have access to perform quantifications of block patch clamp data relative to our in vitro experiments. At the time of creation of these simulations, there had been no recent quantification of IC50 data for the administrations of CQ, HCQ, AZM, and/or their combined administrations within in vitro experiments. The pharmacological data employed in this study was based on quantified guinea pig SAN cells and human ion channels heterologously expressed in human embryonic kidney HEK293 cells. Although not ideal, these utilized values provided a similar response in our virtual population relative to recently published clinical trials. The "naive" approach assumed in this work, regarding the additive effects of the two agents on the ion channel blocking effect, can be revised as emerging research becomes available.

Another potential limitation of this employed framework could be its high computational cost. Computational time availability on commercial and research HPC infrastructures, however, is increasing constantly, but will likely not be a limiting factor for important in silico human heart trials. In the future, we are aiming to produce simulations that may aid in the reduction of human clinical trials, according to the established three R's. Therefore, providing an adequate platform and computational time accessibility, these workflows may be integrated within the drug development pipeline. The results from our simulations reflect the behaviors of both normal and hypokalemic populations, without any additional risk factors (i.e., ischemia, other electrolyte disturbances, infarction, and/or established cardiac genetic disorders). Also note that the assessments of the drug effects on a more broad populations remain part of the future work. Furthermore, we plan to study the assessments of pharmacological agents associated with pro-arrhythmic risks in a greater variety of human cardiac anatomies. Nevertheless, we consider that the assumptions tested in this study are able to provide timely critical information concerning phenotypic characteristics of the subset of our virtual subjects that have a higher predisposition to drug-induced QT prolongations. The use of more detailed ion channel models of hERG can be employed in future work.

5 Conclusion

The analyses of a virtual gender-specific cohort yielded a range of electrocardiographic phenotypes that resemble a normal human subject cohort. The developed and described computational framework was capable of reproducing the effects of one or more administered cardiotoxic drugs, eliciting a remarkable outcome resembling the published clinical studies. Our simulations were capable of reproducing the complexities of the cardiotoxic responses in the human population. Furthermore, these methodologies may be employed in future urgent clinical applications to provide primary information about the dosages and the combined effects of pharmacological agents, most importantly such as applications when clinical guidance are unavailable. The minimal data required for such computational models are the plasma concentrations of the drug or drugs in question and the IC50 values for each and identification of the cardiac ion channels they affect. An in silico clinical trial framework like the one proposed in this work could be capable of providing evidence of the proarrhythmic risks of QTc-interval prolonging agents in normal or diseased populations using high-performance computing within a 24 hour period.

The developed virtual normal human heart cohort can further be interrogated relative to any population variants that produce distinct, arrhythmogenic outcomes after the administrations of one or pharmacological combinations. Importantly, this can be exploited to identify more accurate clinical markers for pro-arrhythmia. The identifications of the most prevalent phenotypes within various populations may further be employed within this computational framework, i.e., to establish predictive bounds of arrhythmogenesis of a relevant, specific sample population.

Funding

Authors JA-S, MV, AKB, and GH were supported by the European Union's Horizon 2020 research and innovation program under grant agreements No 823712 (Comp Biomed project, phase 2) and No 327777204 (SilicoFCM project). https://ec.europa.eu/

programmes/horizon2020/en/home. JA-S was awarded with computation time from PRACE-COVID-19 fast track pandemic response (project COVID1933) at the Joliot-Curie Rome supercomputer hosted by GENCI at CEA, France. https://prace-ri.eu/. JA-S was funded by a Ramon y Cajal fellowship (RYC-2017-22532), Ministerio de Ciencia e Innovacion, Spain. CB was funded by the Torres Quevedo Program (PTQ2018–010290), Ministerio de Ciencia e Innovacion, Spain. MV, AKB, GH, and CB are funded by the Spanish Neotec project EXP - 00123159/SNEO-20191113 Generador de corazones virtuales. The Visible Heart® Laboratories received research support from the Institute for Engineering in Medicine and Medtronic Inc. The funders had no role in study design, data collection and analysis, decision to publish, or preparation of the manuscript.

References

1. Cavalcanti AB et al (2020) Hydroxychloroquine with or without azithromycin in mild-to-moderate covid-19. N Engl J Med 383(21): 2041–2052
2. White NJ (2007) Cardiotoxicity of antimalarial drugs. Lancet Infect Dis 7(8):549–558
3. Ray WA et al (2012) Azithromycin and the risk of cardiovascular death. N Engl J Med 366(20):1881–1890
4. Salama G, Bett GC (2014) Sex differences in the mechanisms underlying long qt syndrome. Am J Phys Heart Circ Phys 307(5):H640–H648
5. Vink AS et al (2018) Effect of age and gender on the QTc-interval in healthy individuals and patients with long-QT syndrome. Trends Cardiovasc Med 28(1):64–75
6. Chen D et al (2020) Assessment of hypokalemia and clinical characteristics in patients with coronavirus disease 2019 in Wenzhou, China. JAMA Netw Open 3(6):e2011122–e2011122
7. Yang PC et al (2020) A computational pipeline to predict cardiotoxicity. Circ Res 126(8): 947–964
8. Bottino D et al (2006) Preclinical cardiac safety assessment of pharmaceutical compounds using an integrated systems-based computer model of the heart. Prog Biophys Mol Biol 90(1–3):414–443
9. Delaunois A et al (2021) Applying the CiPA approach to evaluate cardiac proarrhythmia risk of some antimalarials used off-label in the first wave of COVID-19. Clin Transl Sci 14(3): 1133–1146
10. Beattie KA et al (2013) Evaluation of an in silico cardiac safety assay: using ion channel screening data to predict QT interval changes in the rabbit ventricular wedge. J Pharmacol Toxicol Methods 68(1):88–96
11. Varshneya M et al (2021) Investigational treatments for COVID-19 may increase ventricular arrhythmia risk through drug interactions. CPT Pharmacometrics Syst Pharmacol 10(2): 100–107
12. Llopis-Lorente J et al (2020) In silico classifiers for the assessment of drug proarrhythmicity. J Chem Inf Model 60(10):5172–5187
13. Passini E et al (2021) The virtual assay software for human in silico drug trials to augment drug cardiac testing. J Computat Sci 52:101202
14. Okada J et al (2018) Arrhythmic hazard map for a 3d whole-ventricle model under multiple ion channel block. Br J Pharmacol 175(17): 3435–3452
15. Sahli Costabal F, Yao J, Kuhl E (2018) Predicting drug-induced arrhythmias by multiscale modeling. Int J Numer Methods Biomed Eng 34(5):e2964
16. Hwang M et al (2019) Three-dimensional heart model-based screening of proarrhythmic potential by in silico simulation of action potential and electrocardiograms. Front Physiol 10:1139
17. Okada JI et al (2021) Chloroquine and hydroxychloroquine provoke arrhythmias at concentrations higher than those clinically used to treat covid-19: a simulation study. Clin Transl Sci 14(3):1092–1100
18. Thomet U et al (2021) Assessment of proarrhythmogenic risk for chloroquine and hydroxychloroquine using the CiPA concept. Eur J Pharmacol 913:174632

19. Uzelac I et al (2021) Quantifying arrhythmic long QT effects of hydroxychloroquine and azithromycin with whole-heart optical mapping and simulations. Heart Rhythm O2 2(4):394–404
20. Passini E et al (2015) Mechanisms of pro-arrhythmic abnormalities in ventricular repolarisation and anti-arrhythmic therapies in human hypertrophic cardiomyopathy. J Mol Cell Cardiol 96:72–81
21. Muszkiewicz A et al (2016) Variability in cardiac electrophysiology: using experimentally-calibrated populations of models to move beyond the single virtual physiological human paradigm. Prog Biophys Mol Biol 120(1–3): 115–127
22. Yang PC, Clancy CE (2012) In silico prediction of sex-based differences in human susceptibility to cardiac ventricular tachyarrhythmias. Front Physiol 3:360
23. Fourcade L et al (2014) Bloc de branche gauche douloureux d'effort associé à la chimioprophylaxie antipaludique par chloroquine. Médecine et Santé Tropicales 24(3):320–322
24. Sacco F (2019) Quantification of the influence of detailed endocardial structures on human cardiac haemodynamics and electrophysiology using HPC. Doctoral thesis, Universitat Pompeu Fabra
25. Doste R et al (2019) A rule-based method to model myocardial fiber orientation in cardiac biventricular geometries with outflow tracts. Int J Numer Methods Biomed Eng 35(4): e3185
26. Santiago A et al (2018) Fully coupled fluid-electro-mechanical model of the human heart for supercomputers. Int J Numer Methods Biomed Eng 34(12):e3140
27. Margara F et al (2021) In-silico human electro-mechanical ventricular modelling and simulation for drug-induced pro-arrhythmia and inotropic risk assessment. Prog Biophys Mol Biol 159:58–74
28. Houzeaux G et al (2009) A massively parallel fractional step solver for incompressible flows. J Comput Phys 228(17):6316–6332
29. Vázquez M et al (2016) Alya: multiphysics engineering simulation toward exascale. J Computat Sci 14:15–27
30. Vázquez M et al (2011) A massively parallel computational electrophysiology model of the heart. Int J Numer Methods Biomed Eng 27(12):1911–1929
31. Uekermann B (2016) Partitioned fluid-structure interaction on massively parallel systems. Doctoral thesis, Technische Universität München
32. Casoni E et al (2015) Alya: computational solid mechanics for supercomputers. Arch Computat Methods Eng 22:557–576
33. O'Hara T et al (2011) Simulation of the undiseased human cardiac ventricular action potential: model formulation and experimental validation. PLoS Comput Biol 7(5):e1002061
34. Karypis G, Kumar V (1998) Multilevelk-way partitioning scheme for irregular graphs. J Parallel Distributed Comput 48(1):96–129
35. Gima K, Rudy Y (2002) Ionic current basis of electrocardiographic waveforms: a model study. Circ Res 90(8):889–896
36. Durrer D et al (1970) Total excitation of the isolated human heart. Circulation 41(6): 899–912
37. Dutta S et al (2017) Electrophysiological properties of computational human ventricular cell action potential models under acute ischemic conditions. Prog Biophys Mol Biol 129:40–52
38. Mirams GR et al (2011) Simulation of multiple ion channel block provides improved early prediction of compounds' clinical torsadogenic risk. Cardiovasc Res 91(1):53–61
39. Yang Z et al (2017) Azithromycin causes a novel proarrhythmic syndrome. Circ Arrhythm Electrophysiol 10(4):e003560
40. Collins KP, Jackson KM, Gustafson DL (2018) Hydroxychloroquine: a physiologically-based pharmacokinetic model in the context of cancer-related autophagy modulation. J Pharmacol Exp Ther 365(3):447–459
41. Capel RA et al (2015) Hydroxychloroquine reduces heart rate by modulating the hyperpolarization-activated current if: novel electrophysiological insights and therapeutic potential. Heart Rhythm 12(10):2186–2194
42. Demšar J et al (2013) Orange: data mining toolbox in python. J Mach Learn Res 14(1): 2349–2353
43. Lindeman RH (1980) Introduction to bivariate and multivariate analysis
44. Grömping U (2007) Relative importance for linear regression in r: the package relaimpo. J Stat Softw 17:1–27
45. Goff RP et al (2016) The novel in vitro reanimation of isolated human and large mammalian heart-lung blocs. BMC Physiol 16:1–9

46. Chinchoy E et al (2000) Isolated four-chamber working swine heart model. Ann Thorac Surg 70(5):1607–1614
47. Schmidt MM, Iaizzo PA (2018) The Visible Heart® project and methodologies: novel use for studying cardiac monophasic action potentials and evaluating their underlying mechanisms. Expert Rev Med Devices 15(7):467–477
48. Saleh M et al (2020) Effect of chloroquine, hydroxychloroquine, and azithromycin on the corrected QT interval in patients with SARS-CoV-2 infection. Circ Arrhythm Electrophysiol 13(6):e008662
49. Mercuro NJ et al (2020) Risk of qt interval prolongation associated with use of hydroxychloroquine with or without concomitant azithromycin among hospitalized patients testing positive for coronavirus disease 2019 (COVID-19). JAMA Cardiol 5(9):1036–1041
50. Zhao PA, Li P (2019) Transmural and rate-dependent profiling of drug-induced arrhythmogenic risks through in silico simulations of multichannel pharmacology. Sci Rep 9(1):1–9

Chapter 15

Effect of Muscle Forces on Femur During Level Walking Using a Virtual Population of Older Women

Zainab Altai, Erica Montefiori, and Xinshan Li

Abstract

Aging is associated with a greater risk of muscle and bone disorders such as sarcopenia and osteoporosis. These conditions substantially affect one's mobility and quality of life. In the past, muscles and bones are often studied separately using generic or scaled information that are not personal-specific, nor are they representative of the large variations seen in the elderly population. Consequently, the mechanical interaction between the aged muscle and bone is not well understood, especially when carrying out daily activities. This study presents a coupling approach across the body and the organ level, using fully personal-specific musculoskeletal and finite element models in order to study femoral loading during level walking. Variations in lower limb muscle volume/force were examined using a virtual population. These muscle forces were then applied to the finite element model of the femur to study the variations in predicted strains. The study shows that effective coupling across two scales can be carried out to study the muscle-bone interaction in elderly women. The generation of a virtual population is a feasible approach to augment anatomical variations based on a small population that could mimic variations seen in a larger cohort. This is a valuable alternative to overcome the limitation or the need to collect dataset from a large population, which is both time and resource consuming.

Key words Personalized musculoskeletal model, Muscle volume and force variation, Virtual population, Body-organ coupling, Femoral neck strain, Personal-specific finite element modeling

1 Introduction

Aging is associated with a combination of both muscle and bone loss. This is particularly relevant in countries such as UK, which faces an increasingly aging population with rising cost of medical care. In particular, osteoporosis-related fragility hip fracture is a major public health issue, which disproportionally affects post-menopausal women due to age- and hormone-related bone loss, making bones weaker and more prone to fracture. This is of particular concern with the femur, which is the largest bone of the body and connects to the hip joint to provide mobility.

Alexander Heifetz (ed.), *High Performance Computing for Drug Discovery and Biomedicine*, Methods in Molecular Biology, vol. 2716, https://doi.org/10.1007/978-1-0716-3449-3_15,

In the past, the mechanical changes of muscle and load-bearing bone due to aging were often studied separately due to limitations in computational power and available data. This limited our ability to understand the mechanical interaction between muscles and load-bearing bones, especially in a large elderly population. Multiple studies have pointed to an association between muscle loss (e.g., sarcopenia) and fall history in epidemiological studies [1–4]. These results demonstrated the strong mechanical interactions between these two structures in order to enable movements.

The ability to explore intra-personal muscle variations across multiple subjects is important in order to investigate how individual anatomical parameters (such as muscle size, length, path of action) affect kinematics and forces resulting at bones and joints. Previously, anthropometric parameters (e.g., body mass or body mass index) have been used to explain the variation in muscle anatomy. A recent study found large variations in muscle volume, length, physiological cross-sectional area in eleven post-menopausal women between body sides and across the cohort [5]. However, more than half of these variations remained unexplained. In addition, elderly individuals are known to experience muscle loss at very different rate and extent, which can substantially affect the ability of the muscles to produce force. Muscle strength loss at an older age has been explained by a number of factors such as a reduction in muscle mass [6] and an increase in the percentage of muscular fat with a reduction in physiological cross-sectional area of the muscle [6–8]. These factors can be generalized by a general reduction in muscle volume and hence the force produced by the muscle (assuming a linear relationship between muscle volume and muscle force). These variations will affect the estimated isometric force and consequently the loads exerted on femur, hip, and knee joints. A series of sensitivity analysis was conducted in order to understand how muscle forces would affect the strain predictions on the femur during level walking in five postmenopausal women [9]. Results from this initial study indicated substantial differences in the predicted strains across the cases. These findings suggest that intra-personal variations (muscle anatomy, force and bone strength) are substantial and should be further investigated in a larger population.

This study *aims to* investigate how muscle variability (muscle volume and hence muscle forces) affects femoral loading in a large virtual population using the previously developed personalized body-organ coupling procedure [9]. The variation in muscle forces was estimated at the body level, using a fully subject-specific musculoskeletal model, and data collected during gait analysis. The effect of muscle volume and force variation on the mechanical response of the femoral neck was then investigated using finite element modeling in order to identify the amount of variation in predicted femoral neck strain and any influential muscles during level walking.

2 Methods

This section is split into several parts. First, the participant information and data acquisition details were described. This then leads to the description of personalized musculoskeletal models at the body level and the creation of a virtual population of musculoskeletal models. Maximum isometric muscle forces were extracted from this virtual population to be used as input for the finite element model. The second half of this section describes the generation of a personal-specific finite element model of the femur. The femur's biomechanical responses were simulated and evaluated at two critical time points of the gait cycle through a sensitivity analysis using different muscles forces from the virtual population. This approach provides an elegant coupling between the body and the organ level using individual-specific data collected from multiple modalities as well as various engineering techniques across different scales. The general workflow is presented in Fig. 1.

2.1 Participants and Data Acquisition

This study used retrospective data collected as part of an EPSRC funded study (MultiSim and MultiSim2, EP/K03877X/1 and EP/S032940/1), which involved eleven post-menopausal women (69 ± 7 year, 159 ± 3 cm, 66.9 ± 7.7 kg). Inclusion criterion was having a bone mineral density T-score at the lumbar spine or total hip (whichever was the lower value) less than or equal to -1. Those who were obese (BMI > 35) or underweight (BMI < 18) were

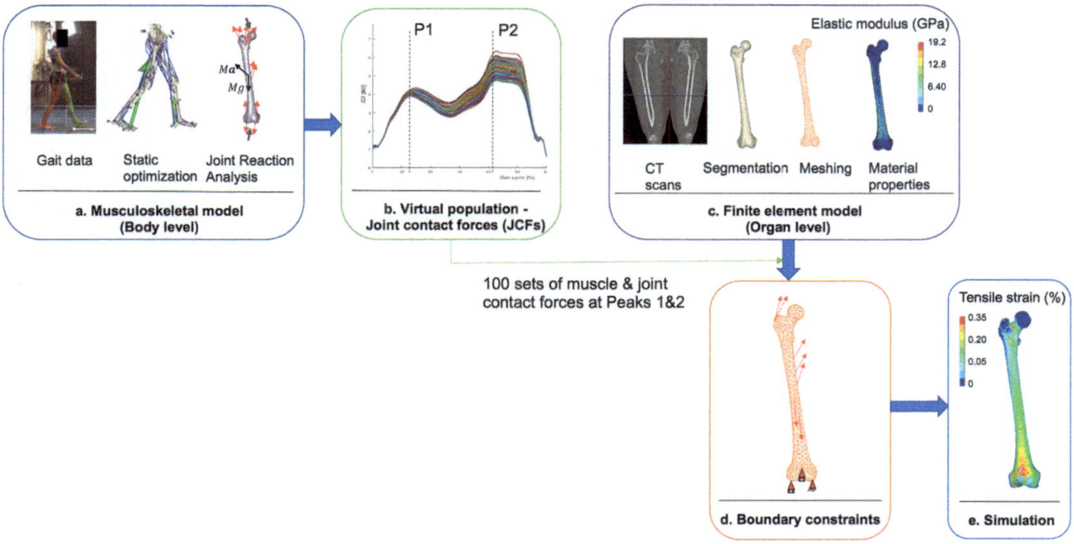

Fig. 1 Body-organ coupling pipeline showing the body-level musculoskeletal model (**a**), the organ-level finite element model of the femur (**c**), an example joint contact forces of the virtual population (**b**), the combined model with boundary conditions applied to the femur, (**d**) and a typical simulation result showing tensile strain distribution (**e**)

excluded. Ethics approval was obtained through the Health Research Authority of East of England (Cambridgeshire and Hertfordshire Research Ethics Committee, reference 16/EE/0049). Each participant was scanned in CT (GE LightSpeed 64 VCT) from the hip to the knee. The CT scan settings were tube current of 120 mA, tube voltage of 100 kVP, and a resolution of $0.742 \times 0.742 \times 0.625$ mm^3. Full lower limb MRIs were collected using a Magnetom Avanto 1.5 T scanner (Siemens). A T1-weighted sequence was used with a voxel size of $1.1 \times 1.1 \times 5.0$ mm^3 for the long bones and $1.1 \times 1.1 \times 3.0$ mm^3 for the joints. Each participant was also invited to the gait lab to collect 3D gait analysis data including marker trajectories and ground reaction forces.

2.2 Baseline Musculoskeletal Models

The 3D gait analysis data and MRI scans of the lower limb were used to build baseline monoliteral musculoskeletal models (Fig. 1a). These included four body segments (pelvis, femur, tibia, foot) articulated by an ideal ball-and-socket joint for the hip, and two ideal hinges, one for the knee and one for the ankle respectively, as well as 43 lower limb muscles. These 43 lower limb muscles corresponded to the lower-limb muscles included in the state-of-the-art OpenSim model gait2392 in the literature [10]. Out of these, 23 muscles can be reliably segmented and an online repository containing personalized muscle volume and lengths has been created [5].[1]

From these eleven women, one participant (70.5 year, 61.4 kg, 164 cm, BMI 22.8, *T*-score of −2.2) was selected to build a fully subject-specific musculoskeletal model (SSMM) and subject-specific finite element model of the femur (described later on). The SSMM was generated using personalized bone geometries and segment inertia generated from MRI [11]. The joint axes were determined via morphological fitting to the articular surface of the segmented bone geometries. The same set of muscles in gait2392 was included in the SSMM; but their origin, insertion, and via points were personalized based on the MRI scans. The personalized muscle information (origin and forces) were later used in the finite element simulation of the femur. Muscle length parameters were linearly scaled from gait2392 values in order to maintain their ratio to musculotendon length. Maximal isometric forces (F_{max}) of 23 lower limb muscles were personalized using MRI-segmented muscle volume (available from the aforementioned online repository). The F_{max} of the remaining 14 muscles (not available as they cannot be repeatedly measured, e.g., gluteus minimus, peroneus longus, etc.) was linearly scaled from gait2392 values based on body mass of the participant.

[1] Available from the online repository (https://doi.org/10.15131/shef.data.9934055.v3), comprising all eleven older women enrolled. Note that in this study, for ease of comparison, each of the adductor magnus, gluteus maximus, and gluteus medius muscles has been split into three bundles.

2.3 Virtual Population

One hundred variations of SSMM representing a virtual population of individuals were generated [12]. First, the mean and standard deviation of 23 muscle forces of the eleven women from the online repository were used to generate normal distributions of F_{max}, representing a virtual population of older women. Independent random sampling of each muscle force distribution was then carried out in order to create 100 sets of F_{max} for each of the 23 muscles. These were then used to characterize muscle properties of each variation of the SSMM described in the previous section. A convergence study was carried out in order to determine the number of sampling points to ensure less than 10% error in the normalized overlap of the resulting joint contact force (JCF) curve bands (example shown in Fig. 1b). This is to remove any anomalies that lead to a JCF pattern outside of the normal range.

2.4 Dynamic Simulations and Data Analysis

Hip, knee, and ankle joint angles and moments were computed from the baseline SSMMs using the OpenSim 3.3 [13] inverse kinematics and inverse dynamics tools relying on MATLAB API (v9.1, R2021b, Mathworks, USA). OpenSim-recommended good practice was followed. One hundred runs of static optimization (where the sum of muscle activations squared was minimized) and joint reaction analysis were carried out in order to estimate the individual muscle (maximum isometric) forces and associated normalized JCFs for each virtual case (Fig. 1b). Ideal moment generators (reserve actuators) were included for each degree of freedom in order to provide joint torque when muscle forces could not balance the external moments, although these ideal moment generators were made unfavorable to recruit by assigning them a unitary maximum force.

Maximum isometric muscle forces and resultant JCFs estimated by the SSMMs were extracted at two specific gait time points, corresponding to the first peak (P1) and the second peak (P2) of hip JCFs during one full gait cycle. These forces were then used as loading conditions to simulate the mechanical response of the femur using finite element modeling, as described below.

2.5 Finite Element Model of the Femur

For the one selected patient in Sect. 2.2, the right full femur was segmented from the CT scans using Mimics 20.0 (Materialise, Belgium). The segmented femur was then automatically meshed with 10-node tetrahedral elements (ICEM CFD 15.0, ANSYS Inc.) using an averaged element size of 3 mm (849,069 degrees of freedom), following the mesh convergence study reported in a previous study [9] using the same MultiSim cohort. Heterogeneous, elastic, and isotropic material properties were estimated from the CT attenuation and mapped onto the finite element mesh following a validated material-mapping protocol (Bonemat v3, Rizzoli Institute) [14–16] (Fig. 1c). The European Spine Phantom was used for bone density calibration. Note that CT scans were required here in order to provide personalized element-based Young's modulus estimation. Such information cannot be obtained through MRI scans.

2.6 Static Femoral Loading During Gait and Data Analysis

Using the finite element model of the femur generated above, muscle isometric forces estimated from the virtual population (representing 100 virtual elderly women) were applied to the model in order to investigate the effect on predicted femoral strain and mechanical behavior. In order to apply muscle isometric forces to the femoral model, the orientation of the muscle and joint forces were transformed from the MRI to the CT scans' reference frame using the Iterative Closet Point (ICP) algorithm [17]. Eighteen muscle forces were applied to the external surface of the finite element model of the femur as point loads (Fig. 1d). More details are described in Altai et al. [9]. Each muscle's attachment point was estimated by the SSMM. Forces were then applied to the nearest surface node on the finite element mesh. Relaxed kinematic constraints were applied at the distal end of the femur to prevent rigid body motion and were chosen to replicate the basic movements involved in walking considering the equilibrium of forces estimated by the SSMM. The most distal node of the medial condyle was completely fixed, while only the anterior-posterior and superior-inferior displacements of the most distal node at the lateral condyle were constrained (Fig. 1d). An extra node in the patella groove was constrained antero-posteriorly [9, 18, 19]. Hip and knee joint reaction forces (predicted by the finite element model) were used to verify that the imposed boundary conditions were appropriate and statically equivalent to applying the hip and knee JCFs estimated from the SSMM.

For each virtual subject, peak principal strains (e1 and e3) at the femoral neck were predicted at two time points (Peak1 and Peak2) corresponding to the first and second peak of the hip JCF curve. The predicted strains were averaged across the surface nodes using a circle of 3 mm radius, to follow the continuum hypothesis avoiding local effects of the load [20, 21]. The predicted strains were then compared to the previously published yield strain limit: 0.73% and 1.04% for tensile and compressive strain, respectively [22]. The location of the peak strains within the femoral neck region was also analyzed. And finally, the peak strain energy density (SED) was computed at Peak1 and Peak2. The femoral neck was chosen as the region of interest as fracture often occurs here during a sideway fall [23, 24]. This area is also away from the subtrochanteric region where hip muscles attach to the femur (and hence the locations of applied muscle forces).

The relation between the muscle forces estimated by the SSMM and the femoral neck strains predicted by the finite element models was investigated using Pearson's product-moment correlation. The correlation was considered to be moderate when correlation coefficient (r) is above 0.3, and strong when r is above 0.5, based on the hypothesis that a p-value below a threshold of 0.05 is significant. The statistical analysis was carried out in MATLAB API (v9.1, R2021b, Mathworks, USA).

All simulations were performed on the high-performance computing cluster at the University of Sheffield (ShARC) using ANSYS Mechanical APDL 19.1 (Ansys Inc., PA, USA). For each virtual subject, the computing time used to solve the static finite element model was less than 1 min for each selected gait time point.

2.7 Results

Across the 100 virtual subjects, the range of hip joint contact forces (normalized by body weight, BW) predicted by baseline SSMMs (using OpenSim) varied by up to 0.8 of the BW at Peak1 and 3.1 of the BW at Peak2, as shown in Fig. 2.

Using the finite element model of the femur, the principal strains from the 100 virtual subjects were predicted at Peak1 and Peak2 after applying muscle forces. At Peak1, the absolute maximum first and third principal strains (median ± SD) at the femoral neck were predicted to be 0.37 ± 0.016% and 0.41 ± 0.016%, respectively. At Peak2, the absolute maximum first and third principal strains were predicted to be 0.22 ± 0.038% and 0.27 ± 0.044%, respectively (Fig. 3). There is a wider variation in the predicted principal strains at Peak2 compared to Peak1. The predicted maximum strain energy density (SED, mean ± SD) at the

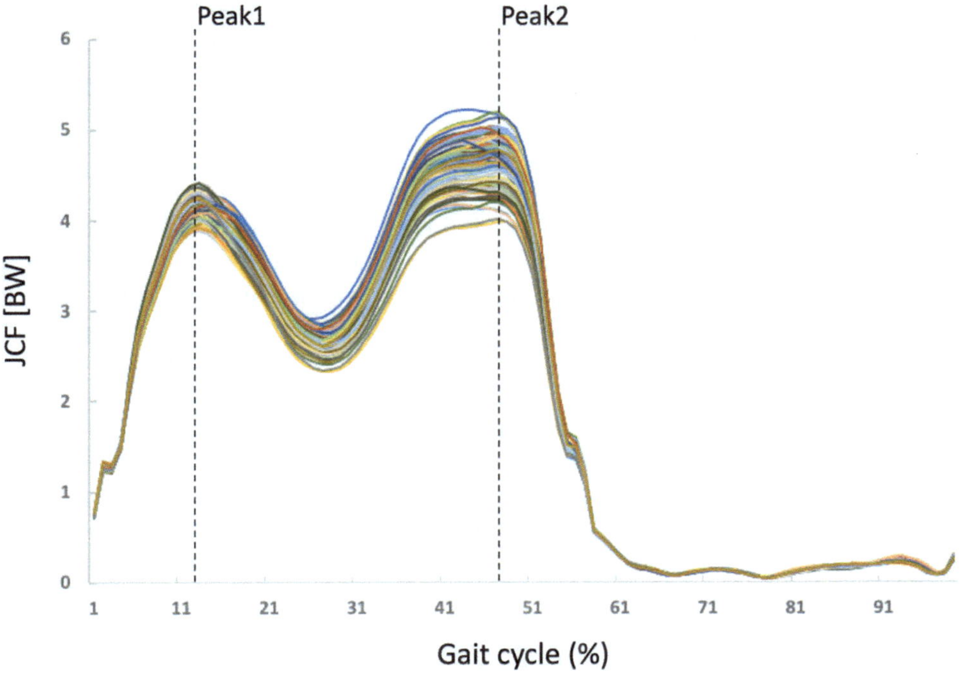

Fig. 2 Hip JCFs as predicted by the SSMM for the 100 virtual subjects. The two selected time points of the level walking cycle (Peak1 and Peak2) are indicated by dashed lines

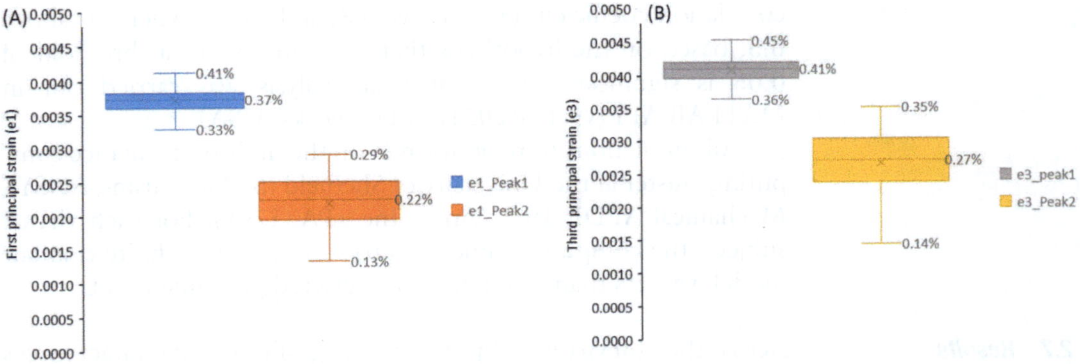

Fig. 3 Distribution of the predicted maximum first principal strains (**a**) and absolute maximum third principal strains (**b**) at Peak1 and Peak2 of one gait cycle for 100 virtual subjects

Table 1
Summary of information predicted at Peak1 and Peak2 of one gait cycle for the 100 virtual subjects. Hip joint contact forces (JCFs) were predicted by the musculoskeletal models. The absolute maximum first (e1) and third (e3) principal strains and strain energy density (SED) were predicted by the finite element models

Mean ± SD	Peak1	Peak2
Hip JCFs	0.8 BW	3.1 BW
e1 (median ± SD)	0.37 ± 0.016%	0.22 ± 0.038%
e3 (median ± SD)	0.41 ± 0.016%	0.27 ± 0.044%
SED (GPa) (mean ± SD)	4.57 ± 0.46	10.73 ± 4.43

BW stands for body weight of the selected subject

femoral neck was 4.57 ± 0.46 GPa at Peak1 and 10.73 ± 4.43 GPa at Peak2. The above information is summarized in Table 1.

For all subjects, potential failure (e.g., bone fracture) was predicted to occur under tension. The maximum first principal strains were consistently predicted at the superior-anterior aspect of the femoral neck region in the finite element model for all cases (Fig. 4).

For most muscles, no statistically significant correlation ($p > 0.05$) was found between muscle volumes (which gave rise to muscle force variations) and the predicted principal strains, as shown in Table 2. However, a few muscles showed significant correlations ($p < 0.05$), which are described here. At Peak1, the gluteus medius muscle bundles 2 and 3 had a moderate to strong positive correlation with the principal strains (r in the range of 0.48 to 0.64 for e1 and e3, $p < 0.001$). The gluteus medius muscle bundle 1 also had a moderate to strong positive correlation ($r = 0.51$ and 0.47 for e1 and e3, respectively, $p < 0.001$) at

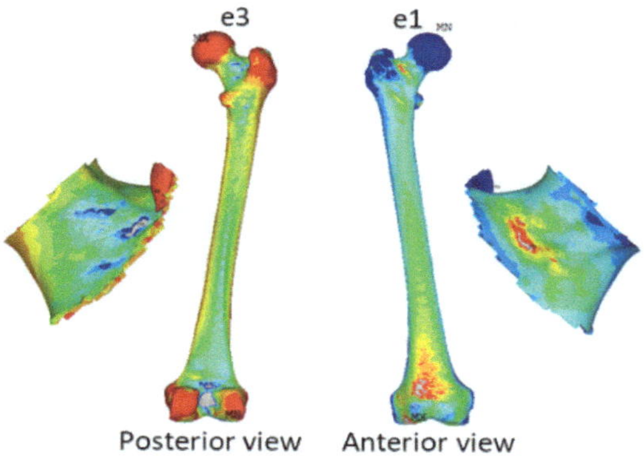

Fig. 4 Example distribution of the absolute maximum first (e1) and third principal (e3) strains predicted by the finite element model, shown in anterior and posterior views, with enlarged views for the femoral neck region (i.e., the region of interest)

Table 2
Correlation coefficients (*r*) between muscle volume (all from the right side) and the predicted absolute maximum first and third principal strains at the two peaks (Peak1 and Peak2) of the gait cycle. Blue: positive correlation. Red: negative correlation. White: no correlation ($p > 0.05$)

Muscle	Peak1		Peak2	
	e1	e3	e1	e3
Adductor brevis	-0.03	-0.03	-0.07	-0.08
Adductor longus	-0.05	-0.01	-0.01	-0.01
Adductor magnus 1	-0.17	-0.20	0.08	0.08
Adductor magnus 2	0.08	0.07	0.27	0.27
Adductor magnus 3	0.04	0.05	-0.06	-0.06
Biceps femoris long head	-0.14	-0.16	-0.14	-0.14
Biceps femoris short head	0.20	0.16	0.26	0.26
Gluteus maximus 1	-0.29	-0.26	-0.13	-0.10
Gluteus maximus 2	-0.04	-0.01	-0.09	-0.10
Gluteus maximus 3	0.04	0.08	-0.17	-0.17
Gluteus medius 1	-0.13	-0.24	0.51	0.47
Gluteus medius 2	0.64	0.63	0.04	0.04
Gluteus medius 3	0.48	0.52	-0.03	-0.02
Gracilis	0.03	0.02	0.11	0.10
Iliacus	-0.06	-0.03	-0.10	-0.09
Gastrocnemius lateralis	-0.04	-0.05	0.06	0.05
Gastrocnemius medialis	-0.12	-0.12	0.01	0.00
Peroneus brevis	-0.11	-0.09	-0.05	-0.06
Rectus femoris	-0.23	-0.19	-0.66	-0.68
Sartorius	-0.13	-0.13	-0.02	-0.03
Semimembranosus	0.09	0.05	0.07	0.07
Semitendinosus	0.20	0.19	0.21	0.22
Soleus	-0.18	-0.17	-0.09	-0.07
Tensor fasciae latae	-0.23	-0.21	-0.44	-0.46
Tibialis anterior	-0.08	-0.02	0.08	0.06
Tibialis posterior	0.14	0.11	0.19	0.21
Vastus intermedius	0.11	0.12	0.06	0.08
Vastus lateralis	0.00	-0.03	-0.19	-0.18
Vastus medialis	0.02	0.01	-0.16	-0.14

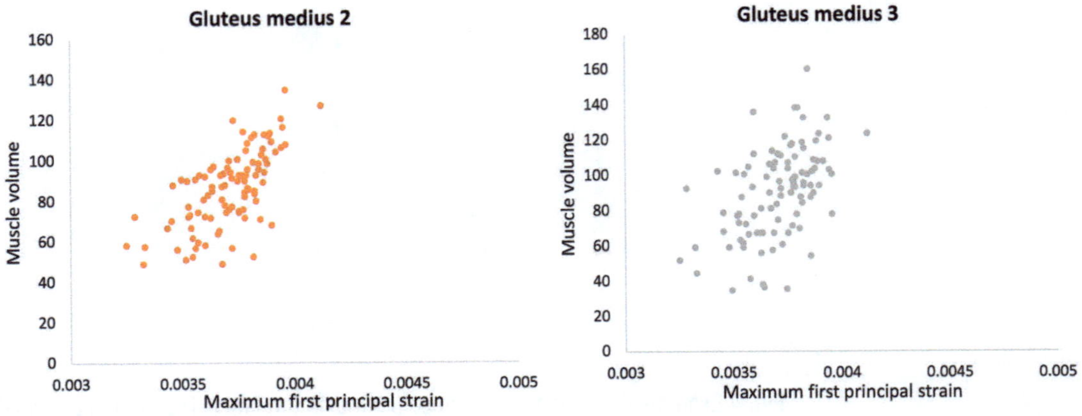

Fig. 5 Plot between the gluteus medius muscle (bundles 2 and 3) volume and maximum first principal strains at Peak1 for 100 virtual subjects

Peak2. At the second peak, two other muscles showed strong and moderate negative correlations with the principal strains: the rectus femoris ($r = -0.66$ and -0.68 for e1 and e3, respectively, $p < 0.001$) and the tensor fasciae latae ($r = -0.44$ and -0.46 for e1 and e3, respectively, $p < 0.001$). For illustration purpose, Fig. 5 shows the gluteus medius muscle volume (bundles 2 and 3) of the 100 virtual subjects plotted against the predicted maximum first principal strains.

2.8 Discussion

This study investigated the effects of muscle variability (muscle volume and hence muscle forces) on femoral loading in a virtual population using a personalized body-organ coupling approach. Multibody dynamic models were used to calculate F_{max} and JCFs, while finite element models of a full femur were used to predict principal strains induced at the femoral neck during one normal walking cycle.

Changes in individual F_{max} caused variations in the estimated JCFs that were broadly similar to those reported by previous studies, with some small differences [11, 25, 26]. As previous studies were mainly based on values derived from cadavers (not live individuals), leading to differences in the definition of the joint axes and muscle paths, and consequently the joint kinematics. In addition, some authors varied all muscle forces simultaneously in the same manner, whereas sampling was carried out in the current study. These differences in approaches could explain the slight differences in results.

The median values of the peak strain (0.22% at Peak1 and 0.37% at Peak2) predicted for the 100 virtual subjects were found to be slightly higher than those found in previous studies [27, 28]. Kersh et al. [27] reported a median peak strain of less than 0.2% at the femoral neck during walking for twenty subjects, and Martelli et al. [29] reported an average tensile strain of 0.25% at the femoral neck during walking based on a single subject. Both the current study and Kersh et al. (2018) study predicted the maximum strains at the femoral neck region. The difference in predicted strains could be explained by the study design. Kersh et al. (2018) reported values of effective strains, which were calculated from the strain energy density and the Young's modulus of each element across the proximal femur. This is different to the peak principal strains reported here on the surface of the femur. The number of subjects used in this study (100 virtual subjects) was much larger than the previous studies: 20 subjects in Kersh et al. (2018) and 1 subject in Martellie et al. (2014). Furthermore, the current study used fully personalized musculoskeletal and finite element models, while Kersh et al. (2018) scaled the musculoskeletal models from the OpenSim generic model. The MultiSim cohort consisted of elderly women in the osteopenia range (T scores ranged from -2.2 to -1.2), while the subjects in Kersh et al. (2018) ranged between healthy and osteopenia.

Although the predicted strain values were different in these studies, the peak strains in all cases were notably below the fracture threshold [22]. This finding supports the theory that, in the absence of trauma, bone fracture is only likely to occur when people with weak bones undertake tasks or suffer from accidents that result in high loads.

As shown in Table 2, strong correlations were found between a few muscles and the predicted peak principal strains. For the rest of the muscles, no strong correlation was found between muscle volumes and the strains, and most correlation coefficients ranged between ± 0.2 and 0 (Table 2). The highest positive correlation coefficient ($r = 0.64$, $p < 0.001$) was observed between the gluteus medius muscles and the first principal strain at Peak1 (Fig. 5). This is in agreement with previous studies where the gluteus muscle group was found to contribute to most of the loading on the femur during walking [9], as well as when carrying out other tasks such as stair ascent, descent, and jumping [27]. In contrast, the highest negative correlation coefficient was observed for the rectus femoris muscle ($r = -0.66$, $p < 0.001$ for e1; and $r = -0.68$, $p < 0.001$ for e3) followed by the tensor fasciae latae ($r = -0.44$, $p < 0.001$ for e1; and $r = -0.46$, $p < 0.001$ for e3) at Peak2. The observed negative correlations suggest that, around the end of the stance phase (toe-off), rectus femoris and tensor fasciae latae muscles play a role in lowering the bending strains within the femoral neck by reducing the bending moment on the proximal femur [30]. A recent study reported similar findings to the current study, where the gluteus medius muscle was found to be the most influential muscle for more than 40% of the gait cycle, followed by the rectus femoris (16%) and tensor fasciae latae (10%) [31]. As the use of musculoskeletal models coupled with finite element models continues to grow, further investigations are necessary to understand the contribution of each particular muscle on loading the femoral neck during walking.

The highest strains were consistently predicted at the femoral neck region for all virtual subjects. This is likely because the cortical bone at the middle of the femoral neck (current region of interest in this study) is thinner than the rest of the proximal femur. For example, the thicker cortex at the trochanteric region can accommodate higher strains than the femoral neck due to the fact that most hip muscles are inserted around this area. This could have implications in hip fractures due to sideway falls because the thinner cortical bone at the femoral neck cannot bear the substantial bending that occurs due to the indirect impact on the greater trochanter.

It is known that muscle forces constitute the largest loads on load-bearing bones (except for case of trauma), which in turn facilitate bone growth, development, and remodeling [32]. The "mechanostat" theory states that if the imposed force by muscles exceeds a particular threshold, then bone formation occurs in favor of bone resorption [32]. This is reflected by the fact that with smaller gluteus medius muscle forces, the predicted strains in femur tend to be lower. This indicates that the femur could be more affected by changes in major hip muscles such as the gluteus medius during aging. A general reduction in muscle size and power during aging will lead to a reduction in typical peak voluntary mechanical loading, and consequently remodeling of the bone with reduced bone strength [32]. This further illustrates the important mechanical interplay between muscles and load-bearing bones.

The current study has a number of limitations. Only 23 of the 43 muscles included in the lower limb model were personalized in this study. This was due to the lack of repeatability found by Montefiori et al. [5] when segmenting the remaining muscles. Automated algorithms based on machine learning or statistical shape modeling approaches could be developed in future to provide faster and more accurate estimations of muscle volumes. This would enable further studies on the role of the remaining 14 muscles (such as the psoas muscle) that were not included in this study.

In an attempt to preserve subject-specificity in the muscle parameters, physiological cross-sectional area was calculated from muscle volume and length, instead of being evaluated directly from higher resolution MRIs [33]. Although muscle volume was altered for each muscle in each virtual subject, the specific muscle path remained unchanged. Anatomical variations in muscle path could lead to changes in moment arm. A change in muscle volume and associated force is also expected to cause variations in resulting kinematics, but this was not accounted for in this study. Only the resulting normalized joint contact forces over one gait cycle were checked to ensure that the results fell within a reasonable range.

Although a large virtual cohort of 100 subjects were used, this data was based on muscle volume measurements obtained from a small cohort of 11 elderly women with no known conditions that affected bone or any neurological disorders. This means the virtual population is only representative of elderly women who are relatively healthy in muscle functions. Therefore, the cohort may not be representative of those who suffer from muscle loss associated with sarcopenia or other musculoskeletal diseases.

All finite element models in the current study were created based on a single femur geometry from one selected subject. Information of this same subject was also used to generate the SSMM of all 100 virtual subjects. Therefore, the variation in femur geometry in combination with muscle volume variations was not considered. Collecting fully personalized data for such a large number of subjects is challenging. Future work can include generating such variations in the finite element models using similar approaches as presented here for the musculoskeletal models.

This study focused on only one gait cycle during level walking, although a ten-meter-long walkway was considered during data collection to ensure a natural cadence of the individual while walking, and hence minimizing variations. It is known that the gait pattern of an individual may differ in two sequential gait cycles [34], producing different joint, muscle, and ground reaction forces. These changes could induce different strain levels and mechanical responses in the femoral neck. The investigation of the gait variability is beyond the scope of the current study. However, future studies should consider gait variations across different physiological loading conditions and quantify the range of changes in predicted femoral strain.

Acknowledgments

All authors received funding from the EPSRC Frontier Engineering Awards, MultiSim and MultiSim2 projects (EP/K03877X/1 and EP/S032940/1). XL has also received funding from the European Commission H2020 program through the CompBioMed and CompBioMed2 Centres of Excellence and the SANO European Centre for Computational Medicine (Grants N. H2020-EINFRA-2015-1/ 675451, H2020-INFRAEDI-2018-1/823712 and H2020-WIDESPREAD-2018-01/857533).

References

1. Clynes MA, Edwards MH, Buehring B et al (2015) Definitions of sarcopenia: associations with previous falls and fracture in a fopulation fample. Calcif Tissue Int 97:445–452
2. Woo N, Kim SH (2013) Sarcopenia influences fall-related injuries in community-dwelling older adults. Geriatr Nurs (Minneap) 35: 279–282
3. Yamada M, Nishiguchi S, Fukutani N et al (2013) Prevalence of sarcopenia in community-dwelling Japanese older adults. J Am Med Dir Assoc 14:911–915
4. Edwards MH, Dennison EM, Aihie Sayer A et al (2015) Osteoporosis and sarcopenia in older age. Bone 80:126–130

5. Montefiori E, Kalkman BM, Henson WH et al (2020) MRI-based anatomical characterisation of lower-limb muscles in older women. PLoS One 15:e0242973
6. Larsson L, Grimby G, Karlsson J (1979) Muscle strength and speed of movement in relation to age and muscle morphology. J Appl Physiol Respir Environ Exerc Physiol 46:451–456
7. Rahemi H, Nigam N, Wakeling JM (2015) The effect of intramuscular fat on skeletal muscle mechanics: implications for the elderly and obese. J R Soc Interface 12:20150365
8. Yoshiko A, Hioki M, Kanehira N et al (2017) Three-dimensional comparison of intramuscular fat content between young and old adults. BMC Med Imaging 17:1–8
9. Altai Z, Montefiori E, van Veen B et al (2021) Femoral neck strain prediction during level walking using a combined musculoskeletal and finite element model approach. PLoS One 16:e0245121
10. Delp SL, Loan JP, Hoy MG et al (1990) An interactive graphics-based model of the lower extremity to study orthopaedic surgical procedures. IEEE Trans Biomed Eng 37:757–767
11. Modenese L, Montefiori E, Wang A et al (2018) Investigation of the dependence of joint contact forces on musculotendon parameters using a codified workflow for image-based modelling. J Biomech 73:108–118
12. Benemerito I, Griffiths W, Allsopp J et al (2021) Delivering computationally-intensive digital patient applications to the clinic: an exemplar solution to predict femoral bone strength from CT data. Comput Methods Prog Biomed 208:106200
13. Delp SL, Anderson FC, Arnold AS et al (2007) OpenSim: open-source software to create and analyze dynamic simulations of movement. IEEE Trans Biomed Eng 54:1940–1950
14. Schileo E, Dall'Ara E, Taddei F et al (2008) An accurate estimation of bone density improves the accuracy of subject-specific finite element models. J Biomech 41:2483–2491
15. Taddei F, Pancanti A, Viceconti M (2004) An improved method for the automatic mapping of computed tomography numbers onto finite element models. Med Eng Phys 26:61–69
16. Morgan EF, Bayraktar HH, Keaveny TM (2003) Trabecular bone modulus-density relationships depend on anatomic site. J Biomech 36:897–904
17. Kjer H, Wilm J (2010) Evaluation of surface registration algorithms for PET motion correction
18. Polgár K, Gill HS, Viceconti M et al (2003) Strain distribution within the human femur due to physiological and simplified loading: finite element analysis using the muscle standardized femur model. Proc Inst Mech Eng H J Eng Med 217:173–189
19. O'Rahilly R, Swenson R, Muller F et al (2008) Chapter 18: Posture and locomotion. In: Dartmouth Medical School (ed) Basic human anatomy
20. Qasim M, Farinella G, Zhang J et al (2016) Patient-specific finite element estimated femur strength as a predictor of the risk of hip fracture: the effect of methodological determinants. Osteoporos Int 27:2815–2822
21. Helgason B, Taddei F, Pálsson H et al (2008) A modified method for assigning material properties to FE models of bones. Med Eng Phys 30:444–453
22. Bayraktar HH, Morgan EF, Niebur GL et al (2004) Comparison of the elastic and yield properties of human femoral trabecular and cortical bone tissue. J Biomech 37:27–35
23. Altai Z, Qasim M, Li X, Viceconti M (2019) The effect of boundary and loading conditions on patient classification using finite element predicted risk of fracture. Clin Biomech 68:137–143
24. Verhulp E, van Rietbergen B, Huiskes R (2008) Load distribution in the healthy and osteoporotic human proximal femur during a fall to the side. Bone 42:30–35
25. Martelli S, Valente G, Viceconti M, Taddei F (2015) Sensitivity of a subject-specific musculoskeletal model to the uncertainties on the joint axes location. Comput Methods Biomech Biomed Eng 18:1555–1563
26. Navacchia A, Myers CA, Rullkoetter PJ, Shelburne KB (2016) Prediction of in vivo knee joint loads using a global probabilistic analysis. J Biomech Eng 138:031002
27. Kersh ME, Martelli S, Zebaze R et al (2018) Mechanical loading of the femoral neck in human locomotion. J Bone Miner Res 33:1999–2006
28. Martelli S, Kersh ME, Schache AG, Pandy MG (2014) Strain energy in the femoral neck during exercise. J Biomech 47:1784–1791
29. Martelli S, Pivonka P, Ebeling PR (2014) Femoral shaft strains during daily activities: implications for atypical femoral fractures. Clin Biomech 29:869–876
30. Taylor M, Tanner KE, Freeman MAR, Yettram A (1996) Stress and strain distribution within the intact femur: compression or bending? Med Eng Phys 18:122–131

31. Benemerito I, Montefiori E, Marzo A, Mazzà C (2022) Reducing the complexity of musculoskeletal models using gaussian process emulators. Appl Sci 12:12932
32. Frost HM (2003) Bone's mechanostat: a 2003 update. Anat Rec Part A Discov Mol Cell Evol Biol 275:1081–1101
33. Handsfield GG, Meyer CH, Hart JM et al (2014) Relationships of 35 lower limb muscles to height and body mass quantified using MRI. J Biomech 47:631–638
34. Beauchet O, Annweiler C, Lecordroch Y et al (2009) Walking speed-related changes in stride time variability: effects of decreased speed. J Neuroeng Rehabil 6:32

Chapter 16

Cellular Blood Flow Modeling with HemoCell

Gabor Zavodszky, Christian Spieker, Benjamin Czaja, and Britt van Rooij

Abstract

Many of the intriguing properties of blood originate from its cellular nature. Bulk effects, such as viscosity, depend on the local shear rates and on the size of the vessels. While empirical descriptions of bulk rheology are available for decades, their validity is limited to the experimental conditions they were observed under. These are typically artificial scenarios (e.g., perfectly straight glass tube or in pure shear with no gradients). Such conditions make experimental measurements simpler; however, they do not exist in real systems (i.e., in a real human circulatory system). Therefore, as we strive to increase our understanding on the cardiovascular system and improve the accuracy of our computational predictions, we need to incorporate a more comprehensive description of the cellular nature of blood. This, however, presents several computational challenges that can only be addressed by high performance computing. In this chapter, we describe HemoCell (https://www.hemocell.eu), an open-source high-performance cellular blood flow simulation, which implements validated mechanical models for red blood cells and is capable of reproducing the emergent transport characteristics of such a complex cellular system. We discuss the accuracy and the range of validity, and demonstrate applications on a series of human diseases.

Key words Blood rheology, Computational Fluid Dynamics, Cellular Blood Simulation, High-Performance Computation, Lattice Boltzmann method, Immersed Boundary Method, Microfluidics

1 The Cellular Properties of Blood

Blood is strongly linked to many of the physiological processes in the human body. Its major three functions can be categorized as transportation, protection, and regulation [1]. It transports oxygen, nutrients, and various wastes of the metabolic functions to maintain normal functioning of tissues all around our body. It also takes a significant role in most of our protective functions, including immune processes and hemostatic mechanisms. Finally, it regulates the balance of body fluids and cellular pH tension, and maintains overall thermobalance.

To be capable of providing its many functions, the composition of blood is far from simple. It is a dense suspension of various cells immersed in blood plasma, such as red blood cells (RBCs), platelets

Alexander Heifetz (ed.), *High Performance Computing for Drug Discovery and Biomedicine*, Methods in Molecular Biology, vol. 2716, https://doi.org/10.1007/978-1-0716-3449-3_16,

(PLTs), and white blood cells (WBCs) [2]. Blood plasma is usually regarded as a Newtonian incompressible fluid containing water and a series of proteins, hormones, nutrients, and gases. Under some circumstances, the effect of these solutes can become significant and lead to non-Newtonian effects such as plasma hardening [3]. In this chapter, we will not consider these plasma effects. RBCs are by far the most numerous of the cells in blood with a normal volume fraction or hematocrit of 45% in adults. They have a biconcave shape that has a large surface to volume ratio, which facilitates the exchange of oxygen by increasing the reaction surface. The diameter of these cells when undeformed is 6–8 μm with a 2–3 μm thickness. Their deformability originates from their structure that consists of a lipid bilayer that can dynamically attach and detach from the underlying supporting cytoskeleton formed by elastic spectrin proteins connected to each other via actin filaments. At low strain, the membrane behaves similarly to elastic solids; however, at high shear deformation, it behaves more like a fluid [4]. The intracellular fluid, the cytosol, is composed of a hemoglobin solution that has a five times higher viscosity compared to blood plasma. A single drop of blood contains approximately 150 million RBCs (*see* Fig. 1).

The dynamics and deformability of RBCs are important, as they give rise to the unique bulk properties of blood. The general behavior of these cells in (homogeneous) shear can be characterized with three separate regimes. At low shear rates, RBCs tumble and flop retaining their original biconcave shape [4]. The increase of shear leads to a swinging motion regime, and finally at higher shear rates, the membrane deforms, (partially) detaches from the

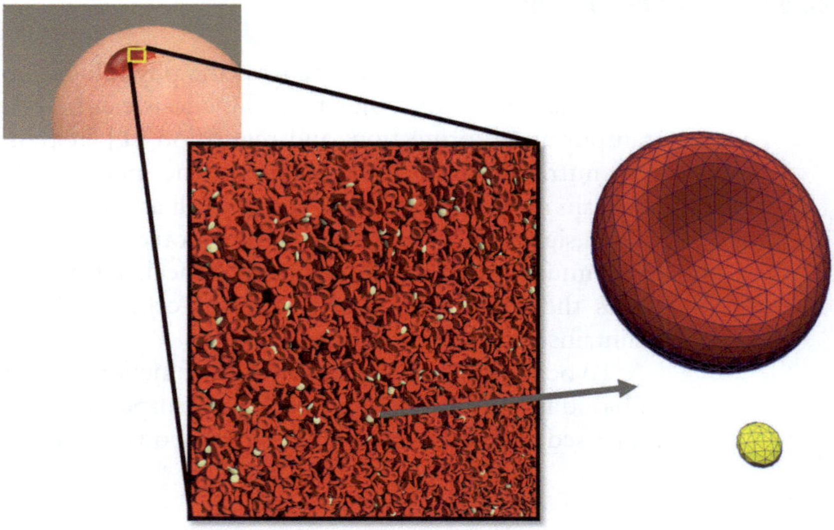

Fig. 1 A drop of blood modeled as a suspension of red blood cells and platelets

cytoskeletal structure, and starts to rotate around like the tread of a tank [5]. In such higher shear flows, the RBCs tend to line up and align with the flow, which in turn leads to a drop in bulk viscosity. This shear-thinning behavior was observed in pure shear (far from walls) in the well-known Chien experiments [6]. These also showed a steep increase in viscosity as the shear rate falls below 10 s^{-1} that later turned out to be caused by the aggregation of RBCs into column-like structures (rouleaux) [7], which has been shown to be induced by external plasma protein pressure [8]. The tendency of RBCs to migrate away from the vessel walls [9] leads to a different variety of shear-thinning. As blood flows through smaller vessels, the apparent viscosity drops, as denoted by the Fåhraeus-Lindqvist effect. As the RBCs vacate the vicinity of the vessel wall, they create a plasma-rich, red cell-free layer (CFL), which can effectively act as a lubrication layer. As we go toward smaller vessels, the relative effect of this layer grows, leading to an overall decreasing viscosity [10]. These observations were followed by decades of experiments that were later aggregated and shown to be consistent in the famous work of Pries [11]. As RBCs move toward the center of the vessel, they create a highly non-homogeneous hematocrit distribution that peaks in the middle of the channel and goes to zero next to the wall (CFL). This results in an uneven viscosity distribution and causes a departure from the Poiseuille profile that characterizes stationary Newtonian channel flows [12].

Given that these processes were studied mainly under synthetic conditions (in straight glass tubes or in pure shear) that are not representative of the human circulation, the full extent of these phenomena is still not completely known. The suspension nature of whole blood therefore needs to be taken into consideration in order to properly study the processes, both physiological and rheological, that occur in blood flow.

2 Methods: Accurate Computational Modeling of Blood Flows

2.1 Simulating Blood on a Cellular Scale

Resolving the complete rheology of blood flows in experimental (in vitro or in vivo) settings has several limitations. These are primarily caused by the limited spatiotemporal resolution of our current measurement technology and by the sheer amount of information necessary to capture the full characteristics of cellular blood. One way to overcome these limitations is by applying cellular flow simulations, that is, moving the experiments in silico. In recent years, many numerical approaches have been developed to simulate the cellular nature of blood [13]. These typically follow the same structure and contain three major components: a fluid model that reproduces the flow of plasma, a mechanical model that describes the deformation of cells, and fluid-structure interaction method that couples these two efficiently. The choice of numerical technique and its implementation varies, and most combinations have specific advantages and disadvantages that can make them suitable for a given research question. Some of the most widely used solutions include discrete particle dynamics (DPD) [14], lattice Boltzmann method (LBM) in combination with finite element (FEM) cells coupled by the immersed boundary method (IBM) [15], or smoothed particle hydrodynamics (SPH) for all three components [16]. In the following, we describe HemoCell, an open-source cellular blood flow simulation code that is designed with the aim of reproducing various microfluidic scenarios to complement experimental measurements. In order to reach this goal, the choice of numerical methods should fulfil a set of requirements:

1. *Enable large-scale flows.* In order to match experimental settings including the geometry (e.g., in a bleeding chip), a large number of cells must be simulated efficiently. Therefore, HemoCell was built with great scalability in mind to allow efficient simulation of a few dozen cells on a laptop up to millions of cells on the largest supercomputers.
2. *Stability at high shear rates.* Many of the investigated phenomena include biomechanical processes as a fundamental component (e.g., initial stages of thrombus formation). To capture these, the simulations need to reproduce realistic flow velocities that often translate to high shear rates locally. A unique feature of HemoCell is that the computational methods are fine-tuned to allow numerical stability at sustained high deformations.
3. *Advanced boundary conditions.* Matching microfluidic scenarios necessitate the support of a series of boundary conditions that can generate constant cell influx, produce pure shear, or enable rotating boundaries in a cellular suspension. Furthermore, the solution needs to be able to represent complex geometric boundaries, such as the surface of a growing thrombus.

To fulfil these requirements, HemoCell is designed with three main components: lattice Boltzmann method for the plasma flow, discrete element method (DEM) for the cellular mechanics, and IBM to create a flexible and accurate coupling between these. The structure of HemoCell is outlined in Fig. 2.

2.2 Simulating Fluid Flow with the Lattice Boltzmann Method

Blood plasma in HemoCell is modeled as an incompressible Newtonian fluid, and the governing equations (Navier-Stokes) are solved with the lattice Boltzmann method. This method is known to be able to accurately capture flow in complex vascular geometries and it is well suited to high-performance parallel execution [17].

Historically, this method is an incremental evolution of the lattice gas automata, and the theoretical bases were formulated by Bhatnagar, Gross, and Krook (BGK) [18]. For a detailed review, the reader is referred to the work of Chen [19]. Using their

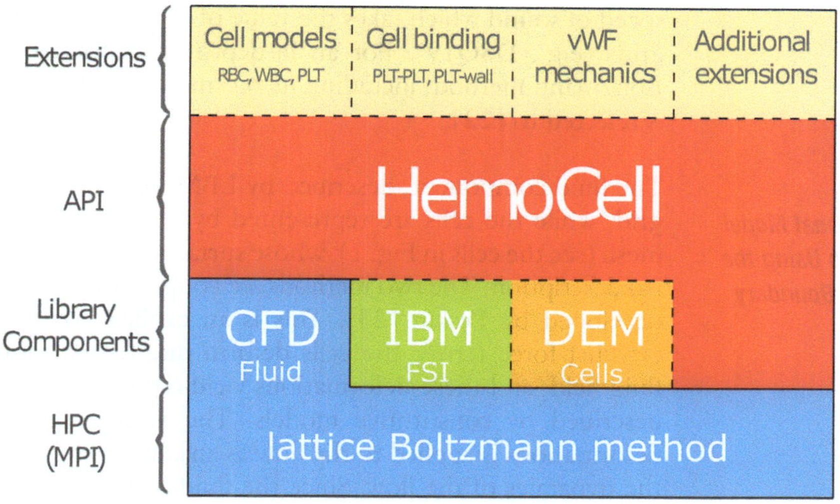

Fig. 2 Outline and structure of the major components of HemoCell

collision operator, the core equation of the LBM method represents distribution functions in lattice space (i.e., discrete time, space, and velocity directions):

$$f_i\left(\vec{x}+\vec{e_i},t+1\right)=f_i\left(\vec{x},t\right)+\frac{1}{\tau}\left(f_i^{\mathrm{eq}}\left(\vec{x},t\right)-f_i\left(\vec{x},t\right)\right),$$

where the i running index denotes the possible discrete velocity directions given by the actual grid model, f_i^{eq} represents the equilibrium distribution function, e_i is the direction of the selected velocity, and τ is the relaxation time of the kinetic system. The fluid density ρ and the macroscopic velocity $\vec{u}$ can be recovered at any lattice site from the first two moments of the distribution function:

$$\rho=\sum\nolimits_i f_i$$

$$\vec{u}=\frac{1}{\rho}\sum\nolimits_i f_i\vec{e_i}$$

The equilibrium distribution function (f_i^{eq}) follows the Boltzmann distribution in the form described by [20]:

$$f_i^{eq}\left(\vec{x},t\right)=w_i\rho\left[1+3\left(\vec{e_i}\vec{u}\right)+\frac{9}{2}\left(\vec{e_i}\vec{u}\right)^2-\frac{3}{2}\overrightarrow{u^2}\right],$$

where w_i denotes the numerical grid-dependent weight values. Applying the expansion described by Chapman and Enskog in the limit of long wavelengths (or low frequencies) [21], from the above-defined system, the Navier-Stokes equation for incompressible flows can be recovered with an ideal equation of state: $p(\rho)=\rho c_s^2$ and a kinematic viscosity of $\nu=c_s^2\left(\tau-\frac{1}{2}\right)$, where $p(\rho)$ stands for the pressure while c_s is the numerical grid-dependent speed of sound which takes the value of $\frac{1}{\sqrt{3}}$ for the most often used grids (e.g., D3Q19). For an in-depth description of the lattice Boltzmann method, including its recent advancements, the reader is referred to [22].

2.3 The Computational Model of the Cells Using the Immersed Boundary Method

The fluid component, described by LBM, operates on an Eulerian grid, while the cells are represented by a triangulated membrane mesh (see the cells in Fig. 1), whose surface vertices have a Lagrangian description. The two methods are coupled together by the IBM developed by Peskin [23]. This is an explicit coupling through external force terms: the cells deform due to the motion of the fluid field, and these deformations yield a non-stress free cell state described by constitutive models. The force response to these deformations is applied to the flow as an external force, influencing the dynamics of the flow. Since the fluid and cellular components are described on different numerical grids, the coupling steps include linear interpolation in both of these steps. A crucial question is how to describe the force response of cells to deformation. There are several existing solutions [24, 25] all based on approximating the mechanics of the cytoskeletal structure of the cells. In HemoCell, this fundamental idea is expanded to differentiate between small deformations that are expected to yield linear reaction based on the response of the bilipid membrane, and large deformations that yield highly nonlinear response that combines contribution from both the membrane and the cytoskeletal structure. The overall force response of this constitutive model comprises four major components:

$$F_{\text{total}} = F_{\text{link}} + F_{\text{bend}} + F_{\text{area}} + F_{\text{volume}}.$$

These force components correspond to a divers set of possible deformations of the cell and its membrane. The first component (F_{link}) relates to the stretching of the spectrin filaments in the cytoskeletal structure. The second one (F_{bend}) accounts for the bending rigidity of the membrane combined with cytoskeletal response for large deformations. The third (F_{area}) maintains the incompressibility of the bilipid membrane, while the last one (F_{volume}) ensures quasi-incompressible volume. The parameters of these forces are derived from limiting behavior of the cells and fitted to experimental measurements (including optical tweezer stretching, Wheeler test). Finally, the free parameters are finetuned for numerical stability. A more in-depth description of these forces and their numerical implementation can be found in [26]. The resulting computational model was thoroughly validated in a series of single-cell and many-cell experiments for both healthy and diabetic blood [27]. To ensure the robustness of the model, it was subjected to detailed sensitivity analysis and uncertainty quantification [28].

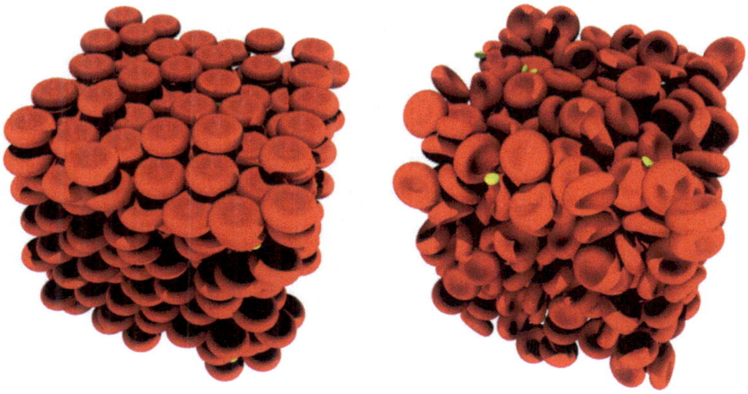

Fig. 3 Two randomized initial conditions. Left, only the positions are randomized; right, both the positions and the orientations are randomized

2.4 Creating Initial Conditions for Cellular Flow

Creating the initial conditions for a cellular suspension is a challenging task. Under physiologic conditions, stationary blood is well mixed, and the position and the orientation of the cells are both random uniform (see Fig. 3).

Naïve randomization routines are only efficient in low-hematocrit regimes. As the volume ratio of cells increases, avoiding overlaps is becoming more difficult. This practically leads to a dense packing problem of irregular cellular shapes. Since HemoCell is designed with large-scale applications in focus, the solution to this problem must also be performant enough to be able to handle up to $O(10^6)$ cells of various types. The way to reduce computational cost efficiently is to utilize surrogate objects that are easier to handle. In this case, the implementation carries out a dense packing of ellipsoidal shapes that each encompass a single cell. These ellipsoids are initialized with random position and orientation, after which an iterative method (the *force-bias method*) computes their overlap and applies repulsive forces between each overlapping pair proportional to their overlapping volume:

$$F_{ij} = \delta_{ij} p_{ij} \frac{\vec{r_j} - \vec{r_i}}{\left| \vec{r_j} - \vec{r_i} \right|},$$

where δ_{ij} equals 1 if there is an overlap between particle i and j and 0 otherwise, while p_{ij} is a potential function proportional to the overlapping volume. This method is capable of efficiently initializing large, well-mixed cellular domains, while reaching high packing density (i.e., hematocrit) values [29].

2.5 Advanced Boundary Conditions

Every computational model that has a finite domain requires boundary conditions. These define the behavior at the borders of the domain or at borders of different regimes within the domain. A trivial example implemented even in simple fluid flow simulations is the no-slip boundary condition. In the case of cell-resolved flow models like HemoCell, additional boundary conditions need to be introduced to cover cell-to-cell, cell-to-fluid, and cell-to-domain boundary interactions. To enable diverse applications for HemoCell simulations, e.g., in more arbitrary geometries such as curved and bifurcated blood vessels, more advanced boundary conditions are necessary. Here, three boundary conditions implemented in the HemoCell framework, namely the periodic pre-inlet, the rotating velocity field boundary, and the Lees-Edwards boundary condition [30], are discussed along with the more standard periodic boundary condition.

A periodic boundary condition is commonly used for cellular flow simulations in domains with high symmetry, such as a straight channel section with a single inlet and outlet (see, e.g., [26]). By mimicking an indefinitely long tube or channel, computational costs can be saved without sacrificing spatial resolution, e.g., through downscaling. For an example application, a periodic boundary condition is applied in two dimensions to mimic the flow profile and cellular interactions at the center of a parallel plate flow chamber (*see* Fig. 4a).

In many cases, a periodic boundary condition cannot be implemented, for example, in case of multiple outlets with only a single inlet, as is the case for a bifurcated vessel section. In situations where periodicity cannot be applied, such as in the case of a curved (or any arbitrarily shaped) vessel section, additional description is required for the in- and outflowing cellular distributions. To allow for a constant influx of cells, a periodic pre-inlet is attached in a

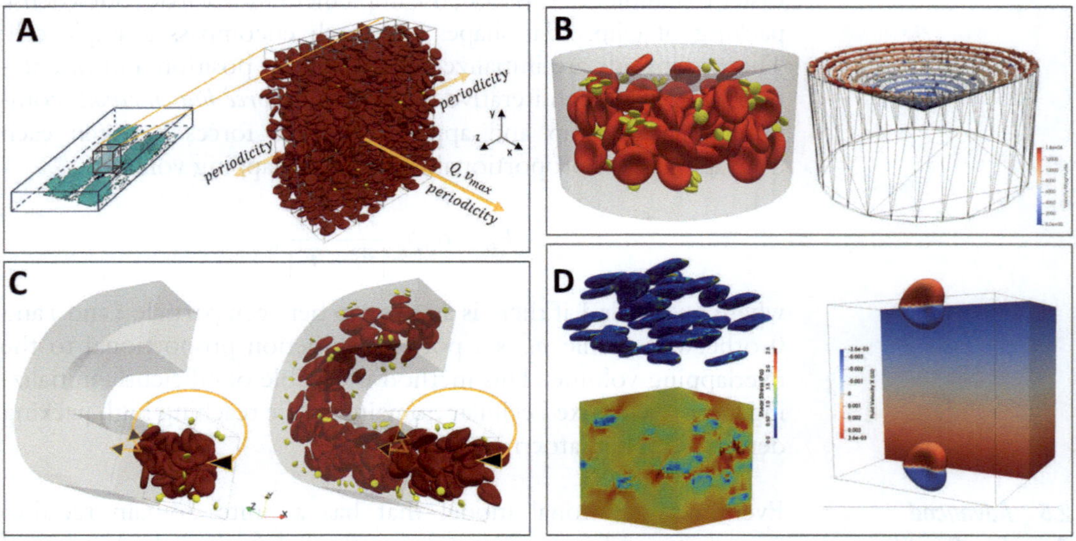

Fig. 4 Boundary conditions in HemoCell. (**a**) Periodic boundary condition. (**b**) Rotating boundary condition—in this case the cone-shaped top lid. (**c**) Pre-inlet section to provide a continuous influx of well-mixed cells. (**d**) Lees-Edwards boundary condition to provide pure shear environment

serial fashion to the inlet of the domain of interest. This periodic pre-inlet is a straight channel with periodic boundaries in the direction of the flow [31]. It can be regarded as a small additional simulation prepending the domain of interest. Cells and fluid propagating across the outlet of the periodic pre-inlet (that is joined to the inlet of the main simulated domain) are duplicated into the main domain. Figure 4c visualizes this process in case of a curved vessel simulation by depicting the initial time-step of the empty main domain and filled pre-inlet and a later stage when the entire domain is filled with cells from the influx.

Many hematological devices include moving or rotating components. One such example is a cylindrical device aiming to create constant shear in a volume to observe the shear-induced behavior of blood. This device allows the analysis of platelet deposition at the device's bottom plate due to a constant shear gradient which is maintained by a rotating cone top lid [32]. A virtual replica of this device can be created in HemoCell, which allows access to more detailed information on the process (*see* Fig. 4b), such as the trajectory of the cells, or local fluid stresses. The top lid of the geometry is a cone shape indented at a desired angle between 0 and 45°. The rotation of this lid is implemented via a rotating velocity field at the surface of the cone shape. This results in a linear velocity profile at the cone surface, which propagates through the cylinder and thus causes rotation of the fluid including the immersed cells.

Finally, the implementation of a uniform shear environment, such as in a Couette flow viscometer, allows for the calculation of bulk viscosity of blood and the diffusivities of the cells. This setup comes with the caveat that the presence of boundaries (wall or velocity boundary) induces lift force on the immersed cells, which in return causes the formation of a CFL [26]. One solution to this problem is to oversize the domain until the effects of the boundaries become negligible. In order to prevent this computationally expensive upscaling, Lees-Edwards boundary conditions are implemented to create a boundary-less uniform shear environment. The simulation is periodic in all dimensions, while imposing constant opposing velocities on two parallel boundaries in the direction of the shear. Figure 4d (left) shows the stress patterns on the immersed cells and the embedding fluid, which are used to calculate the bulk viscosity. Figure 4d (right) shows a single red blood cell crossing the Lees-Edwards boundary and hence being copied onto the other side of the domain, while experiencing forces from two opposing directions.

The advanced boundary conditions implemented in HemoCell widen the range of potential applications. The 3D periodic pre-inlet, the rotating velocity field boundary, and the Lees-Edwards boundary condition are novel implementations for cellular simulations, which enable their application in complex geometries mimicking rheological devices and realistic vessels.

2.6 Performance and Load-Balancing

The largest scale deployment of HemoCell at the time of writing demonstrated the execution of a single simulation on over 330,000 CPU cores, while maintaining over 80% weak-scaling efficiency. Given the tightly coupled nature of the code involving both structured (LBM) and unstructured (DEM) grids, this performance scaling can be regarded as excellent. To achieve this, several advanced techniques are applied.

The computation of the constitutive model of the cells is very costly (at the typical resolution, a single RBC is simulated via a system of 5000 equations). Since the flow field and the cellular components utilize different numerical methods, their stability w.r.t. the time-step size also differs. The equations of cells are integrated by either the Euler or Adams-Bashforth method depending on the choice of the user. Both of these allow larger integration steps compared to LBM. For this reason, the computation of the cells is separated in time from that of the fluid field, allowing the constitutive model to be evaluated less frequently and save significant computational cost. The separation of integration time scales can be set to a constant value or can be set adaptively during the simulation depending on the scale of stresses in the system [29].

Another challenge that arises at scale is the change in computational load distribution. Cells move around relative to the stationary geometry. This means that they can vacate certain region or aggregate in others. The initial domain decomposition is based on the starting homogeneous state of the simulated domain; however, during the simulation this can increasingly become non-homogeneous. If this happens, the original domain decomposition is no longer sufficient, since regions with no or very few cells compute much faster than regions with high cell density. This means that CPUs with less work will be idle, reducing the parallel efficiency significantly. To circumvent this problem, HemoCell has load-balancing capability that monitor the fractional load-imbalance of the simulation, and when it surpasses a given threshold, the simulation is paused (checkpointing), redistributed, and continued with the new domain decomposition [33]. This load-balancing step necessarily incurs additional computational cost; therefore, the choice of the imbalance threshold is important.

3 Applications of HemoCell

3.1 Cellular Trafficking and Margination

A major reason to apply microscopic blood simulation is to gain access to high-resolution information that is not possible through experimental techniques. For example, simulations can provide the detailed trajectories of cells, or correlate the cell positions and the local flow environment with their deformations. This information is used to investigate cellular interactions and emergent properties,

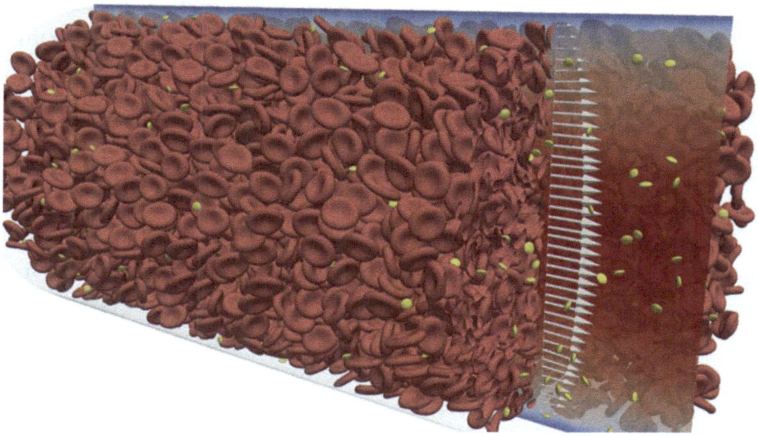

Fig. 5 Outline of cellular flow in a straight pipe geometry. The flow velocity is denoted by arrows, showing the typical "plug" profile arising from the increased hematocrit toward the center of the vessel

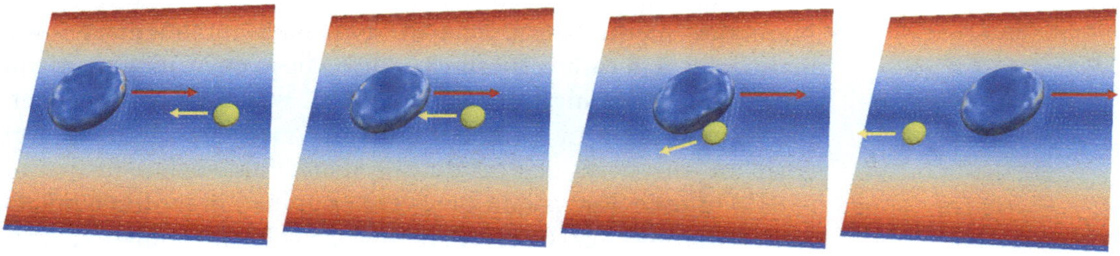

Fig. 6 Displacement of a platelet during a collision with a red blood cell. Time progresses from left to right. The platelet is displaced by approx. 1 μm during the collision

such as the formation of a cell free layer or the margination of platelets. For instance, even a simple flow in a straight vessel segment (e.g., *see* Fig. 5) has various gradients that cannot be tracked accurately in an experimental setting [34]. These include gradients in the cell distribution, gradients in velocity, and gradients in shear-rate (note: this flow is not Newtonian).

These gradients have significant effect on the trafficking of the cells and therefore they influence the distribution of the cells and the overall rheology. A well-known implication is the emergent phenomenon of platelet margination. RBCs move away from the vessel walls primarily due to the effect of wall-induced lift and they create a hematocrit distribution that increases toward the center of the channel. This in turn pushes platelets out of the center of the vessel toward the wall through shear-induced cell-cell collisions. Every time a platelet bumps into an RBC, it is displaced a little (*see* Fig. 6). This displacement mechanism favors the direction toward the wall, since the RBC density is lower there.

After a long sequence of collisions, platelets tend to get captured in the CFL at the wall. In an evolutionary sense, this is a very important mechanism, since it drives PLTs next to the vessel wall, where they need to act in case of an injury to stop the bleeding. When this mechanism is impaired, it leads directly to bleeding disorders as it was demonstrated recently using HemoCell in combination with a series of bleeding experiments [35].

3.2 Cellular Flow in Microfluidic Devices

Biochemical processes of blood are often investigated in microfluidic devices (also called "bleeding chips"). In recent years, the role of the biomechanical environment is gaining focus as an inseparable component acting along biochemistry. See for instance the mechanism of shear-induced thrombus formation [36] that produces the majority of stroke and heart infarct cases. While the role of the mechanical forces is becoming obvious through mechanosensing components, such as the von Willebrand factor (VWF) [37], their quantitative description is lacking. The primary reason for this is the limitation in current measurement technologies to capture detailed stresses in fluid and cells. Numerical methods can yield the missing complimentary information by replicating the microfluidic flow environment of the measurements in high detail. HemoCell was used to investigate the starting point of forming high-shear rate thrombi to explore what are the necessary conditions that allow such a formation [38]. Multiple flow conditions were examined in multiple microstenosis geometries. Cross-matching the cellular simulations with the experimental results revealed that at the initial location of the thrombus, multiple conditions need to be present: (1) availability of platelets (the high availability is facilitated by margination); (2) suitable "collision-free" volume that can be provided for example behind a stenosis by the mechanism of cell-free layer formation; and (3.) (in case of a shear-induced thrombus) high shear rate values that can uncoil and activate VWF. Apart from such a suitable mechanical environment, the components of Virchow's triad, including the thrombogenic surface, are still necessary for the development of the complete thrombus.

The cellular nature of blood significantly influences the mechanical responses, for instance, the emerging stresses. Current empirical blood models that use a continuum approximation cannot include the effects of non-homogeneous cell distributions. When investigating the biomechanics of the vessel wall (for wall remodeling, inflammation, endothelial layer disruption, etc.), wall shear stress is one of the key quantities of interest. Continuum approaches cannot resolve the CFL, and as a consequence, they underestimate the wall shear rate, and at the same time, they tend to significantly overestimate wall stresses. The reason for this misprediction is that the CFL (forming next to the wall) is a pure plasma layer that has a viscosity three times lower than the average whole blood viscosity. This natural lubrication layer influences the local flow dynamics significantly that cellular models can capture successfully [39]. It is also worth noting that most microfluidic experiments are characterized based on wall shear rate or stress calculated using continuum theories (such as Hagen-Poiseuille or Poiseuille flow). These calculations likely have the same shortcoming since they neglect the CFL.

3.3 Flow in a Curved Micro-Vessel Section

Traditionally, cellular blood flow mechanics during platelet adhesion and aggregation are studied in straight channels, sometimes including local geometric variations (e.g., stenoses) [39], while the influence of more complex vessel geometries on blood flow characteristics remains understudied. The effects arising from more complicated geometry are discussed here focusing on initial platelet adhesion and aggregation, via a combination of in silico (i.e., HemoCell) and in vitro approaches. The flow behavior in regard to shear rate and rate of elongation as well as cellular distributions is investigated in the simulations of a curved channel and compared to the results of complementary microfluidic aggregation experiments in a similarly shaped flow chamber. The simulations reveal the occurrence of high elongational flow at the inner arc of the curvature which corresponds to the site of increased platelet aggregation in the experimental results.

The simulations are performed in a U-shaped square duct geometry with a $25 \times 25\ \mu m^2$ cross-section, and the curvature has an inner diameter of 25 μm as well. The domain is initialized with RBCs to result in a discharge hematocrit of 30%. In order to ensure a constant inflow from a straight vessel section, a periodic pre-inlet is utilized (*see* Fig. 7a). The blood flow is simulated for 2 s at two different flow velocities that are indicated by their initial wall shear rates: 300 s^{-1} and 1600 s^{-1}. For the microfluidic aggregation experiments, a similarly shaped U-chamber is designed at a larger scale. The chamber is coated with collagen and perfused for 4 min with hirudinated human whole blood at the same two flow velocities as used in the simulations. More detailed setup methodology can be found in [40].

In order to capture effects of the curvature, cell distributions, width of the CFL, and cross-sectional flow profiles are quantified in three sections: close to the inlet, at the center of the curvature, and close to the outlet. While differences in CFL width remain insignificant and only a slight shift of RBC concentration toward the inner wall of the curvature is observed, the flow profiles display significant differences when comparing the curvature to the inlet region. Increased shear rate gradients which cause sites of high elongational flow are observed at the inner arc of the curvature (*see* Fig. 7b, c). When comparing the two flow velocities, we observe qualitative similarity with different magnitudes in each case.

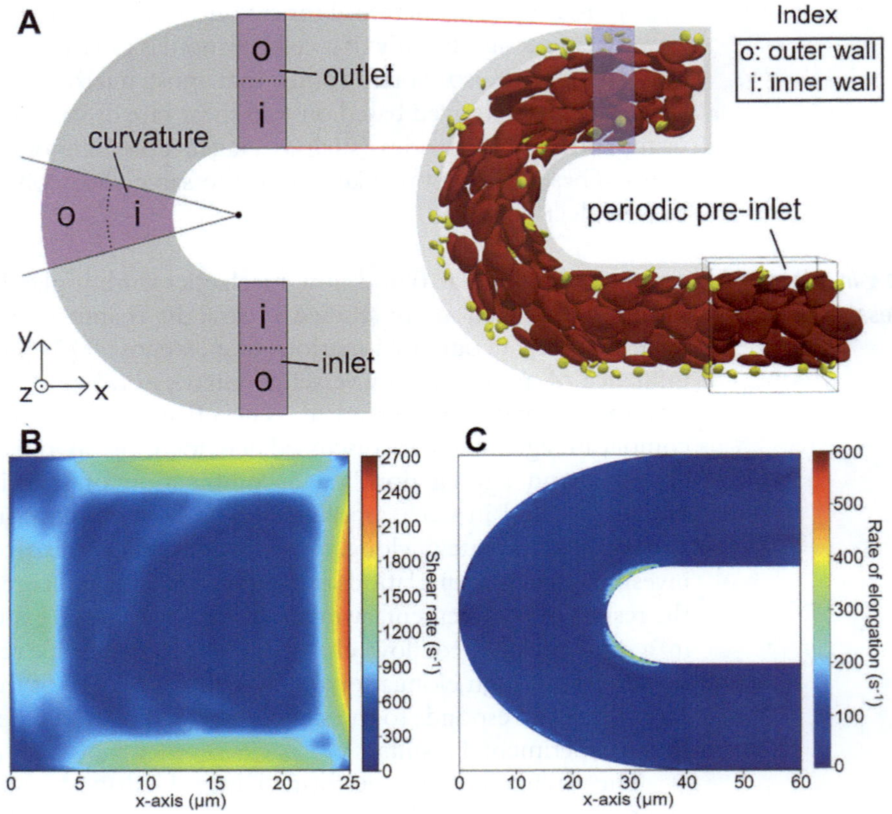

Fig. 7 Curved channel simulations. (**a**) Setup of curved channel domain with flow in negative x-direction from periodic pre-inlet. The regions of interest are highlighted with the corresponding inner and outer wall division. (**b**) Cross-sectional shear rate profile in curvature section and (**c**) top view of elongational flow magnitude across channel at initial wall shear rate of 1600 s^{-1}

To allow for a comparison of the simulations to the experiments, the aggregate sizes accumulated on the microfluidic chamber after perfusion are quantified in the same respective regions, denoted inlet, curvature, and outlet. The results show different observations for the two flow velocity cases. While the low-flow velocity chamber displays no significant difference in aggregate size between the different regions, the high-flow velocity chamber depicts elevated aggregate formation at the inner arc of the curvature.

The increased aggregate area of the microfluidic experiments is located at the sites of high shear and elongational flows, observed in the simulations. As the blood used in the experiments is treated with the anticoagulant hirudin, only the initial steps of platelet adhesion and aggregation are observed, which highly depend on the plasma suspended molecule von Willebrand factor as a mediator [41]. While shear-induced platelet aggregation occurs at shear rates much higher than observed in the simulations [42], elongational flows are found to enable the von Willebrand factor-mediated adhesion process at comparatively low shear flow [43]. Since the peak rate of elongation in the simulations reaches this critical range only in the high-flow velocity case, it is hypothesized to be responsible for the elevated aggregate formation, which is also only observed at the higher flow velocity in the experiments. In summary, the results highlight the role of elongational flows as well as the importance of vessel geometry in initial platelet adhesion and aggregation.

3.4 Flow of Diabetic Blood in Vessels

Several diseases can alter the deformability of cells significantly and, therefore, impact the overall rheology of whole blood. The deformability of a red blood cell can be impeded, and stiffened, by various pathologies such as sickle cell disease [44], diabetes [45], malaria [46], and Parkinson's disease [47] for example. In diabetic flow, RBCs are stiffer and less capable of deformations, which have a significant effect on bulk viscosity as well as on platelet margination. A novel stiffened red blood cell model was applied to study the effects of changing red blood cell deformability. This model was developed and validated in HemoCell by matching the deformation indices from ektacytometry measurements in vitro with the deformation indices computed from single-cell shearing numerical experiments in silico [27]. Stiffening of RBCs membranes was induced in vitro by incubating RBCs with tert-butyl hydroperoxide (TBHP), which induces oxidative stress on the cell membrane. TBHP can be used as a general model for cell membrane perturbations resulting from oxidative disorders [48]. The stiffened numerical RBC model was achieved by scaling the mechanical parameters, in particular the link force coefficient and the internal viscosity ratio of the original validated RBC model [26, 49].

The behavior of stiffened RBCs in bulk rheology was studied by simulating whole blood flowing through a periodic pipe of radius $R = 50$ μm with a tank hematocrit of 30% driven by a body force resulting in a wall shear rate of $1000\ s^{-1}$. The fraction of stiff/healthy RBCs (0/100, 30/70, 50/50, 70/30, and 100/0) was varied in each simulation, maintaining a total hematocrit of 30%. The main observation from this study is that the RBC free layer decreases because of an increase of stiffened RBC fraction in flowing blood, shown in the left panel of Fig. 8. The more rigid RBCs experience decreased lift force leading to reduction in the CFL size. Furthermore, a decrease of platelet localization at the vessel wall as the fraction of stiffened RBCs increase is also observed (right panel of Fig. 8). Platelet margination is likely altered, on the one hand, by the decrease in size of the CFL as there is limited volume next to the wall for platelets to be trapped in, and, on the other hand, by a reduced shear induced dispersion that would drive the PLTs toward the CFL. Note that these changes might have an influential effect on the hemostatic processes, where the high availability of PLTs is one of the necessary conditions for physiologic operation.

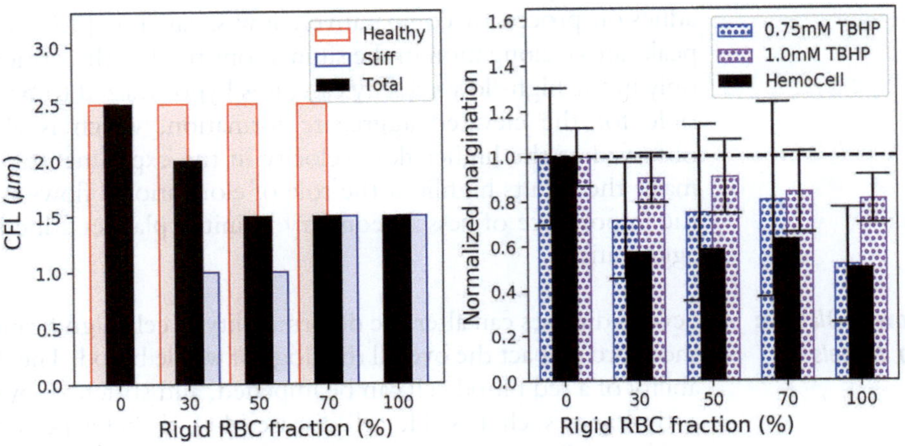

Fig. 8 Red blood cell-free layer (left panel) and platelet margination (right panel) as a function of rigid RBC fractions. The computed CFL from HemoCell is shown in black, with each of the RBC components shown in red for healthy and blue for 1.0 mM TBHP. Platelet concentration at the wall is computed in the volume 4 μm from the wall normalized to the concentration of the 100% healthy RBC case (HemoCell:black and in vitro results for 0.75 mM TBHP:blue and 1.0 mM TBHP:purple)

The work presented in this section proposes a general model for altered RBC deformability as a result of disease, both experimental and computational, and offers evidence of the detrimental effect rigid RBCs have on physiological blood flow and the ability of platelets to localize to the vessel wall. The diseased, stiffened, red blood cell model from this section has further been applied to investigate whole blood flow through a patient-specific segmented retinal microaneurysm [50].

References

1. Boron WF, Boulpaep EL (eds) (2017) Medical physiology, 3rd edn. Elsevier, Philadelphia
2. Caro CG (2012) The mechanics of the circulation, 2nd edn. Cambridge University Press, Cambridge
3. Varchanis S, Dimakopoulos Y, Wagner C, Tsamopoulos J (2018) How viscoelastic is human blood plasma? Soft Matter 14(21):4238–4251. https://doi.org/10.1039/C8SM00061A
4. Dupire J, Socol M, Viallat A (2012) Full dynamics of a red blood cell in shear flow. Proc Natl Acad Sci U S A 109(51):20808. https://doi.org/10.1073/pnas.1210236109/-/DCSupplemental; www.pnas.org/cgi/doi/10.1073/pnas.1210236109
5. Skotheim JM, Secomb TW (2007) Red blood cells and other nonspherical capsules in shear flow: oscillatory dynamics and the Tank-Treading-to-Tumbling Transition. Phys Rev Lett 98(7):078301. https://doi.org/10.1103/PhysRevLett.98.078301
6. Chien S (1970) Shear dependence of effective cell volume as a determinant of blood viscosity. Science 168(3934):977–979. https://doi.org/10.1126/science.168.3934.977
7. Samsel RW, Perelson AS (1984) Kinetics of rouleau formation. II. Reversible reactions. Biophys J 45(4):805–824. https://doi.org/10.1016/S0006-3495(84)84225-3
8. Brust M et al (2014) The plasma protein fibrinogen stabilizes clusters of red blood cells in microcapillary flows. Sci Rep 4:1–6. https://doi.org/10.1038/srep04348
9. Secomb TW (2017) Blood flow in the microcirculation. Annu Rev Fluid Mech 49 (August):443–461. https://doi.org/10.1146/annurev-fluid-010816-060302

10. Fåhræus R, Lindqvist T (1931) The viscosity of the blood in narrow capillary tubes. Am J Physiol-Leg Content 96(3):562–568. https://doi.org/10.1152/ajplegacy.1931.96.3.562
11. Pries AR, Neuhaus D, Gaehtgens P (1992) Blood viscosity in tube flow: dependence on diameter and hematocrit. Am J Physiol 263(6 Pt 2):H1770–H1778
12. Carboni EJ et al (2016) Direct tracking of particles and quantification of margination in blood flow. Biophys J 111(7):1487–1495. https://doi.org/10.1016/j.bpj.2016.08.026
13. Freund JB (2014) Numerical simulation of flowing blood cells. Annu Rev Fluid Mech 46(1):67–95. https://doi.org/10.1146/annurev-fluid-010313-141349
14. Müller K, Fedosov DA, Gompper G (2014) Margination of micro- and nano-particles in blood flow and its effect on drug delivery. Sci Rep 4:4871. https://doi.org/10.1038/srep04871
15. Krüger T, Gross M, Raabe D, Varnik F (2013) Crossover from tumbling to tank-treading-like motion in dense simulated suspensions of red blood cells. Soft Matter 9(37):9008–9015. https://doi.org/10.1039/C3SM51645H
16. Hosseini SM, Feng JJ (2009) A particle-based model for the transport of erythrocytes in capillaries. Chem Eng Sci 64(22):4488–4497. https://doi.org/10.1016/j.ces.2008.11.028
17. Závodszky G, Paál G (2013) Validation of a lattice Boltzmann method implementation for a 3D transient fluid flow in an intracranial aneurysm geometry. Int J Heat Fluid Flow 44:276–283. https://doi.org/10.1016/j.ijheatfluidflow.2013.06.008
18. Bhatnagar PL, Gross EP, Krook M (1954) A model for collision processes in gases. I. Small amplitude processes in charged and neutral one-component systems. Phys Rev 94(3):511–525. https://doi.org/10.1103/PhysRev.94.511
19. Chen S, Doolen GD (1998) Lattice Boltzmann method for fluid flows. Annu Rev Fluid Mech 30(1):329–364. https://doi.org/10.1146/annurev.fluid.30.1.329
20. Qian Y, D'Humières D, Lallemand P (1992) Lattice BGK models for Navier-Stokes equation. EPL Europhys Lett 479 [Online]. Available: http://iopscience.iop.org/0295-5075/17/6/001. Accessed 12 Jul 2014
21. Rosenau P (1989) Extending hydrodynamics via the regularization of the Chapman-Enskog expansion. Phys Rev A 40(12):7193–7196. https://doi.org/10.1103/PhysRevA.40.7193
22. Krüger T, Kusumaatmaja H, Kuzmin A, Shardt O, Silva G, Viggen EM (2017) The Lattice Boltzmann method: principles and practice in graduate texts in physics. Springer International Publishing, Cham. https://doi.org/10.1007/978-3-319-44649-3
23. Peskin CS (2002) The immersed boundary method. Acta Numer 11:479–517. https://doi.org/10.1017/S0962492902000077
24. Hansen JC, Skalak R, Chien S, Hoger A (1996) An elastic network model based on the structure of the red blood cell membrane skeleton. Biophys J 70(1):146–166. https://doi.org/10.1016/S0006-3495(96)79556-5
25. Li J, Dao M, Lim CT, Suresh S (2005) Spectrin-level modeling of the cytoskeleton and optical tweezers stretching of the erythrocyte. Biophys J 88(5):3707–3719. https://doi.org/10.1529/biophysj.104.047332
26. Závodszky G, van Rooij B, Azizi V, Hoekstra A (2017) Cellular level in-silico modeling of blood rheology with an improved material model for red blood cells. Front Physiol 8. https://doi.org/10.3389/fphys.2017.00563
27. Czaja B, Gutierrez M, Závodszky G, de Kanter D, Hoekstra A, Eniola-Adefeso O (2020) The influence of red blood cell deformability on hematocrit profiles and platelet margination. PLOS Comput Biol 16(3):e1007716. https://doi.org/10.1371/journal.pcbi.1007716
28. de Vries K, Nikishova A, Czaja B, Závodszky G, Hoekstra AG (2020) Inverse uncertainty quantification of a cell model using a Gaussian process metamodel. Int J Uncertain Quantif 10(4). https://doi.org/10.1615/Int.J.UncertaintyQuantification.2020033186
29. Zavodszky G, van Rooij B, Azizi V, Alowayyed S, Hoekstra A (2017) Hemocell: a high-performance microscopic cellular library. Procedia Comput Sci 108:159–165. https://doi.org/10.1016/j.procs.2017.05.084
30. Lees AW, Edwards SF (1972) The computer study of transport processes under extreme conditions. J Phys C Solid State Phys 5(15):1921. https://doi.org/10.1088/0022-3719/5/15/006
31. Azizi Tarksalooyeh VW, Závodszky G, van Rooij BJM, Hoekstra AG (2018) Inflow and outflow boundary conditions for 2D suspension simulations with the immersed boundary lattice Boltzmann method. Comput Fluids 172:312–317. https://doi.org/10.1016/j.compfluid.2018.04.025
32. Varon D et al (1997) A new method for quantitative analysis of whole blood platelet interaction with extracellular matrix under flow conditions. Thromb Res 85(4):283–294.

https://doi.org/10.1016/S0049-3848(97)00014-5

33. Alowayyed S, Závodszky G, Azizi V, Hoekstra AG (2018) Load balancing of parallel cell-based blood flow simulations. J Comput Sci 24:1–7. https://doi.org/10.1016/j.jocs.2017.11.008
34. Závodszky G, van Rooij B, Czaja B, Azizi V, de Kanter D, Hoekstra AG (2019) Red blood cell and platelet diffusivity and margination in the presence of cross-stream gradients in blood flows. Phys Fluids 31(3):031903. https://doi.org/10.1063/1.5085881
35. Kimmerlin Q et al (2022) Loss of α4A- and β1-tubulins leads to severe platelet spherocytosis and strongly impairs hemostasis in mice. Blood 140(21):2290–2299. https://doi.org/10.1182/blood.2022016729
36. Casa LDC, Ku DN (2017) Thrombus formation at high shear rates. Annu Rev Biomed Eng 19(1):415–433. https://doi.org/10.1146/annurev-bioeng-071516-044539
37. Gogia S, Neelamegham S (2015) Role of fluid shear stress in regulating VWF structure, function and related blood disorders. Biorheology 52(5–6):319–335. https://doi.org/10.3233/BIR-15061
38. van Rooij BJM, Závodszky G, Azizi Tarksalooyeh VW, Hoekstra AG (2019) Identifying the start of a platelet aggregate by the shear rate and the cell-depleted layer. J R Soc Interface 16(159):20190148. https://doi.org/10.1098/rsif.2019.0148
39. van Rooij BJM, Závodszky G, Hoekstra AG, Ku DN (2021) Haemodynamic flow conditions at the initiation of high-shear platelet aggregation: a combined in vitro and cellular in silico study. Interface Focus 11(1):20190126. https://doi.org/10.1098/rsfs.2019.0126
40. Spieker CJ et al (2021) The effects of micro-vessel curvature induced Elongational flows on platelet adhesion. Ann Biomed Eng 49(12):3609–3620. https://doi.org/10.1007/s10439-021-02870-4
41. Ruggeri ZM, Orje JN, Habermann R, Federici AB, Reininger AJ (2006) Activation-independent platelet adhesion and aggregation under elevated shear stress. Blood 108(6):1903–1910. https://doi.org/10.1182/blood-2006-04-011551
42. Casa LDC, Deaton DH, Ku DN (2015) Role of high shear rate in thrombosis. J Vasc Surg 61(4):1068–1080. https://doi.org/10.1016/j.jvs.2014.12.050
43. Sing CE, Alexander-Katz A (2010) Elongational flow induces the unfolding of Von Willebrand factor at physiological flow rates. Biophys J 98(9):L35–L37. https://doi.org/10.1016/j.bpj.2010.01.032
44. Chirico EN, Pialoux V (2012) Role of oxidative stress in the pathogenesis of sickle cell disease. IUBMB Life 64(1):72–80. https://doi.org/10.1002/iub.584
45. Shin S, Ku Y-H, Ho J-X, Kim Y-K, Suh J-S, Singh M (2007) Progressive impairment of erythrocyte deformability as indicator of microangiopathy in type 2 diabetes mellitus. Clin Hemorheol Microcirc 36(3):253–261
46. Tan JSY, Závodszky G, Sloot PMA (2018) Understanding malaria induced red blood cell deformation using data-driven Lattice Boltzmann simulations. In: Computational science – ICCS 2018, Y Shi, H Fu, Y Tian, VV Krzhizhanovskaya, MH Lees, J Dongarra, PMA Sloot (eds.), in Lecture Notes in Computer Science. Cham: Springer International Publishing, pp. 392–403. https://doi.org/10.1007/978-3-319-93698-7_30
47. Jenner P (2003) Oxidative stress in Parkinson's disease. Ann Neurol 53(S3):S26–S38. https://doi.org/10.1002/ana.10483
48. Rice-Evans C, Baysal E, Pashby DP, Hochstein P (1985) t-butyl hydroperoxide-induced perturbations of human erythrocytes as a model for oxidant stress. Biochim Biophys Acta BBA 815(3):426–432. https://doi.org/10.1016/0005-2736(85)90370-0
49. De Haan M, Zavodszky G, Azizi V, Hoekstra AG (2018) Numerical investigation of the effects of red blood cell cytoplasmic viscosity contrasts on single cell and bulk transport behaviour. Appl Sci 8(9):9. https://doi.org/10.3390/app8091616
50. Czaja B et al (2022) The effect of stiffened diabetic red blood cells on wall shear stress in a reconstructed 3D microaneurysm. Comput Methods Biomech Biomed Engin 25:1–19. https://doi.org/10.1080/10255842.2022.2034794

Chapter 17

A Blood Flow Modeling Framework for Stroke Treatments

Remy Petkantchin, Franck Raynaud, Karim Zouaoui Boudjeltia, and Bastien Chopard

Abstract

Circulatory models can significantly help develop new ways to alleviate the burden of stroke on society. However, it is not always easy to know what hemodynamics conditions to impose on a numerical model or how to simulate porous media, which ineluctably need to be addressed in strokes. We propose a validated open-source, flexible, and publicly available lattice-Boltzmann numerical framework for such problems and present its features in this chapter. Among them, we propose an algorithm for imposing pressure boundary conditions. We show how to use the method developed by Walsh et al. (Comput Geosci 35(6):1186–1193, 2009) to simulate the permeability law of any porous medium. Finally, we illustrate the features of the framework through a thrombolysis model.

Key words Stroke treatment, Porous media, Computational fluid dynamics, Lattice-Boltzmann, Partial bounce-back, Mesoscopic modeling

1 Introduction

Stroke is a leading cause of disability and death worldwide, and developing effective treatments is a critical medical challenge. State-of-the-art treatments consist of the use, alone or in combination, of thrombolysis, sonothrombolysis, thrombectomy, thrombus aspiration in the case of ischemic strokes; and coiling, stenting, or slowing down blood flow for hemorrhagic strokes. Blood flow modeling can thus provide valuable insights into the mechanisms underlying stroke and guide the development of new therapies. Modeling is one of the pillars of the scientific method. To increase our knowledge on natural phenomena, the scientific approach tells us to observe, make hypotheses, and model the process, and re-iterate until the model satisfyingly replicates or predicts the observations. The development of high-performance computing (HPC) has brought new possibilities in the field of biomedical research, enabling a wider range of spatial and temporal scales for numerical

Alexander Heifetz (ed.), *High Performance Computing for Drug Discovery and Biomedicine*, Methods in Molecular Biology, vol. 2716, https://doi.org/10.1007/978-1-0716-3449-3_17,

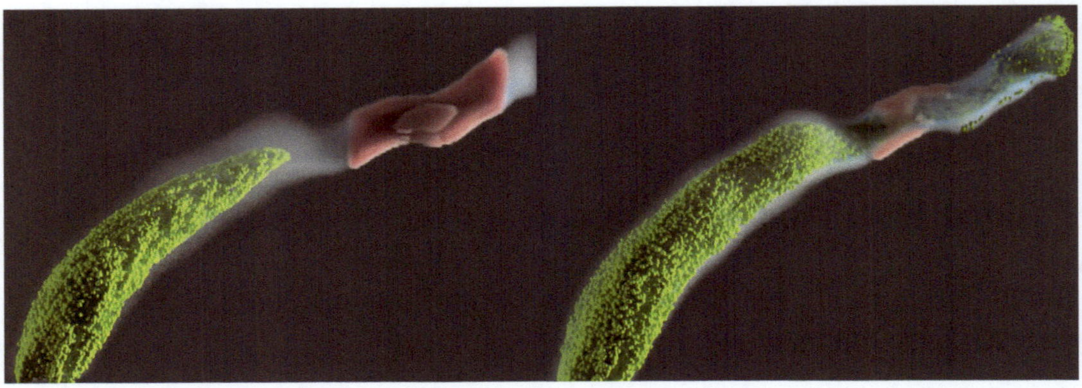

Fig. 1 Snapshots of a fibrinolysis simulation with a 3D model in a patient-specific artery. The thrombus (red) initially blocks part of the flow (blue), which makes it hard for the profibrinolytic drug (green) to reach it. When the drug finally reaches the thrombus and initiates the lysis reaction, the artery can be recanalized

models, and the analysis of larger experimental and clinical datasets. Porous media are central to stroke problems, may it be the thrombi that clog brain arteries, or the flow-diverting stents that will prevent aneurysm-rupture-induced hemorrhagic strokes. Even brain tissue can be modeled as a porous medium that needs oxygenation [1, 2]. We present here an HPC-compatible methodology that allows the simulation of porous media at a mesoscale, using the lattice-Boltzmann (LB) method, through the example of the simulation of a thrombus for a previously developed thrombolysis model (https://doi.org/10.1101/2023.05.09.539942), depicted in Fig. 1. Finally, we simulate the lysis of the thrombus and the subsequent hemodynamic changes through a minimal toy thrombolysis model. The presented methodology can be applied to any porous medium, provided that one knows a priori the permeability law of the medium.

2 Methods

In our thrombolysis example, we consider blood flow as a laminar Newtonian fluid [2–6]. We simulate blood flow using the open-source LB library Palabos [7], which offers high control and flexibility to solve a wide variety of problems.

2.1 The Lattice-Boltzmann Method

Inspired by Lattice Gas Automata [8], the LB method is an efficient alternative to the Navier–Stokes equations for computational fluid dynamics [9–13] that has already been widely used to address biomedical problems [14–17]. In a few words, it solves the Boltzmann equation, which describes the evolution of the so-called "fluid" particles, in a regularly discretized lattice. This spatial discretization enforces a set of discrete velocities c_i, $i = 0, 1, ..., q - 1$ for the fluid particle densities f_i, also called *populations*, such that f_1 is the density of fluid particles going with velocity c_1. Macroscopic variables such as density ρ and momentum $\rho\mathbf{u}$ are obtained by computing the moments of f_i:

$$\rho(\mathbf{x}, t) = \sum_i f_i(\mathbf{x}, t), \tag{1}$$

$$\rho(\mathbf{x}, t)\mathbf{u}(\mathbf{x}, t) = \sum_i \mathbf{c}_i f_i(\mathbf{x}, t). \tag{2}$$

The temporal evolution of the populations $f_i(\mathbf{x}, t)$ is done in two steps. First, during the collision step, we compute at each lattice position $\mathbf{x}$ the scattered populations $f_i^*(\mathbf{x}, t)$. Then, the streaming step propagates the populations $f_i^*(\mathbf{x}, t)$ to the neighboring lattice sites. Formally, the scattered populations are given by

$$f_i^*(\mathbf{x}, t) = f_i(\mathbf{x}, t) + \Omega_i(f), \tag{3}$$

where $\Omega_i(f)$ is the collision operator. In this chapter, we consider a relaxation-type collision process between populations, the Bhatnagar–Gross–Krook (BGK) collision operator [18]:

$$\Omega_i(f) = -\frac{f_i(\mathbf{x}, t) - f_i^{eq}(\mathbf{x}, t)}{\tau}\Delta t, \tag{4}$$

where Δt is the temporal discretization, and τ is the model's relaxation time associated with the fluid's kinematic viscosity. The equilibrium populations $f_i^{eq}(\mathbf{x}, t)$ are expressed as follows [19]:

$$f_i^{eq}(\mathbf{x}, t) = w_i\, \rho\left(1 + \frac{\mathbf{u}\cdot\mathbf{c_i}}{c_s^2} + \frac{(\mathbf{u}\cdot\mathbf{c_i})^2}{2c_s^4} - \frac{\mathbf{u}\cdot\mathbf{u}}{2c_s^2}\right), \tag{5}$$

where w_i are weighting factors that depend on the dimension and the number of discretized velocities. c_s is the speed of sound in the lattice, defined by the lattice topology [12]. Finally, the streaming step propagates the scattered populations:

$$f_i(\mathbf{x} + \mathbf{c_i}\Delta t, t + \Delta t) = f_i^*(\mathbf{x}, t). \tag{6}$$

2.2 Appropriate Boundary Conditions

A common problem that comes beforehand of the simulation is the choice of boundary conditions (BCs) to apply to the boundaries of the simulation domain. This choice usually depends on available knowledge of the process to simulate or on the predicted evolution of the system. Unfortunately, detailed patient-specific data are hardly available in the case of strokes, due to the emergency of the condition. This leaves us to make a choice for the BCs. A standard [5, 6, 20] choice is the pressure-flow BCs, which specify the pressure and flow rate at the inlet and outlet of the system. Another option is the pressure–pressure BCs, where the pressure at the inlet and outlet are specified. We focus on the latter because it

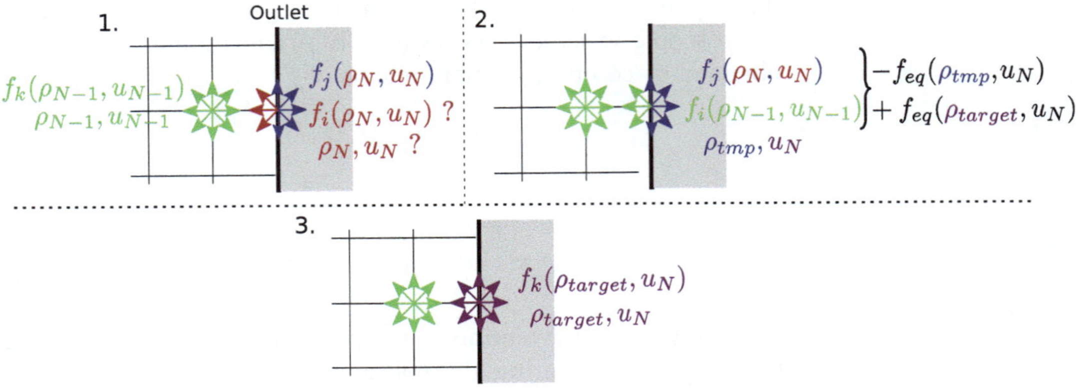

Fig. 2 Pressure boundary condition used for inlet and outlet, with ρ_N the density and u_N the velocity at node N. The computation steps are the following (here showing outlet): (**1**) Pre-computation state: At node N, populations f_j are known from the streaming of neighbor nodes at previous iterations, and populations f_i, density, and velocities are unknown. At node $N-1$, f_k, density, and velocities are known. It is chosen as the previous node in the direction normal to the boundary. (**2**) Unknown populations at node N are copied from populations at node $N-1$. From populations f_j and f_i, a temporary density ρ_{tmp} and the velocities are computed. Two equilibrium functions are then computed, one with ρ_{tmp}, and the other with the target density ρ_{target}. We subtract the former and add the latter to the populations at node N. (**3**) Node N now has target density ρ_{target} and populations f_k

allows the flow rate to vary based on the system's dynamics, which is particularly useful in the simulation of thrombolysis, where the dynamics of blood flow in the cerebral vasculature can be highly complex and challenging to predict. During the recanalization of the artery, we furthermore need to be able to adapt the imposed pressure gradient. This can be done with the following boundary condition for LB, which yields stable simulations and can also simulate pulsatility by applying a time modulation to the inlet pressure.

As a foreword, it is essential to know that in LB, pressure is imposed through the density of the fluid. Let us consider the situation displayed in Fig. 2, where we impose the fluid density at the outlet boundary node N. After having copied the unknown populations f_i from node $N-1$, the density at node N can be computed as usually in LB:

$$\rho_{\text{tmp}} = \sum_j f_j(\mathbf{x_N}, t) + \sum_i f_i(\mathbf{x_N}, t) = \sum_k f_k(\mathbf{x_N}, t), \tag{7}$$

where the f_j designates the populations that are known from the streaming step, at node N. The equilibrium function f_k^{eq} at node N, used in the standard BGK collision, reads [12]

$$f_k^{eq}(\mathbf{x_N}, t) = w_k\, \rho_{\text{tmp}} \left(1 + \frac{\mathbf{u_N} \cdot \mathbf{c_k}}{c_s^2} + \frac{(\mathbf{u_N} \cdot \mathbf{c_k})^2}{2c_s^4} - \frac{\mathbf{u_N} \cdot \mathbf{u_N}}{2c_s^2}\right), \tag{8}$$

where w_k are the weighting factors, $\mathbf{c_k}$ the lattice discrete velocities, and c_s the speed of sound magnitude in the lattice. By using the property that $\sum_k f_k^{eq}(\mathbf{x_N}, t) = \sum_k f_k(\mathbf{x_N}, t) = \rho_{\text{tmp}}$ at node N, we subtract and add the following equilibrium functions to the populations f_k at node N:

$$\begin{aligned}\rho' &= \sum_k (f_k(\mathbf{x_N}, t) - f_k^{eq}(\rho_{\text{tmp}}, \mathbf{u_N}) + f_k^{eq}(\rho_{\text{target}}, \mathbf{u_N}))\\ &= \sum_k (f_k(\mathbf{x_N}, t) - \sum_l f_l^{eq}(\rho_{\text{tmp}}, \mathbf{u_N}) + \sum_m f_m^{eq}(\rho_{\text{target}}, \mathbf{u_N}))\\ &= \rho_{\text{tmp}} - \rho_{\text{tmp}} + \rho_{\text{target}}\\ &= \rho_{\text{target}}.\end{aligned}$$

Combining these equilibrium functions, we can thus impose the density ρ_{target} at the boundary. It is worth noting that the velocities $\mathbf{u_N}$ remain unchanged.

2.3 Porous Medium Simulation

When the populations f_i hit an obstacle during a standard collision, their velocities get reversed after the streaming. This leads to a zero fluid velocity relative to the boundary of the obstacle (no-slip) and is called solid bounce-back boundary in LB:

$$f_i(\mathbf{x_b}, t + \Delta t) = f_{\hat{i}}^*(\mathbf{x_b}, t), \tag{9}$$

where the subscript b denotes the location of a boundary and $\hat{i}$ is the opposite lattice direction of i. The partial bounce-back (PBB) method is an extension of the solid bounce-back boundary that can simulate porosity. The idea is to assign a fraction of bounce-back to each volume element, whether none, part, or all of the fluid is to be reflected at the boundary. The outgoing populations in direction i are calculated as a weighted sum of the scattered populations $f_i^*(\mathbf{x}, t)$ and the bounced populations $f_{\hat{i}}(x, t)$ [21]:

$$f_i(\mathbf{x} + \mathbf{c_i}\Delta t, t + \Delta t) = (1 - \gamma) f_i^*(\mathbf{x}, t) + \gamma f_{\hat{i}}(\mathbf{x}, t), \tag{10}$$

where γ is the bounce-back fraction, ranging between 0 (completely fluid voxel) and 1 (completely solid voxel). Note that for $\gamma = 0$, we recover the usual streaming of a fluid node (Eq. 6), while for $\gamma = 1$, we have the equation of the solid bounce-back boundary (Eq. 9). Consistently, the macroscopic velocity $\mathbf{u_f}$ is computed as [21]

$$\mathbf{u_f}(\mathbf{x}, t) = \frac{(1 - \gamma)}{\rho(\mathbf{x}, t)} \sum_i \mathbf{c_i} f_i(\mathbf{x}, t). \tag{11}$$

Now, let us recall Darcy's law for laminar flow [22]:

$$u_f = -\frac{k}{\rho \nu} \frac{\Delta P}{L}, \tag{12}$$

with ν the kinematic viscosity, ρ the fluid density, ΔP the pressure drop over length L, and k the *permeability*. In parallel, Walsh et al.

show that the permeability k_{PBB} in m^2 of PBB voxels is related to γ and the simulation time step Δt as follows:

$$k_{\mathrm{PBB}} = \frac{(1-\gamma)\nu\,\Delta t}{2\gamma}. \tag{13}$$

By combining 11, 12, and 13, we can link Darcy's law to the PBB formalism:

$$u_{\mathrm{f}} = -\frac{\Delta t}{2\gamma}\frac{\Delta P}{L\rho}. \tag{14}$$

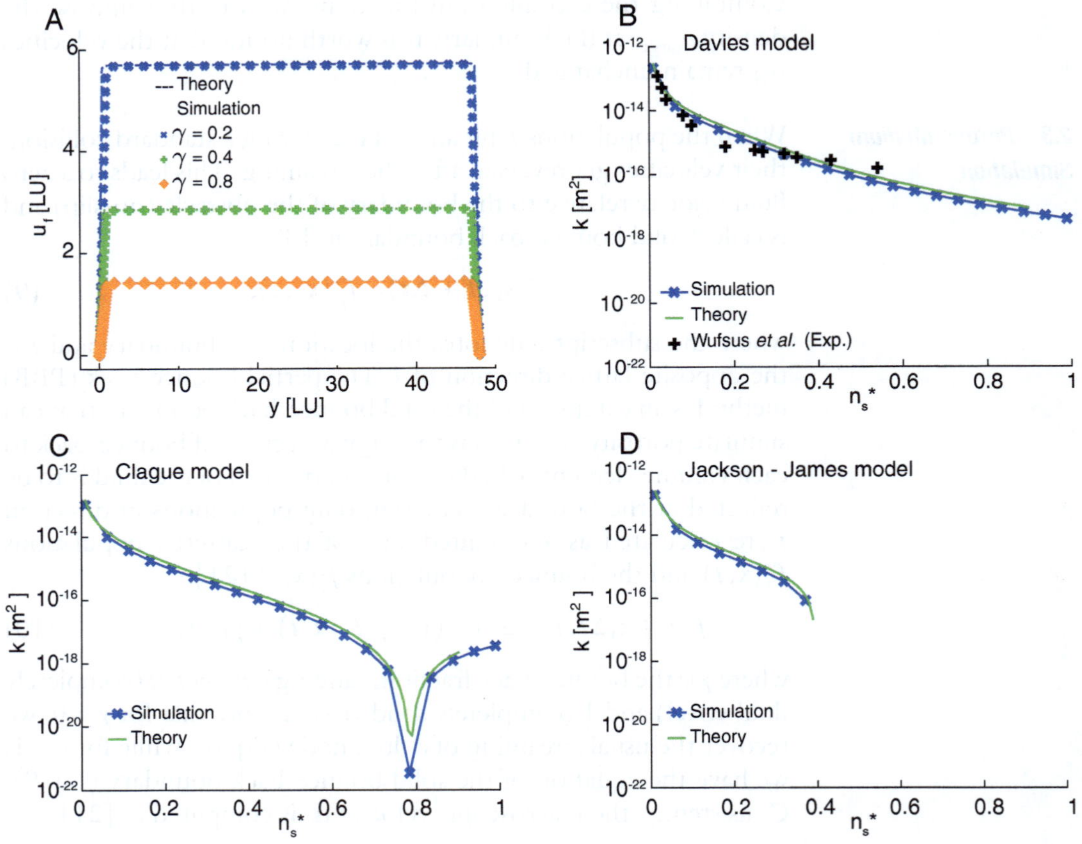

Fig. 3 (**a**) Comparison of fluid velocity of a transverse slice of a clot in a rectangular duct, between Darcy's law and our implementation of the PBB method, for three bounce-back fractions $\gamma = 0.2, 0.4, 0.8$. Velocity and position are given in lattice units (LU). Space and time steps were $\Delta x = 1.13 \cdot 10^{-4}$ m and $\Delta t = 1.83 \cdot 10^{-5}$ s, $\nu = 3 \cdot 10^{-6}\ \mathrm{m}^2 \cdot \mathrm{s}^{-1}$, and the pressure gradient $\nabla P = 7700\ \mathrm{Pa}\cdot\mathrm{m}^{-1}$. The domain size was $N_x = N_y = N_z = 48$ lattice nodes. (**b–d**) Fibrous thrombus intrinsic permeability k as a function of fiber solid fraction n_s^*. Wufsus et al. [26] in vitro fibrin gels measurements (black +), Davies' equation (B) [23], Clague (C) [24], and Jackson and James (D) [25], according to Eqs. 16, 17, and 18, respectively. Simulations were made in a tube of radius 4.5 mm (23 nodes), a length 1 cm (46 nodes), with a fully occluding homogeneous thrombus (i.e., same dimensions as the tube). The space–time discretization was $\Delta x = 2.17 \cdot 10^{-4}$ m and $\Delta t = 1.58 \cdot 10^{-3}$ s, and the pressure gradient $\nabla P = 100\ \mathrm{Pa}\cdot\mathrm{m}^{-1}$. The hydrated fiber radius ranges from $R_f = 51$ to 60 nm for formulae and simulations, as in [26]

We validate the use of the PBB method by measuring the values of the fluid velocity u_f in numerical simulations and comparing with the analytical solution of Darcy and Walsh Eq. 14 in Fig. 3a. Flow was simulated in a 3D rectangular duct filled with a porous medium with bounce-back fractions of $\gamma = 0.2, 0.4, 0.8$, respectively. The velocity profile was measured along the y-axis (transversal) in the center of the duct. In the analytical solution, we imposed no-slip conditions, i.e., $u_f = 0$ at the walls.

2.4 Permeability Laws

The PBB method is a mesoscopic method, which means that it acts as an interface between macroscopic quantities, such as permeability, and microscopic quantities of interest, such as fiber radius or fiber arrangement. One advantage is the possibility of simulating spatially or temporally macroscopic systems with a non-trivial description and a granularity that can reproduce heterogeneity in realistic computational times. Another advantage is that it does not need the knowledge of microscopic details, such as pore topology, provided that the permeability law of the medium is known. The permeability is a macroscopic quantity that measures the ability of a porous medium to let a fluid flow through it. It depends on the material's nature, the shape of the pores, and their local arrangements. Consequently, a given material's permeability law is generally not known a priori. Thus, it has to be measured experimentally through flow measurements and estimated with, for instance, Darcy's law. Using Eq. 13, we can express the bounce-back fraction in terms of the permeability of the porous medium to simulate

$$\gamma = \frac{1}{1 + \frac{2k}{\nu \, \Delta t}}. \tag{15}$$

In our test case, it implies that in order to simulate a porous material with the characteristic permeability of a thrombus, we have to incorporate in our model a permeability law that provides a value of k, knowing the radius of the fibers (R_f) and the relative volume occupied by the fibers (solid fraction n_s^*). Several studies established permeability laws of fibrous media, either experimentally [23] or numerically, using randomly organized fibers [24], or two-dimensional periodic square arrays of cylinders [25]. These different laws yield to similar results, but Wufsus et al. [26] show that the permeability of in vitro fibrin clots is best described by Davies' permeability law [23]:

$$k_D(R_f, n_s^*) = R_f^2\left(16 n_s^{*1.5}\left(1 + 56 n_s^{*3}\right)\right)^{-1}. \tag{16}$$

Finally, we obtain the relationship $\gamma(n_s^*, R_f)$ by replacing in Eq. 15 the value of k_D given by Eq. 16.

Other common laws to describe fibrous porous media permeabilities are Clague's equation [24], a numerical solution of permeability of randomly organized fibrous media:

$$k_C(R_f, n_s^*) = 0.50941 R_f^2 \left(\left(\frac{\pi}{4n_s^*} \right)^{0.5} - 1 \right)^2 e^{-1.8042 n_s^*}, \tag{17}$$

and Jackson–James' equation [25], the weighted average of the solution to the Stokes equation for flow parallel to or normal to 2D periodic square arrays of cylinders:

$$k_{JJ}(R_f, n_s^*) = R_f^2 \frac{3}{20 n_s^*} (-\ln(n_s^*) - 0.931). \tag{18}$$

Permeabilities measured with the PBB model as a function of input solid fractions are shown in Fig. 3b–d; they correctly replicated their empirical counterparts Eqs. 16, 17, and 18.

3 Proof-of-Concept: Minimal Thrombolysis

We illustrate the results of implementing the presented methodology through a minimal example of thrombolysis. In this example, a thrombus of length 4.75 mm initially fully occludes a 2D tube of width 5.8 mm. We impose a pulsatile pressure at the inlet. The pressure drop magnitude is linearly interpolated throughout the recanalization, depending on the current blood velocity:

$$\Delta P(t) = \Delta P(0) + u(t) \frac{\Delta P_{\text{empty}} - \Delta P(0)}{u_{\text{empty}} - u(0)}, \tag{19}$$

where the subscript *empty* designates the situation where the thrombus is fully lysed. u_{empty} is imposed to the physiologically realistic value of 20 cm·s^{-1}, and ΔP_{empty} is the corresponding pressure drop, obtained with Poiseuille's formula using the dimensions of the simulation tube. The initially imposed pressure drop $\Delta P(0)$ is taken from measurements in ischemic stroke patients during thrombolytic therapy [27] and is set at 3000 Pa ($\nabla P(0) \sim$ 45 mmHg·cm^{-1}). Finally, $u(t)$ is a result of the LB computation.

We simulate the lysis of the thrombus by reducing at each iteration the fibrin concentration F of lattice cells that are at the front of lysis by an arbitrary amount proportional to the ratio $u(t)/u_{\text{empty}}$ and to the current fibrin concentration:

$$F(t + \Delta t) = F(t) - C \frac{u(t)}{u_{\text{empty}}} F(t) \Delta t, \tag{20}$$

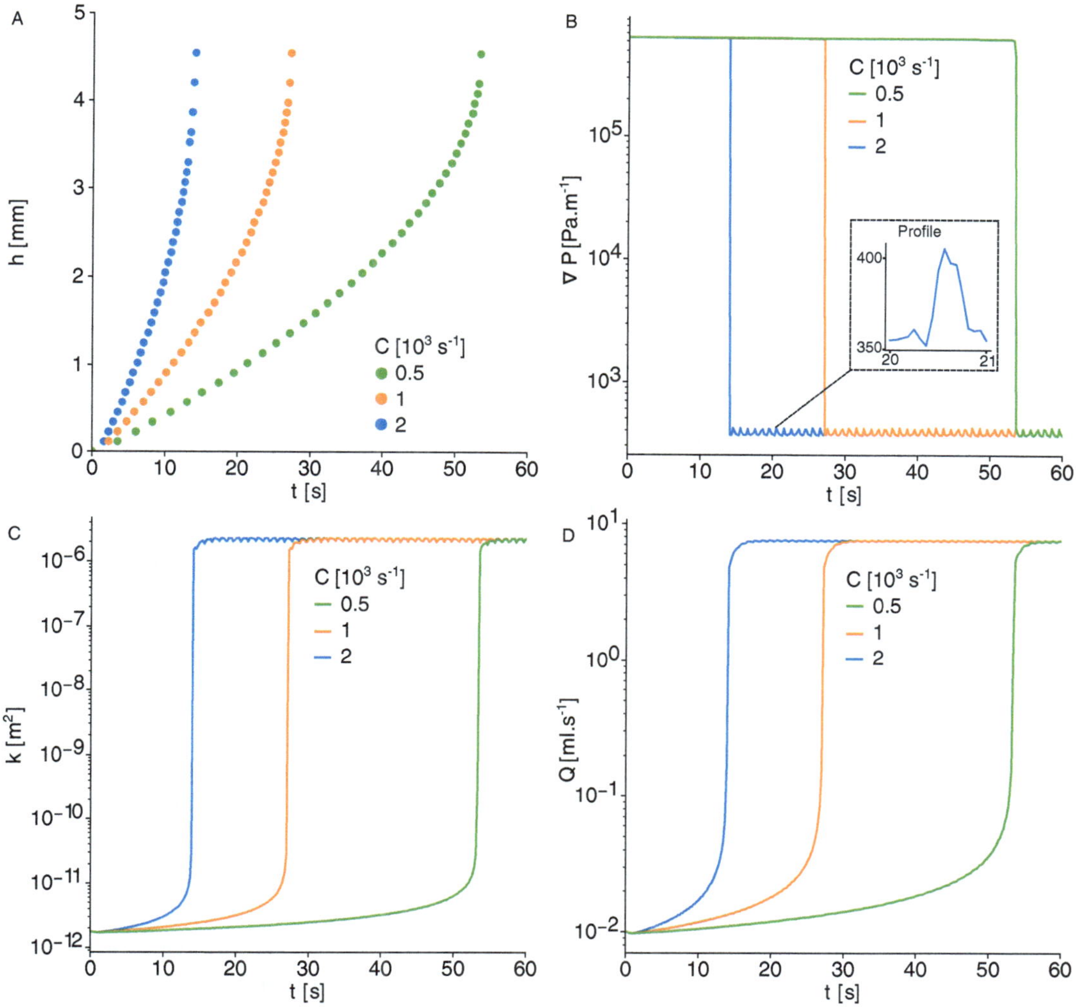

Fig. 4 Hemodynamic quantities evolution during a mock thrombolysis simulation with pulsatile flow. The initial fibrin concentration is set to 2 mg·ml^{-1}, $R_{f_0} = 140$ nm, and using Davies' law. The reaction rate *C* is varied. (**a**) The position of the lysis front *h*. (**b**) The pressure gradient applied, which is imposed as in eq. 19. (**c**) The permeability of the thrombus. When completely lysed ($\sim$ 28 s), the permeability shows the tube's hydraulic resistance. It oscillates at the rate imposed in the inlet density, because the hydraulic resistance depends on the fluid density. (**d**) The flow rate. Recanalization occurs mostly at the end of the lysis, when the thrombus is totally dissolved

where C is a constant that controls the reaction rate. We define the lysis front h as the position where 10% or less of initial fibrin concentration remains. The permeability and bounce-back fraction are subsequently recomputed using Davies' law, and computing the solid fraction n_s^* as described in [28], with $R_{f_0} = 140$ nm. With this minimalistic model, we observe, as could be expected, a faster-than-linear evolution of the front position Fig. 4a. Figure 4b–d shows the evolution of the hemodynamic conditions during the lysis. Note that when the clot is completely lysed (~ 28 s), the permeability on Fig. 4c becomes the hydraulic resistance of the tube. As a result, we showed that the framework presented can work in both the physiological and pathological regimes of stroke treatment.

4 Notes

1. Although widely used, one must be careful with the approximation of laminar blood flow. A recent study reports physiological blood flow as turbulent [29] even at low-mean Reynolds number (<400). Turbulence has previously been associated with the initiation of vascular diseases such as intracranial aneurysms or atherosclerosis [30, 31], due to the mechanosensory function of endothelial cells, which produces a biological response to the wall shear stress applied [32]. It might therefore be necessary to include blood flow turbulence and wall shear stress when designing new therapies for vascular diseases.
2. The open-source library Palabos offers interesting features such as: its flexibility due to being open-source, the built-in possibility to import custom geometries, and being continuously improved and maintained by an active community. Computation using graphical processing units (GPU) has already been showcased [17] and is underway for stable release, including multi-GPU computation.
3. A minimal working example of our thrombolysis model with particles acting as a profibrinolytic agent can be found in Palabos, under `examples/showCases/partialBounceBack/fibrinolysis.cpp`.
4. The pressure boundary condition presented here is implemented in Palabos, under the names `FluidPressureInlet3D` and `myFluidPressureOutlet3D` in `fibrinolysis.cpp`.
5. The PBB dynamics is implemented in Palabos, under `src/boundaryCondition/partialBBdynamics.hh`.
6. More detailed versions of the thrombolysis model can be found here https://gitlab.com/remy.pet/insist-2020 (3D model) and here https://gitlab.com/remy.pet/thrombolysis2d (2D model).
7. If one wants to simulate heterogeneous porous media at the mesoscopic scale where the permeability law of individual entities is known, but the overall permeability is not, we suggest computing an estimated equivalent permeability of the

medium. This was done for instance in [26] for fibrin–platelets thrombi. It can be shown [33] that the equivalent permeability of two media organized in series is given by the harmonic mean, and by the arithmetic mean if they are organized in parallel. Yet another option could be to increase the spatial resolution so that there is only one entity per lattice node.

8. It is also possible to use the PBB dynamics to simulate down to the microscopic scale, if one knows the detailed topology of the medium, by imposing a bounce-back fraction of 1 to the solid parts of the medium. The PBB method is a one-size-fits-all solution for porous media simulations with LB.
9. The permeability values for the physiological range of low solid fractions (0.003–0.04, i.e., 1–4 mg.ml^{-1} fibrinogen concentration) do not vary substantially from one model to another; thus, if the clots studied are highly porous, the choice of model has little impact. Clague's law predicts an increase of permeability with the increase of solid fraction in the 0.8–1 range, and Jackson–James' law asymptotically approaches zero as n_s^* approaches 0.4. Thus, in order to be able to simulate the whole range of solid fractions, Davies' permeability law was chosen for the model.
10. To date, the work of Wufsus et al. [26] serves as a reference for quantifying thrombi permeability. However, recent experiments [34] estimated permeabilities in vivo two to three orders of magnitude higher than those observed by Wufsus et al.. Possible interpretations of this discrepancy are the magnification of the thrombus' pores due to pulsatile flow, thrombus histology, structural composition and heterogeneity, and vessel compliance, which creates paths for flow to seep through. This is a point to consider when transferring in vitro experiments, or models, to the in vivo context. One might need to choose a more appropriate permeability law that depends on parameters such as shear stress or diastole–systole pressure difference. Quantitative data are needed on that matter but are hardly available.
11. In the case of turbulent flows, the quadratic flow term from Darcy–Forchheimer's law might be needed. However, a study giving the permeability law for flow-diverting stents [35] supports that in the physiological blood flow regime, the linear relationship should be sufficient.
12. One can also model flow-diverting stents not as porous media, but as 2D screen deflectors, as done in [36].

Acknowledgements

This research has been developed under the umbrella of the CompBioMed Consortium and the INSIST project [37], which has received funding from the European Union's Horizon 2020 research and innovation programme under grant agreement No. 777072.

References

1. Wagner A, Ehlers W (2008) A porous media model to describe the behaviour of brain tissue. PAMM 8(1):10201–10202. ISSN: 1617-7061. https://doi.org/10.1002/pamm.200810201
2. Padmos RM et al (2021) Coupling one-dimensional arterial blood flow to three-dimensional tissue perfusion models for in silico trials of acute ischaemic stroke. Interface Focus 11(1):20190125. ISSN: 2042–8901, 2042–8901. https://doi.org/10.1098/rsfs.2019.0125
3. Piebalgs A et al (2018) Computational simulations of thrombolytic therapy in acute ischaemic Stroke. Sci Rep 8(1). ISSN: 2045–2322. https://doi.org/10.1038/s41598-018-34082-7
4. Piebalgs A, Yun Xu X (2018) Towards a multiphysics modelling framework for thrombolysis under the influence of blood flow. J R Soc Interface 12(113):20150949. ISSN: 1742-5689, 1742–5662. https://doi.org/10.1098/rsif.2015.0949
5. Sun C, Munn LL (2008) Lattice Boltzmann simulation of blood flow in digitized vessel networks. Comput Math Appl 55(7):1594–1600. ISSN: 0898-1221. https://doi.org/10.1016/j.camwa.2007.08.019
6. Závodszky G (2015) Hemodynamic investigation of arteries using the lattice Boltzmann method. PhD thesis. https://doi.org/10.13140/RG.2.1.4110.5448
7. Latt J et al (2020) Palabos: parallel lattice Boltzmann solver. Comput Math Appl ISSN: 08981221. https://doi.org/10.1016/j.camwa.2020.03.022
8. Frisch U, Hasslacher B, Pomeau Y (1986) Lattice-gas automata for the Navier–Stokes equation. Phys Rev Lett 56(14):1505–1508. https://doi.org/10.1103/PhysRevLett.56.1505
9. He X, Luo L-S (1997) Lattice Boltzmann model for the incompressible Navier–Stokes equation. J Stat Phys 88(3/4):927–944. ISSN: 0022-4715. https://doi.org/10.1023/B:JOSS.0000015179.12689.e4
10. Xiaoyi He and Li-Shi Luo (1997) Theory of the lattice Boltzmann method: From the Boltzmann equation to the lattice Boltzmann equation. Phys Rev E **56**(6):6811–6817. ISSN: 1063-651X, 1095–3787. https://doi.org/10.1103/PhysRevE.56.6811
11. Chopard B et al (2002) Cellular automata and lattice Boltzmann techniques: an approach to model and simulate complex systems. Adv Complex Syst 5(02n03):103–246. ISSN: 0219-5259, 1793–6802. https://doi.org/10.1142/S0219525902000602
12. Krüger T et al (2017) The lattice Boltzmann method: principles and practice. Graduate texts in physics. Springer, Cham. ISBN: 978-3-319-44647-9; 978-3-319-44649-3. https://doi.org/10.1007/978-3-319-44649-3
13. Succi S (2018) The lattice Boltzmann equation: for complex states of flowing matter. Oxford University Press, Oxford. ISBN: 978-0-19-959235-7. https://doi.org/10.1093/oso/9780199592357.001.0001
14. Malaspinas O et al (2015) A spatio-temporal model for spontaneous thrombus formation in cerebral aneurysms. https://doi.org/10.1101/023226
15. Li R et al (2014) Lattice Boltzmann modeling of permeability in porous materials with partially percolating voxels. Phys Rev E 90(3). ISSN: 1539-3755, 1550-2376. https://doi.org/10.1103/PhysRevE.90.033301
16. Li S, Chopard B, Latt J (2019) Continuum model for flow diverting stents in 3D patient-specific simulation of intracranial aneurysms. J Comput Sci 38:101045. ISSN: 1877-7503. https://doi.org/10.1016/j.jocs.2019.101045
17. Kotsalos C, Latt J, Chopard B: Bridging the computational gap between mesoscopic and continuum modeling of red blood cells for fully resolved blood flow. J Comput Phys 398:

108905. ISSN: 0021-9991. https://doi.org/10.1016/j.jcp.2019.108905

18. Bhatnagar PL, Gross EP, Krook M (1954) A model for collision processes in gases. I. Small amplitude processes in charged and neutral one-component systems. Phys. Rev 94(3): 511–525. https://doi.org/10.1103/PhysRev.94.511
19. Shan X, Yuan X-F, Chen H (2006) Kinetic theory representation of hydrodynamics: a way beyond the Navier–Stokes equation. J Fluid Mech 550:413–441. ISSN: 1469-7645, 0022-1120. https://doi.org/10.1017/S0022112005008153
20. Xiu-Ying K et al (2005) Simulation of blood flow at vessel bifurcation by lattice Boltzmann method. Chin Phys Lett 22(11):2873. ISSN: 0256-307X. https://doi.org/10.1088/0256-307X/22/11/041
21. Walsh SD, Burwinkle H, Saar MO (2009) A new partial-bounceback lattice-Boltzmann method for fluid flow through heterogeneous media. Comput Geosci 35(6): 1186–1193. ISSN: 00983004. https://doi.org/10.1016/j.cageo.2008.05.004
22. Darcy H (1856). Les fontaines publiques de la ville de Dijon. https://gallica.bnf.fr/ark:/12148/bpt6k624312 (visited on 12/18/2021)
23. Davies CN (1952) The separation of airborne dust and particles. Inst Mech Eng B1:185–213
24. Clague DS et al (2000) Hydraulic permeability of (un)bounded fibrous media using the lattice Boltzmann method. Phys. Rev. E 61(1): 616–625. ISSN: 1063-651X, 1095–3787. https://doi.org/10.1103/PhysRevE.61.616
25. Jackson GW, James DF (1986) The permeability of fibrous porous media. Can J Chem Eng 64(3):364–374. ISSN: 00084034, 1939019X. https://doi.org/10.1002/cjce.5450640302
26. Wufsus AR, Macera NE, Neeves KB (2013) The hydraulic permeability of blood clots as a function of fibrin and platelet density. Biophys J 104(8):1812–1823. ISSN: 00063495. https://doi.org/10.1016/j.bpj.2013.02.055
27. Sorimachi T et al (2011) Blood pressure measurement in the artery proximal and distal to an intra-arterial embolus during thrombolytic therapy. J NeuroIntervent Surg 3(1): 43–46. ISSN: 1759–8478, 1759–8486. https://doi.org/10.1136/jnis.2010.003061
28. Diamond SL, Anand S (1993) Inner clot diffusion and permeation during fibrinolysis. Biophys J 65(6):2622–2643. ISSN: 00063495. https://doi.org/10.1016/S0006-3495(93)81314-6
29. Saqr KM et al (2020) Physiologic blood flow is turbulent. Sci Rep 10(1):15492. issn: 2045-2322. https://doi.org/10.1038/s41598-020-72309-8
30. Chiu J-J, Chien S (2011) Effects of disturbed flow on vascular endothelium: pathophysiological basis and clinical perspectives. Physiol. Rev. 91(1):327–387. ISSN: 0031-9333. https://doi.org/10.1152/physrev.00047.2009
31. Saqr KM et al (2020) What does computational fluid dynamics tell us about intracranial aneurysms? A meta-analysis and critical review. J Cerebral Blood Flow Metabol 40(5): 1021–1039. ISSN: 0271-678X. https://doi.org/10.1177/0271678X19854640
32. Rashad S et al (2020) Epigenetic response of endothelial cells to different wall shear stress magnitudes: a report of new mechano-miRNAs. J Cell Physiol. **235**(11): 7827–7839 ISSN: 1097–4652. https://doi.org/10.1002/jcp.29436
33. Ali MA, Umer R, Khan KA (2019) Equivalent permeability of adjacent porous regions. In: 2019 Advances in science and engineering technology international conferences (ASET), pp. 1–4. https://doi.org/10.1109/ICASET.2019.8714226
34. Terreros NA et al (2020) From perviousness to permeability, modelling and measuring intra-thrombus flow in acute ischemic stroke. J Biomech 111:110001. ISSN: 00219290. https://doi.org/10.1016/j.jbiomech.2020.110001
35. Csippa B et al (2020) Hydrodynamic resistance of intracranial flow-diverter stents: measurement description and data evaluation. Cardiovasc Eng Technol 11(1):1–13. ISSN: 1869–4098. https://doi.org/10.1007/s13239-019-00445-y
36. Li S (2019) Continuum model for flow diverting stents of intracranial aneurysms. https://archive-ouverte.unige.ch/unige:115538
37. INSIST (2017) In Silico Clinical Trials for treatment of acute ischemic stroke (INSIST) H2020 project. https://insist-h2020.eu/

Chapter 18

Efficient and Reliable Data Management for Biomedical Applications

Ivan Pribec, Stephan Hachinger, Mohamad Hayek, Gavin J. Pringle, Helmut Brüchle, Ferdinand Jamitzky, and Gerald Mathias

Abstract

This chapter discusses the challenges and requirements of modern Research Data Management (RDM), particularly for biomedical applications in the context of high-performance computing (HPC). The FAIR data principles (Findable, Accessible, Interoperable, Reusable) are of special importance. Data formats, publication platforms, annotation schemata, automated data management and staging, the data infrastructure in HPC centers, file transfer and staging methods in HPC, and the EUDAT components are discussed. Tools and approaches for automated data movement and replication in cross-center workflows are explained, as well as the development of ontologies for structuring and quality-checking of metadata in computational biomedicine. The CompBioMed project is used as a real-world example of implementing these principles and tools in practice. The LEXIS project has built a workflow-execution and data management platform that follows the paradigm of HPC–Cloud convergence for demanding Big Data applications. It is used for orchestrating workflows with YORC, utilizing the data documentation initiative (DDI) and distributed computing resources (DCI). The platform is accessed by a user-friendly LEXIS portal for workflow and data management, making HPC and Cloud Computing significantly more accessible. Checkpointing, duplicate runs, and spare images of the data are used to create resilient workflows. The CompBioMed project is completing the implementation of such a workflow, using data replication and brokering, which will enable urgent computing on exascale platforms.

Key words High-performance computing, Biomedicine, Research data management, FAIR principles, Resilient distributed workflows, Exascale

1 Introduction

Biomedical simulations cover a wide range of research areas from molecular medicine over the cardiovascular and respiratory systems to neuro-musculoskeletal diseases. Understanding and searching for cures of these diseases with simulation and modeling applications have quickly evolved in methodology and accuracy over recent years. At the same time, the associated computational needs have grown exponentially. Therefore, biomedical computing

Alexander Heifetz (ed.), *High Performance Computing for Drug Discovery and Biomedicine*, Methods in Molecular Biology, vol. 2716, https://doi.org/10.1007/978-1-0716-3449-3_18,

has entered the realm of high-performance computing (HPC) and strives toward deployment on the computers of the exascale era. Along with the computational needs, also the amounts of data that are processed, generated, and/or stored have grown correspondingly and can reach multiple petabytes. The corresponding datasets do not only stick out by their mere size, but also due to special requirements concerning security and privacy, particularly, if they stem from patients. Furthermore, they are a significant asset for the research community, since a lot of expertise and computational power went into their generation.

Accordingly, biomedicine on HPC systems requires modern Research Data Management (RDM) to meet all the challenges of size, compliance, and full scientific re-usability. These requirements are included in the so-called FAIR principles (Findable, Accessible, Interoperable, Reusable [1]), which have become universally endorsed by many communities. Libraries, funding agencies, and publishers are forefront players in this, while domain-specific research communities are on their way to technically and methodically adopt these practices.

It is important to emphasize that RDM in HPC is quite a challenging topic: Several standard approaches, which are used both to make existing data or the output of new projects follow the FAIR principles and to make the research process transparent, fail here. In particular, the approach to create a data repository for outputs and intermediate-step (i.e., partially processed) data is not feasible. Output data of HPC applications reside on huge, fast, parallel file systems, which are tightly integrated in the supercomputers and are a substantial cost factor. The ingestion in other systems (such as institutional or general-purpose research data repositories) would usually bring those beyond their limits up to which they can administratively and technically accommodate larger amounts of data. In addition, the data transfer itself may take a long time and can be extremely cumbersome. Thus, an affordable compromise for making the data directly available from storage or archival systems close to the HPC systems has usually to be found. Accordingly, approaches to publish data, while keeping them in place, have been developed [2]. As an example, EUDAT [3], with its components to manage data with metadata (EUDAT B2SAFE [4]), and to assign PIDs (EUDAT B2HANDLE [5]), has the B2STAGE component in order to efficiently exchange data between HPC file systems and EUDAT-driven data systems.

The CompBioMed[1] [6–9] Centre of Excellence is a European collaborative project developing and applying computational methods to advance understanding of human physiology and disease. The project focuses on stand-alone applications and complex

[1] https://www.compbiomed.eu.

workflows that integrate multiple simulation tools and data sources to enable ever more accurate and comprehensive simulations of biological systems. In the project, researches from many universities and institutes, as well as scientists at large HPC facilities, team up to meet these challenges, particularly, managing datasets at many different sites. Therefore, RDM is inherent to the project.

Beyond the mere administration of data, data management in such a project also covers the movement of data, which, ideally, will work in an automated fashion. The need for automated data management tools became particularly apparent when the CompBioMed consortium engaged in the research on the coronavirus, due to the outbreak of the COVID-19 pandemic. Here, drug discovery simulations were conducted at different sites from large throughput screening simulations to highly accurate docking simulations on promising drug candidates [10, 11]. Because these simulations build on each other, results from one site had to be forwarded and served at the next site as input data. This had to be mainly initiated by hand, which is time-consuming and error-prone.

To address the challenges of multi-center workflows in the future, CompBioMed has teamed up with two other Horizon 2020 initiatives, DICE[2] (Data infrastructure capacity for the European Open Science Cloud, GA No. 101017207 [12]) and LEXIS[3] (Large EXecution for Industry and Science, GA No. 825532 [13]). Here, the goal was to integrate their RDM tools for in the CompBioMed RDM and workflows. The LEXIS platform brokers HPC and IaaS-Cloud infrastructure from European supercomputing centers and provide orchestration and distributed data management tools on top for running simulation workflows. This enables researchers to collaborate and to share data more effectively and to access the best-matching resources easily. The LEXIS distributed data management builds on EUDAT's collaborative data infrastructure, as mentioned above, and augments it by APIs for automated data movement in workflows [14]. The results are workflows (also involving different cloud-computing/HPC systems) that inherently fulfill the FAIR principles to a certain degree.

On the domain science and data curation side, these workflow-automation efforts are complemented by approaches to develop ontologies for structuring and quality-checking metadata, e.g., for simulations from engineering [15] or other HPC-centric topics in computational biomedicine [7, 10, 16–19].

In this chapter, we discuss the founding aspects of HPC-related biomedical RDM as well as their implementation in the

[2] https://dice-eosc.eu.

[3] https://lexis-project.eu.

CompBioMed project. A special focus will be the challenges associated with data movement and replication and how corresponding tools can be used to build resilient cross-center workflows.

2 BioMedical Research Data Management

RDM for biomedical HPC data is a special use case of modern RDM, which follows FAIR principles [1]. This requires appropriate file formats, data formats, and metadata schemes, as well as publication platforms. The following discussion of these components will pay special attention to specific biomedical and HPC aspects.

2.1 The FAIR Principles

The strong push toward FAIR principles and the idea of reproducible research in general has shaped novel methods in RDM during the last decade. These go far beyond the idea of optimum disk, tape, or semiconductor-based storage and management systems on top of them: Concepts of semantic nature, from metadata over persistent, unique identifiers (PIDs) to provenance tracking, and semantics-aware data management (e.g., in triple stores) have added a new quality to the business of managing scientific data. All this is aimed at making the research process transparent and reproducible. To a major degree, such ideas have been picked up by research stakeholders, such as funding agencies and editorial houses. This often implies serious requirements on the data management of nowadays' researchers. For example, output data must be provided in a FAIR (and often open data) manner, and data management plans (DMPs) have to be provided in order to obtain funding of research.

In practice, basic Findability, Accessibility, Interoperability, and Re-usability can be implemented with relatively simple methods, while optimal FAIRness needs more work, the description or assessment of which is out of scope of this chapter. The framework of Digital Object Identifiers (DOIs [20]—a special form of PIDs, also used for research articles, for example), as they are provided by DataCite [21], gives a good example of an adaptable, low-threshold implementation in particular of "F" and "A." The "minting" of a DOI with DataCite requires the collection of fundamental metadata according to the DataCite standard [22], and the provisioning of a landing (web-)page with a data description, making metadata and data accessible. DataCite provides an own search facility making the data findable on the Web once they are registered [23]. Clearly, it is subject to vivid community debates if, and when, DOIs or other PIDs (e.g., B2HANDLE PIDs, [5]) should be used for datasets. These debates are beyond the scope of this chapter, which merely lays out the principles.

Interoperability and re-usability involve important aspects, such as formatting of the (meta-) data. However, also legal issues are important for re-usability: Wilkinson et al. [1] clearly state that data must come with an appropriate license. This implies, somewhat contrary to naive intuition, that FAIR data do not necessarily have to be open data (while the respective metadata must be open [1]). Since the FAIR principles have first been stated, a plethora of technical and methodical efforts, from the setup of repositories and metadata stores, to the development of metadata standards and ontologies, have been developed to apply the FAIR principles to larger parts of the research data produced. Currently, a consolidation phase can be observed in which registries are built to collect, e.g., all the various repositories in a list [24], and thus give orientation to scientists.

2.2 Data Formats

In biomedical applications, the need to store and organize large amounts of data has become increasingly important. To ensure that data are effectively shared and used by the scientific community, it is important to choose a suitable file format that embodies the FAIR principles. Some practical considerations for choosing a file format include:

- Self-describing formats contain all the necessary information for interpreting the data within the file, such as data type, size, and organization. This eliminates the need for external documentation and makes it easier to share and use the data.
- Extensible file formats allow for the addition of new data types or structures as needed, without requiring major modifications to the format. This ensures that the format can adapt to changing data requirements over time.
- Meta-standards define how different data standards and formats can be integrated and used together. By using meta-standards, it is possible to ensure interoperability between different data formats and systems.

Biomedical applications based on computational fluid dynamics (CFD), or structural mechanics, produce the largest datasets in the community. Here, file formats such as the Hierarchical Data Format (HDF5 [25]), the Network Common Data Format (NetCDF [26, 27]), the CFD General Notation System (CGNS [28]), the Adaptable IO System (ADIOS [29]), and others, are commonly used. These formats follow similar design principles, aimed at providing portability and ease of access. For example, HDF5 is a self-describing format that allows for the hierarchical organization of data, and it supports extensibility through the addition of user-defined data types. NetCDF is another self-describing format that is widely used in the meteorological community, and it provides a flexible data model that allows for the representation of complex data structures.

A second large type of data comes from HPC-based molecular medicine, where structures for biological macro-molecules, or molecular agents, have to be stored, along with the trajectories out of molecular dynamics (MD) simulations. Although seemingly every large community-based MD program such as Gromacs,[4] Amber,[5] or NAMD[6] [32] seems to have its own input and output file formats, a large set of conversion, analysis, and visualization tools exits that make these file formats interoperable.

In conclusion, choosing a suitable file format is important to ensure that data are effectively shared and used by the scientific community. Self-describing formats, extensibility, and the use of meta-standards are all important considerations when choosing a file format for biomedical applications. File formats such as HDF5 and NetCDF have been proposed to address the challenge of storing and organizing big amounts of data in scientific computing and can be adapted to meet the specific needs of biomedical research.

2.3 Publication Platforms

Appropriate data curation of large biomedical datasets requires a reliable and distributed infrastructure. Such an infrastructure is provided by the EUDAT [3] services for proper FAIR data management and curation, namely the B2SAFE [4] and the B2SHARE [33] repository services. In collaboration with the DICE project, the CompBioMed project set up B2SAFE federation among the CompBioMed Core Partners, the Co-operative University Computing Facilities in the Netherlands (SURF), Barcelona Supercomputing Center (BSC), and University College London (UCL). The B2SAFE service is based on the Integrated Rule-Oriented Data System (iRODS, [34]) and facilitates transfer of large amounts of data as well as data management aspects such as annotating data with PID (persistent identifier). The B2SAFE instance in SURF is connected to tape storage in the backend and offers long-term preservation of the data stored via B2SAFE.

As a next step in collaboration with DICE, CompBioMed plans to deploy an instance of the B2SHARE service, as the community repository for publishing computational biomedical data and results. In this repository, it is possible to implement the different metadata schemas including up-to-date metadata fields that are being used by the respective disciplines. Also, a community responsible role can be assigned to define curation regimes and review the data to protect and amplify the value of the data.

To achieve these targets, CompBioMed has put in place policies and processes. Core Partners are expected to adhere to a minimum

---

[4] https://manual.gromacs.org/current/index.html [30, 31].

[5] https://ambermd.org.

[6] http://www.ks.uiuc.edu/Research/namd/.

set of metadata, enforced by the EUDAT B2SHARE platform. To promote reuse, community-specific metadata schemas are used, which can be implemented automatically in the B2SHARE platform. In line with data curation strategies, this helps to cater both to the immediate and future needs of computational biomedical researchers.

There are, of course, other publication platforms for biomedical applications that use the FAIR principles:

- Zenodo[7] is a free, open-access publication platform that allows researchers to publish and share their research data and software. Zenodo is compatible with the FAIR principles and allows users to assign persistent identifiers to their research outputs, making them easily discoverable and citable.
- Dataverse[8] is a research data repository that allows researchers to share, store, and discover research data. Dataverse is compatible with the FAIR principles and provides tools for data citation, versioning, and sharing. Dataverse also provides metadata templates that help ensure data are well-described and discoverable.
- The Open Science Framework[9] (OSF) is a free, open-source research management platform that allows researchers to share and collaborate on research projects. The OSF is compatible with the FAIR principles and provides tools for managing research data, including version control, data sharing, and collaborative workflows.
- BioSharing[10] is a curated, searchable database of biomedical data standards, databases, and policies. BioSharing is compatible with the FAIR principles and provides tools for discovering and using data standards and resources. BioSharing also provides tools for tracking data usage and impact, making it a valuable resource for both researchers and funders.
- FAIRsharing[11] is a curated, searchable database of data standards, databases, and policies in the life sciences. FAIRsharing is compatible with the FAIR principles and provides tools for discovering and using data standards and resources. FAIRsharing also provides tools for tracking data usage and impact, making it a valuable resource for both researchers and funders.

These are just a few examples of publication platforms that use the FAIR principles in biomedical research. By using these platforms, researchers can ensure that their data and resources are

[7] https://zenodo.org.
[8] https://dataverse.org.
[9] https://osf.io.
[10] https://bio-sharing.org.
[11] https://fairsharing.org.

effectively shared and used by the scientific community, promoting collaboration and advancing scientific knowledge. The CompBioMed project chose the EUDAT services because it is a technologically sound, pan-European solution following essential European policies (from technological to ethical aspects), and it offers integration with the European Open Science Cloud (EOSC, [35]) via DICE.

2.4 Annotation Schemata

The CompBioMed data storage system is built from components of the EUDAT Collaborative Data Infrastructure, including the data storage and publishing service B2SHARE. By default, B2SHARE uses the EUDAT Core metadata schema,[12] which is derived from the DataCite schema [22] for bibliographic information. However, unlike many other data repositories, B2SHARE is extensible, allowing registered communities to add their own domain-specific metadata.

One of the purposes of the schema is to support targeted filtering of large volumes of data. This becomes particularly important with the growing number of big data and ML applications. Since it is difficult to anticipate the future kinds of queries and filters submitted by a plurality of parties, including clinicians and biomedical researchers, a continuous improvement process based upon the logged user queries would be beneficial. These should be reviewed for potential new labels and other possible improvements of the search tool (e.g., B2FIND [36]). Newly proposed labels should also be applied to already existing datasets. However, this might not be possible in case the metadata was not (properly) preserved in the first place, no responsible data owner can be found, or the resources needed to support this kind of activity are unavailable.

As discussed above, B2SHARE and other data publication frameworks can extend the metadata with domain-specific annotations. This has two important benefits: the increase in interoperability due to the metadata being provided in a standardized, structured way; and the possibility to query, filter, and ultimately retrieve datasets, using the domain-specific metadata fields as search input.

The selection of the annotation schema and a matching ontology not only depends on the scientific domain, but also on the target audience for future data usage. For CompBioMed, such users would predominantly be other computational and data scientists at the moment. But as the methods will become more widely available, also clinical researchers, or, in the future, even patients, will be a target audience for the data published. On the community website https://schema.org/, associated problems are discussed: "*There is a great deal of high-quality health and medical information*

---

[12] https://www.eudat.eu/meta/kernel-20.

on the web. Today it is often difficult for people to find and navigate this information, as search engines and other applications access medical content mostly by keywords and ignore the underlying structure of the medical knowledge contained in that content. Moreover, high-quality content can be hard to find using a search engine if the content is not optimized to map the content's concepts to the keywords that users tend to use in search. And while the medical community has invested significant effort in building rich structured ontologies to describe medical knowledge, such structure is today typically available only 'behind the scenes' rather than shared in the Web using standard markup" (https://schema.org/docs/meddocs.html, retrieved March 31, 2023).

Clearly, these groups are interested in different aspects and rely on different vocabularies. CompBioMed, for example, uses specific schemas for computational workflows,[13] for software,[14] and for a large collection of biomedical ontologies.[15] These are used for data published by CompBioMed researchers.

Standardized annotation schemes and terminology ensure consistency and interoperability across different datasets. Standards such as the Dublin Core Metadata Initiative (DCMI) or the Data Documentation Initiative (DDI) can be used to provide a framework for annotation. Clear and descriptive labels help users to understand the meaning and context of the data. Metadata provides additional information about the dataset, such as its source, format, and date of creation. Including metadata in the annotation helps to provide additional context and makes the dataset more discoverable and usable. This can be done using a variety of file formats, such as Comma-Separated Values Format (CSV), the JavaScript Object Notation (JSON), or Extensible Markup Language (XML). In addition to annotations, it is helpful to provide detailed documentation about the dataset, including any preprocessing steps, data cleaning procedures, and other relevant information.

Effective data annotation is essential for ensuring that data are usable and understandable by others. By following these general annotation strategies, organizations can improve the quality and usability of their data and promote data sharing and collaboration.

3 Automated Data Management and Staging

An overview of systems, methods, and tools for automated data management and staging is required first, before the HPC-centric workflows of the CompBioMed collaboration [10, 17, 18] can be

[13] https://bioschemas.org/profiles/ComputationalWorkflow/1.0-RELEASE.

[14] https://schema.org/SoftwareSourceCode.

[15] https://bioportal.bioontology.org/ontologies.

discussed. Here, data management methods such as workflow control [37, 38] and remote access [39] and/or staging [40] in a general computer science context beyond their application in HPC are out of scope for this work. The following discussion rather focuses on a few necessary aspects and on EUDAT and LEXIS methods to implement distributed data management in the scope of cross-site and cross-system workflows. These concepts are applied to workflows in CompBioMed to fully automatize them (*see* Subheading 4).

3.1 Data Infrastructure in HPC Centers

Traditional HPC systems have often been running Unix operating systems or the like; nowadays, Linux is a typical choice. File systems with massive storage and parallel-writing capabilities are attached to these HPC clusters via their internal network. A typical installation of a HPC cluster (e.g., SuperMUC-NG at the Leibniz Supercomputing Centre (LRZ), or Karolina at IT4Innovations) involves a backed-up file system with small quotas for each user for important files (e.g., home directories), and one or more file systems for huge input and output datasets. These split into file systems keeping files on the long-term storage, and file systems for volatile data, where on the latter, a lot of storage space can be offered, often at the cost of rolling file deletion. In the context of longer-term data archival and FAIR RDM, the systems keeping files permanently are most important, as are the tape archives attached to most HPC systems. All these systems are often only accessible on the HPC cluster itself; sometimes parts of the system can be exported, e.g., as Network File System (NFS) ([41] and references therein) mounts to other machines, at least within the same computing center.

3.2 File Transfer and Staging Methods in HPC

Due to security requirements, access to supercomputers is strictly controlled. This is implemented via firewalls and through rigorous cluster operation security concepts. The user typically connects to a "login node" of the HPC cluster, where access is granted via a secure shell (SSH) [42]. On the login node, the user can compile code and run basic tests, submit compute jobs, and perform simple forms of data analysis. Established approaches to access or share files on a supercomputing system, or to copy them to other computing facilities (including the home institute), are based on a few simple methods.

For basic file transfer, `scp`, `sftp`, and `sshfs` (secure copy, secure file transfer protocol, secure-shell filesystem, [39]) are widely used and very well-established tools. It is straightforward to access files with these tools. `scp` [39] is used for encrypted transfer of a file set between a local machine and remote machines, such as users work station, storage services mentioned above, and HPC platforms. The `sftp` and `sshfs` expand these possibilities and allow for more convenient file management over SSH connections. Also, the remote file synchronization tool `rsync` [43] builds on ssh and allows efficient backup and replication, since it only transfers the changes between local and remote file sets.

In addition to this, many computing centers offer data transfer nodes, which run server end points for advanced broadband data transfer methods in the HPC field, such as GridFTP [44], UFTP [45], or GLOBUS [46]. GridFTP is a high-performance data transfer protocol that is based on the file transfer protocol FTP. It is optimized to work with high-bandwidth wide-area networks and should be considered if very large files are to be transferred to and from HPC platforms. Many HPC platforms have GridFTP deployed at an end point. It uses the available bandwidth much more effectively than many other data transfer tools, by using multiple simultaneous TCP streams. It is also resilient against transfer failures.

GridFTP was specified by the Open Grid Forum, and its most widely used implementation is that provided by the GLOBUS Project/Alliance [44]. However, GLOBUS, with its massive online file transfer service GLOBUS Online, has since shifted to a proprietary file transfer setup. A scheduling mechanism (like the LEXIS approach discussed later) is used in GLOBUS Online for asynchronous execution of file transfers, for retrying failed transfers, and for queuing the users' requests, as well as notifying them about the results.

Meanwhile, the GridFTP server and other parts of the original Globus toolkit continue to be maintained as "Grid Community Toolkit" by the Grid Community Forum [47]. Also, alternatives such as UNICORE's[16] UFTP [45] are increasingly considered by the HPC community.

Data transfers to and from object storage systems (as implemented at SURF in the Netherlands as a CompBioMed2 partner) are typically supported by the Amazon Simple Storage Service API (`Amazon S3 REST API` [48]), which can also be set up on top of existing storage solutions using the MinIO framework [49]. Graphical data browsers such as Cyberduck[17] can be used to visually access the data stored. The S3 API has been strongly established in cloud-native computing facilities and may also see increasing use in HPC in the coming years.

3.3 EUDAT Components

The HPC-centric approaches to data movement described above are nicely complemented by EUDAT services, systems, and tools. In particular, the B2SAFE system of EUDAT, based on iRODS [34], offers distributed data and metadata management. iRODS is a middleware installed on servers and implements a layer on top of raw storage systems, effectively making data on all involved

[16] https://www.unicore.eu/.

[17] https://cyberduck.io/.

systems—potentially located at multiple sites—appearing in one file tree. It implements a metadata store for each file and directory, rule-based data management (where rules can implement storage policies, such as "all files larger than one Terabyte go to storage X"), and a wealth of possibilities to react on events (e.g., file creation or deletion). One so-called zone, based on one iRODS "provider server" or "iCAT server"—i.e., a metadata store with a file system and customized metadata—can integrate multiple back-end storage systems. Multiple zones can then be federated across geographical sites, while access rights management remains with the zone owners. Clearly, users from other zones can be given access. iRODS supports a fine-granular rights model based on iRODS access-control lists. A wealth of plugins are available, e.g., for authentication (via OpenID-connect [50]), for serving iRODS as a NFS [41] storage, or for using different storage backends including S3-compliant object stores.

In HPC contexts, iRODS is usually not installed on top of the parallel file system, where the HPC codes read and write, but on top of other backends. Relevant data are then staged between the "raw-usage" HPC cluster file systems and the "managed-data" systems based on iRODS. B2SAFE, on top of the basic iRODS system, provides automated capabilities for cross- and intra-zone data replication, and for minting and managing persistent identifiers (PIDs) for original data and their replica. This gives the iRODS capabilities for cross-site data transfer (via its own single- and multi-stream protocol) an extra edge. If adequate metadata are stored together with the PIDs (as in LEXIS, see below), the data are ready for FAIR publication.

In order to obtain globally unique PIDs and assign them to data, B2SAFE (as other EUDAT components) employs EUDAT's B2HANDLE service. This service is based on the Handle System [51], an international system used also by the Digital Object Identifier (DOI [20]) community. B2HANDLE PIDs are matching the idea of automated data management well, as (e.g., in contrast to DOIs) they can be assigned to any file or directory without fulfilling any special requirements on the dataset (such as the presence of certain required metadata or the long-term availability of (meta-) data. B2HANDLE can be installed on-site (such as B2SAFE), or a central instance can be used.

These tools and services are complemented by further EUDAT services [3] to enhance (meta-)data accessibility. In the context of this chapter, in particular, EUDAT-B2STAGE is worth mentioning, since it has implemented a GridFTP-based method for data staging between iRODS and HPC file systems, and B2SHARE, a (meta-)data portal and repository system. Via B2SHARE, datasets—including those not directly uploaded, but residing on other B2SAFE instances or HPC cluster file systems—can be published and made findable.

3.4 The LEXIS Platform and Distributed Data Infrastructure

The LEXIS [13] platform is an advanced, user-friendly platform that enables users in research and industry to run advanced simulations and computational workflows with automated data management. It uses advanced data management and orchestration techniques deployed across geographically distributed computing centers to provide a seamless experience for users. Thus, it has also been chosen to enable integrated workflows and RDM in CompBioMed2.

A portal [52] provides easy access for up- and downloading data as well as workflow specifications and to start workflows across appropriate target systems. One of the key components of the LEXIS platform in the context of this chapter is the Distributed Data Infrastructure (DDI, for a deep discussion, *see* refs. 14, 52), making heavy use of EUDAT components as described above. On the computing/orchestration side, this is complemented by a Distributed Compute Infrastructure (DCI, *see* also Subheading 4).

The execution of workflows across different data centers constantly requires movement of input and output data. The LEXIS DDI leverages the iRODS system [34] to provide a unified view on data across the data centers. Each center administers their own iRODS zone (as discussed above) on top of their backend data systems. The communication between zones is established natively in the iRODS protocol. On top of iRODS, EUDAT B2SAFE and B2HANDLE have been deployed for data replication across zones and persistent identifier (PID) assignment. The current iRODS federation includes three data centers: Leibniz Supercomputing Centre (LRZ) in Germany, IT4Innovations National Supercomputing Center (IT4I) in the Czech Republic, and SURF in the Netherlands. Figure 1 shows the federation setup.

In the scope of LEXIS, data are divided into three visibility scopes: user data, project data, and public data. Here, project refers to a computational project and defines a user group on the LEXIS platform that shares a common scientific goal and the associated data. The access is managed through the integration of the iRODS-OpenID plugin [53] with the LEXIS Authentication and Authorization Infrastructure (AAI) based on Keycloak [54]. Each user has private access to their user directory within a computational project. Moreover, all users within the computational project have read/write access to the project directory, while only (computational) project managers have write access to the public directory to allow the publishing of data. All users of the platform have read access to data within the public directory, which is also used for data publication.

In order to allow automated data handling on the LEXIS DDI by the workflow-orchestration system, and to standardize data management patterns via the LEXIS portal, all communication of the LEXIS platform with its iRODS subsystem takes place through RESTful APIs. This includes data search, upload, and download

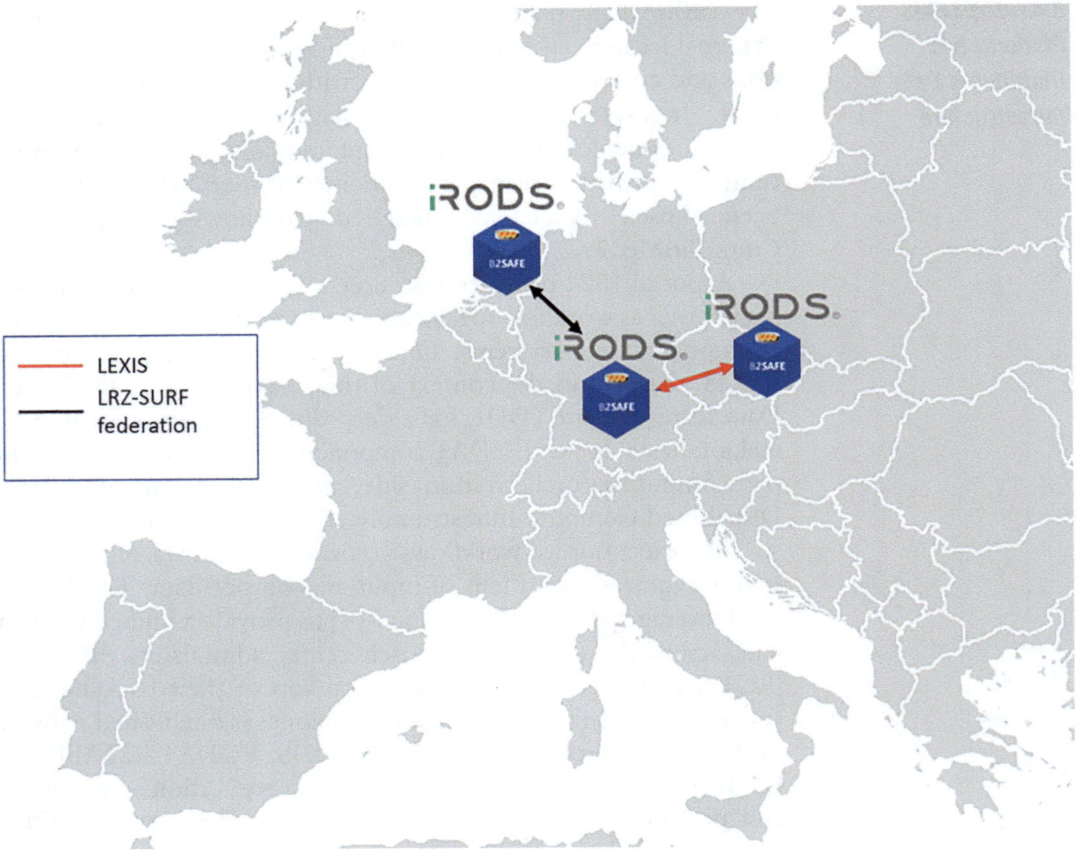

Fig. 1 The Figure shows the LEXIS iRODS/EUDAT-B2SAFE cross-site federation, complemented by the LRZ-SURF federation. HPC centers in Amsterdam (SURF), Munich (LRZ), and Ostrava (IT4I) are connected. Logos/software are subject to BSD-style licensing as evident from and detailed in LICENSE files as of 2022 under https:/github.com/irods/irods and https:/gitlab.eudat.eu/b2safe/B2SAFE-core

APIs, and a data staging API, which are discussed below, as they are relevant examples to understand the basic DDI concept. Further APIs may also be involved, e.g., the Replication API [14].

The LEXIS Staging API is an asynchronous HTTP-REST API that allows the orchestrator to copy, delete, and replicate data between different data sources and HPC infrastructures across the different data centers. Such operations can be triggered via simple HTTP requests to appropriate end points of the API and are executed via a queuing system. The API is based on Django [55] and makes use of the RabbitMQ [56] queuing system and Celery [57]. Once a staging request is triggered, a task ID is assigned. The orchestrator can query the status of the request through the ID and can retrieve the exact location of the staged/replicated data.

The data search, upload, and download APIs allow users not only to transfer their files, but also to assign DataCite-like metadata to their data [22], and to perform metadata search through the LEXIS portal. Very basic metadata assignment is made mandatory through the portal. The metadata management makes use of the iRODS capability to assign key-value information to datasets, which is stored in the iCAT database.

4 Resilient HPC Workflows with Automated Data Management

HPC workflows can connect various data sources, compute sites, and data targets for storage and publication. In complex applications, such workflows typically have multiple source, compute, and target sites, each of which may be unique or redundant. For example, the drug search against the coronavirus conducted in the CompBiomed consortium involved multiple compute sites that provided different hardware capabilities, and the workflow applied in this search involved dependencies and data movement between the sites. Commonly, in such complex scenarios, data staging, data processing, and computation are triggered manually, but for large computational campaigns, it becomes apparent that partial or full automation of such workflows is required.

4.1 Resilient Workflows

Leading edge HPC machines feature very high node counts. Individual nodes have a reasonable mean time to failure (MTF); however, when one utilizes thousands or ten thousands of nodes together in a single highly parallel application, the overall MTF of a job becomes significant. This is mainly due to the MPI programming paradigm that used in most modern HPC codes. A typical MPI simulation will abort if a single MPI task fails and takes the whole application down with it, although modern MPI provides first features for resiliency. Computational biomedical simulations employing these exascale platforms may well employ time- and safety-critical simulations, where results are required at the operating table in faster than real time. Given the increased threat of node failure, one mitigation is to employ resilient HPC workflows.

HPC applications often implement a series of strategies to recover from hardware and software errors, to ensure that the application can be continued when such errors occur.

A simple strategy used in most HPC applications is checkpointing, where the state of the application is stored periodically to the file system, so that on a hardware error, or a termination of the application due to reaching the requested duration, the application can be restarted from the last checkpoint.

Duplicate runs increase the reliability and resilience of a workflow. This involves running multiple copies of the same simulation in parallel, such that at least one run finishes correctly. If more runs return results, they can be checked for consistency. Such strategies are often used in large distributed campaigns such as Rosetta@-home or Folding@home, where larger failure probabilities are expected [58]. For large, data-centric workflows, spare images of the input data provide the corresponding strategy to run duplication. Ideally, copies of the datasets are available from different sources, such that the inaccessibility of a data source can be mitigated. This might, however, involve additional actions in the workflow, such as staging the data at a different site.

4.2 CompBioMed Workflows on the LEXIS Platform

The LEXIS project has built a workflow-execution and data management platform following the paradigm of HPC–Cloud convergence for demanding Big Data applications. It orchestrates workflows with YORC [59], utilizing the DDI, and the distributed computing resources within the platform (DCI), and is accessed by a user-friendly LEXIS portal for workflow and data management. This makes HPC, and Cloud Computing, a significantly lower-threshold endeavor than before. The portal allows end users to manage their workflows and data without deep-diving into technical details.

Using this platform, the CompBioMed project is currently completing the implementation of a new workflow, which is resilient to catastrophic failure of an upcoming exascale platform, using a novel evocation of data replication and brokering.

LEXIS already offers a resilient HPC workflow, wherein the same simulation is launched concurrently on multiple HPC platforms: the chances of all the platforms failing are far less than that of any individual platform failing. However, such replicated computation can prove expensive, especially when employing millions of cores, as is to be expected at the exascale. This novel workflow employs replicated data rather than replicated computation.

The biomedical use case is an example of urgent computing: the application HemoFlow (University of Amsterdam) will simulate the flow of blood through a brain aneurysm with the goal of optimally inserting a stent. In the operating theatre, a surgeon will use the simulation results, produced remotely using live patient data, to help determine the best stent alignment.

The LEXIS DCI and DDI have both been used and extended in the course of the collaboration between CompBioMed and LEXIS. The DCI includes nodes from HPC- and Cloud-Computing platforms at LRZ and IT4I and has been extended to include EPCC's HPC system Cirrus at the University of Edinburgh; the DDI has been extended to include a data node at SARA.

The use case runs as follows: the application will launch at the initial HPC center, currently the Leibniz Supercomputing Centre (LRZ). The application often writes data snapshots and, less frequently, checkpoint files. As soon as these checkpoint files are created, the LEXIS Platform will automatically duplicate these files to all relevant data nodes in the CompBioMed/LEXIS DDI. These other platforms are designed to be located in different countries to mitigate against a center-wide failure, e.g., power outage, at the initial HPC center.

To test the resiliency of the workflow, a single MPI task can be aborted. This causes the entire simulation to fail, thus mimicking a node failure. This failure triggers the LEXIS Platform to restart the simulation on one of the remote HPC platforms in the CompBioMed/LEXIS DCI, where it will stage the latest restart file at all the HPC nodes in the DCI from its nearest data node. The choice of HPC platform is then handled by the LEXIS Platform's orchestration system.

The duplication of data for exascale simulations must consider not only the amount of data that needs to be copied and the bandwidth required, but also ensure that both the duplication and the subsequent staging follow appropriate data management principles, given that biomedical simulations can contain patient-sensitive data. The amount and bandwidth can be addressed by employing the existing GEANT2 network. For the publication of open results from workflows, the FAIR data principles are endorsed by the public parts of the LEXIS DDI.

At present, the target application has been installed at EPCC, LRZ, and IT4I, and the staging mechanism is currently being enabled at EPCC. The workflow manager has been written and is undergoing testing using the target simulation at LRZ and IT4I, and EPCC will be incorporated in the near future.

5 Summary

Especially when large-scale biomedical simulations are applied, and collaborations between multiple research centers are involved, efficient and reliable data management is crucial. With a suitable file format (e.g., HDF5 or NetCDF), the right choice of metadata, and persistent identifiers (e.g., DOIs provided by DataCite), the FAIR principles can be followed. This ensures effective data sharing and usage within the scientific community, as it helps to make the research process transparent, reproducible, and enables quality-checking, which is important not only in the context of biomedical applications.

Publication platforms can play an important role in sharing research findings, best practices, and facilitate collaboration among researchers working with HPC systems in the biomedical field. This can ultimately contribute to more efficient and reliable data management.

Annotation schemata guide the metadata collection and help in structuring and labeling the data, making it easier for researchers and HPC systems to process and analyze the information effectively. This, in turn, leads to more accurate and reliable results. Automated data movement and replication can be implemented, even in cross-center workflows on HPC systems that use high-end parallel file systems. For example, iRODS can be installed, which implements a layer on top of raw storage systems, effectively making data on all involved systems (potentially located at multiple sites) appearing in one file tree.

Moving data to and from HPC centers can be cumbersome when using traditional tools such as scp and rsync. Transferring very large amounts of data requires more sophisticated tools like GridFTP. The services from the EUDAT project can help to make access and transfer easier and more transparent. For example, the B2SAFE system of EUDAT is based on iRODS and offers distributed data and metadata management. EUDAT's B2HANDLE service helps to obtain globally unique PIDs and assign them to data. To enhance (meta-)data accessibility, EUDAT's B2STAGE has implemented a GridFTP-based method for data staging between iRODS and HPC file systems, and B2SHARE is a (meta-)data portal and repository system that also allows to publish data and make it findable.

The CompBioMed project is using the LEXIS platform for its workflow execution and data management. LEXIS has built a workflow-execution and data management platform that follows the paradigm of HPC–Cloud convergence for demanding Big Data applications. The CompBioMed project is currently completing the implementation of a new workflow on the LEXIS platform, which is resilient to (catastrophic) failure of upcoming exascale platforms, using a novel evocation of data replication and brokering. The collaboration between CompBioMed and LEXIS has led to the extension of the LEXIS Distributed Data Infrastructure (DDI) and Distributed Computing Infrastructure (DCI), including additional nodes and data sources.

The CompBioMed project is completing the implementation of a resilient workflow, using data replication and brokering, which will enable urgent computing on upcoming exascale platforms.

Acknowledgements

This work has been conducted within the EU Horizon 2020 projects CompBioMed2 (GA No. 823712), LEXIS (GA No. 825532, giving workflow and data management support in the scope of the LEXIS Open Call), and DICE (Data infrastructure capacity for the European Open Science Cloud, GA No. 101017207).

References

1. Wilkinson MD et al (2016) The FAIR Guiding Principles for scientific data management and stewardship. Sci Data 3:1–9 (2016)
2. Götz A, Weber T, Hachinger S (2019) Let the data sing—a scalable architecture to make data silos FAIR—poster from RDA plenary 14 (2019). https://doi.org/10.5281/zenodo.3497321
3. EUDAT Collaborative Data Infrastructure: EUDAT—Research Data Services, Expertise & Technology Solutions (2023). https://www.eudat.eu, Cited 20 Feb 2023
4. EUDAT Collaborative Data Infrastructure: B2SAFE-EUDAT (2023). https://www.eudat.eu/services/b2safe, Cited 20 Feb 2023
5. EUDAT Collaborative Data Infrastructure: B2HANDLE-EUDAT (2023). https://www.eudat.eu/services/b2handle, Cited 20 Feb 2023
6. compbiomed.eu: CompBioMed2 Project (2019). https://doi.org/10.3030/823712
7. Alowayyed S, Groen D, Coveney PV, Hoekstra AG (2017) Multiscale computing in the exascale era. J Comput Sci 22:15–25 (2017). https://doi.org/10.1016/j.jocs.2017.07.004
8. Coveney PV (2020) Computational biomedicine. Part 1: molecular medicine. Interface Focus 10(6):20200047. https://doi.org/10.1098/rsfs.2020.0047
9. Coveney PV, Hoekstra A, Rodriguez B, Viceconti M (2020) Computational biomedicine. Part II: organs and systems. Interface Focus 11(1):20200082. https://doi.org/10.1098/rsfs.2020.0082
10. Saadi AA, Alfe D, Babuji Y, Bhati A, Blaiszik B, Brace A, Brettin T, Chard K, Chard R, Clyde A, Coveney P, Foster I, Gibbs T, Jha S, Keipert K, Kranzlmüller D, Kurth T, Lee H, Li Z, Ma H, Mathias G, Merzky A, Partin A, Ramanathan A, Shah A, Stern A, Stevens R, Tan L, Titov M, Trifan A, Tsaris A, Turilli M, Van Dam H, Wan S, Wifling D, Yin J (2021) Impeccable: integrated modeling pipeline for covid cure by assessing better leads. In: Proceedings of the 50th international conference on parallel processing, ICPP '21. Association for Computing Machinery, New York. https://doi.org/10.1145/3472456.3473524
11. Bhati AP, Wan S, Alfè D, Clyde AR, Bode M, Tan L, Titov M, Merzky A, Turilli M, Jha S, Highfield RR, Rocchia W, Scafuri N, Succi S, Kranzlmüller D, Mathias G, Wifling D, Donon Y, Di Meglio A, Vallecorsa S, Ma H, Trifan A, Ramanathan A, Brettin T, Partin A, Xia F, Duan X, Stevens R, Coveney PV (2021) Pandemic drugs at pandemic speed: infrastructure for accelerating covid-19 drug discovery with hybrid machine learning- and physics-based simulations on high-performance computers. Interface Focus 11:20210018. https://doi.org/10.1098/rsfs.2021.0018
12. dice eosc.eu: DICE Project (2021). https://doi.org/10.3030/101017207
13. Scionti A et al (2020) HPC, Cloud and Big-Data Convergent Architectures: The LEXIS Approach. In Barolli L, Hussain F, Ikeda M (eds.) CISIS 2019, Advances in intelligent systems and computing, vol. 993, pp. 200–212. Springer, Cham. https://doi.org/10.1007/978-3-030-22354-0_19
14. Munke J, Hayek M, Golasowski M, García-Hernández RJ, Donnat F, Koch-Hofer C, Couvee P, Hachinger S, Martinovič J (2022) Data System and Data Management in a Federation of HPC/Cloud Centers. CRC Press, Boca Raton, pp 59–77. https://doi.org/10.1201/9781003176664-4
15. Schembera B, Iglezakis D (2020) EngMeta: metadata for computational engineering. Int J Metadata Semant Ontol 14(1):26–38. https://doi.org/10.1504/IJMSO.2020.107792
16. Pérez A, Martínez-Rosell G, De Fabritiis G (2018) Simulations meet machine learning in structural biology. Curr Opin Struct Biol 49: 139–144. https://doi.org/10.1016/j.sbi.2018.02.004. Theory and simulation • Macromolecular assemblies
17. Alowayyed S, Piontek T, Suter J, Hoenen O, Groen D, Luk O, Bosak B, Kopta P, Kurowski K, Perks O, Brabazon K, Jancauskas V, Coster D, Coveney P, Hoekstra A (2019) Patterns for high performance multiscale computing. Fut Gener Comput Syst 91: 335–346. https://doi.org/10.1016/j.future.2018.08.045
18. Lee H, Merzky A, Tan L, Titov M, Turilli M, Alfe D, Bhati A, Brace A, Clyde A, Coveney P, Ma H, Ramanathan A, Stevens R, Trifan A, Van Dam H, Wan S, Wilkinson S, Jha S (2021) Scalable HPC & AI infrastructure for covid-19 therapeutics. In: Proceedings of the platform for advanced scientific computing conference, PASC '21. Association for Computing Machinery, New York. https://doi.org/10.1145/3468267.3470573
19. Benemerito I, Mustafa A, Wang N, Narata AP, Narracott A, Marzo A (2023) A multiscale computational framework to evaluate flow alterations during mechanical thrombectomy for treatment of ischaemic stroke. Front

Cardiovasc Med 10 (2023). https://doi.org/10.3389/fcvm.2023.1117449

20. DOI Foundation: Home Page (2022). https://www.doi.org/, Cited 23 Mar 2023
21. DataCite - International Data Citation Initiative e.V.: Welcome to DataCite (2023). https://datacite.org, Cited 20 Feb 2023
22. DataCite Metadata Working Group.: DataCite Metadata Schema Documentation for the Publication and Citation of Research Data and Other Research Outputs. Version 4.4. (2021). https://doi.org/10.14454/3w3z-sa82. https://datacite.org, Cited 20 Feb 2023
23. DataCite - International Data Citation Initiative e.V.: DataCite Search (2023). https://search.datacite.org/, Cited 20 Feb 2023
24. re3data.org: Registry of Research Data Repositories (2023). https://doi.org/10.17616/R3D, Cited 20 Feb 2023
25. The HDF Group: Hierarchical data format version 5 (2000–2010). http://www.hdfgroup.org/HDF5
26. Rew R, Davis G (1990) NetCDF: an interface for scientific data access. IEEE Comput Graph Appl 10(4):76–82. https://doi.org/10.1109/38.56302
27. Brown SA, Folk M, Goucher G, Rew R (1993) Software for portable scientific data management. Comput Phys 7(3):304–308. https://doi.org/10.1109/TETC.2020.3019202
28. Poinot M, Rumsey CL (2018) Seven keys for practical understanding and use of CGNS. In: 2018 AIAA aerospace sciences meeting. AIAA, pp. 1–14. https://doi.org/10.2514/6.2018-1503
29. Godoy WF, Podhorszki N, Wang R, Atkins C, Eisenhauer G, Gu J, Davis P, Choi J, Germaschewski K, Huck K, Huebl A, Kim M, Kress J, Kurc T, Liu Q, Logan J, Mehta K, Ostrouchov G, Parashar M, Poeschel F, Pugmire D, Suchyta E, Takahashi K, Thompson N, Tsutsumi S, Wan L, Wolf M, Wu K, Klasky S (2020) ADIOS 2: the adaptable input output system. A framework for high-performance data management. SoftwareX 12:100561 (2020). https://doi.org/10.1016/j.softx.2020.100561
30. Páll S, Abraham M, Kutzner C, Hess B, Lindahl E (201) Tackling exascale software challenges in molecular dynamics simulations with GROMACS. In Markidis S, Laure E. (eds.) Research and advanced technology for digital libraries. Springer, Berlin, pp 3–27. https://doi.org/10.1007/978-3-319-15976-8
31. Abraham MJ, Murtola T, Schulz R, Páll S, Smith JC, Hess B, Lindahl E (2015) GROMACS: high performance molecular simulations through multi-level parallelism from laptops to supercomputers. SoftwareX 1–2:19–25. https://doi.org/10.1016/j.softx.2015.06.001
32. Phillips JC, Hardy DJ, Maia JDC, Stone JE, Ribeiro JV, Bernardi RC, Buch R, Fiorin G, Hénin J, Jiang W, McGreevy R, Melo MCR, Radak BK, Skeel RD, Singharoy A, Wang Y, Roux B, Aksimentiev A, Luthey-Schulten Z, Kalé LV, Schulten K, Chipot C, Tajkhorshid E (2020) Scalable molecular dynamics on CPU and GPU architectures with NAMD. J Chem Phys 153(4):044130. https://doi.org/10.1063/5.0014475
33. EUDAT Collaborative Data Infrastructure: B2SHARE-EUDAT (2023). https://www.eudat.eu/services/b2share, Cited 30 Mar 2023
34. Xu H, Russell T, Coposky J, Rajasekar A, Moore R, de Torcy A, Wan M, Shroeder W, Chen SY (2017) iRODS primer 2: integrated rule-oriented data system. Morgan & Claypool Publishers, Williston. https://doi.org/10.2200/S00760ED1V01Y201702ICR057
35. Schouppe M, Burgelman JC (2018) Relevance of EOSC and FAIR in the realm of open science and phases of implementing the EOSC. In: Kalinichenko LA, Manolopoulos Y, Stupnikov SA, Skvortsov NA, Sukhomlin V (eds) Selected papers of the XX international conference on data analytics and management in data intensive domains (DAMDID/RCDL 2018), Moscow, Russia, October 9–12, 2018, CEUR Workshop Proceedings, vol 2277. CEUR-WS.org, pp 1–4
36. EUDAT Collaborative Data Infrastructure: B2FIND-EUDAT (2023). https://www.eudat.eu/services/b2find, Cited 20 Feb 2023
37. Carnero J, Nieto FJ (2018) Running simulations in HPC and cloud resources by implementing enhanced TOSCA workflows. In: 2018 international conference on high performance computing & simulation (HPCS), pp 431–438. https://doi.org/10.1109/HPCS.2018.00075
38. Ilyushkin A, Bauer A, Papadopoulos AV, Deelman E, Iosup A (2019) Performance-feedback autoscaling with budget constraints for cloud-based workloads of workflows. arXiv:1905.10270. https://doi.org/10.48550/arXiv.1905.10270
39. Thelin J (2011) Accessing Remote Files Easily and Securely. Linux Journal (2011). https://www.linuxjournal.com/content/accessing-remote-files-easy-and-secure, Cited 20 Feb 2023

40. Jin T, Zhang F, Sun Q, Romanus M, Bui H, Parashar M (2020) Towards autonomic data management for staging-based coupled scientific workflows. J Parallel Distrib Comput 146: 35–51. https://doi.org/10.1016/j.jpdc.2020.07.002
41. Haynes T, Noveck D (2015) Network file system (NFS) version 4 protocol. RFC 7530. https://doi.org/10.17487/RFC7530
42. Ylonen T (1996) SSH—secure login connections over the Internet. In: 6th USENIX Security Symposium (USENIX Security 96). USENIX Association, San Jose. https://www.usenix.org/conference/6th-usenix-security-symposium/ssh-secure-login-connections-over-internet
43. Davison W (2023) Rsync community: Rsync. https://rsync.samba.org, Cited 20 Feb 2023
44. Allcock W, Bresnahan J, Kettimuthu R, Link M (2005) The globus striped GridFTP framework and server. In: SC '05: proceedings of the 2005 ACM/IEEE conference on supercomputing, pp 54–54. https://doi.org/10.1109/SC.2005.72
45. Schuller BT, Pohlmann T (2011) UFTP: high-performance data transfer for UNICORE. In: Proceedings of the 2011 UNICORE summit, Torun, Poland, IAS Series, Forschungszentrum Jülich GmbH Zentralbibliothek, Jülich, vol. 9. pp 135–142
46. Foster I (2011) Globus online: accelerating and democratizing science through cloud-based services. IEEE Internet Comput 15:70–73
47. Grid Community Forum: Overview — #Grid Community Forum (2023). https://www.eudat.eu, Cited 20 Feb 2023
48. Amazon Web Services, Inc. and affiliates: Cloud Object Storage – Amazon S3 – Amazon Web Services (2023). https://aws.amazon.com/de/s3/, Cited 23 Mar 2023
49. MinIO, Inc. MinIO — High Performance, Kubernetes Native Object Storage (2023). https://min.io, Cited 19 Mar 2023
50. Sakimura N, Bradley J, Jones MB, de Medeiros B, Mortimore C (2014). OpenID Connect Core 1.0 incorporating errata set 1 (2014). https://openid.net/specs/openid-connect-core-1_0.html, Cited 20 Feb 2023
51. Lannom L, Boesch LCBP, Sun S (2003) Handle system overview. RFC 3650. https://doi.org/10.17487/RFC3650
52. Hachinger S, Golasowski M, Martinovič J, Hayek M, García-Hernández RJ, Slaninová K, Levrier M, Scionti A, Donnat F, Vitali G, Magarielli D, Goubier T, Parodi A, Parodi A, Harsh P, Dees A, Terzo O (2022) Leveraging high-performance computing and cloud computing with unified big-data workflows: the LEXIS project. In: Curry E, Auer S, Berre AJ, Metzger A, Perez MS, Zillner S (eds) Technologies and applications for big data value. Springer, Cham, pp 159–180. https://doi.org/10.1007/978-3-030-78307-5_8
53. García-Hernández RJ, Golasowski M (2020) Supporting Keycloak in iRODS systems with OpenID authentication (2020). Presented at CS3—workshop on cloud storage synchronization & sharing services. https://indico.cern.ch/event/854707/contributions/3681126, Cited 6 Nov 2020
54. JBoss: Keycloak (2023). https://www.keycloak.org, Cited 20 Feb 2023
55. Django Software Foundation: Django: The web framework for perfectionists with deadlines (2023). https://www.djangoproject.com, Cited 20 Feb 2023
56. Pivotal Software: Messaging that just works—RabbitMQ (2023). https://www.rabbitmq.com, Cited 20 Feb 2023
57. Ask Solem & contributors; GoPivotal: Celery - Distributed Task Queue—Celery 5.2.7 documentation (2023). https://docs.celeryq.dev/en/stable, Cited 20 Feb 2023
58. Snow CD, Nguyen H, Pande VS, Gruebele M (2002) Absolute comparison of simulated and experimental protein-folding dynamics. Nature 420:102–106
59. Atos: Ystia Suite (2023). https://ystia.github.io, Cited 20 Feb 2023

Chapter 19

Accelerating COVID-19 Drug Discovery with High-Performance Computing

Alexander Heifetz

Abstract

The recent COVID-19 pandemic has served as a timely reminder that the existing drug discovery is a laborious, expensive, and slow process. Never has there been such global demand for a therapeutic treatment to be identified as a matter of such urgency. Unfortunately, this is a scenario likely to repeat itself in future, so it is of interest to explore ways in which to accelerate drug discovery at pandemic speed. Computational methods naturally lend themselves to this because they can be performed rapidly if sufficient computational resources are available. Recently, high-performance computing (HPC) technologies have led to remarkable achievements in computational drug discovery and yielded a series of new platforms, algorithms, and workflows. The application of artificial intelligence (AI) and machine learning (ML) approaches is also a promising and relatively new avenue to revolutionize the drug design process and therefore reduce costs. In this review, I describe how molecular dynamics simulations (MD) were successfully integrated with ML and adapted to HPC to form a powerful tool to study inhibitors for four of the COVID-19 target proteins. The emphasis of this review is on the strategy that was used with an explanation of each of the steps in the accelerated drug discovery workflow. For specific technical details, the reader is directed to the relevant research publications.

Key words Machine learning, ML, Artificial intelligence, AI, Novel drug design, Molecular dynamics, ESMACS, TIES, Affinity prediction, Structure-based drug design, SBDD

1 Introduction

The COVID-19 pandemic is caused by the novel severe acute respiratory syndrome coronavirus 2 (SARS-CoV-2), a member of the coronavirus family. Structure-based drug design (SBDD) has been deployed by many research groups to identify candidate therapeutics for the treatment of COVID-19 [1]. Although SARS-CoV-2 is beginning to be considered an endemic virus, the risk of the virus mutating into a vaccine-resistant variant persists. As a result, the demand of efficacious drugs to treat COVID-19 is still pertinent. To this end, scientists continue to identify and repurpose marketed drugs for this new disease [1–5]. However, the

Alexander Heifetz (ed.), *High Performance Computing for Drug Discovery and Biomedicine*, Methods in Molecular Biology, vol. 2716, https://doi.org/10.1007/978-1-0716-3449-3_19,

identification of conclusive drug molecules has been slowed down by the huge chemical space that needs to be explored. It is clearly not possible to synthesize and test each of these experimentally, especially since it is known that, as with any high-throughput screen, the vast majority of these ligands will not bind to the target protein. This is where in silico methods can play an important role in virtual screening (VS) and can support the de novo design of novel anti-COVID-19 drugs.

Physics-based (PB) techniques like molecular dynamics (MD) are one of the traditional techniques for in silico drug design. MD simulations can provide not only plentiful dynamical structural information on target protein but also a wealth of energetic information about protein and ligand binding, interactions, and affinity. Such information is very important to understanding the structure-function relationship of the target and the essence of protein-ligand interactions and to guiding the drug discovery and design process. Thus, MD simulations have been applied widely and successfully in each step of modern drug discovery [6–8]. However, MD-based methods require known ligand structures and do not possess de novo design capabilities needed to identify novel active compounds. Among these capabilities are included essential optimization goals such as activity on primary target, selectivity against off-targets, and physicochemical and ADMET properties.

Several attempts have been made to create synergies between PB and artificial intelligence (AI) methods to compensate for these limitations [9–15]. AI tools find increasing application in drug discovery, supporting every stage of the Design-Make-Test-Analyze (DMTA) cycle [16]. The application of AI and machine learning (ML) approaches promises to revolutionize the DMTA cycle, accelerating the drug design process and, therefore, reducing cost. Over the past few years, the field of AI/ML has moved from largely theoretical studies to real-world applications [9, 14, 15]. Much of that explosive growth has been enabled by the advances in AI/ML algorithms, such as deep learning (DL). The use of neural networks in DL algorithms enables computers to imitate human intelligence by learning from data. The application of these approaches can be exploited across multiple aspects of drug design, allowing us to generate novel drug-like molecules by sampling a significant subset of the chemical space of relevance. DL techniques are computationally far cheaper than experimental methods and enable a quick turnaround of results that allows millions to billions of compounds to be explored.

However, the accuracy of ML/DL methods is very much dependent on the training data and the accuracy of the tools used to evaluate protein-ligand affinity. The predictive capability of ML/DL methods can be dramatically improved by providing them with reliable data and by curating them with accurate tools for identification of drug–protein interactions (DPIs) and/or affinity [6, 7, 17]. DPIs are crucial in drug discovery, and several ML methods trained on a curated industry-scale benchmark data set have been developed to predict them. These new approaches are promising; however, the existing methods are still dependent on the use of unrealistic data sets that contain hidden bias, which limits the accuracy of VS methods based on DL/ML [17]. Combining the capability of MD methods to calculate protein–ligand affinity and DPIs with ML algorithms represents a significant advancement towards automated and efficient drug design. This approach holds even more potential when implemented on HPC systems. This review showcases one example of how this approach has been effectively adopted to screen for SARS-CoV-2 inhibitors during the COVID-19 pandemic.

2 Methods and Results

2.1 MD-ML-HPC Workflow

The proof-of-concept for a successful integrated MD-ML approach has been demonstrated by Prof. Coveney's team [5]. The MD-based approaches for docking pose refinement and affinity prediction were integrated with de novo ML design of novel ligands and adopted for HPC, as described in Fig. 1. This protocol was applied for the exploration of ligands for four of the SARS-CoV-2 target proteins, specifically, 3C-like protease (3CLPro; also known as the main protease), papain-like protease (PLPro), ADP-ribose phosphatase (ADRP; a macrodomain of NSP3), and non-structural protein 15 (NSP15). This protocol aimed to predict/design ligand structures with improved binding potency toward a given target protein.

2.1.1 Docking and MD-Based Refinement of Docked Poses

The pre-docking procedure was implemented to generate an initial docking pose and filter out all obvious non-binders. The Zinc database, a collection of commercially available chemical compounds, was docked into binding sites of the 4 target proteins, and 100 top-ranked ligands (according to docking score) were selected for further MD exploration. This is because while docking programs are generally good at pose prediction, they are less effective in predicting binding free energy of the compounds. An additional problem is encountered with the standard docking protocols because the protein is kept rigid and its intrinsic flexibility (in response to the ligand) is ignored. To overcome these docking issues, each pre-docked protein-ligand complex was subject to further refinement with MD to optimize the docked structure by allowing flexibility for both protein and ligand.

2.1.2 MD-Based Binding Affinity Prediction (with ESMACS or TIES)

Two MD-based methods, enhanced sampling with approximation of continuum solvent (ESMACS) [18, 19] and thermodynamic integration with enhanced sampling (TIES) [20], have previously

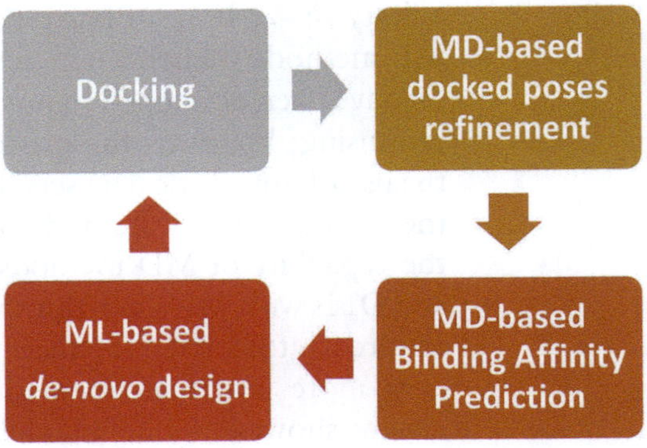

Fig. 1 MD-ML-HPC workflow [ref]

been reported to deliver accurate and reproducible binding affinity predictions. Their excellent scalability allows them to calculate binding affinities for a large number of protein-ligand complexes in parallel, utilizing the great power of the modern supercomputers.

ESMACS [18, 19] involves performing an ensemble of MD simulations followed by free energy estimation using a semi-empirical method called molecular mechanics Poisson-Boltzmann surface area (MMPBSA). The free energies for the ensemble of conformations are analyzed in a statistically robust manner, yielding precise free energy predictions for any given complex. The use of ensembles is particularly important because the usual practice of performing MMPBSA calculations on conformations generated using a single MD simulation does not give reliable binding affinities.

TIES (as described in detail in Chap. 11) is based on an alchemical free energy method called thermodynamic integration (TI). Alchemical methods involve calculating free energy along a non-physical thermodynamic pathway to get relative free energy between the two endpoints. This knowledge is highly important in lead-optimization (LO) stage of drug design where we want to know if a given modification in ligand structure can improve affinity or not. TIES provides an excellent tool to do so with confidence. A positive value of relative of binding affinities ($\Delta\Delta G$) predicted by TIES indicates a diminished relative binding potency for the "modified" compound, whereas a negative value means that the modification studied is favorable. Such information then informs ML predictive model about the desirable as well as undesirable chemical modifications to be introduced into the selected lead compounds. In this way, it improves the predictive accuracy of the ML models, progressively leading to quicker convergence toward the region of main interest in the huge chemical space.

2.1.3 ML-Based De Novo Design

ML is used to gather and accumulate information from the MD stages described above in order to quickly locate the most interesting region(s) in the chemical space, in terms of assessing the potential of a lead compound to bind strongly. The ML model is trained using data from both docking and MD-based binding affinity predictions to enable it to actively relate structural/chemical features with corresponding binding potencies. This allows the ML model to make progressively more accurate predictions of ligand structures that can be classified as potential protein binders. These are then fed into the MD component of the drug discovery workflow to filter them, first using docking and then by ESMACS and TIES, to finally select those that bind the most effectively. This is repeated iteratively, with the ML model improving after each iteration.

3 Results

According to the ESMACS prediction for most of the molecular systems studied, between 4% and 19% of the compounds showed promising binding affinities corresponding to K_D values of 10 nM (−10.98 kcal/mol), 100 nM (−9.61 kcal/mol), and 1 μM (−8.24 kcal/mol). These ligands are currently being subjected to extensive experimental validation. This study also showed that the known binders used as the test set for the ML model were selected more reliably using the ESMACS prediction than the docking scores, supporting the idea that multiple methods need to be used for SBDD.

TIES was performed on a set of 19 compound modifications (that is chemically modifying the “original” compound into a “new” compound) to study the effect of small structural changes on a compound’s binding potency with ADRP, one of the four target proteins, and only one modification appeared to be favorable.

The calculations described in this accelerated approach to COVID-19 drug discovery were performed on a variety of supercomputers including Leibniz Rechenzentrum’s SuperMUC-NG, Hartree Centre’s ScafellPike, Oak Ridge National Laboratory’s Summit, and Texas Advanced Computing Center’s Frontera. The ESMACS calculations were accelerated with OpenMM as the MD engine on GPUs. TIES required longer wall-clock times than ESMACS as only CPUs were employed to obtain the data [5].

4 Conclusion

As reported by Prof. Coveney’s et al. [3, 5], this heterogeneous 4 steps workflow has the potential to accelerate any drug discovery process substantially by coupling machine learning with physics-based methods such that each compensates for the weakness of the other. Molecules filtered at one stage proceed to the next to be filtered once again using a more accurate and computationally thorough method. A refined set of lead compounds emerges at the end of this multistage process that can then be used for experimental studies. With information relating structural features to energetics and binding potencies being fed into the ML model at each stage, the model is able to identify how to improve the prediction of the next generation of compounds. This iterative process, along with the upstream and downstream flow of information, allows it to accelerate the sampling of relevant chemical space much more rapidly than with conventional methods. For efficient implementation at extreme scale, exactly the circumstances under which pandemic drug discovery is performed, a dedicated workflow manager is needed to handle the large number of heterogeneous computational tasks. With technological advances on the horizon in exascale, quantum, and analogue computing, this hybrid MD-ML-HPC approach offers the potential in the long term to deliver novel pandemic drugs at pandemic speed.

Acknowledgements

Corresponding author is grateful to Profs Peter Coveney and Andrea Townsend-Nicholson from University College London for their support, to the European Commission for the EU H2020 CompBioMed2 Centre of Excellence (grant no. 823712), and to Dr. Tim Holt, senior publishing editor of Interface Focus that allowed the author to describe the findings originally published in Interface Focus [5] as a source of information for the current review.

References

1. Asselah T, Durantel D, Pasmant E, Lau G, Schinazi RF (2021) COVID-19: discovery, diagnostics and drug development. J Hepatol 74:168–184. https://doi.org/10.1016/j.jhep.2020.09.031
2. Monteleone S, Kellici TF, Southey M, Bodkin MJ, Heifetz A (2022) Fighting COVID-19 with artificial intelligence. Methods Mol Biol 2390:103–112. https://doi.org/10.1007/978-1-0716-1787-8_3
3. Wan S, Bhati AP, Wade AD, Alfè D, Coveney PV (2022) Thermodynamic and structural insights into the repurposing of drugs that bind to SARS-CoV-2 main protease. Mol Syst Des Eng 7:123–131. https://doi.org/10.1039/d1me00124h
4. Chilamakuri R, Agarwal S (2021) COVID-19: characteristics and therapeutics. Cell 10. https://doi.org/10.3390/cells10020206
5. Bhati AP, Wan S, Alfè D, Clyde AR, Bode M, Tan L, Titov M, Merzky A, Turilli M, Jha S, Highfield RR, Rocchia W, Scafuri N, Succi S, Kranzlmüller D, Mathias G, Wifling D, Donon Y, Di Meglio A, Vallecorsa S, Ma H, Trifan A, Ramanathan A, Brettin T, Partin A, Xia F, Duan X, Stevens R, Coveney PV (2021) Pandemic drugs at pandemic speed: infrastructure for accelerating COVID-19 drug discovery with hybrid machine learning- and physics-based simulations on high-performance computers. Interface Focus 11:20210018. https://doi.org/10.1098/rsfs.2021.0018

6. Wright DW, Hall BA, Kenway OA, Jha S, Coveney PV (2014) Computing clinically relevant binding free energies of HIV-1 protease inhibitors. J Chem Theory Comput 10:1228–1241. https://doi.org/10.1021/ct4007037
7. Wan S, Bhati AP, Zasada SJ, Coveney PV (2020) Rapid, accurate, precise and reproducible ligand-protein binding free energy prediction. Interface Focus 10:20200007. https://doi.org/10.1098/rsfs.2020.0007
8. Hollingsworth SA, Dror RO (2018) Molecular dynamics simulation for all. Neuron 99:1129–1143. https://doi.org/10.1016/j.neuron.2018.08.011
9. Muller C, Rabal O, Diaz Gonzalez C (2022) Artificial intelligence, machine learning, and deep learning in real-life drug design cases. Methods Mol Biol 2390:383–407. https://doi.org/10.1007/978-1-0716-1787-8_16
10. Clyde A (2022) Ultrahigh throughput protein-ligand docking with deep learning. Methods Mol Biol 2390:301–319. https://doi.org/10.1007/978-1-0716-1787-8_13
11. Isert C, Atz K, Schneider G (2023) Structure-based drug design with geometric deep learning. Curr Opin Struct Biol 79:102548. https://doi.org/10.1016/j.sbi.2023.102548
12. Anighoro A (2022) Deep learning in structure-based drug design. Methods Mol Biol 2390:261–271. https://doi.org/10.1007/978-1-0716-1787-8_11
13. Potterton A, Heifetz A, Townsend-Nicholson A (2022) Predicting residence time of GPCR ligands with machine learning. Methods Mol Biol 2390:191–205. https://doi.org/10.1007/978-1-0716-1787-8_8
14. James T, Hristozov D (2022) Deep learning and computational chemistry. Methods Mol Biol 2390:125–151. https://doi.org/10.1007/978-1-0716-1787-8_5
15. Paul D, Sanap G, Shenoy S, Kalyane D, Kalia K, Tekade RK (2021) Artificial intelligence in drug discovery and development. Drug Discov Today 26:80–93. https://doi.org/10.1016/j.drudis.2020.10.010
16. Patronov A, Papadopoulos K, Engkvist O (2022) Has artificial intelligence impacted drug discovery? Methods Mol Biol 2390:153–176. https://doi.org/10.1007/978-1-0716-1787-8_6
17. Wang P, Zheng S, Jiang Y, Li C, Liu J, Wen C, Patronov A, Qian D, Chen H, Yang Y (2022) Structure-aware multimodal deep learning for drug-protein interaction prediction. J Chem Inf Model 62:1308–1317. https://doi.org/10.1021/acs.jcim.2c00060
18. Wan S, Bhati AP, Skerratt S, Omoto K, Shanmugasundaram V, Bagal SK, Coveney PV (2017) Evaluation and characterization of Trk kinase inhibitors for the treatment of pain: reliable binding affinity predictions from theory and computation. J Chem Inf Model 57:897–909. https://doi.org/10.1021/acs.jcim.6b00780
19. Wan S, Bhati AP, Zasada SJ, Wall I, Green D, Bamborough P, Coveney PV (2017) Rapid and reliable binding affinity prediction of bromodomain inhibitors: a computational study. J Chem Theory Comput 13:784–795. https://doi.org/10.1021/acs.jctc.6b00794
20. Bhati AP, Wan S, Hu Y, Sherborne B, Coveney PV (2018) Uncertainty quantification in alchemical free energy methods. J Chem Theory Comput 14:2867–2880. https://doi.org/10.1021/acs.jctc.7b01143

Chapter 20

Teaching Medical Students to Use Supercomputers: A Personal Reflection

Andrea Townsend-Nicholson

Abstract

At the "Kick Off" meeting for CompBioMed (compbiomed.eu), which was first funded in October 2016, I had no idea that one single sentence ("I wish I could teach this to medical students") would lead to a dedicated program of work to engage the clinicians and biomedical researchers of the future with supercomputing. This program of work which, within the CompBiomed Centre of Excellence, we have been calling "the CompBioMed Education and Training Programme," is a holistic endeavor that has been developed by and continues to be delivered with the expertise and support from experimental researchers, computer scientists, clinicians, HPC centers, and industrial partners within or associated with CompBioMed. The original description of the initial educational approach to training has previously been published (Townsend-Nicholson Interface Focus 10:20200003, 2020). In this chapter, I describe the refinements to the program and its delivery, emphasizing the highs and lows of delivering this program over the past 6 years. I conclude with suggestions for feasible measures that I believe will help overcome the barriers and challenges we have encountered in bringing a community of users with little familiarity of computing beyond the desktop to the petascale and beyond.

Key words High-performance computing, University education, Medical student, Undergraduate, Molecular biosciences, Experimental-computational workflow, Metagenomics, Next-generation sequencing, Computational biology, Computational biomedicine

1 Introduction

In 2016, we had started the program by designing training in computational biomedicine with two different modalities: (1) as part of a credit bearing unit within a taught undergraduate degree program and (2) as an extracurricular training course of shorter duration. As part of a taught program of study, we focused on medicine because we wanted to engage clinicians with computational predictions that were sufficiently accurate that they could use them to inform their clinical practice. We included the molecular biosciences (biochemistry and molecular biology) to encourage the translation of basic research into clinical applications and to

Alexander Heifetz (ed.), *High Performance Computing for Drug Discovery and Biomedicine*, Methods in Molecular Biology, vol. 2716, https://doi.org/10.1007/978-1-0716-3449-3_20,

facilitate personalized medicine initiatives. Also, as Head of Teaching, I was responsible for the design and delivery of the Molecular Biosciences degree program and could ensure the provision of appropriate domain-level knowledge to facilitate the integration of relevant computational methods into the curriculum. The extra-curricular training course was something that we ran as part of the PRACE Advanced Training Courses (PATC) delivered at the Barcelona Supercomputing Center as a Winter School in Computational Biomedicine [2]. The trainees on the PATC course were primarily MSc students of varying backgrounds, mostly computer science but with some from biological disciplines.

The taught programs of study were first delivered at UCL to both medical and molecular biosciences students. Medical student course delivery was through a type of course called a "student selected component (SSC)". These are mandatory courses within the curriculum that provide medical students with the opportunity to learn about specific areas in medicine that are of interest to them. Molecular biosciences course delivery was through a third-year Specialist Research Project. Both the SSC and the Specialist Research Project allowed students to conduct a research project in metagenomics based on the Human Microbiome Project [3]. These courses are consistently well received by the students taking them, and the feedback is used to tune the courses to make engagement with supercomputing easier for those unfamiliar with the technology.

2 Course Developments

It quickly became clear, in our first year of implementation, that although students were able to upload images, audio and video to a number of different social media platforms, the vast majority of them had no experience with the command line, with connecting to a remote system or with running job scripts on a machine. To address this in all subsequent years, we have ensured that the taught content of the medical students' SSC includes an introduction to the command line and to high-performance computing (HPC) before beginning the metagenomic analyses. For the molecular biosciences students, this introductory material is provided in their first year of study. Currently, our undergraduate students in molecular biosciences engage with HPC in their first year, learning to connect to a remote system and to transfer files to and from this system. They learn how to submit jobs to a scheduler and to retrieve and visualize their results.

Our next great challenge came with the COVID-19 pandemic. We had been running metagenomics courses that integrated experimental and computational work. In 2020, we (and the rest of the planet) found ourselves unable to ensure that the experimental component was able to be achieved during lockdown. However, this provided an opportunity to obtain data from public repositories for analysis. Between 2017 and 2020, we had noted that a number of students were finding metagenomic sequence data to include in their analyses beyond that which they had generated themselves experimentally and we took the opportunity to provide next-generation sequence datasets and links to data repositories during the pandemic. This worked exceptionally well for the course and has had the added benefit of helping to expand our students' interest in data science. It also has allowed us to run our metagenomics courses online as a purely computational course with no experimental component. Although this was not what we had originally envisaged, this computational-only offering has actually helped us to expand our teaching delivery beyond UCL and into other universities with the establishment of new courses.

3 Course Delivery

3.1 Location Location Location

It takes a great deal of effort to establish a course or module within a university curriculum, but once it is running, it is likely to be within the curriculum for a period of time that can be measured in years, if not decades. In the summer of 2020, we ported our medical student SSC from UCL to the University of Sheffield. This wasn't an obvious "lift and shift" because SSCs have different formats in different medical schools. The UCL SSC 1 block format is a course with 3 h per week for 8 weeks in total, and it takes place in either the first or second year of study; students complete either one double block or two single block SSCs per year. The Sheffield SSC format is a research SSC that comprises 5 weeks of dedicated study with an academic supervisor and takes place in year 2. Despite these differences, we have successfully run the metagenomics SSC at both UCL and Sheffield since 2020 (*see* Table 1).

What has been particularly notable about the transfer is that although primarily a computational project, a domain/subject specialist is clearly required for effective teaching delivery of this SSC. From 2020, we have needed to have an academic in each institution collaborating in the delivery of the module: for this particular SSC, the domain specialist is at UCL (where the SSC was originally developed) and the computational expert is at Sheffield. The successful establishment of computational biomedicine courses will be greatly facilitated by identifying educators possessing both domain and computational knowledge. We have, from 2021, expanded to include a taught course at the University Pompeu Fabra (UPF; Barcelona, Spain) in our CompBioMed curriculum. In this case, the course organizer is both domain and computationally expert and no co-delivery of content is required. We are currently developing a Sheffield SSC project with significant computational

Table 1
Medical students on HPC. The location and delivery mode of medical student SSC modules delivered from 2017 as part of the CompBioMed Education and Training Programme. The University Pompeu Fabra (UPF) course is multidisciplinary and integrates students from medicine, biology, biomedical engineering, and data science degrees

Academic year	Medical student course	University (Year)	Medicine students	Delivery
2017–2018	SSC	UCL (Year 1)	20	Face to face
	SSC	UCL (Year 2)	20	
2018–2019	SSC	UCL (Year 1)	20	Face to face
2019–2020	SSC	UCL (Year 1)	20	Face to face
2020–2021	SSC	UCL (Year 1)	20	Online
	SSC	USFD (Year 2)	20	
2021–2022	SSC	UCL (Year 1)	8	Face to face
	SSC	UCL (Year 2)	30	Face to face
	SSC	USFD (Year 2)	17	Hybrid
	Inter-departmental	UPF (Year 5)	0/24	Face to face
2022–2023	SSC	UCL (Year 1)	12	Face to face
	SSC	UCL (Year 2)	60	Face to face
	SSC	USFD (Year 2)	12	Hybrid
	Inter-departmental	UPF (Year 5)	tba	Face to face
		TOTAL	259*	

biomedicine content and will port it to UCL to see what training and support are required to run it as a standalone SSC module there. We are also working with our Core and Associate Partners in CompBioMed to find training modalities that will allow us to expand our SSC training to medical schools across Europe, particularly in the EU13 (member states admitted to the EU since 2004) and in HPC-under-represented countries (*see* Fig. 1).

3.2 HPC Resource

We had originally been using the training allocation of the CompBioMed grant to provide access to compute for our students. This was provided by a number of HPC centers, including the Edinburgh Parallel Computing Centre [4] in the UK and SURF [5] in the Netherlands. However, the pandemic also provided us with the opportunity to explore different methods of course delivery. Specifically, we looked at cloud-based methods, whereas we had previously been using federated HPC resources. In the summer of 2020, together with our colleagues in Alces Flight and one of our undergraduate students, we built nUCLeus—a proof of concept cloud HPC education environment [6] that we were able to showcase at SC20. We have reverted to federated resources at present, but we can reactivate nUCLeus whenever needed and we envisage using it for training delivery in the future. We have also explored the use of

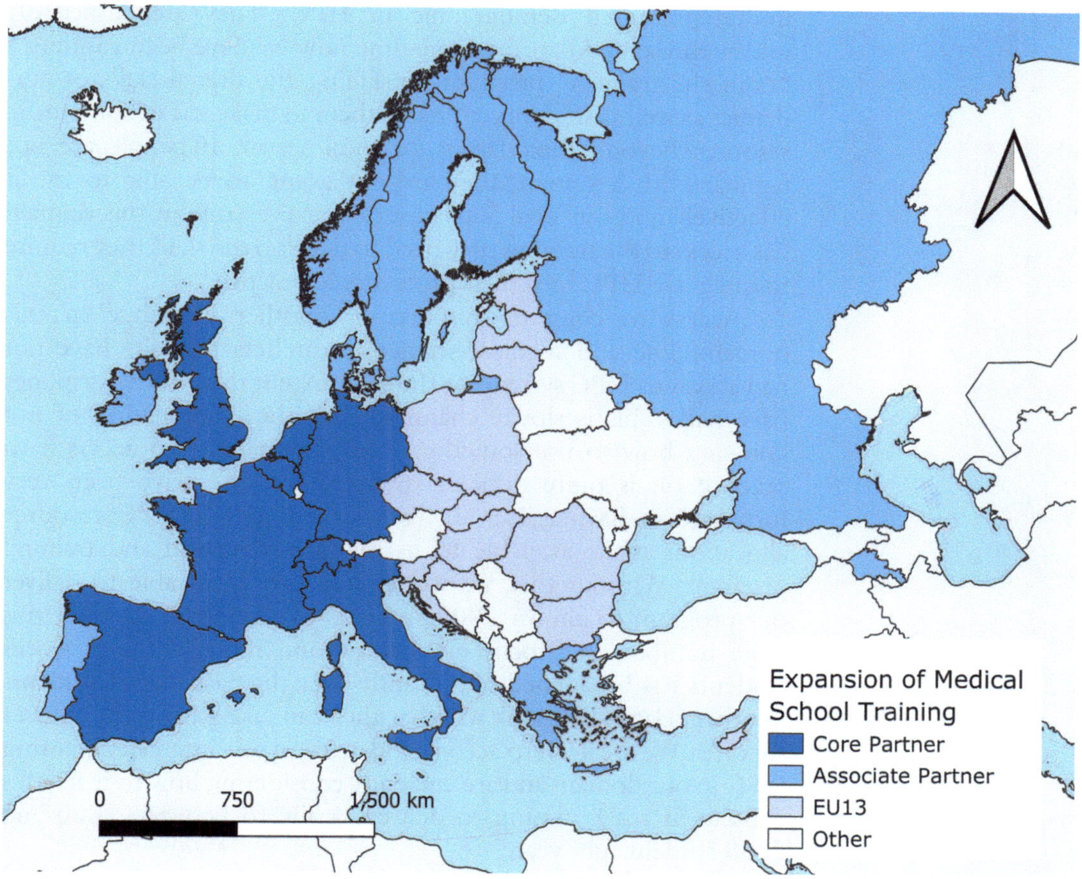

Fig. 1 Expanding CompBioMed SSC delivery across Europe. The geographic location of countries with CompBioMed core partners, associate partners, and EU13 countries with medical schools facilitates opportunities for teaching computational biomedicine in medical schools across Europe

Google Colaboratory to explore open datasets with cloud computing platforms [7].

4 Challenges and Barriers

There are two major challenges that we have collectively faced in embedding computational biomedicine training in the educational curriculum of medical and biomedical undergraduate degrees. The first is access to HPC resources and the second is funding to support the teaching program.

In 2016, a mere 10% of the UK's national supercomputers were being used for research in the life and medical sciences and only 0.2% of this was being used for medical applications [1]. This access was for research and none of it was being used for teaching at an undergraduate level. Life and medical sciences are primarily experimental/clinical rather than computational and are an underrepresented demographic in HPC. They are sufficiently underrepresented that for some time now we have been running a "train the trainers" program, upskilling the digital skills of academics as well as students to enable them to make use of computing resource beyond their laptop or local server. If people are not familiar with a system, they are not going to be able to use it effectively and our goal is to ensure that everyone in this domain has access to the training they need to develop the skills they require to integrate HPC into their professional practice.

Access to compute resource is another challenge in this domain. Life and medical science researchers typically have not had access to HPC systems to the same extent that other disciplines have had. This is slowly changing, but the combination of not knowing how to use something and the inability to access it to practice on is quite difficult to overcome. We have been very fortunate in having Tier 0, Tier 1, and Tier 2 HPC resource allocations made available to us for our education and training program. Without this, we would not have been able to deliver our program to almost 2000 students (*see* Table 2). Despite this large number, our focus on medical and molecular biosciences students has broadened significantly over the past year as students in other related fields ask whether they can take part in the courses we offer. We are keen to accommodate everyone interested in using HPC in our domain and are currently considering how to teach this content at scale, aiming to deliver locally to between 1000 and 2000 students per year.

Table 2
Trainees in the CompBioMed Education and Training Programme. The number and origin of students trained annually from 2017 are shown. The increase in numbers from 2021 is due to embedding HPC training in all 3 years of the molecular biosciences undergraduate degree program

Academic year	Medicine students	Biosciences students	Extracurricular course	Total number of students
2017–2018	40	89	40	169
2018–2019	20	104	38	162
2019–2020	20	108	31	159
2020–2021	40	86	35	161
2021–2022	55	370	40	465
2022–2023	84	472	16	572
	259	1229	200	1688

5 Future Directions

It was not at all clear at the start of the program whether there would be the support needed to foster this kind of innovation. It turned out to be possible within both the grant or the higher education system: engaging domain-specific practitioners with HPC is something that was very much supported by the grant and, as a result, the CompBioMed Education and Training Programme became an important part of our undergraduate teaching program. This new knowledge has been well-received by our students, too, many of whom are completely taken with computational insights into the molecular aspects of their subject. There is a significant demand for more access, more courses, and more training, and I see students coming back again for new opportunities to engage with HPC. There are so many examples, but to highlight a few: a first year medical student who enjoyed the metagenomics SSC has gone on to do a third research project that involves molecular dynamics simulations of protein-protein interactions involved in neurological disorders; a first year medical student who chose to do an intercalated BSc degree in mathematics, computers and medicine; and, several fourth year MSci project students have gone on to work in the area of computational biomedicine as part of their PhD studies. The appetite for knowledge and experience in this area is significant and wonderful to see!

Having watched the program grow to the point where we are easily teaching almost 600 students per year and knowing how we came to be able to do this, there are some things that I think we will need to consider going forward:

1. *There is an urgent need to support programs that will create a culture of engagement with and integration of computational methodologies into professional practice for domains that are primarily experimental/clinical.*
2. *Local institutions should be encouraged to provide HPC for teaching or, at the very least, to implement schedulers that can be used to book a partition on their machine(s) for teaching events like classes and workshops, so that jobs can be run during the teaching session without getting stuck in the queue.*
3. *If local HPC systems do not operate under a "free to use" model, there is the potential for inequality of access between students at institutions that make the resource freely available and those enrolled at institutions that charge to use it; for academics delivering HPC-based courses in the curriculum at institutions that charge to use local HPC resource, the teaching budget will need to include HPC access costs.*

4. *You don't need to deliver digital upskilling through dedicated taught courses. Key skills training, short workshops, week-long summer schools are all good ways of bringing this content to students. The advantage of doing it through a taught curriculum is that it is easier to keep track of what has been taught, to whom and when.*
5. *There will always be growth in the program. We are now building more computationally intensive workflows for teaching our medical and undergraduate students. These will require resource beyond that of a local cluster.*

Acknowledgments

I would like to specifically thank Carlos Teijeiro Barjas, Marco Verdicchio, Gavin Pringle, Andrew Narracott, Guillaume Hautbergue, Alberto Marzo, Oscar Camara Rey, Mariano Vazquez and Peter Coveney for their advice, support and contributions to the development of this novel education and training programme. I am indebted to Dr Claire Ellul (Civil, Environmental and Geomatic Engineering, UCL) for generating the map shown in Figure 1, which was sourced from the European Commission, Eurostat (ESTAT) using the following sources of data: Euro Geographics, TurkStat and UN FAO. I am grateful to the European Commission (grants 675451 and 823712) and to EPSRC (EP/X019446/1 and EP/Y008731/1) for funding support.

References

1. Townsend-Nicholson A (2020) Educating and engaging new communities of practice with high performance computing through the integration of teaching and research. Interface Focus 10:20200003. https://doi.org/10.1098/rsfs.2020.0003
2. https://www.bsc.es/education/training/other-training/online-short-course-hpc-based-computational-bio-medicine
3. https://www.hmpdacc.org
4. https://www.epcc.ed.ac.uk
5. https://www.surf.nl/en
6. Townsend-Nicholson A, Gregory D, Hoti A, Merritt C, Franks S (2020) Demystifying the Dark Arts of HPC – Introducing biomedical researchers to supercomputers. https://sc20.supercomputing.org/proceedings/sotp/sotp_files/sotp104s2-file2.pdf
7. Poolman TM, Townsend-Nicholson A, Cain A (2022) Teaching genomics to life science undergraduates using cloud computing platforms with open datasets. Biochem Mol Biol Educ 50(5): 446–449. https://doi.org/10.1002/bmb.21646

Index

Alexander Heifetz (ed.), *High Performance Computing for Drug Discovery and Biomedicine*, Methods in Molecular Biology, vol. 2716, https://doi.org/10.1007/978-1-0716-3449-3,

C

D

M

N

R

S

X

Y

Z